ARTIFICIAL INTELLIGENCE IN ENERGY AND RENEWABLE ENERGY SYSTEMS

ARTIFICIAL INTELLIGENCE IN ENERGY AND RENEWABLE ENERGY SYSTEMS

SOTERIS KALOGIROU
EDITOR

Nova Science Publishers, Inc.
New York

For permission to use material from this book please contact us:
Telephone 631-231-7269; Fax 631-231-8175
Web Site: http://www.novapublishers.com

Library of Congress Cataloging-in-Publication Data

Kalogirou, Soteris.
Artificial intelligence in energy and renewable energy systems / Soteris Kalogirou.
p. cm.
Includes index.
ISBN 1-60021-261-1
1. Power (Mechanics) 2. Power resources. 3. Renewable energy sources. 4. Artificial intelligence. I. Title.
TJ163.9.K33 2006
621.042--dc22 2006016748

Published by Nova Science Publishers, Inc. ✣New York

CONTENTS

PREFACE

Artificial intelligence is a relatively new subject. It deals with computer programs capable of making predictions and classifications based on past history data. The purpose of this book is to give an overview of the various artificial intelligence techniques and to present various applications in the field of energy and renewable energy. The various techniques comprise artificial neural networks, evolutionary and genetic algorithms, fuzzy logic systems and various hybrid systems which employ more than one of the above techniques.

As editor I had the responsibility and at the same time the challenge to specify the principal categories for the chapters, to identify the foremost scientists in the respective areas, to review the chapters for clarity and homogeneity, to overview the assembly of the book including adherence to deadlines and to coordinate matters among the authors and the publisher. What motivated me was the prospect of emerging with an invaluable reference book for any serious student or scientist entering one of the areas of artificial intelligence. This type of book concentrates a wealth of accumulated wisdom and experience and would save considerable effort in literature review. It attempts to gain an understanding of the basic concepts and demonstrate possible areas in which artificial intelligence techniques can be applied.

The book comprises thirteen chapters. The first chapter is introductory in which the various artificial intelligence techniques are briefly introduced. The major purpose of this chapter is to ease the understanding of the subsequent chapters. Each of the other chapters cover a major application of artificial intelligence as applied to energy and renewable energy systems. In most of the chapters there is an introductory review of applications of the technique under consideration, followed by the particular application/s in more detail.

Chapter 2 reviews various applications in meteorology, which comprise solar radiation, temperature and wind. These are the input source to all renewable energy systems and some of the most influencing parameters in the general area of energy systems. Chapters 3 to 6 present applications in renewable energy systems. Chapter 3 reviews applications in solar thermal systems including the development of drift detection systems based on neural networks. Chapter 4 shows how artificial neural networks and genetic algorithms can be used to optimize solar water heating systems. Chapter 5 deals with the drawing of solar radiation maps constructed for the purpose of identifying suitable locations to install solar systems and various applications of neural networks in photovoltaics. Finally, chapter 6 deals with the prediction of wind speed and show the development of control algorithms for active power generation of wind energy systems.

Chapters 7 to 13 present applications in the general area of energy, which includes applications in fuel cells, buildings, acoustics, internal combustion engines, combustion and refrigeration. Chapter 7 presents a neuro-fuzzy hammerstein model which is used for the predictive control of a solid oxide fuel cell. Chapter 8 deals with the application of fuzzy systems for the modeling and control of building systems. Related to buildings but to a different energy source, that of sound is chapter 9, which deals with the applications of artificial intelligence techniques for the architectural and building acoustics. Chapter 10 shows active learning strategies for the neural estimation of engine maps. Chapter 11 reviews the application of artificial intelligence techniques in combustion systems whereas chapter 12 presents learning control of fluidized-bed combustion processes for power plants. Finally, chapter 13 describes the use of artificial intelligence techniques to reduce the energy consumption of refrigeration systems.

If the readers find the list of topics incomplete is not necessarily because I overlooked or belittled other application areas but because either suitable authors willing to contribute have not been identified or the topic is just under development. It should be noted that the application of artificial intelligence techniques in energy and renewable energy systems, is a relatively new field of research.

I would like to thank all contributors for the effort they put into preparing their chapters and I hope the readers of this book will extend a profound gratitude to all of them. We all hope that this book will initiate research in areas of the fields covered in this book but also to other fields of science and technology.

Soteris A. Kalogirou, Editor
Higher Technical Institute
Nicosia, Cyprus
E-mail address: SKalogirou@hti.ac.cy

List of Authors

Jorge Aguilera
Research Group "IDEA"
Department of Electronics Engineering
Polytechnics School. Jaén. University of Jaén
Campus de las Lagunillas. 23071- Jaén
Spain

Florencia Almonacid
Research Group "IDEA"
Department of Electronics Engineering
Polytechnics School. Jaén. University of Jaén
Campus de las Lagunillas. 23071- Jaén
Spain

Alex Z. S. Chong
Intelligent Systems Research Centre
School of Technology
University of Glamorgan
Pontypridd CF37 1DL
UK

Debarati Datta
School of Engineering and Design
Brunel University
Uxbridge, Middlesex, UB8 3PH
UK

Giuseppe De Nicolao
Department of Computer Engineering and Systems Science
University of Pavia
Via Ferrata 1, 27100 Pavia
Italy

Patxi Flores
The University of the Basque Country (Polytechnic Faculty)
Plaza de Europa, 1, 20018 Donostia-San Sebastián
Spain

Fergus Fricke
School of Architecture, Design Science and Planning
University of Sydney, NSW 2006
Australia

Ian N. Grace
School of Engineering and Design
Brunel University
Uxbridge, Middlesex, UB8 3PH
UK

Leocadio Hontoria
Research Group "IDEA"
Department of Electronics Engineering
Polytechnics School. Jaén. University of Jaén
Campus de las Lagunillas. 23071- Jaén
Spain

Enso Ikonen
Systems Engineering Laboratory
Department of Process and Environmental Engineering
4PYOSYS, PL 4300, FIN-90014 University of Oulu
Finland

Francisco Jurado
University of Jaén, Department of Electrical Engineering
23700 Alfonso X, nº 28, EPS Linares (Jaén)
Spain

Soteris Kalogirou
Higher Technical Institute
P. O. Box 20423, Nicosia 2152
Cyprus

Jeno Kovács
Systems Engineering Laboratory
Department of Process and Environmental Engineering
4PYOSYS, PL 4300, FIN-90014 University of Oulu
Finland

Sylvain Lalot
Laboratory of Mechanics and Energetics (LME)
University of Valenciennes and Hainaut-Cambresis
Le Mont Houy
59313 Valenciennes Cedex 9
France

Silas C. Michaelides
Meteorological Service,
28 Nikis Avenue, Nicosia 1418
Cyprus

Joseph Nannariello
School of Architecture, Design Science and Planning
University of Sydney, NSW 2006
Australia

Marta Neve
Department of Computer Engineering and Systems Science
University of Pavia
Via Ferrata 1, 27100 Pavia
Italy

Gustavo Nofuentes
Research Group "IDEA"
Department of Electronics Engineering
Polytechnics School. Jaén. University of Jaén
Campus de las Lagunillas. 23071- Jaén
Spain

Chee-Keong Tan
Intelligent Systems Research Centre
School of Technology
University of Glamorgan
Pontypridd CF37 1DL
UK

Arantxa Tapia
The University of the Basque Country (Polytechnic Faculty)
Plaza de Europa, 1, 20018 Donostia-San Sebastián
Spain

Gerardo Tapia
The University of the Basque Country (Polytechnic Faculty)
Plaza de Europa, 1, 20018 Donostia-San Sebastián
Spain

Savvas A. Tassou
School of Engineering and Design
Brunel University
Uxbridge, Middlesex, UB8 3PH
UK

Filippos S. Tymvios
Meteorological Service,
28 Nikis Avenue, Nicosia 1418
Cyprus

Chris P. Underwood
School of the Built Environment,
Northumbria University, Newcastle upon Tyne
UK

John Ward
Intelligent Systems Research Centre
School of Technology
University of Glamorgan
Pontypridd CF37 1DL
UK

Steven J. Wilcox
Intelligent Systems Research Centre
School of Technology
University of Glamorgan
Pontypridd CF37 1DL
UK

Jingfeng Xu
School of Architecture, Design Science and Planning
University of Sydney, NSW 2006
Australia

Pedro Zufiria
Research Group "Neural Networks"
Department of Applied Mathematics to Information Technologies
Telecommunications School. Polythecnics University of Madrid
Ciudad Universitaria s/n. 28040 Madrid
Spain

In: Artificial Intelligence in Energy and Renewable Energy Systems ISBN 1-60021-261-1
Editor: Soteris Kalogirou, pp. 1-46

Chapter 1

INTRODUCTION TO ARTIFICIAL INTELLIGENCE TECHNOLOGY

Soteris A. Kalogirou
Higher Technical Institute, Nicosia, Cyprus

ABSTRACT

Artificial intelligence is widely accepted as a technology offering an alternative way to tackle complex and ill-defined problems. Artificial intelligence (AI) systems comprise areas like, artificial neural networks (ANN), genetic algorithms (GA), fuzzy logic (FL) and various hybrid systems, which combine two or more techniques. AI systems can learn from examples, are fault tolerant, i.e., they are able to handle noisy and incomplete data, are able to deal with non-linear problems, and once trained can perform prediction and generalization at very high speed. They have been used in diverse applications in pattern recognition, forecasting, optimization, signal processing, medicine, power systems, manufacturing, control, robotics, social/psychological sciences and many more. They are particularly useful in system modeling such as in implementing complex mappings and system identification. The major objective of this introductory chapter is to outline an understanding of how artificial intelligence systems operate and how they are set up in order to be used in the different disciplines of energy and renewable energy engineering.

Keywords: Artificial intelligence, neural networks, genetic algorithms, fuzzy logic, hybrid systems.

NOMENCLATURE

A(R)	Radial basis function
b_1	Constant term (bias)
E	RMS error
E_n	Normalized mean squared error
I	Number of input parameters

k	Smoothing factor
m	Momentum factor
o	Desired output vectors over all pattern (p)
O	Number of output parameters
p	Patterns
	Constant parameter
P_i	Number of training patterns
q	Constant parameter
r	Constant parameter
R	Radial basis centre
S	Binary operator representing the addition of membership functions $\mu_A(x)$ and $\mu_B(x)$
T	Binary operator representing the multiplication of membership functions $\mu_A(x)$ and $\mu_B(x)$
t	Network output (target)
x_{max}	Maximum value contained in the data set
x_{min}	Minimum value contained in the data set
w	Weight values

Greek

α	Activation for each node
β_i	Weighted average obtained by combining all input numerical information from upstream nodes
δ	Error term
ε	Learning rate which determines the size of the weight adjustments during each training iteration
μ	Membership function
μ_i	Centre location of each basis
σ_i	Scaling of the width of each basis

Abbreviations

AI	Artificial Intelligence
ANN	Artificial Neural Networks
BP	BackPropagation
FL	Fuzzy Logic
GA	Genetic algorithm
GMDH	Group Method of Data Handling
GRNN	General Regression Neural Network
IMQE	Inverse Multi-Quadratic Equation
MLP	Multi-Layer Perceptron
PID	Proportional-Integral-Derivative

RBF	Radial Basis Function
RMS	Root Mean Squared
SOM	Self-Organizing Map
TSP	Travelling Salesman Problem

1. INTRODUCTION

The possibility of developing a machine that would "think" has intrigued human beings for many centuries. The French philosopher-mathematician Rene Descartes in 1637 predicted that it would never be possible to make a machine that thinks as humans do. After three centuries however (1950), the British mathematician and computer pioneer Alan Turing predicted that one day there would be a machine that could duplicate human intelligence in every way.

Artificial Intelligence (AI) is a term that in its broadest sense would indicate the ability of a machine or artifact to perform similar kind of functions that characterize human thought. The term AI has also been applied to computer systems and programs capable of performing tasks more complex than straightforward programming, although still far from the realm of actual thought. Artificial intelligence is the part of computer science concerned with the design of intelligent computer systems, i.e., systems that exhibit the characteristics we associate with intelligence in human behavior, understanding, language, learning, reasoning, solving problems and so on.

However, solving a computation does not indicate understanding, something a person who solved a problem would have. This is because human reasoning is not based solely on rules of logic but involves perception, awareness, emotional preferences, values, evaluating experience, the ability to generalize and weigh options, and many more.

In a similar way that machinery can outperform humans physically, computers can outperform mental functions in limited areas such as in the speed of mathematical calculations. The fastest computers developed today are able to perform roughly 10 billion calculations per second. Making computers that are more powerful however, will probably not be the way to create a machine capable to think. Computer programs operate according to algorithms, which are set procedures and logic steps. Additionally, most computers do serial processing and operations of recognition and computation are performed one at a time. The human brain works in a manner called parallel processing, performing a number of operations simultaneously. To achieve simulated parallel processing, recently some supercomputers have been made with multiple processors to follow several algorithms at the same time.

Artificial intelligence comprise four major branches, i.e., artificial neural networks, genetic algorithms, fuzzy logic and various hybrid systems, which are combinations of two or more of the branches mentioned previously.

Artificial neural networks (ANNs) are collections of individually interconnected processing units. Information is passed between these units along interconnections. An incoming connection has two values associated with it, an input value and a weight. The output of the unit is a function of the summed value of the inputs multiplied by the weights. ANNs are implemented on computers but are not programmed to perform specific tasks. Instead, they are trained with representative data sets until they learn the patterns used as inputs and their association with the output/s. Once they are trained, new patterns may be presented to them for prediction or classification. ANNs can automatically learn to recognize patterns in data from real systems or from physical models, computer programs, or other sources such as data acquisition systems. They can handle many inputs and produce answers that can be meaningful to designers or end users.

Genetic algorithms are inspired by the way living organisms adapt to the harsh realities of life in a hostile world, i.e., by evolution and inheritance. The algorithm imitates in the process the evolution of a population by selecting only fit individuals for reproduction. Thus, a genetic algorithm is an optimum search technique based on the concepts of natural selection and survival of the fittest. It works with a fixed-size population of possible solutions of a problem, called individuals, which are evolving in time. A genetic algorithm utilizes three principal genetic operators: selection, crossover, and mutation.

Fuzzy logic is based on fuzzy logic reasoning which employs linguistic rules in the form of IF-THEN statements. Fuzzy logic and fuzzy control feature a relative simplification of a control methodology description which allows the application of a "human language" to describe the problems and their fuzzy solutions. In many control applications, the model of the system is unknown or the input parameters are highly variable and unstable. In such cases, fuzzy controllers can be more easily applied than the conventional PID controllers as they are more robust and cheaper. It is also easier to understand and modify fuzzy controller rules, which use a human operator's strategy and are expressed in natural linguistic terms.

Hybrid systems combine one or more of the technologies outlined above, either as part of an integrated method of problem solution, or to perform separate tasks of the same problem. Typical examples include the neuro-fuzzy controllers which use neural networks and fuzzy logic for the same task, i.e., to control a process; and systems which utilize a neural network to derive some parameters and a genetic algorithm employed subsequently to find an optimum solution to a problem.

For the modeling, prediction of performance and control of energy and renewable energy processes, analytic computer codes are often used. The algorithms employed are usually complicated involving the solution of complex differential equations. These programs usually require powerful computers and need a considerable amount of time to give accurate predictions. Instead of complex rules and mathematical routines, artificial intelligence systems are able to learn the key information patterns within a multidimensional information domain. In addition, many of the artificial intelligence systems like, neural networks are fault tolerant, robust, and noise immune (Rumelhart *et al.*, 1986). Data from energy and renewable energy processes being inherently noisy are good candidate problems to be handled with artificial intelligence systems.

When dealing with research and design associated with energy and renewable energy processes, there are often difficulties encountered in handling situations where there are many variables involved. To adequately model and predict the behavior of energy and renewable energy processes requires consideration of nonlinear multivariate inter-relationships, often in a 'noisy' environment. For example, in the prediction of performance, modeling or control of a solar system, a building or any other energy system, from the point of view of energy efficiency or prediction of system long-term performance, there are numerous variables involved. Many of these variables depend on weather conditions and the precise interactions of these variables on the system are not fully understood or cannot easily be modeled.

Analytical techniques have been used successfully in the study of the behavior of engineering systems such as heat transfer, thermal processes, and other areas. While the analytical models have been valuable in understanding principles and useful where less than optimal designs were acceptable with the advent of digital computers, numerical methods became much more attractive than analytical solutions, as they could handle more complex and realistic situations. Numerical methods have their limitations as well. They cannot easily account for practical limitations, they tend to perform well at analyzing a situation but not so well as a designer's tool for quickly looking at options. Additionally, the number of variables that can be considered is still limited and numerical solutions usually cannot be obtained directly. Nowadays with the rapid development of engineering, complex systems for which there is no exact model of behavior need to be designed. Furthermore, designers have to design or deal with complex systems whose expected performance is completely unknown. Much of this complexity is due to the multi-parameter and multi-criteria aspects of the design of a system, which are not easily handled using rules of thumb, analytical methods, physical models or numerical methods.

Many of the energy and renewable energy systems are exactly the types of problems and issues for which artificial intelligence approach appear to be most applicable. In these models of computation, an attempt is made to simulate the powerful cognitive and sensory functions of the human brain and to use this capability to represent and manipulate knowledge in the form of patterns. A pattern is a set of input and output data. Based on these patterns, neural networks for example, model input-output functional relationships and can make predictions about other combinations of unseen inputs. Many of the artificial intelligence techniques have the potential for making better, quicker and more practical predictions than other traditional methods.

Artificial intelligence (AI) analysis is based on past history data of a system (represented as a set of patterns) and it is therefore likely to be better understood and appreciated by designers than other theoretical or empirical methods. AI may be used to provide innovative ways of solving design issues and will allow designers to get an almost instantaneous expert opinion on the effect of a proposed change in a design.

The scope of this book is to demonstrate the possibilities of applying artificial intelligence to energy and renewable energy processes, which will show the capability of artificial intelligence techniques as tools for the prediction, modeling and control of these processes. The objective of this chapter is to introduce briefly the various artificial intelligence techniques so as it will make the understanding of the applications presented in the following chapters easier.

2. Artificial Neural Networks

Although the concept of artificial neural network (ANN), or simply neural network, analysis has been discovered nearly 50 years ago, it is only in the last 20 years or so that application software has been developed to handle practical problems. In this section, a brief overview of how neural networks operate and the basic features of some of the mostly used neural network architectures are presented.

ANNs are good for some tasks while are not so effective in some others. Specifically, they are good for tasks involving incomplete data sets, fuzzy or incomplete information, and highly complex and ill-defined problems, where humans usually decide on an intuitional basis. They can learn from examples, and are able to deal with non-linear problems. Furthermore, they exhibit robustness and fault tolerance (Kalogirou, 2003). Tasks that ANNs cannot handle effectively are those requiring high accuracy and precision as in logic and arithmetic. ANNs have been applied successfully in a number of application areas. Some of the most important ones are (Kalogirou, 2003):

(1) *Function approximation.* Mapping of a multiple input to a single output is established. Unlike most statistical techniques, this can be done with adaptive model-free estimation of parameters.
(2) *Pattern association and pattern recognition.* This is a problem of pattern classification. ANNs can be effectively used to solve difficult problems in this field, like for instance in sound, image, or video recognition. This task can even be made without an *a priori* definition of the pattern. In such cases, the network learns to identify totally new patterns.
(3) *Associative memories.* This is the problem of recalling a pattern when given only a subset clue. In such applications, the network structures used are usually complicated, composed of many interacting dynamical neurons.
(4) *Generation of new meaningful patterns.* This general field of application is relatively new. Some claims are made that suitable neuronal structures can exhibit rudimentary elements of creativity.

ANNs have been applied successfully in various fields of mathematics, engineering, meteorology, economics, medicine, psychology, neurology, and many others. Some of the most important ones are; in the analysis of electromyographs and other medical signatures, in pattern, sound and speech recognition, in the identification of military targets and in a number of safety uses such as the identification of explosives in passenger suitcases. They have also been used in the prediction of mineral exploration sites, in weather and market trends forecasting, in adaptive and robotic control, in electrical and thermal load prediction and many others. Neural networks have also been used for process control in energy and renewable energy fields because they can build predictive models of the process from multidimensional data routinely collected from sensors.

Neural networks obviate the need to use complex mathematically explicit formulas, computer models, and impractical and costly physical models. Some of the characteristics that support the success of artificial neural networks and distinguish them from the conventional computational techniques are (Nannariello and Fricke, 2001):

- The direct manner in which artificial neural networks acquire information and knowledge about a given problem domain (learning linear and possibly non-linear relationships) through the 'training' phase.
- Neural networks can work with numerical or analogue data that would be difficult to deal with by other means because of the form of the data or because there are so many variables.

- Neural network analysis can be conceived of as a 'black box' approach and the user does not require sophisticated mathematical knowledge.
- The compact form in which the acquired information and knowledge is stored within the trained network and the ease with which it can be accessed and used.
- Neural network solutions can be robust even in the presence of 'noise' in the input data.
- The high degree of accuracy reported when artificial neural networks are used to generalize over a set of previously unseen data (not used in the 'training' process) from the problem domain.

While neural networks can be used to solve complex problems they do suffer from a number of shortcomings. The most important of them are (Kalogirou, 2003):

- The data used to train neural nets should contain information, which ideally, is spread evenly throughout the entire range of the system.
- There is limited theory to assist in the design of neural networks.
- There is no guarantee of finding an acceptable solution to a problem.
- There are limited opportunities to rationalize the solutions provided.

The following sections describe briefly how from a biological neuron the artificial one is developed and the steps required to set-up an artificial neural network. Additionally, the characteristics of some of the most popularly used neural network architectures are described.

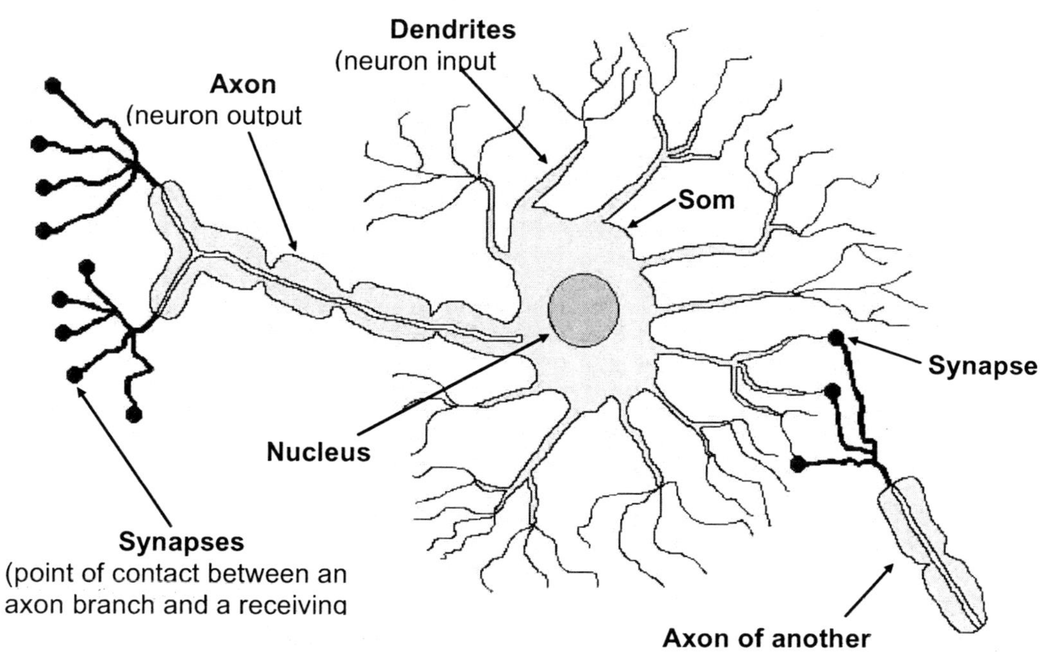

Fig. 1. A simplified illustration of a biological neuron

2.1. Biological and Artificial Neurons

A simplified illustration of a biological neuron as can be found in the human brain is shown in Fig. 1. The neuron consists of the soma with nucleus, the axon, dendrites and synapses. The axon is the neuron output device and the dendrites are the input devices. The point of contact between an axon branch and a receiving neuron is called a synapse. In this, the flow of coded information from the synapses towards the axon is done using electrochemical media, the so-called neurotransmitters. As can be seen the axon of each neuron transmits information to a number of other neurons. Also, the neuron receives information at the synapses from a large number of other neurons. It is estimated that each neuron may receive stimuli from as many as 10,000 other neurons. Groups of neurons are organized into subsystems and the integration of these subsystems forms the brain. It is estimated that the human brain contains around 100 billion interconnected neurons.

Computing networks are far simpler than their biological counterparts are. The first simplified neuron model was proposed by McCulloch and Pitts in 1943. A highly simplified model of an artificial neuron, which may be used to stimulate some important aspects of the real biological neuron, is shown in Fig. 2.

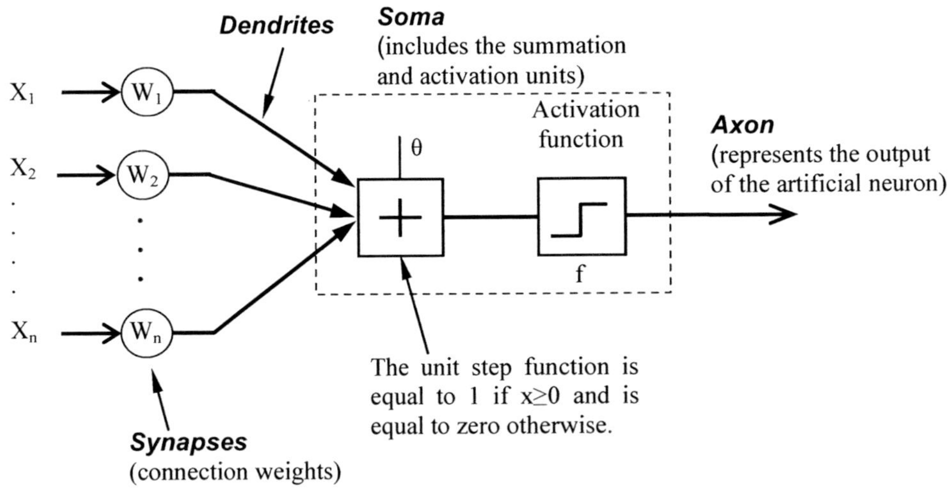

Fig. 2. A simplified model of an artificial neuron

The artificial neuron has all the features of a biological neuron, i.e., the synapses are the connection weights, the soma includes the summation and activation units whereas the axon represents the output of the artificial neuron as in the biological counterpart. The unit step function is equal to 1 if $x \geq 0$ and is equal to zero otherwise. The neuron is considered to be a binary threshold device with n inputs $x_1 \ldots x_n$, one output y, a threshold level θ, and n synaptic weights $w_1 \ldots w_n$. In such a system, excitation is applied to the input of the neuron. Following some suitable operation, it results in a desired output.

An ANN consists of a group of interconnected artificial neurons, interacting with one another in a concerted manner. In such a system, excitation is applied to the input neurons of the network. Following some suitable operation, it results in a desired output. At the synapses, there is an accumulation of some potential, which in the case of the artificial neurons is modeled as a connection weight. These weights are continuously modified during training, based on suitable learning rules.

2.2. Artificial Neural Network Principles

A neural network is a massively parallel distributed processor that has a natural propensity for storing experiential knowledge and making it available for use. It resembles the human brain in two respects (Haykin, 1994):

- the knowledge is acquired by the network through a learning process, and
- inter-neuron connection strengths known as synaptic weights are used to store the knowledge.

Artificial neural networks (ANNs) mimic somewhat the learning process of a human brain. They operate like a "black-box model", requiring no detailed information about the system. Instead, they learn the relationship between the input parameters and the controlled and uncontrolled variables by studying previously recorded data, similar to the way a non-linear regression might perform. Another advantage of using ANNs is their ability to handle large and complex systems with many interrelated parameters. They seem to simply ignore excess input parameters that are of minimal significance and concentrate instead on the more important inputs. Thus, ANN models may be used as an alternative method in engineering analysis and predictions. Additionally, ANNs can reveal non-linear relations between the variables used and connect them to improve the accuracy of the model in a way that is almost impossible with traditional methods (Haykin, 1994).

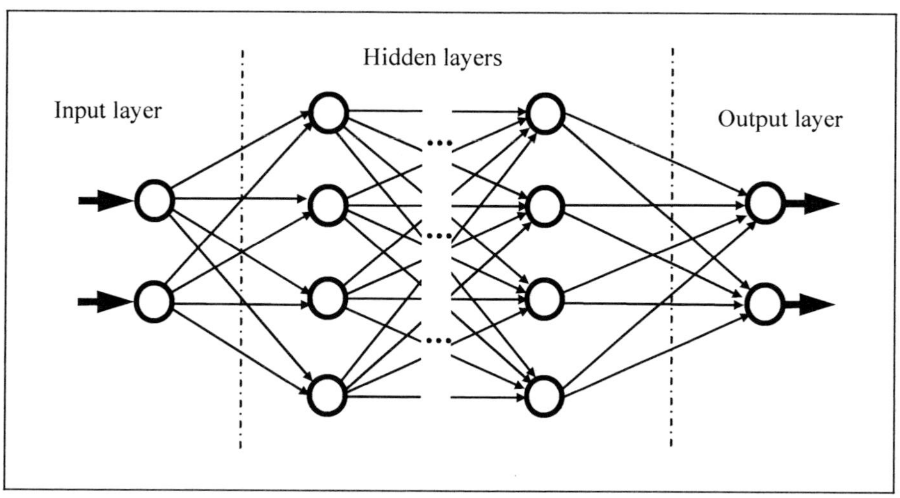

Fig. 3. Schematic diagram of a multilayer feed forward neural network

A schematic diagram of a typical multilayer feedforward neural network architecture is shown in Fig. 3. The network usually consists of an input layer, one, two or more hidden layers and an output layer. In its simple form, each single neuron is connected to other neurons of a previous layer through adaptable synaptic weights. Knowledge is usually stored as a set of connection weights (presumably corresponding to synapse efficacy in biological neural systems). Training is the process of modifying the connection weights in some orderly fashion using a suitable learning method. During learning or training of the network an input vector is presented to the network along with the desired output vector and the weights are adjusted so that the network attempts to produce the desired output. The weights before training are random and have no meaning whereas after training contain meaningful information.

The way that information is processed through a single node is shown in Fig. 4. As can be seen, the node receives weighted activation of other nodes through its incoming connections. First, these are added up (summation). The result is then passed through an activation function; the outcome of which is the activation of the node. For each of the outgoing connections, this activation value is multiplied with the specific weight and transferred to the next node.

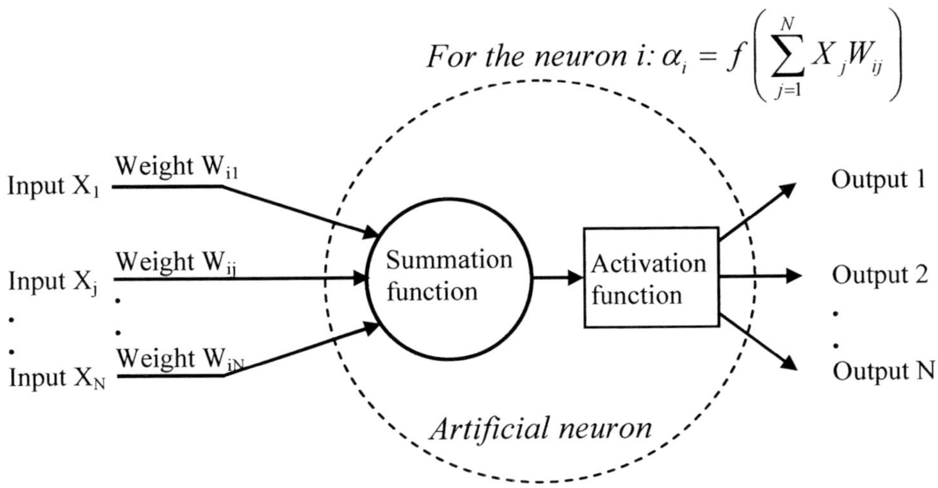

Fig. 4. Information processing in a neural network unit

A training set is a group of matched input and output patterns used for training the network, usually by suitable adaptation of the synaptic weights. The output is one or more (depending on the problem) dependent variable/s that the network produces for the corresponding input. It is important that all the information the network needs to learn is supplied to the network as a data set. When each pattern (a set of input and output parameters) is read, the network uses the input data to produce an output, which is then compared to the correct or desired output. If there is a difference, the connection weights (usually but not always) are altered in such a way that the error is decreased. After the network has run through all the input patterns, if the error is still greater than the maximum desired tolerance, the ANN runs through all the input patterns repeatedly until all the errors are within the required tolerance. When the training reaches a satisfactory level, the network is frozen, i.e., it holds the weights constant, and the trained network can be used to make decisions, identify patterns, or define associations in new input data sets not used to train it.

By learning, it is meant that the system adapts (usually by changing suitable controllable parameters) in a specified manner so that, some parts of the system suggest a meaningful behavior, projected as output. The controllable parameters have different names such as synaptic weights, synaptic efficacies, free parameters and others (Kalogirou, 2003).

The classical view of learning is well documented and interpreted in approximation theories. In these theories, learning may be considered as finding a suitable hypersurface that fits known input/output data points in such a manner that the mapping is acceptably accurate. Such a mapping is usually accomplished by employing simple non-linear functions that are used to compose the required function (Pogio and Girosi, 1990).

A more general approach of learning is adopted by Haykin (1994) who stated that learning is a process by which the free parameters of a neural network are adapted through a continuing process of simulation by the environment in which the network is embedded. The type of learning is determined by the manner in which the parameters' change take place.

An even more general approach is suggested by Neocleous (1998) who indicated that learning is achieved through any change, in any characteristic of a network, so that meaningful results are achieved. Thus, learning could be achieved, among others, through synaptic weight modification, network structure modifications, through appropriate choice of activation functions and many others. A procedure for choosing the appropriate network parameters to facilitate learning is presented in section 2.3.

Achievement of meaningful results means that a desired objective is met with a satisfactory degree of success. The objective is usually quantified by a suitable criterion or cost function. Therefore, this is usually a process of minimizing an error function or maximizing a benefit function. In this respect, learning is similar to optimization. That is why a genetic algorithm, which is an optimum search technique (see section 3), can also be used to train an artificial neural network.

Several algorithms are commonly available which can be used to achieve the minimum error in the shortest possible time. There are also many alternative forms of neural networking systems and, indeed, many different ways in which they may be applied to a given problem. The suitability of an appropriate paradigm and strategy for application is very much dependent on the type of problem to be solved.

The most popular learning algorithm is the back-propagation and its variants (Werbos, 1974; Rumelhart *et al.*, 1986). The Back-Propagation (BP) is one of the most powerful learning algorithms in neural networks. Back-propagation training is a gradient descent algorithm. It tries to improve the performance of the neural network by reducing the total error by changing the weights along its gradient. The error is expressed by the root-mean-square value (RMS), which can be calculated by:

$$E = \frac{1}{2}\sqrt{\sum_p \sum_i \left| t_{ip} - o_{ip} \right|^2} \tag{1}$$

Where (E) is the RMS error, (t) the network output (target), and (o) the desired output vectors over all pattern (p). An error of zero would indicate that all the output patterns computed by the ANN perfectly match the expected values and the network is well trained. The training set has to be a representative collection of input-output examples. The training of all patterns of a training data set is called an epoch.

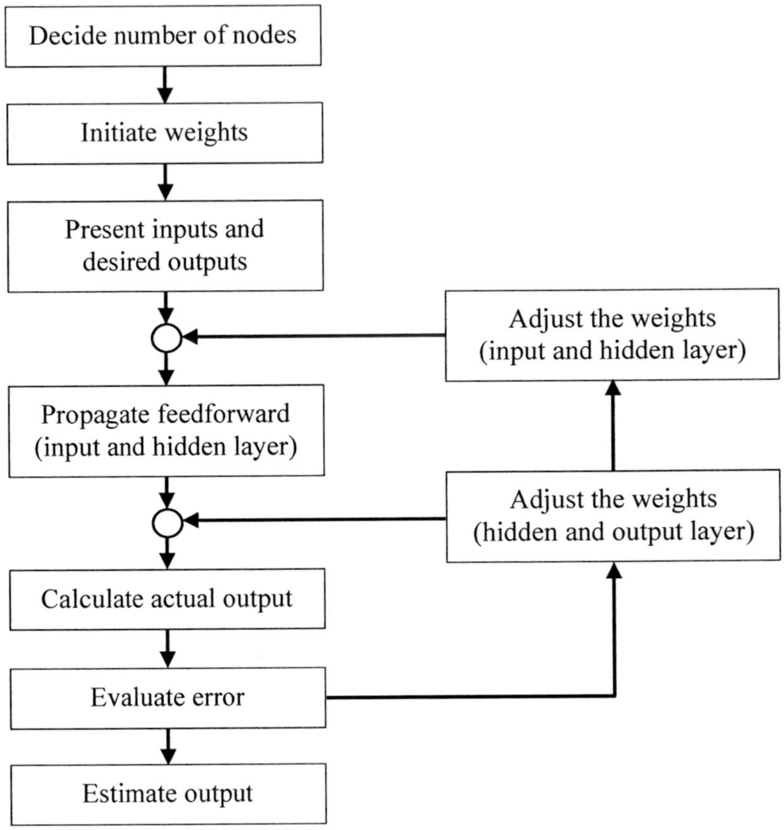

Fig. 5. Flowchart of the backpropagation learning algorithm

The flow chart of backpropagation learning algorithm is shown in Fig. 5. In brief, backpropagation training is performed by initially assigning random values to the weight terms $(w_{ij})^1$ in all nodes (Kalogirou, 2003). Each time a training pattern is presented to the ANN, the activation for each node, α_{pi}, is computed. After the output of the layer is computed the error term, δ_{pi}, for each node is computed backwards through the network. This error term is the product of the error function, E, and the derivative of the activation function and hence is a measure of the change in the network output produced by an incremental change in the node weight values. For the output layer nodes and for the case of the logistic-sigmoid activation, the error term is computed as:

$$\delta_{pi} = (t_{pi} - \alpha_{pi})\alpha_{pi}(1 - \alpha_{pi}) \tag{2}$$

For a node in a hidden layer:

$$\delta_{pi} = \alpha_{pi}(1 - \alpha_{pi})\sum_k \delta_{pk} w_{kj} \tag{3}$$

[1] The j subscript refers to a summation of all nodes in the previous layer of nodes and the i subscript refers to the node position in the present layer.

In the latter expression, the k subscript indicates a summation over all nodes in the downstream layer (the layer in the direction of the output layer). The j subscript indicates the weight position in each node. Finally, the (δ) and (α) terms for each node are used to compute an incremental change to each weight term via:

$$\Delta w_{ij} = \varepsilon\left(\delta_{pi}\alpha_{pj}\right) + mw_{ij}(old) \tag{4}$$

The term (ε) is referred to as the learning rate and determines the size of the weight adjustments during each training iteration. The term (m) is called momentum factor. Momentum prevents the network to get trapped at a shallow local minimum of the error surface, leading to a failure to converge and subsequently in reaching a global minimum. Momentum is applied to the weight change used in the previous training iteration, $w_{ij}(old)$. Both learning rate and momentum are constant terms, specified at the start of the training cycle and determine the speed and stability of the network.

2.3. Network Parameters Selection

The neural network architecture refers to the arrangement of neurons into layers and the connection patterns between layers, activation functions and learning methods. While most scholars are concerned with the techniques to define an artificial neural network architecture, engineers want to apply the ANN architecture to the model and obtain quick results. The neural network model and the architecture of a neural network determine how a network transforms the input into output. This transformation is in fact a computation. Often the success depends upon a clear understanding of the problem regardless of the type of the network architecture applied. However, in determining which neural network architecture provides the best prediction it is necessary to build a good model. It is essential to be able to identify the most important variables in a process, which can be used as input to the neural network. The way to identify and define the best model is very controversial.

Although there are fundamental differences between traditional approaches and neural networks, both methods require the preparation of the model. The classical approach is based on the precise definition of the problem domain as well as the identification of a mathematical function or functions to describe it. It is however very difficult to identify an accurate mathematical function when the system is non-linear and there are parameters that vary with time and season or operate under unstable weather or process conditions. Control programs often lack the capability to adapt to parameter changes. Neural networks on the other hand are used to learn the behavior of the system and subsequently to simulate and predict its behavior. In defining the artificial neural network model, first the process and the process control constrains have to be understood and identified. Then the model must be defined and validated.

When using a neural network for prediction, the following steps are crucial:

- A neural network needs to be built to model the behavior of the process and the values of the output are predicted based on the model.

- Based on the neural network model obtained on the first phase, the output of the model is simulated using different scenarios, and
- The control variables are modified such that to control and optimize the output.

When building an artificial neural network model of a process the input and output variables that characterize the process has to be identified. The inputs may include measurements of the physical dimensions, measurements of the variables specific to the equipment or environment in which the process operates and controlled variables modified by the operator. Variables that do not have any effect on the variation of the measured output are discarded or should not be used as they may defocus the network from obtaining a correct mapping. These variables are estimated by the contribution factors of the various input parameters. These factors indicate the contribution of each input parameter to the learning of the neural network and are usually estimated, depending on the software employed, after the network training or at any intermediate stage when the user wants to check if the input parameters are correctly selected.

The selection of training data has also a vital role in the performance and convergence of the neural network model. A very important step is the analysis of historical data for identification of variables that are important to the process. Plotting charts to check whether the curves of the various variables reflect what is known about the process from operating experience and for discovery of errors in data is also very helpful.

The input and output values need to be normalized. All input and output values are usually scaled individually such that the overall variance in the data set is maximized. This is necessary as it leads to faster learning. The input and output parameters can be scaled either in the range –1 to 1 or in the range 0 to 1 depending on the type of data and the activation function used. For example to scale the data from –1 to 1, the following equation may be used:

$$y_i = \frac{2(x_i - x_{min})}{x_{max} - x_{min}} - 1 \quad (5)$$

Where x_{max} and x_{min} are the maximum and minimum values contained in the data set.

The basic operation that has to be followed to successfully handle a problem with ANNs, is to select the appropriate architecture and the suitable learning rate, momentum, number of neurons in each hidden layer and the activation functions. This is a laborious and time-consuming work but as experience is gathered, some parameters can be predicted easily, thus shortening tremendously the time required. The procedure for finding the best architecture and the other network parameters is shown graphically in Fig. 6.

The first step is to collect the required data and prepare them in a spreadsheet format with various columns representing the input and output parameters. A number of architectures with default values of network input parameters (usually given by the software employed) are tried and the one which gives the more accurate results is chosen. The results are evaluated with the validation data set. If these results are not completely satisfactory then the various network parameters are modified as shown in Fig. 6. If all possible variations are tried and the

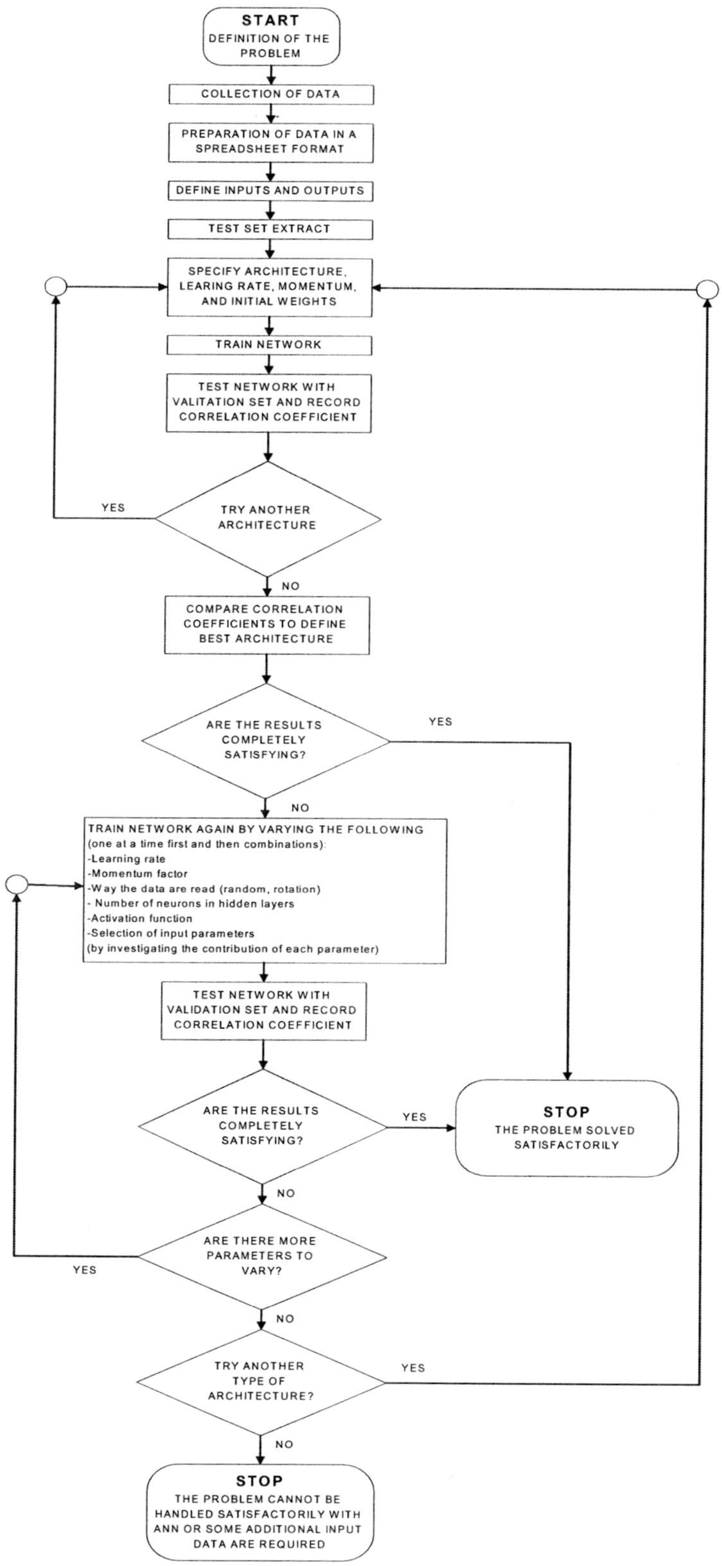

Fig. 6. Network parameters selection procedure

results are still not satisfactory then another architecture is chosen whose network parameters are again modified until satisfactory results are obtained. It should be noted that it is not guaranteed that the problem can be solved with ANNs.

If a large number of sequences/patterns are available in the input data file, to avoid long training times at the beginning, when a trial and error procedure is followed in order to select the best network parameters, a smaller training file may be created, containing as much as possible representative samples of the whole problem domain. Then, when the network parameters are selected, the complete data set can be used for the final training of the optimum network. This optimum network can be further refined, if required, when the results obtained with the complete data set are not as good as when the smaller files were used.

Three types of data files are required; a training data file, a test data file and a validation data file. The former and the latter should contain representative samples of all the cases the network is required to handle, whereas the test file may contain about 10% of the cases contained in the training file. During training, the network is tested against the test file to determine accuracy and training should be stopped when the mean average error remains unchanged for a number of epochs[2]. This is done in order to avoid overtraining where, the network learns perfectly the training patterns but is unable to make predictions when an unknown data set is presented to it.

In back-propagation networks, the number of hidden neurons determines how well a problem can be learned. If too many are used, the network will tend to memorise the problem, and thus its ability to generalise well is reduced. If too few are used, the network will generalise well but may not have enough "power" to learn the patterns satisfactorily. Getting the right number of hidden neurons is done by trial and error, since there are no rules that can be applied and this strongly depends on the variability of the data. A general good approximation, which seems to give good results, is to use the following empirical formula in order to determine the number of hidden neurons (N) (Neuroshell, 1996):

$$N = \frac{I + O}{2} + \sqrt{P_i} \qquad (6)$$

where I is the number of input parameters, O is the number of output parameters and P_i is the number of training patterns available.

2.4. Artificial Neural Network Architectures

There are two basic types of neural networks; supervised and unsupervised. Supervised networks build models, which classify patterns or make predictions according to other patterns or inputs and outputs they have learned. They give the most reasonable answer based upon the variety of patterns used for their training. In supervised networks, the network gives predictions, classifications or decisions by training them with a large number of correct classifications or predictions from which it can learn. Unsupervised networks can classify a set of training patterns into a specified number of categories without being shown in advance

[2] An epoch is the complete run through all data of the training data file.

how to categorize them. This is done by clustering the patterns by their proximity in N dimensional space, where N is the number of inputs. The patterns are clustered into a number categories specified by the user.

There are a number of architectures that have been applied by various researchers. A short description of the most important ones, which are suitable for engineering systems, is given in this section. The architectures described include various backpropagation (BP) networks, general regression neural networks (GRNN), group method of data handling (GMDH), radial basis function (RBF) networks and unsupervised Kohonen architecture.

2.4.1. Standard and Jump Connections Architectures

This category of architectures belongs to the backpropagation networks. Backpropagation networks are known for their ability to generalize well on a wide variety of problems. Backpropagation networks are a supervised learning type of networks, i.e., training is done with both inputs and outputs. Backpropagation networks are used in a large number of working applications, as they tend to generalize well. The various alternative standard and jump connection architectures that may be considered are shown in Fig. 7.

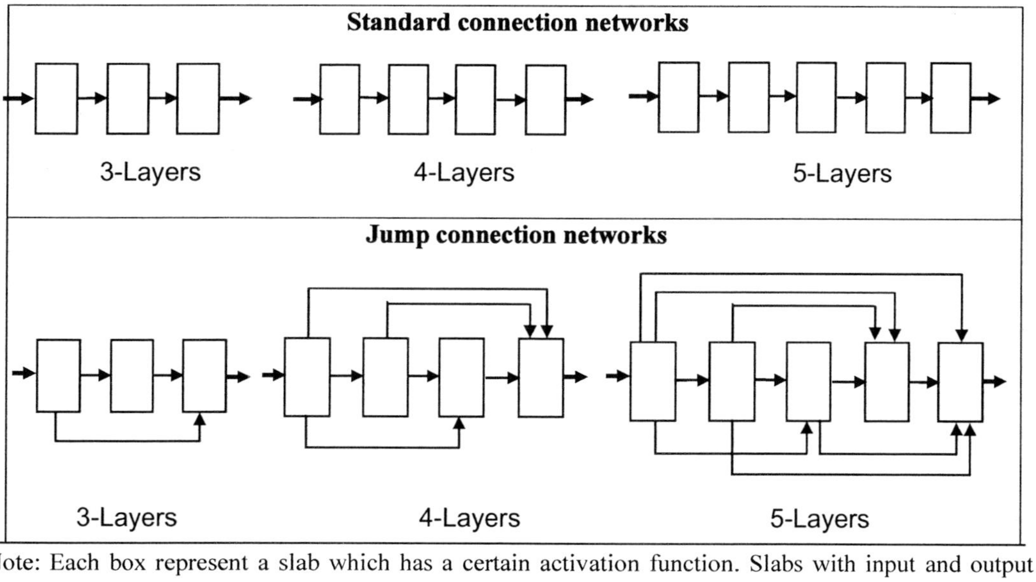

Note: Each box represent a slab which has a certain activation function. Slabs with input and output arrows are input and output slabs. All others are hidden

Fig. 7. Types of standard and jump connection neural network architectures

The standard connections neural network architecture, also known as multi-layer perceptron (MLP), is the one where each layer is connected to the immediately previous layer. Generally, three layers (an input, a hidden and an output layer) are sufficient for the majority of problems to be handled. A three-layer Backpropagation network with standard connections is suitable for almost all problems as it is considered a universal approximator. One, two or three hidden layers architecture can be used however, depending on the problem characteristics. Use of more than 5 layers in total generally offers no benefit and should be avoided.

Variations of standard connection architectures are the jump connection ones where various layers are connected to previous layers. These network architectures are found to be useful when working with very complex patterns, i.e., when it may be very difficult for a human to define the different patterns that are inherent in the data.

2.4.2. Recurrent Type Architecture

Another category of backpropagation architectures is the recurrent with dampened feedback from either the input, hidden, or output layer. This type of architecture is excellent for time series data. The various topologies that can be used are shown in Fig. 8.

The recurrent architecture shown in the middle of Fig. 8 is shown schematically and in more detail in Fig. 9. It is composed of four slabs; one of which is hidden and one is used for dampened feedback. In this case the extra slab is connected to the hidden layer. This architecture is commonly called "Jordan Elman recurrent network". It holds the contents of one of the layers as it existed when the previous pattern than the current one was trained. In this way the network uses previous knowledge which was acquired from the previous inputs. This extra slab is sometimes called the network's "long term" memory.

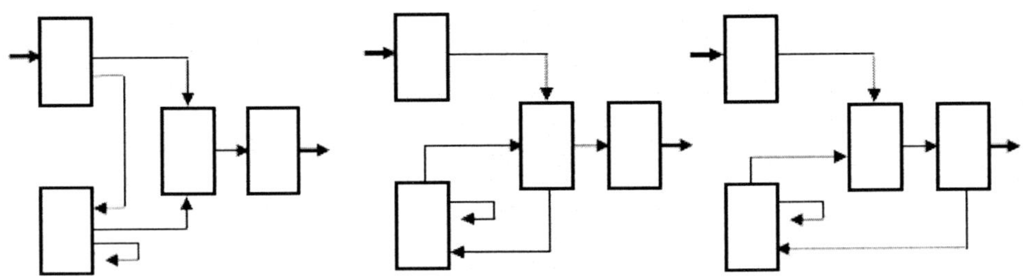

Fig. 8. Types of Recurrent networks with dampened feedback architectures

The long term memory remembers the hidden layer and contains features detected in the raw data of previous patterns. Recurrent neural networks are particularly suitable for prediction of sequences so they are particularly effective for time series data.

One such application for the prediction of the energy consumption of a passive solar building is shown in Kalogirou and Bojic (2000). The difference of this architecture from the networks of the previous category is that a back-propagation network with standard connections responds to a given input pattern with exactly the same output pattern every time the input pattern is presented whereas a recurrent network may respond to the same input pattern differently at different times, depending on the pattern of the inputs presented to the network just previously. Thus, the sequence of the patterns is as important as the input pattern itself. Recurrent networks are trained similarly as the standard back-propagation networks except that patterns must always be presented in the same order. The difference in the structure of this architecture is that there is one extra slab in the input layer that is connected to the hidden layer just like the normal input slab. This extra slab holds the contents of one of the layers (the hidden layer in this case) as it existed when the previous pattern was trained. In this way the network sees previous knowledge it had about previous inputs.

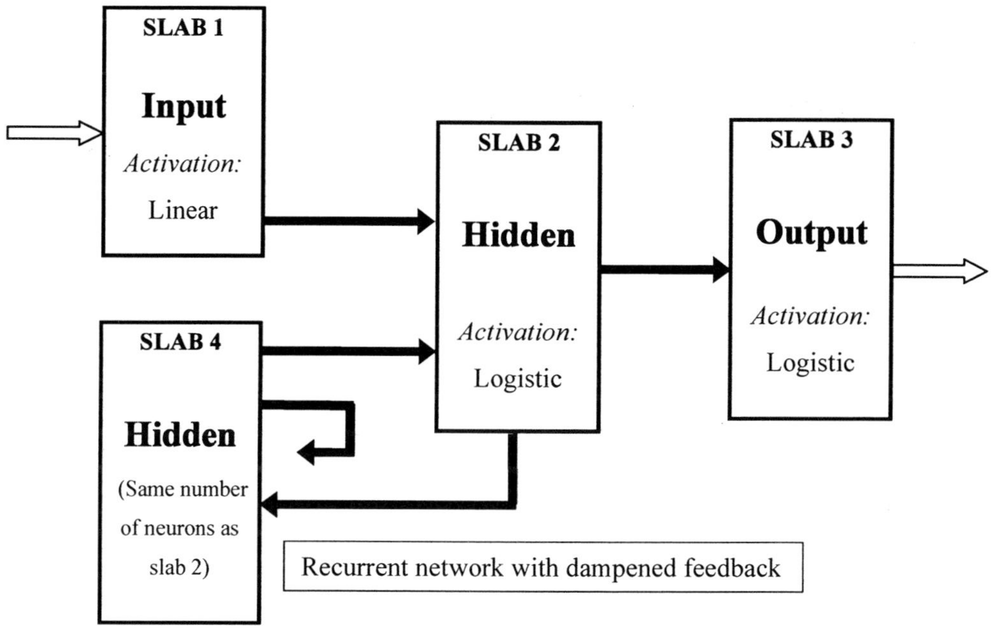

Fig. 9. Recurrent type architecture

The activation function that may be used for each slab is also shown in Fig. 9. The activation function used in the input slab is linear (i.e., of the form y=x) whereas in the hidden and output slabs is the sigmoid given by the logistic form:

$$y = \frac{1}{1+e^{-x}} \tag{7}$$

The learning procedure in this network may be implemented by using the back-propagation algorithm. In this case the learning rate, the initial value of the weights and the momentum factor need to be specified by the user at the beginning.

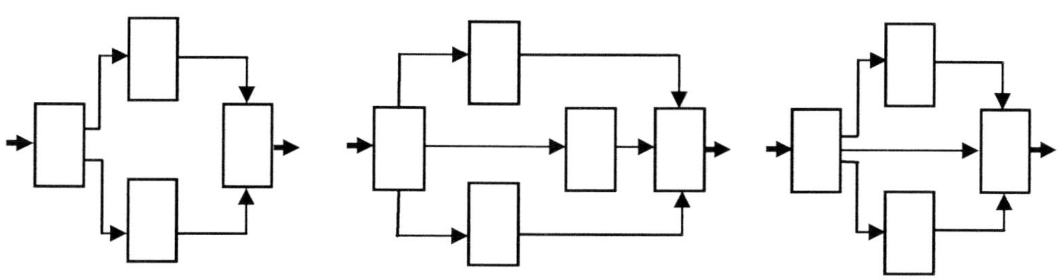

Fig. 10. Types of feedforward architectures with multiple hidden slabs and different activation functions

2.4.3. Feedforward with Multiple Hidden Slabs Architecture

Another category of the backpropagation networks is the feedforward with multiple hidden slabs. These network architectures are very powerful to detect different features of the input vectors when different activation functions are employed in the hidden slabs. The various topologies that can be used are shown in Fig. 10. This type of architecture has been used in a number of engineering problems for modeling and prediction with very good results (Kalogirou *et al.*, 1996; 1997; 1999; Kalogirou, 1996).

Figure 11 shows in detail the feedforward architecture shown in the middle of Fig. 10 and has three hidden slabs. The information processing at each node site is performed by combining all input numerical information from upstream nodes in a weighted average of the form:

$$\beta_i = \sum_j w_{ij}\alpha_{pj} + b_1 \quad (8)$$

where $\alpha_{(pi)}$ is the activation for each node and b_1 is a constant term referred to as the bias.

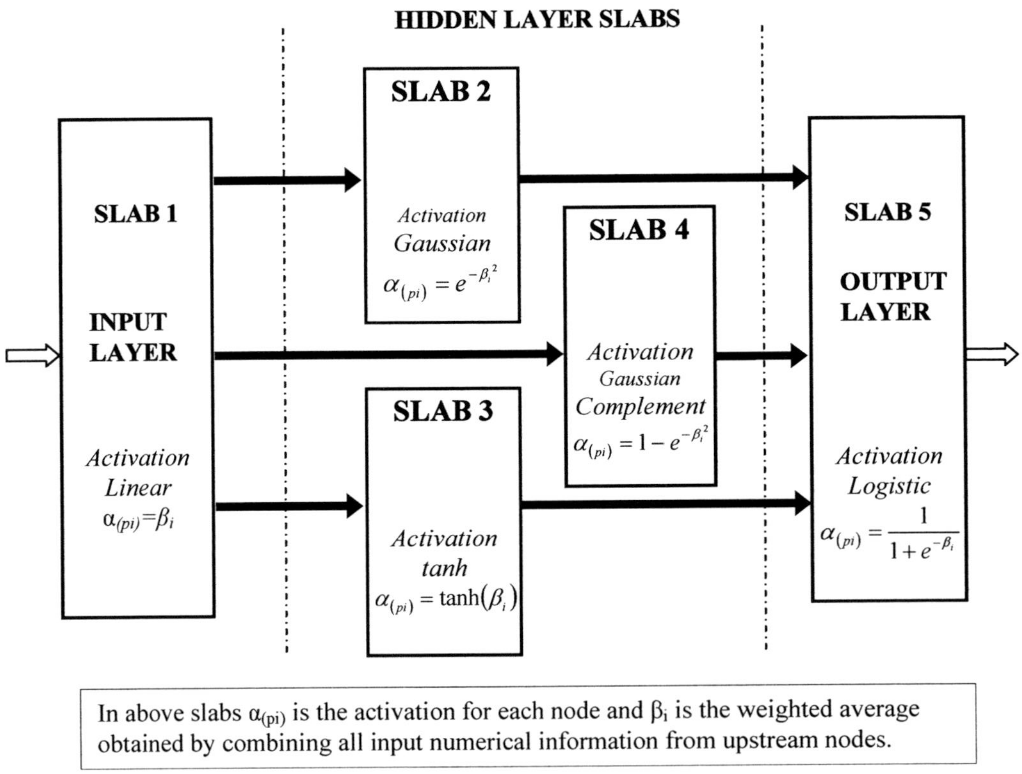

Fig. 11. Feedforward with multiple hidden slabs architecture

The final nodal output is computed via the activation function. The characteristics of this architecture is that it has different activation functions in each slab. The activation functions that may be used in the various slabs are also shown in Fig. 11. Thus, the input slab activation function is linear, i.e., $\alpha_{(pi)}=\beta_i$, (where β_i is the weighted average obtained by combining all input numerical information from upstream nodes), while the activations used in the other slabs are:

SLAB 2-Gaussian given by: $$\alpha_{(pi)} = e^{-\beta_i^2} \quad (9)$$

SLAB 3-Tanh given by: $$\alpha_{(pi)} = \tanh(\beta_i) \quad (10)$$

SLAB 4- Gaussian complement given by: $$\alpha_{(pi)} = 1 - e^{-\beta_i^2} \quad (11)$$

OUTPUT SLAB-Logistic. This is similar to Eq (7) but in the new notation is given by:

$$\alpha_{(pi)} = \frac{1}{1+e^{-\beta_i}} \quad (12)$$

The reason for applying different activation functions to hidden layer slabs is to detect different features in a pattern processed through a network. The same problem of finding the right number of hidden layer neurons applies here as before and the number of them may also be calculated with Eq (6). However, depending on the complexity of the data, an increased number of hidden layer neurons may be used in order to get more "degrees of freedom" and allow the network to learn patterns that are more complex. This is usually done when the input data are highly non-linear. It is highly recommended in this architecture to use Gaussian function on one hidden slab to detect features in the mid-range of the data and Gaussian complement in another hidden slab to detect features from the upper and lower extremes of the data. Thus combining the two feature sets in the output layer usually leads to a better prediction.

2.4.4. General Regression Neural Network (GRNN) Architecture

General regression neural networks (GRNN) have the ability to train quickly on sparse data sets. In numerous trials, it was found that GRNN responds much better than back-propagation in many types of problems, although this is not a rule. GRNN is especially useful for continuous function approximation. It can have multidimensional input, and it will fit multidimensional surfaces through the data. GRNN work by measuring how far a given sample pattern is from patterns in the training set in N dimensional space, where N is the number of inputs in the problem. For this purpose the Euclidean distance is usually adopted.

A GRNN architecture is shown in Fig. 12. As can be seen a GRNN is a four-layer feed-forward neural network architecture based on the nonlinear regression theory consisting of the input layer; the pattern layer, the summation layer and the output layer. There are no training parameters, such as learning rate and momentum used in back-propagation networks, but there is a smoothing factor that is applied after the network is trained. The smoothing factor determines how well the predictions of the network match the data in the training patterns. The neurons in the first three layers are fully connected while each output neuron is connected only to some processing units in the summation layer. The summation layer has two different types of processing units, i.e., the summation units and a single division unit. The number of the summation units is always equal to the number of the GRNN output units whereas the division unit only sums the weighted activations of the pattern units of the hidden layer. This is done without using any activation function. Each of the GRNN output units is connected only to its corresponding summation unit and to the division unit (there are no weights in these connections). The function of the output units is to perform a simple division of the signal coming from the summation unit by the signal coming from the division unit. The summation and output layers together basically perform a normalization of the output vector, thus making GRNN much less sensitive to the proper choice of the number of pattern units. More details on GRNN can be found in (Tsoukalas and Uhrig, 1997; Ripley, 1996).

All the data used, as in all neural networks, are scaled from their numeric range into the numeric range that the neural network deals with efficiently. For the GRNN an activation function is only required in the input layer slab.

If a lot of patterns are available in the training data set (perhaps more than 2000), then GRNN may become too slow to be feasible unless a very fast computer is available. The reason for the slow learning is that applying a GRNN network requires a comparison between a new pattern and each of the training patterns.

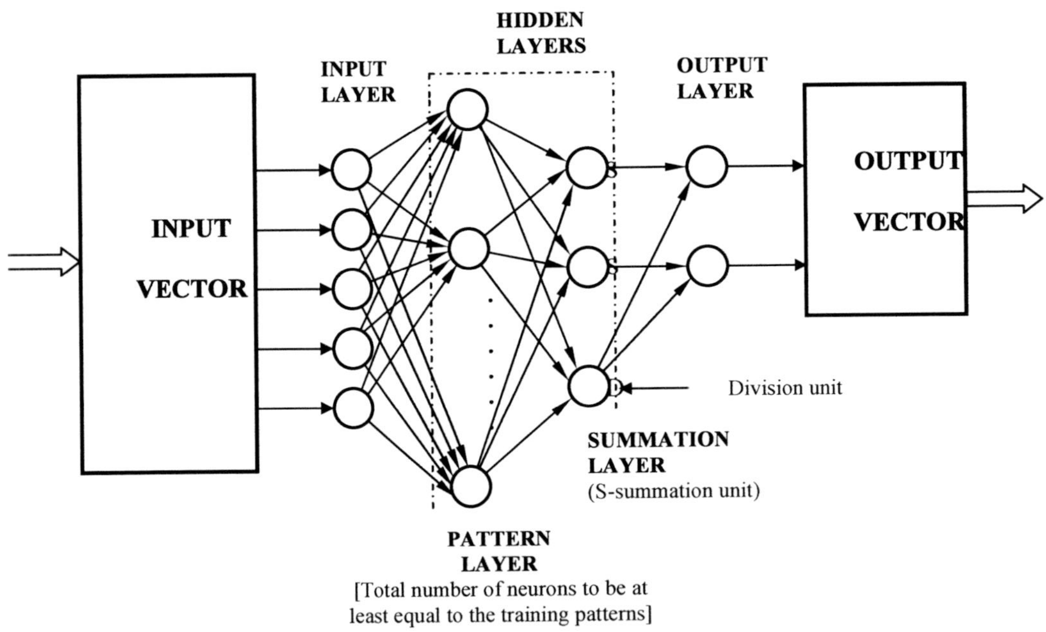

Fig. 12. General regression neural network architecture

In GRNN networks, the number of neurons in the hidden layer is usually equal to the number of patterns available in the training set because the hidden layer consists of one neuron for each pattern in the training set. This number can be made larger if one may want to add more patterns later, but it cannot be made smaller. As usual, the number of neurons in the input layer is equal to the number of input parameters and the number of neurons in the output layer corresponds to the number of actual outputs. If the number of input, output, or hidden neurons is changed, the network must be retrained. This for example may occur when more training patterns are added.

The GRNN is usually trained using genetic algorithms (GA). The genetic algorithm is used to find the appropriate individual smoothing factors for each input as well as an overall smoothing factor. Details on genetic algorithms are given in Section 3 but generally a GA uses a "fitness" measure to determine which of the individuals in the population survive and reproduce. Thus, survival of the fittest causes good solutions to progress. A genetic algorithm works by selective breeding of a population of "individuals", each of which could be a potential solution to the problem. In the case of GRNN, a potential solution is a set of smoothing factors, and the genetic algorithm is seeking to breed an individual that minimizes the mean squared error of the test set, which is calculated by:

$$E = \frac{1}{p} \sum_{p} \left(t_p - o_p\right)^2 \tag{13}$$

Where (E) is the mean squared error, (t) the network output (target), and (o) the desired output vectors over all pattern (p) of the test set.

The input smoothing factor is an adjustment used to modify the overall smoothing and provides a new value for each input. When the network training is completed, the individual smoothing factors can be used as a sensitivity analysis tool that can be used to find the most important inputs , i.e., the bigger the factor for a given input, the more important that input is to the model, as far as the test set is concerned. Inputs with low smoothing factors are candidates for removal for a latter trial if the overall results are not satisfactory. Individual smoothing factors are unique to each network. The numbers are relative to each other within a given network and they cannot be used to compare inputs from different networks.

In genetic algorithms the larger the breeding pool size, the greater is its potential of producing a better individual. However, as the smoothing factor produced by every individual must be applied to the test set on every reproductive cycle, larger breeding pools take longer time. After testing all the individuals in the breeding pool, a new "generation of individuals" is produced for testing. The training of the GRNN is quite different from the training used in other neural network architectures. Oppositely with back-propagation algorithm, which propagates the error through the neural network many times seeking a lower mean squared error between the network's output and the actual output, GRNN training is completed after each input-output vector pair from the training data set is presented to the GRNN input layer once.

2.4.5. Group Method Data Handling Neural Network (GMDH) Architecture

One type of neural network, which is very suitable for modelling energy and renewable energy systems, is the Group Method of Data Handling (GMDH). The GMDH technique was invented by A. G. Ivakhnenko from the Institute of Cybernetics in Ukrainian Academy of Sciences (Ivakhnenko, 1968; 1971), but enhanced by others (Farlow, 1984). This network is also known as "polynomial networks" as it results in a polynomial equation that can represent the modelled system. GMDH is a feature-based mapping network. Ivakhenenko developed GMDH for the purpose of building more accurate predictive models of fish populations in rivers and oceans. He found that GMDH worked well for modelling fisheries but it can also be used for many other modelling applications (Hecht-Nielsen, 1991).

GMDH works by building successive layers with links that are simple polynomial parameters. These polynomial parameters are created by using both linear and non-linear regression. The first layer is simply the input layer. The second layer created is made by computing regressions of the input variables from which the best ones are chosen. The third layer is created by computing regressions of the terms in the second layer along with the input variables. Only the best regressions are chosen by the algorithm, called survivors. This process continues until the network stops getting better, according to a prespecified selection criterion. Finally the resulting network is represented as a complex polynomial description of the model of a system which is in the form of a mathematical equation. The complexity of the resulting polynomial depends on the variability and complexity of the training data. In some respects GMDH, is very similar to regression analysis, but it is far more powerful than the latter. GMDH can build very complex models while avoiding overfitting of the data. An additional advantage of GMDH is that as it trains it recognizes the best variables and for problems with many variables, the ones with low contribution are discarded during the training process.

A typical GMDH architecture is shown in Fig. 13. The GMDH network is not like a regular feedforward network and in fact it was not originally represented as a neural network. The GMDH network is implemented with polynomial terms in the links and can be used with a genetic algorithm component to decide how many layers are built during the training process. The result of training at the output layer is represented as a polynomial function of all or some of the inputs (for problems with many inputs those with low contribution are discarded).

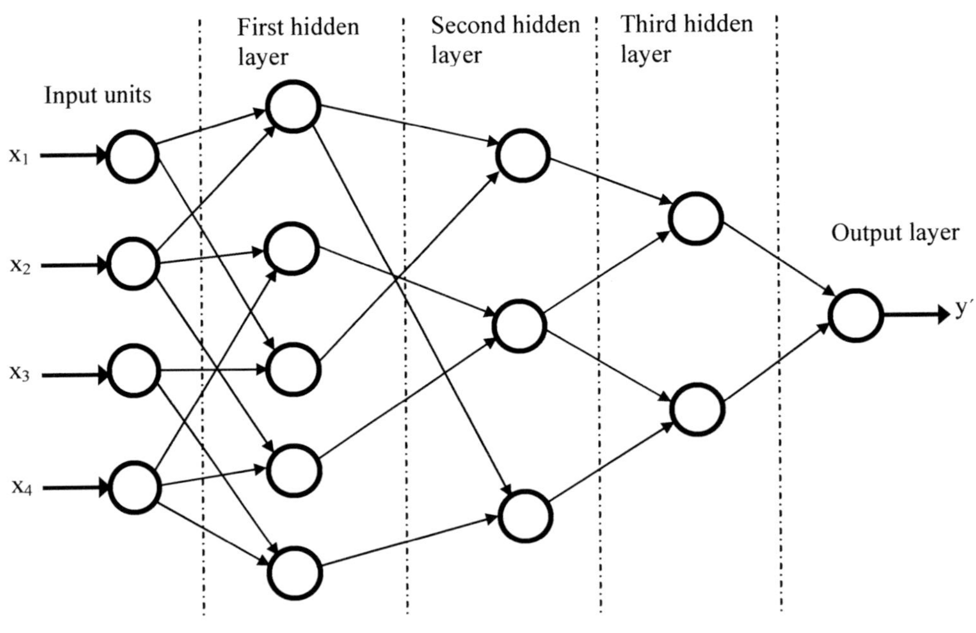

Fig. 13. GMDH neural network architecture

With reference to Fig. 13, the network maps a vector input **x** to a scalar output y′. Each processing element has two inputs, which are combined in a general quadratic form with six weights to yield the output signal. Layers in these units are combined to implement the mapping. The main idea behind GMDH is that it is trying to build a function (i.e., a polynomial model), which would behave in such a way that the actual and predicted values of the output parameters would be as close as possible.

For many end users, like energy engineers, it may be more convenient to have a model, which is able to make predictions using polynomial formulas, which are widely understood, instead of normal neural networks that operate like "black box" models. The most common method to solve such models is to use regression analysis. Thus, the first step is to decide the type of polynomial that regression should find. A good idea is to use powers of input variables along with their covariants and trivariants as terms of the polynomial as for example:

$$\{x_1, x_2, x_3, ..., x_1^2, x_2^2, x_3^2, ..., x_1x_2, x_1x_3, ...,x_{n-1}x_n, x_1x_2x_3, ...\} \quad (14)$$

The next step is to construct a linear combination of all of the polynomial terms with variable coefficients. Then the algorithm determines the values of these coefficients by minimizing the squared sum of the differences between actual values and model predictions, over all samples.

The main problem when utilizing regression is the correct choice of the set of polynomial terms. In addition, decisions need to be made about the degree of the polynomial such as how complex the terms should be or whether the model should evaluate for example terms such as x^8, or maybe limit the consideration to terms up to x^5 and lower.

The decision about the quality of each model must be made using a numeric criterion. The simplest criterion, which is also used in linear regression analysis, is the sum, over all samples, of the squared differences between the actual output (y_a) and the model's prediction (y_p) divided by the sum of the actual squared. This is also called the normalized mean squared error (E_n) and is given by:

$$E_n = \frac{\sum_{i=1}^{N}\left(y_a - y_p\right)^2}{\sum_{i=1}^{N} y_a^{\,2}} \quad (15)$$

If however only the normalized mean squared error is used there is a problem as its value gets smaller as long as extra terms are added to the model. This is because, as it is obvious, the more complex the model, the more exact it is. This result in an "overcomplex" model or model "overfit", which means that the model, does not generalize well because it pays too much attention on noise in the training data. This is similar to "overtraining" other neural networks. This is true if normalized mean squared error is used alone as it determines the quality of the model by evaluating the same information already used to build the model. To avoid this danger, a more powerful criterion is needed, which could be based on information other than the one which was used to build the evaluated model. There are several ways to define such criteria. One of them is to use the squared sum of differences between the actual values and model prediction over some other set of experimental data, i.e., a test set. Another way to avoid overfitting is to introduce a penalty for model complexity, called the predicted squared error criterion. Theoretical considerations show that increasing model complexity should be stopped when the selection criterion reaches a minimum value.

2.4.6. Radial Basis Function Neural Networks

Radial basis function (RBF) networks also belong to supervised learning category. Radial basis functions networks have a very strong mathematical foundation rooted in regularization theory for solving ill-conditioned or ill-defined problems. RBFs have been used for data modeling, such as curve fitting, by many researchers (Powel, 1987; Poggio and Girossi, 1989). Recently these functions have been put in a neural network paradigm in what is called RBF neural network (Broomhead and Lowe, 1988).

RBF networks have the same feed-forward layered architecture, as the multi-layer perceptrons but the weighting function W and the activation function α are different. With RBF networks, there are only weights between the output and hidden layers (see Fig. 14). Each node in the hidden layer has a unique function, called the basis function. For the simple network of Fig. 14 with a single input, single output and two basis functions, the output is given by the sum of the two basis functions; each multiplied with its own weighting factor. In principle, any type of function can be used to act as basis function. For example, spline functions are used by Kavli (1992), but the identification RBF network, applies only if radial basis functions are used.

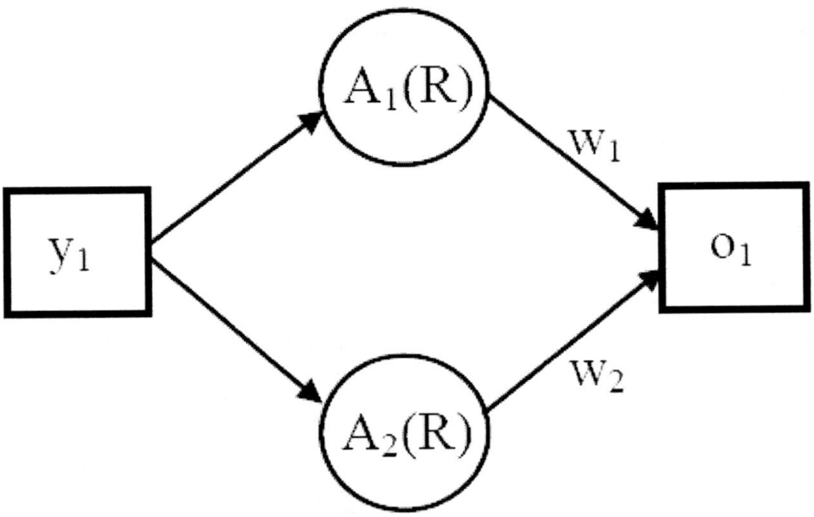

Fig. 14. Schematic representation of Radial Basis Function network (single input, two basis functions and one output)

Radial basis functions give local support to data points. The output of the hidden nodes, peaks when the input is near the centroid of the node and then falls of symmetrically as the Euclidean distance between input and centroid of a node increases (see Fig. 15). The consequence of this behavior is that RBF networks are good for data interpolation, but not good for data extrapolation.

Several different radial basis functions are in use, with the Gausian function (Fig. 15a), being the most widely used. If the radial basis centre R is defined as:

$$R = \sqrt{\sum_{i=1}^{L} \frac{(y_i - \mu_i)^2}{\sigma_i^2}} \tag{16}$$

where μ_i represents the centre location of each basis and σ_i indicates a scaling of the width of each basis, then the Gaussian activation function is given by:

$$\alpha_R = \exp\left(-\frac{R^2}{2}\right) \tag{17}$$

Multiplication of the activation function α_R with a weighting factor w then yields the output o (Fig. 14).

Another widely used RBF function is the so-called Inverse Multi-Quadratic Equation (IMQE, Fig. 15b), defined as:

$$\alpha_R = \frac{1}{\sqrt{R + k^2}} \tag{18}$$

where k is an empirically determined smoothing factor (usually 0.5).

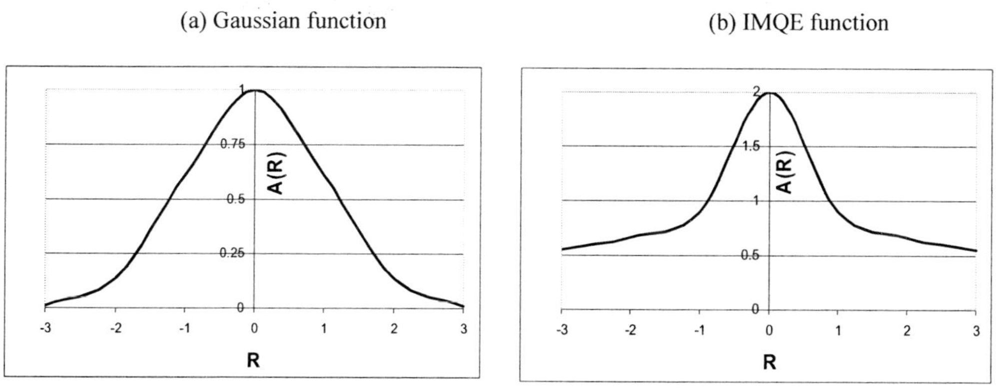

Fig. 15. Activation functions for RBF networks (a) Gaussian function (b) IMQE function

It should be noted that the widths in RBF functions are specified independently from each input dimension, making the functions elliptic rather than spherical. Also, no bias is included in the RBF functions.

Centre locations are typically determined by randomly selecting training examples from a large set of training data. The smoothing parameters and the number of nodes are typically adjusted empirically during training. One important aspect when comparing RBF networks with multi-layer percentrons is the training speed. RBF networks can be trained within a fraction of the time that is required for training multi-layer percentrons. RBF networks however, generally require more nodes to obtain similar performance.

2.4.7. Unsupervised Training Networks

Unsupervised networks have the ability to learn without being shown correct outputs in sample patterns. These networks are able to separate data into a specified number of categories.

These networks are called self-organizing map (SOM) or Kohonen networks from the name of the inventor (Kohonen, 1990; 1995). These networks are using an unsupervised learning algorithm where the input patterns are freely distributed over the output node matrix. The weights are adapted without supervision in such a way, so that the density distribution of the input data is preserved and represented on the output nodes. This mapping of similar input patterns to output nodes, which are close to each other, represents a discretisation of the input space, allowing visualization of the distribution of the input data. The output nodes are usually ordered in a two dimensional grid and the weights of the output nodes in the neighborhood of the winning output node are adapted simultaneously. The neighborhood around the winning output node decreases in size over time. The weights of the output nodes in the neighborhood are adapted with:

$$w(t+1) = w(t) + \varepsilon h(t)\left(x - w(t)\right) \tag{19}$$

where ε is the learning rate ($0<\varepsilon\leq 1$) and h(t) is a decreasing function which decreases with time.

At the end of the training phase, the output nodes are labelled with the class of the majority of the input patterns assigned to each node. In the evaluation phase, an input pattern is said to belong to the class label of the winning output node where it has been assigned.

Unsupervised learning is particularly well suited to perform feature extraction. There are a lot of problems where it is not known *a priori* the desired signal, nor what are the most important features of a given signal. This happens in data compression, in signal and noise separation and in feature extraction problems.

Unsupervised learning networks are normally very simple topologies that utilize the idea of correlation between input and output to process signals. It turns out that correlation can be easily implemented in a neural network through Hebbian learning. In fact most of the unsupervised learning rules, in one way or another, are based on Hebbian learning.

In Hebbian type learning, the weights are modified according to the product of the input and output at the weight, implying that there is no desired signal. Hence, it is not necessary to propagate the signal forward and the error backward, as in gradient descent learning.

The Kohonen self-organizing map network contains only two layers; an input layer and an output layer which has one neuron for each possible category. The maximum number of output categories can be specified by setting the number of neurons in the output layer. Thus, if it required that the patterns are categorized into N clusters, then the output layer should contain N neurons. The Kohonen network may decide that the patterns have less than N categories, but N will be the maximum it can find.

The training patterns are presented to the input layer, then propagated to the output layer and evaluated. One output neuron is the "winner". The network weights are adjusted during training. This process is repeated for all patterns for a number of epochs chosen in advance.

This network is very sensitive to learning rate. This is better to be lowered slightly but steadily as the training progresses, thus, causing smaller and smaller weight changes. This causes the network to stabilize.

The network adjusts the weights of the neurons in a neighborhood around the winning neuron. The neighborhood size is variable, starting off fairly large (sometimes even close to the number of categories) and decreasing with learning until during the last training events the neighborhood is zero, meaning that by then only the winning neuron's weights are changed. By that time the learning rate is very small and the clusters have been defined.

Kohonen networks are usually trained for a fixed number of epochs. Small problems may train in as little as 50 epochs, but for really large problems the number of epochs can be set to very large numbers (e.g. 10,000).

How well a Kohonen network classifies data is dependent upon how well the learning rate, initial weights and neighborhood size are set. Learning may not be successful if a large enough neighborhood size is not available or the network is not trained for enough epochs, or if a too small learning rate is used at the beginning.

3. Genetic Algorithms

A genetic algorithm (GA) is a model of machine learning, which derives its behavior from a representation of the processes of evolution in nature. This is done by the creation within a computer of a population of individuals, which are represented by chromosomes. Essentially these chromosomes are a set of character strings that are analogous to the ones that can be found in the DNA of human beings. The individuals in the population then go through a process of evolution.

It should be noted that evolution as occurring in nature or elsewhere is not a purposive or directed process. For example in the animal kingdom the processes of nature seem to end to different individuals competing for resources in the environment. Some are better than the others and those that are better are more likely to survive and propagate their genetic material. In fact in nature, the encoding for the genetic information is done in a way that admits asexual reproduction typically resulting in offspring that are genetically identical to their parents. Sexual reproduction allows the creation of genetically radically different offspring that are still of the same general species.

In an over simplified way, what happens at the molecular level is that a pair of chromosomes bump into one another, exchange chunks of genetic information and drift apart. This is called the recombination operation, which in GAs is generally referred to as "crossover" because of the manner in which the genetic material crosses over from one chromosome to another. The crossover operation takes place in an environment where the selection of who mate is a function of the fitness of the individual. The fitness is a parameter which shows how good the individual is at competing in its environment. Some GAs use a simple function of the fitness measure to select individuals probabilistically who undergo genetic operations such as crossover or asexual reproduction. It should be noted that in asexual reproduction the propagation of genetic material remains unaltered. This is called the fitness-proportionate selection. Other implementations use a model in which certain randomly selected individuals in a subgroup compete in order to select the fittest. This is called tournament selection. The two processes that contribute mostly to evolution are crossover and fitness based selection/reproduction. Mutation also plays a role in this process.

GAs are used for a number of different application areas such as multidimensional optimization problems in which the character string of the chromosome can be used to encode the values for the different parameters to be optimized. In practice, this genetic model of computation can be implemented by having arrays of bits or characters to represent the chromosomes. Simple bit manipulation operations allow the implementation of crossover, mutation and other operations.

When a GA is executed, the following cycle is usually followed:

- Evaluate the fitness of all of the individuals in the population.
- Create a new population by performing operations such as crossover, fitness-proportionate reproduction and mutation on the individuals whose fitness has just been measured.
- Discard the old population and iterate using the new population.

One iteration of this loop is referred to as a generation (epoch in neural networks).

The structure of the standard genetic algorithm is shown in Fig. 16 (Zalzala and Fleming, 1997). With reference to Fig. 16, in each generation individuals are selected for reproduction according to their performance. This is done with respect to the fitness function of each individual. In essence, selection works in a way that gives a higher chance of survival to better individuals. Subsequently the genetic operations are applied to form new and possibly better offspring. The first generation (generation 0) of this process operates on a population of randomly generated individuals. From there on, the genetic operations, in concert with the fitness measure, operate to improve the population. The algorithm is terminated either after a certain number of generations or when a target value representing the optimal solution is found. More details on genetic algorithms can be found in Michalewicz (1996), Goldberg (1989) and Davis (1991).

Begin (1)
    n = 0 [start with an initial time]
    Initialize Population P(n) [initialize a usually random population of individuals]
    Evaluate fitness of Population P(n) [evaluate fitness of all individuals in population]
    While (Generations < Total Number) do begin (2)
    n = n + 1 [increase the time counter]
    Select Population P(n) out of Population P(n-1) [select sub-population for offspring production]
    Apply Crossover on Population P(n)
    Apply Mutation on Population P(n)
    Evaluate fitness of Population P(n) [evaluate new fitness of population]
    end (2)
end (1)

Fig. 16. Standard genetic algorithm

During each step in the reproduction process, the individuals in the current generation are evaluated by the fitness function value, which is a measure of the accuracy with which the individual solves the problem. Then each individual is reproduced in proportion to its fitness. Thus the higher the fitness of an individual the higher is its chance to participate in mating (crossover) and to produce an offspring. Then a small number of newborn offspring undergo the action of the mutation operator. Therefore, after many generations, only those individuals who have the best genetics (with respect to their fitness function) survive. The individuals that emerge from this 'survival of the fittest' process are the ones that represent the optimal solution to the problem as specified by the fitness function and the constraints.

As can be understood from the above discussion genetic algorithms are suitable for finding the optimum solution in problems were a fitness function is present or can be derived. For example, a fitness function can be the square of the summation of the difference of the actual to the predicted values over all samples in a dataset. Genetic algorithms use the "fitness" measure to determine which of the individuals in the population survive and reproduce. Thus, survival of the fittest causes good solutions to progress. A genetic algorithm works by selective breeding of a population of "individuals", each of which could be a potential solution to the problem. The genetic algorithm is seeking to breed an individual, which either maximizes, minimizes or it is focused on a particular solution of a problem. In the case of the above example, the optimum is found when the fitness function is zero, i.e., all the predicted values are equal to the actual values.

The processes of genetic searches, which are selection, crossover and mutation, can be described as follows:

1. Selection

The selection operator biases the search process in favor of the more fit members, based on their fitness value. The fitness value of the ith member in a population can participate in this operation based on its probability. The probability of the ith member in a population can be computed from:

$$P_i = \frac{f_i}{\sum_{k=1}^{m} f_k} \tag{20}$$

The selection operation helps to increase the participation of the more fit members of the population while less fit members may be completely suppressed that leads to an increase in average fitness.

2. Crossover

Crossover operation allows for an exchange of design characteristics among the mating members. The crossover operation is performed by picking two mating parents, then randomly selecting two sites on each of the chromosomal strings and swapping strings between the sites among the pair based on a crossover probability p_c. The crossover operation can be illustrated is as follows:

Parent 1 = 1000101011 → Child 1 = 1000000111
Parent 2 = 1010010111 → Child 2 = 1010110111

3. Mutation

The mutation operator is invoked with a low probability (p_m) at a randomly picked site on a chromosomal string of a randomly chosen design. The operation consists of switching a 0 value to 1 and vice versa. The mutation operation can be illustrated as follows:

Design = 10001010110011100
New design = 10000010110111100

A genetic algorithm is not gradient based, and uses an implicitly parallel sampling of the solutions space. The population approach and multiple sampling indicates that it is less subject to becoming trapped to local minima that can happen in traditional direct approaches, and can navigate a large solution space with a highly efficient number of samples. Although not guaranteed to provide the globally optimum solution, GAs have been shown to be highly efficient at reaching a very near optimum solution in a computationally efficient manner.

It should be noted that the larger the breeding pool size, the greater is its potential to produce a better individual. However, the fitness value produced by every individual must be compared with all other fitness values of all other individuals on every reproductive cycle, therefore larger breeding pools take longer time. After testing all of the individuals in the pool, a new "generation" of individuals is produced for testing.

During the setting up of the GA the user has to specify the adjustable chromosomes, i.e. the parameters that would be modified during evolution to obtain the maximum, minimum or a set value of the fitness function. Additionally the user has to specify the ranges of these values called constraints. The genetic algorithm is terminated after best fitness remained unchanged for a number of generations or when the optimum solution is reached. The genetic algorithm parameters that need to be specified by the user are:

1. Population size

Population size determines the size of the genetic breeding pool, i.e., the number of individuals contained in the pool. If the population size is set to a low value, there would not be enough different kinds of individuals to solve the problem satisfactorily. On the other hand, if there are too many in the population, a good solution will take longer to be found because the fitness function must be calculated for every individual in every generation.

2. Crossover Rate

Crossover rate determines the probability that the crossover operator will be applied to a particular chromosome during a generation. This parameter is usually set near 90%.

3. Mutation Rate

Mutation rate determines the probability that the mutation operator will be applied to a particular chromosome during a generation. This parameter is usually set to a very small number, near 1%. Low values of mutation probability are recommended, as higher values tend to be disruptive to the conversion process.

4. Generation Gap

Generation gap determines the fraction of those individuals that should not go into the next generation. It is sometimes desirable that individuals in the population be allowed to go into the next generation. This is especially important if the individuals selected are the most fit ones in the population. This parameter is usually set near 95%.

5. Chromosome Type

Populations are composed of individuals, and individuals are composed of chromosomes, which are equivalent to the variable parameters of a problem/system. Chromosomes are composed of smaller units called genes. There are two types of chromosomes, continuous and enumerated as shown in Fig. 17.

Continuous chromosomes are implemented in the computer as binary bits. The two distinct values of a gene, called alleles, can take values of 0 or 1. Multiple chromosomes make up each individual. Each partition is one chromosome, each binary bit is a gene, and the value of each bit (1, 0, 0, 1, 1, 0) is an allele. The genes in a chromosome can take on a wide range of values between the minimum and maximum values of the associated variable. One variation of continuous chromosomes is the 'integer chromosomes'. These are used in problems that require to take only integer values of chromosomes and genes (no decimals allowed).

1	1	0	0	1	0

A. Continuous chromosomes

2	3	5	4	1	6

B. Enumerated chromosomes

Fig. 17. Continuous and enumerated chromosomes

Enumerated chromosomes consist of genes, which can have allele values that include values more than just 0 and 1, and these values, depending on the software, are usually visible to the user. This kind of chromosomes is suitable for a category of problems, called combinatorial problems. The most famous example of combinatorial problems is the Travelling Salesman Problem (TSP). In the TSP, a salesman must visit a number of cities, say 8, and he only wants to visit each city once. The objective of this problem is to find the shortest route through all cities in order to minimize his travelling expenses. The values of chromosomes range from a minimum to a maximum value, and the order of their genes in the chromosome is important. In the TSP example with 8 cities, there will be 8 genes, each with values ranging from 1 to 8. The entire chromosome represents the path through the cities. Therefore each individual has only one chromosome, comprising the entire list of cities. For example, if a particular chromosome in the population is 7, 3, 4, 5, 2, 8, 1, 6, this represents the solution where the salesman will go first to city 7, then to city 3, and so on. Each partition (city) is a gene, and each value (7, 3, 4, 5, 2, 8, 1, 6,) is an allele.

It is obvious that the chromosomes in the TSP problem should not contain duplicate genes, i.e., genes with the same allele values otherwise the salesman may travel to the same city more than once. Two different types of enumerated chromosomes are available with either 'repeating genes' or 'unique genes'. For the TSP, unique genes have to be used but in some other problems repeating genes where chromosomes can be 5,2,3,6,4,7,5,1,5,8 or even 4,4,4,4,4,4,4 may be required.

4. Fuzzy Logic

Fuzzy logic was initialled in the United States by Professor Lofti Zateh (1973). In fact, Zadeh's theory not only offered the theoretical basis for fuzzy control, but also established a bridge connecting artificial intelligence to control engineering. Fuzzy logic was initially developed as a tool for controlling household and entertainment electronics, as well as industrial processes, diagnosis systems and other expert systems. Fuzzy logic is basically a multi-valued logic that allows intermediate values to be defined between conventional evaluations like large/small, black/white, true/false, yes/no, etc. With this method notions like "pretty cold" or "rather warm" can be formulated mathematically and processed in computers. Thus, a more human-like way of thinking is applied in the programming of computers.

Fuzzy logic is all about the relative importance of precision, i.e., how important is to be exactly right when a rough answer will work. Additionally fuzzy logic is almost synonymous with the theory of fuzzy sets, a theory that relates to classes of objects without sharp boundaries in which membership is a matter of degree. Fuzzy logic is a logical system, which can be considered as an extension of multi-valued logic. Fuzzy inference systems have been applied successfully in various fields such as automatic control, computer vision, expert systems, decision analysis and data classification. Fuzzy logic is a convenient way to map an input space to an output space, as for example, according to how hot the water is required for a process adjust the steam supply valve to the right setting, or according to the fluid outlet temperature required adjust the fuel flow in the boiler's burner. Thus, it is clear from the above examples that fuzzy logic mainly deals with the design of controllers.

Conventional control is based on the derivation of a mathematical model of a plant from which the controller's mathematical model can be obtained. When a mathematical model of a process or plant cannot be developed then there is no way through classical control to create a controller. Other limitations of conventional control are (Zadeh, 1973):

- Nonlinear process or plant models are computationally intensive and have complex stability problems.
- Accurate models of a process or plant can not be created due to uncertainty and lack of perfect knowledge.
- Multi-variable and multi-loop systems have complex environmental constrains and dependencies.
- Uncertainty in measurements due to noise.
- Plants, controllers, environments and their constrains vary with time (temporal behaviour) and time delays are difficult to model.

The advantages of fuzzy control are (Zadeh, 1973):

- Fuzzy controllers are more robust than PID controllers as they can cover a much wider range of operating conditions and can operate with noise and disturbances.
- Development of fuzzy controllers is cheaper than that of model-based controllers to do the same job.
- Fuzzy controllers are customisable since it is easier to understand and modify their rules as they are expressed in natural linguistic terms.
- It is easy to learn how fuzzy controllers operate and how to design and apply them in a particular application.
- Fuzzy controllers can model nonlinear functions of arbitrary complexity.
- A fuzzy controller can be built based on the experience of experts.
- A fuzzy controller can be blended with conventional control techniques.

As a rule fuzzy control should not be used when conventional control theory yields satisfactory results and when an adequate and solvable mathematical model already exists or can easily be developed.

A fuzzy controller design process contains the same steps as any other design process. Initially one needs to choose the structure and parameters of a fuzzy controller, test the model or the controller itself and change the structure and/or parameters based on the test results (Zadeh, 1973). A basic requirement for implementing fuzzy control is the availability of a control expert who would provide the necessary knowledge for the control problem (Nie and Linkens, 1997). More details on fuzzy control and practical applications can be found in (Reznik, 1997). The linguistic description of the dynamic characteristics of a controlled process can be interpreted as a fuzzy model of the process. In addition to the knowledge of a human expert, a set of fuzzy control rules can also be derived by using knowledge obtained from the experimental behaviour of the system. A fuzzy controller avoids precise mathematical models and is more robust than the classical approaches in cases which the process or plant cannot be or are with great difficulties precisely modelled mathematically. Fuzzy rules are used to describe in linguistic terms the quantitative relationship between two or more variables. Processing of the fuzzy rules provides the mechanism with which the rules can be used to compute the response of a given fuzzy controller input.

The basis of a fuzzy or any fuzzy rule system is the inference engine. This is responsible for the inputs fuzzification, fuzzy processing and defuzzification of the output. A schematic of the inference engine is shown in Fig. 18. Fuzzification generally means that the actual inputs are fuzzified and thus fuzzy inputs are obtained. Fuzzy processing means that the inputs are processed according to the various fuzzy rules and produces fuzzy outputs. Finally defuzzification generally means to produce a crisp real value for a fuzzy output which is also the output of the controller.

The goal of a fuzzy logic controller is to achieve a satisfactory control of a process or plant. Based on the input parameters, the operation of the controller (output) can be determined. The typical design scheme of a fuzzy logic controller is shown in Fig. 19 (Zadeh, 1973). The design of such a controller usually contains the following steps:

(1) Define the inputs and the control variables.
(2) Define the condition interface and express inputs as fuzzy sets.
(3) Design the rule base.

(4) Design the computational unit (many readymade programs are available for this purpose).
(5) Determine the rules for defuzzification, which will be used to transform fuzzy control output to crisp control action.

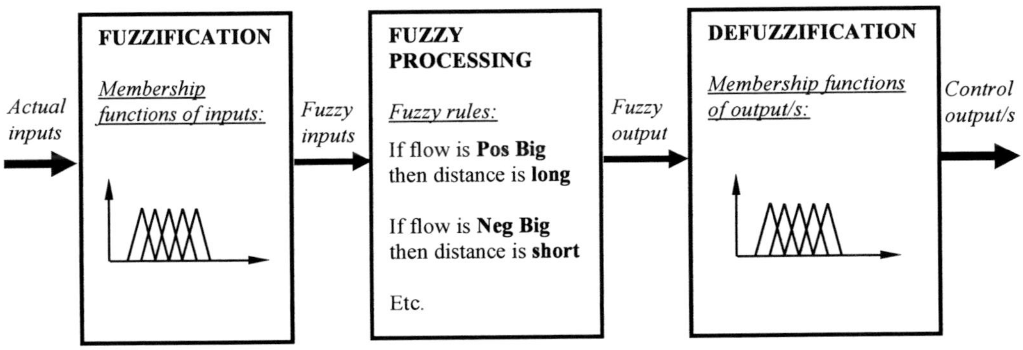

Fig. 18. Fuzzy controller operation

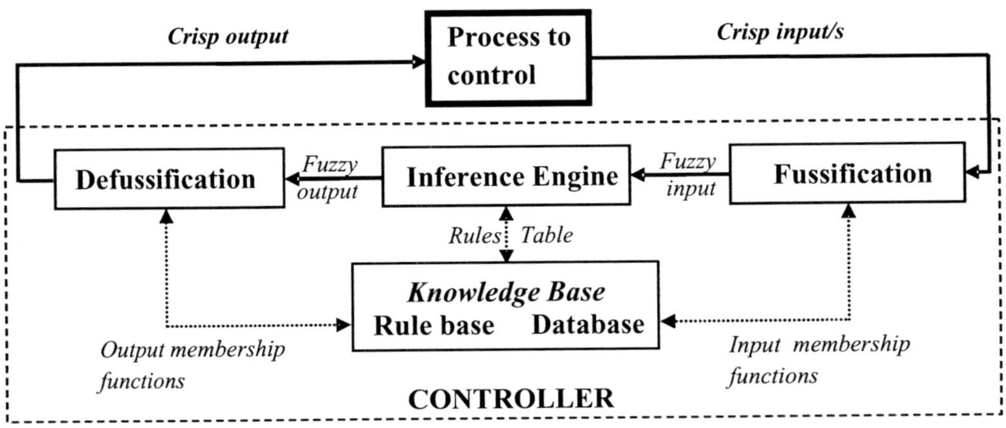

Fig. 19. Basic configuration of fuzzy logic controller

4.1. Membership Functions

A membership function is a curve that defines how each point in the input space (sometimes referred to as the universe of discourse) is mapped to a membership value, or degree of membership, between 0 and 1. The only condition that a membership function must actually satisfy is that it must vary between 0 and 1. Additionally it is possible in a fuzzy set to have partial membership functions, e.g. the weather is rather cold. The function itself can be an arbitrary curve whose shape can be defined as a function that suits the particular problem from the point of view of simplicity, convenience, speed and efficiency. For this purpose a variety of curves like the triangular, trapezoidal, quadratic, Gaussian (exponential), cos-function and many others can be used.

Based on signals obtained from sensors and common knowledge, membership functions for the input and output variables need to be defined. The inputs are described in terms of linguistic variables such as, Very High, High, Okay, Low and Very Low as shown in Fig. 20 or Positive Big, Positive, Positive Small, Neutral, Negative Small, Negative and Negative Big. Depending on the problem, different sensors could be used representing different parameters like slope, angle, distance, resistance, etc.

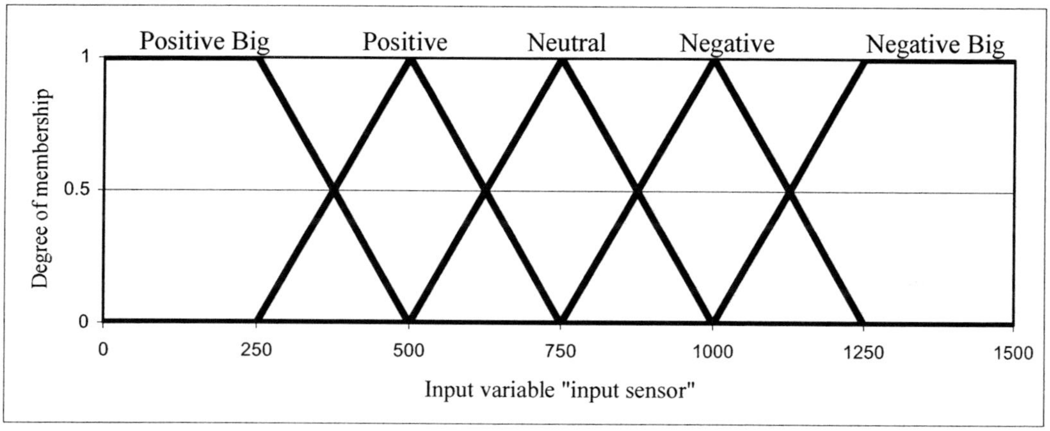

Fig. 20. Membership functions for linguistic variables describing an input sensor

Similarly, the output can be determined in a similar way according to some membership functions as for example the ones presented in Fig. 21. As was indicated before in both cases membership curves other than the triangular can be used.

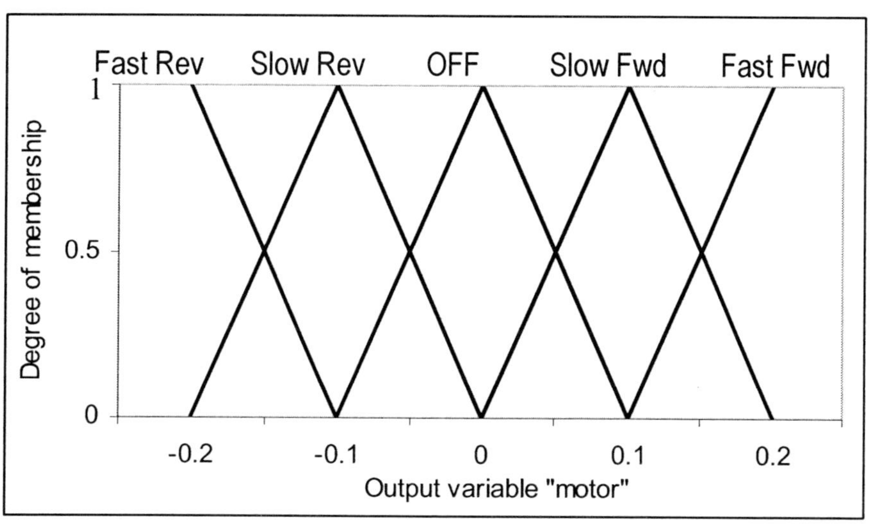

Fig. 21. Membership functions for linguistic variables describing output motor operation

4.2. Logical Operations

The first and most important thing to note about fuzzy logical reasoning is that it is a superset of standard Boolean logic. This means that if fuzzy values are at their extremes of 1 (completely true) and 0 (completely false), standard logical operations will apply. In fuzzy logic however, the truth of any statement is a matter of degree. The input values can be real numbers between 0 and 1. Therefore the result of the statement A AND B, where A and B are limited to the range (0,1) can be determined by using *min* (A,B). Similarly OR operation can be replaced with the *max* function so that A OR B becomes equivalent to *max* (A,B) and the operation NOT A is equivalent to the operation 1-A. Given these three functions any statement can be determined using fuzzy sets and the fuzzy logical operation of AND, OR and NOT. An example of the operations on fuzzy sets are shown in Fig. 22.

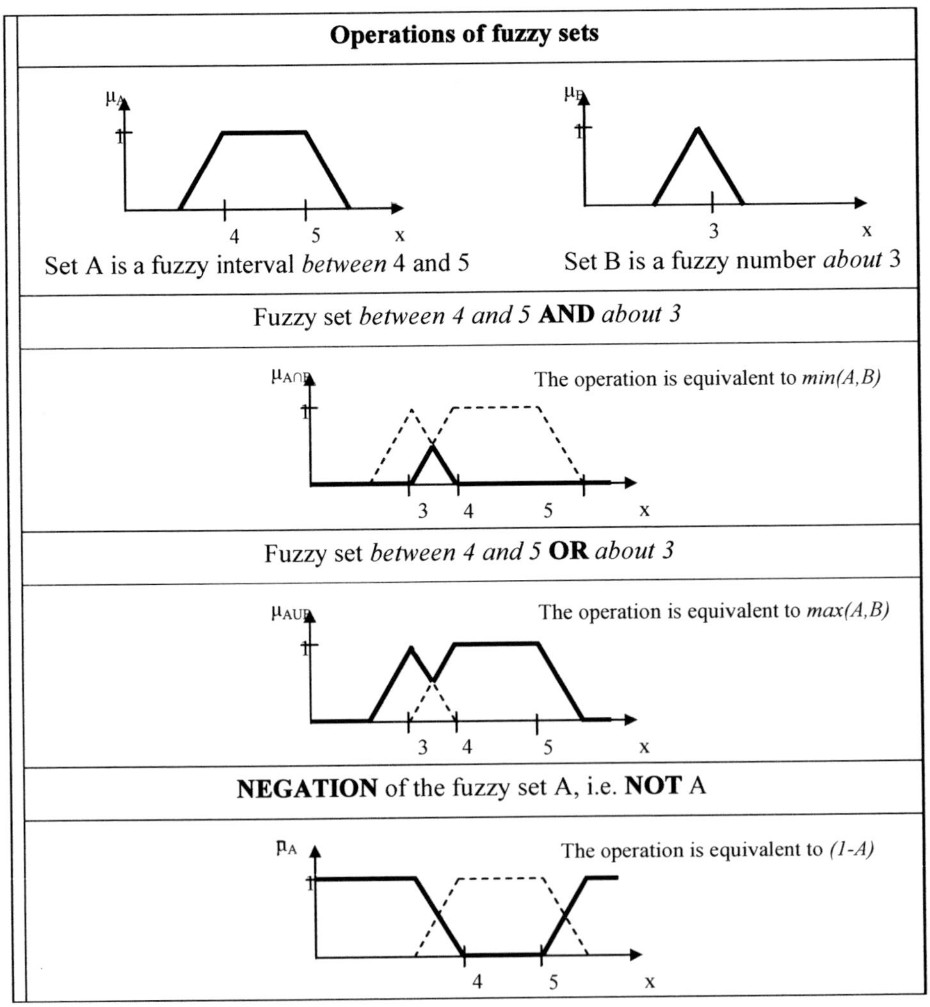

Fig. 22. Operations on fuzzy sets

In Fig. 22 only one particular correspondence between two-valued and multivalued logical operations for AND, OR and NOT is defined. This correspondence is by no means unique.

In more general terms, what is known as the fuzzy intersection or conjunction (AND), fuzzy union or disjunction (OR) and fuzzy complement (NOT) can be defined. As seen before the classical operators for these functions are defined as: AND=min, OR=max and NOT=additive complement.

The intersection of two fuzzy sets A and B is specified in general by a binary mapping T, which aggregates two membership functions as (Kalogirou, 2003):

$$\mu_{A \cap B}(x)=T(\mu_A(x), \mu_B(x)) \tag{21}$$

The binary operator T may represent the multiplication of $\mu_A(x)$ and $\mu_B(x)$. These fuzzy intersection operators are usually refined as T-norm (triangular norm) operators. Similarly to fuzzy intersection, the fuzzy union operator is specified in general by a binary mapping S as (Kalogirou, 2003):

$$\mu_{AUB}(x)=S(\mu_A(x), \mu_B(x)) \tag{22}$$

The binary operator S may represent the addition of $\mu_A(x)$ and $\mu_B(x)$. These fuzzy union operators are usually referred to as T-conorm (or S-norm) operators.

4.3. If-Then Rules

While the differential equations are the language of conventional control, IF-THEN rules, which determine the way a process is controlled, are the language of fuzzy control. In fact fuzzy sets and fuzzy operators are the subjects and verbs of fuzzy logic. The fuzzy rules serve to describe the quantitative relationship between variables and this is done in linguistic terms. These if-then rule statements are used to formulate the conditional statements that comprise fuzzy logic. Depending on the problem, several rule bases of different complexity can be developed. For example for a solar tracker the following rules may be used (Kalogirou, 2002):

IF Sensor #A is Very Low AND Sensor #B is Very Low THEN Motor is Fast Reverse
IF Sensor #A is High AND Sensor #B is Low THEN Motor is Slow Reverse
IF Sensor #A is Okay AND Sensor #B is Okey THEN Motor Off
IF Sensor #A is Low AND Sensor #B is High THEN Motor is Slow Forward
IF Sensor #A is Very Low AND Sensor #B is Very High THEN Motor is Fast Forward

In general, a single fuzzy IF-THEN rule is of the form:

IF m is A and n is B THEN x is C (23)

Where A, B and C are linguistic values defined by fuzzy sets on the ranges (universe of discourse) M, N and X, respectively. The general terminology in IF-THEN rules is to call the term following the IF statement the *premise* or antecedent and the term following THEN the *consequent.*

In Eq. (23) A and B are represented as numbers between 0 and 1, and therefore the antecedent is an interpretation that returns a single number between 0 and 1. On the contrary C is represented as a fuzzy set, and so the consequent is an assignment that assigns the entire fuzzy set C to the output variable x. In the if-then rule the word "is" is used in two entirely different ways depending on whether it appears in the antecedent or in the consequent part. In general the input to an if-then rule (m and n in Eq. 23) is the current value of an input variable and the output (x in Eq. 23) is an entirely fuzzy set. This will need to be defuzzified later, assigning a unique value to the output.

Interpreting an if-then rule involves two distinct parts:

(1) Evaluate the antecedent, which involves fuzzifying the input and applying any necessary fuzzy operators, and
(2) Apply the result to the consequent, known as implication.

In the case of two-valued or binary logic, if-then rules are interpreted quite easily, i.e., if the premise is true, then the conclusion is true. In the case of a fuzzy statement, if the antecedent is true to some degree of membership, then the consequent is also true to that same degree, that is:

In binary logic: p → q (p and q are either both true or both false).
In fuzzy logic: 0.5p → 0.5q (partial antecedents provide partial implication).

Generally both the antecedent and the consequent parts of a rule can have multiple components. For example, as was also seen in Eq. (23), the antecedent part can be:

IF steam temperature is high and sun is shining and pressure is rising THEN...

In this case all parts of the antecedent are calculated simultaneously and resolved to a single number using the logical operators described before.

The consequent of a rule can also have multiple parts, as for example:

....THEN boiler supply valve is shut and make-up water valve is open

In this case all consequents are affected equally by the result of the antecedent. The consequent specifies a fuzzy set assigned to the output. The implication function then modifies that fuzzy set to the degree specified by the antecedent. The most common way to modify the output set is by truncation using the *min* function.

In general interpreting if-then fuzzy rules is a three-part process:

(1) *Fuzzify inputs.* All fuzzy statements in the antecedent are resolved to a degree of membership between 0 and 1.
(2) *Apply fuzzy operator to multiple part antecedents.* If there are multiple parts to the antecedent, apply fuzzy logic operators and resolve the antecedent to a single number between 0 and 1.

(3) *Apply implication method.* The degree of support for the entire rule is used to shape the output fuzzy set. The consequent of a fuzzy rule assigns an entire fuzzy set to the output. This fuzzy set is represented by a membership function that is chosen to indicate the quantities of the consequent. If the antecedent is only partially true (it has a value of more than zero and less than one), then the output fuzzy set is truncated according to the implication method.

4.4. Fuzzy Inference System

Fuzzy inference is a method that interprets the values of the input vector and, based on some sets of rules, assigns values to the output vector. As already seen, in fuzzy logic the truth of any statement becomes a matter of a degree.

Fuzzy inference is the process of transforming the mapping from a given input to an output using fuzzy logic. The mapping then provides a basis with which decisions can be made, or patterns discerned. The process of fuzzy inference involves all the items that are described so far, i.e., membership functions, fuzzy logic operators and if-then rules. There are mainly two types of fuzzy inference systems that can be implemented, Mamdani-type (Mamdani, 1974; 1977) and Sugeno-type (Sugeno, 1985). Basically the only variation of these two types of inference systems is in the way outputs are determined.

Mamdani's fuzzy inference method is the most commonly used fuzzy methodology. This method was among the first control systems built using fuzzy set theory and was performed in an attempt to control a steam engine and boiler combination. Mamdani-type inference expects the output membership functions to be fuzzy sets. After the aggregation process, there is a fuzzy set for each output variable that needs defuzzification. It is possible, and usually more efficient, to use a single spike as the output membership function rather than a distributed fuzzy set. This is called a singleton output membership function and it can be considered as a pre-defuzzified fuzzy set. The method enhances the efficiency of the defuzzification process because it greatly simplifies the computation required compared to the more general Mamdani method, which finds the centroid of a two-dimensional function. Thus, the weighted average of a few data points can be used to find the centroid instead of integrating across the two-dimensional function.

Sugeno method of fuzzy inference has many similarities to the Mamdani method. In fact the first two parts of the fuzzy inference process, i.e., fuzzifying the inputs and applying the fuzzy operator, are exactly the same. The main difference between the two fuzzy inference methods is that the output membership functions are only linear or constant for the Sugeno-type fuzzy inference. A typical fuzzy rule in a first-order Sugeno fuzzy model is of the form:

$$\text{If } m \text{ is } A \text{ and } n \text{ is } B \text{ then } x = pm + qn + r \quad (24)$$

where A and B are fuzzy sets in the antecedent, while p, q and r are all constants. Second or higher order Sugeno fuzzy models can be used, but they introduce significant complexity with little obvious merit. The Sugeno method is ideal for acting as an interpolating supervisor of multiple linear controllers that are applied to different operating conditions of a dynamic nonlinear system because of the linear dependence of each rule on the system's input variables. A Sugeno fuzzy inference system is very well suited to the task of smoothly interpolating the linear gains that would be applied across the input space, thus, it is a natural and efficient gain scheduler. Similarly, nonlinear systems can be easily modeled with the Sugeno model by interpolating multiple linear models.

5. Hybrid Systems

With the term "hybrid systems" it is meant, systems that combine two or more artificial intelligence techniques in order to solve a problem. The typical example of a hybrid system is the neuro-fuzzy control whereas other types combine genetic algorithms and fuzzy logic, or artificial neural networks and genetic algorithms as part either of an integrated problem solution or to perform specific separate tasks of the same problem. As most of these systems are problem specific more details are given here for the neuro-fuzzy systems.

As was seen in the previous section, a fuzzy system possesses great power in representing linguistic and structured knowledge by fuzzy sets and performing fuzzy reasoning and fuzzy logic in a qualitative manner. It usually relies however in domain experts to provide the necessary knowledge for a specific problem. As was also seen in section 2, neural networks are particularly effective at representing nonlinear mappings in a computational fashion and they are "constructed" through training procedures presented to them as patterns. The behaviour of fuzzy systems can be understood easily due to their logical structure and step-by-step inference procedures, while a neural network acts generally as a "black-box" model. Quite recently the possibility of integrating the two technologies into a new kind of system called neuro-fuzzy control was considered. Thus, several strengths of both systems are utilised and combined appropriately.

More specifically by neuro-fuzzy control it is meant:

(1) The controller has a structure which combines fuzzy systems and artificial neural networks,
(2) The resulting control system consists of fuzzy systems and neural networks as independent components performing different tasks of the same problem, and
(3) The design methodologies for constructing respective controllers are originated from ideas in both fuzzy and neural control.

In the case of a neuro-fuzzy controller a trained neural network can be viewed as a means of knowledge representation. Particularly, instead of representing knowledge by IF-THEN localised associations (as in fuzzy systems), a neural network stores knowledge through its structure, (more specifically in its connection weights and local processing units) in a distributed or localized manner. Many commercial software (like Matlab) include routines for neuro-fuzzy modelling.

The basic structure of fuzzy inference system is described in previous section. As was seen this is a model that maps the input membership functions, the input membership functions to rules, the rules to a set of output characteristics, then output characteristics to output membership functions, and finally the output membership function to a single-valued output associated with the output. In this procedure, the membership functions are fixed. Therefore in this way, fuzzy inference can be applied to modeling systems whose rule structure is essentially predetermined by the user's interpretation (who acts as an expert) of the characteristics of the variables of the process or plant.

There are however some modeling situations in which the shape of the membership functions cannot be determined from the data characteristics. In this case, instead of choosing the parameters associated with a given membership function arbitrarily, these parameters could be chosen so as to adjust the membership functions to the input/output data characteristics which could account for the variations in the data values. In case that past history input/output data are available for a system these can be used to determine the membership functions of the fuzzy inference. Thus a neuro-fuzzy system can be developed which, using a given input/output data set, a fuzzy inference system is constructed, whose membership function parameters are tuned or adjusted by using a neural network. Therefore, the basic idea behind a neuro-fuzzy technique is to provide a method for the fuzzy modeling system to learn information about a data set, in order to compute the membership function parameters that best allows the associated fuzzy inference system to track the given input/output data. Thus, a neural network can be used to interpret the input/output map. This is done by mapping inputs through input membership functions and associated parameters, and then through output membership functions and associated parameters to outputs. The parameters associated with the membership functions are adjusted through the learning process and the procedure followed is similar to that applied to any other neural network described in section 2.

Finally, it needs to be pointed out that this type of modeling works well if the training data, presented to a neuro-fuzzy system for training and estimating the membership function parameters, is representative of the features of the data that the trained fuzzy inference system is intended to model. However, this is not always the case, as usually the data are collected using noisy measurements or because of the peculiarities of the process training data cannot be representative of all features of the model. For this purpose, model validation needs to be used as in any other neural network system. This is the process by which input vectors from the input/output data sets, which the neuro-fuzzy system has not seen before, are presented to the trained model to check how well it predicts the corresponding output values.

6. Conclusions

A number of artificial intelligence techniques have been described in this chapter. Artificial neural networks are collections of small individually interconnected processing units. Information is passed between these units along interconnections. ANNs while implemented on computers are not programmed to perform specific tasks. Instead, they are trained with respect to data sets until they learn the patterns presented to them. Once they are trained, new patterns may be presented to them for prediction or classification. Genetic algorithms are inspired by the way living organisms adapt to the harsh realities of life in a hostile world, i.e., by evolution and inheritance. The algorithm imitates in the process the evolution of population by selecting only fit individuals for reproduction. Therefore, a genetic algorithm is an optimum search technique based on the concepts of natural selection and survival of the fittest. A genetic algorithm utilizes three principal genetic operators: selection, crossover, and mutation. Fuzzy logic is used mainly in control engineering. It is based on fuzzy logic reasoning which employs linguistic rules in the form of IF-THEN statements. Fuzzy logic and fuzzy control feature a relative simplification of a control methodology description. This allows the application of a "human language" to describe the problems and their fuzzy solutions. Hybrid systems combine more than one of the technologies described above, either as part of an integrated method of problem solution, or to perform a particular task that is followed by a second technique, which performs some other task.

Like all other approximation techniques, artificial intelligence techniques have relative advantages and disadvantages. There are no rules as to when this particular technique is more or less suitable for an application. It is believed that artificial intelligence techniques offer an alternative method, which should not be underestimated.

Various applications of artificial intelligence in energy and renewable energy systems are presented in this book. From these one can see that artificial intelligence techniques have been applied in a wide range of fields for modeling, prediction and control in energy and renewable energy processes. What is required for setting up such a system is data that represents the past history and performance of the real system and a selection of a suitable model. The selection of this model is done empirically and after testing various alternative solutions. The performance of the selected models is tested with data of the past history of the real system.

REFERENCES

Broomhead D.S., Lowe D., 1988. Multivariable functional interpolation and adaptive networks, *Complex Systems*, Vol. 2, pp. 231-355.

Davis L., 1991. *Handbook of Genetic Algorithms*, New York, Van Nostrand.

Farlow S.J. (Ed.), 1984. *Self-Organizing Methods in Modeling*, Marcel Dekker, New York.

Goldberg D.E., 1989. *Genetic Algorithms in Search Optimization and Machine Learning*, Reading, MA, Addison-Wesley.

Haykin S., 1994. *Neural Networks: A Comprehensive Foundation*. Macmillan, New York.

Hecht-Nielsen R., 1991. *Neurocomputing*, Addison-Wesley Publishing Company.

Ivakhenko A.G., 1971. Polynomial theory of complex systems. *IEEE Transactions,* Systems, Man and Cyber., SMC-12, pp. 364-378.

Ivakhenko A.G., 1968. The group method of data handling-A rival of stochastic approximation, (In Russian), *Soviet Automatic Control*, Vol. 1, pp. 43-55.

Kalogirou S., Neocleous C., Schizas C., 1996. Artificial neural networks in modelling the heat-up response of a solar steam generation system. *Proceedings of the Engineering Applications of Neural Networks (EANN'96) Conference*, London, UK, pp. 1-4.

Kalogirou S., 1996. Artificial neural networks for estimating the local concentration ratio of parabolic trough collectors. *Proceedings of the EuroSun'96 Conference*, Freiburg, Germany, Vol. 1, pp. 470-475.

Kalogirou S., Neocleous C., Schizas C., 1997. Artificial neural networks used for estimation of building heating load, *Proceedings of the CLIMA 2000 International Conference*, Brussels, Belgium, Paper Number, P159.

Kalogirou S., Panteliou S., Dentsoras A., 1999. Modelling of solar domestic water heating systems using artificial neural networks. *Solar Energy,* Vol. 65, No. 6, pp. 335-342.

Kalogirou S., Bojic M., 2000. Artificial neural networks for the prediction of the energy consumption of a passive solar building. *Energy*, Vol. 25, No. 5, pp. 479-491.

Kalogirou S., 2002. Design of a fuzzy single-axis sun tracking controller, *International Journal of Renewable Energy Engineering*, Vol. 4, No. 2, pp. 451-458.

Kalogirou S., 2003. Artificial intelligence for the modelling and control of combustion processes: A review, *Progress in Energy and Combustion Science*, Vol. 29, No. 6, pp. 515-566.

Kavli T.O., 1992. Learning principles in dynamic control, PhD Thesis, University of Oslo, ISBN No. 82-411-0394-8.

Kohonen T., 1990. The self-organizing map, *Proceedings of the IEEE,* Vol. 78, No. 9, pp. 1464-1480.

Kohonen T., 1995. *Self-Organizing Maps*, Springer Series in Information Sciences, Berlin.

Nannariello J., Frike F.R., 2001. Introduction to neural network analysis and its applications to building services engineering. *Building Services Engineering Research and Technology*, Vol. 22, No. 1, pp. 58-68.

Neocleous C., 1998. A neural network architecture composed of adaptively defined dissimilar single-neurons, *PhD Thesis*, Brunel University, UK.

Neuroshell, 1996. *Neuroshell-2 Program Manual*, Ward Systems Group, Inc., MD.

Nie J., Linkens D.A., 1995. *Fuzzy-Neural Control: Principles, Algorithms and Applications*. Prentice Hall.

Mamdani E.H., 1974. Application of fuzzy algorithms for control of simple dynamic plant. *Proceedings of IEEE*, Vol. 121, pp. 1585-1588.

Mamdani E.H., 1997. Applications of fuzzy set theory to control systems: a survey. In Fuzzy Automata and Decision Process, Gupta M.M. *et al.*, Eds, Amsterdam: North-Holland, pp. 77-88.

Michalewicz Z., 1996. *Genetic Algorithms + Data Structures = Evolution Programs*, (3rd Ed.), Springer, Berlin.

Pogio T., Girosi F., 1989. A theory of networks for approximation and learning, *Technical Report*, Artificial Intelligence Laboratory, Massachusetts Institute of Technology, Jul. 1989.

Pogio T., Girosi F., 1990. Networks for approximation and learning. *Proceedings of IEEE*, Vol. 78, pp. 1481-1497.

Powel M.J.D, 1987. Radial basis functions for multivariable interpolation: A review, *Algorithms and Approximation*, Editors: Mason J.C. and Cox M.G., Clarendon Press, London.

Reznik L., 1997. *Fuzzy Controllers*, Newnes, Oxford.

Ripley B.D., 1996. *Pattern Recognition and Neural Networks*, Cambridge University Press.

Rumelhart D.E., Hinton G.E., Williams R.J., 1986. Learning internal representations by error propagation, *Parallel Distributed Processing: Explorations in the Microstructure of Cognition*, Vol. 1, Chapter 8, Cambridge, MA, MIT Press.

Sugeno M., 1985. *Industrial Applications of Fuzzy Control*,. Amsterdam: North-Holland.

Tsoukalas L.H., Uhrig R.E., 1997. *Fuzzy and Neural Approaches in Engineering*. John Wiley and Sons, Inc.

Werbos P.J., 1974. Beyond regression: New tools for prediction and analysis in the behavioral science, *PhD Thesis*, Harvard University, Cambridge, MA.

Zadeh L.A., 1973. Outline of a new approach to the analysis of complex systems and decision processes. *IEEE Trans. Syst. Man Cyber.*, Vol. 3, pp. 28-44.

Zalzala A., Fleming P., 1997. *Genetic Algorithms in Engineering Systems*, The Institution of Electrical Engineers, London, UK.

In: Artificial Intelligence in Energy and Renewable Energy Systems ISBN 1-60021-261-1
Editor: Soteris Kalogirou, pp. 47-82

Chapter 2

ARTIFICIAL NEURAL NETWORKS FOR METEOROLOGICAL VARIABLES PERTAINED TO ENERGY AND RENEWABLE ENERGY APPLICATIONS

***Silas C. Michaelides*[*], *Filippos S. Tymvios*[*] and *Soteris A. Kalogirou*[**]**

[*]Meteorological Service, Nicosia, Cyprus
[**]Higher Technical Institute, Nicosia, Cyprus

ABSTRACT

This chapter is a broad survey of published research on the use of Artificial Neural Networks (ANNs) in estimating values of three meteorological parameters which are related to energy and renewable energy related issues, namely solar radiation, temperature and wind. The respective research tasks are reviewed, focusing on the neural methodology that was adopted, the database that was employed and the validation method that was subsequently performed. The survey presented in this chapter aims at assisting the reader to obtain a good understanding of the extent and variability of the areas that can be treated with the neural network methodology. Hopefully, the information presented in this chapter will trigger research in more diverse areas related to energy and renewable energy issues that make direct or indirect use of meteorological variables.

Keywords: Artificial neural networks, solar radiation, temperature, wind, forecasting, prediction.

NOMENCLATURE

AP	Atmospheric pressure (hPa or mb)
DB	Dry-bulb temperature (°C)
DL	Day length (hours)

G	Total solar radiation (W/m^2)
$\hat{G}_{(n)}$	Solar radiation of a day ($Wh/m^2/day$)
G_0	Available radiation at the top of the atmosphere (W/m^2)
LC	Low cloud amount (in oktas)
PM	Observed minimum temperature of the previous night (°C)
PW	Present weather condition (from 0 to 99, see Table 2)
TC	Total cloud amount (in oktas)
VI	Visibility (m)
WB	Wet-bulb temperature (°C)
W1	Wind speed eastward component (m/s)
W2	Wind speed northward component (m/s)
S	Measured sunshine duration (hours)
S_0	Theoretical available sunshine duration (hours)

Greek

α	Constant in Ångström-Prescott relationship
β	Constant in Ångström-Prescott relationship

Abbreviations

ANFIS	Adaptive Neuro-Fuzzy Inference Systems
ANN	Artificial Neural Networks
ECMWF	European Centre for Medium-Range Weather Forecasts
LR	Linear Regression
GR	Global Radiation
IIR	Impulse Infinite Response
MLP	Multilayer Perceptron
MTM	Markov Transition Matrices
RBF	Radial Basis Function
RMSE	Root Mean Squared Error

1. Introduction

In designing renewable energy systems (that make use of renewable environmental resources), a detailed long-term series of meteorological parameters such as solar radiation, temperature or wind data is normally required. The effort is to design systems that can make an efficient conversion and utilization of these renewable energy resources.

However, one of the problems that designers of any such system are often confronted with is the acquisition of sufficiently long series of meteorological variables for direct utilization. Such measurements are often non-existent for most locations or, wherever they are available, they suffer from a number of shortcomings (e.g. poor data quality, missing data, insufficiently long series etc). In order to overcome this serious problem, several attempts have been made by researchers in many parts of the world to fill in the gap of missing real series of meteorological data by ***artificial*** ones. The term artificial is invariably used here to describe ***not real*** time series generated with various methods.

It is fundamental to recognize that the "artificial" time series, regardless of the methodologies employed, must have similar statistical properties with the real ones (see for example Aguiar and Collares-Pereira, 1991; Hontoria *et al.*, 2002). Such an appreciation is very important because, when used in specific applications, the artificial data should produce comparable results to those the real time series yield.

Artificial Neural Networks (ANNs) have long been widely used in many disciplines, such as forecasting, pattern recognition, classification, optimization, control etc (see Jain *et al.*, 1996). The use of the ANNs for modeling and prediction purposes has increasingly becoming popular. Also, ANNs have been trained to overcome known limitations of conventional approaches in solving complex problems.

Meteorological variables, in the broad sense of the term, have long been considered as potential candidates for study by employing ANNs, both from a diagnostic and a prognostic viewpoint. Such studies date back in the early 80's. The easiness in using ANNs is partly dictated by one of their major characteristics; they can be used to build data-driven models, as opposed to other statistical techniques that can be considered as model-driven.

A large part of the conceivable atmospheric processes are neither fully comprehensible nor explicitly expressible in deterministic terms. Commonly, intuition and empiricism (including stochastic treatment) have long been employed in an attempt to model such processes. ANNs are considered as a more modern methodology, exploiting more deeply the available mathematical and computational resources. Also, the generally non-linear character of the meteorological processes renders ANNs as a suitable tool for their study. Sometimes, the highly complex nature underlying such processes goes beyond human comprehension. In this respect, ANNs appear to be appropriate in an attempt to by-pass a total understanding of the physical processes involved.

ANNs have extensively been used in atmospheric sciences and the reader is referred to publications reviewing past applications of ANNs in this broader area. For example, Gardner and Dorling (1998) present a general introduction and discussion of applications of the multilayer perceptron in the atmospheric sciences. Also, Hsieh and Tang (1998) present a general discussion of issues related to ANN applications in meteorology and oceanography.

The broad review of ANN applications to atmospheric related variables is beyond the scope of this review. Rather, in this chapter, the applications of ANNs in the study of meteorological variables that are directly related to energy and renewable energy issues are considered in more detail. More specifically, studies on three meteorological elements, namely, ***solar radiation***, ***temperature*** and ***wind*** are surveyed and reviewed. In particular, we focus our attention on the various modeling approaches that have been adopted in the estimation of these variables for subsequent use in different applications.

The majority of the studies that will be presented in the following simply adopt ANNs in the modeling of the meteorological variables for general-purpose energy applications. However, some studies have incorporated the meteorological elements' modeling into multi-modular schemes that make direct use of meteorological elements estimations in purpose-specific energy-related applications.

In this review, no extensive discussion will be made of the various architectures used in each particular endeavor (the reader can refer to the original articles for more detail). The main effort is to provide the reader with a global view of the extent of the research that has been undertaken, outlining the problem confronted in each case and describing the major steps taken to tackle it.

The presentation of the various research efforts is stripped of mathematical or theoretical treatments for which the reader can consult existing textbooks or the original papers. Rather, this review attempts to provide the reader with a wider view of the diversity of ANNs adopted, the input data and the results obtained. Compiling the existing literature and focusing on these three aspects, namely, reference to the models used, data used as input and results obtained, the reader can assess in a fast-track approach the various techniques adopted.

Not all the examples from the scientific literature surveyed in the following have exclusively been used in energy and renewable energy applications. However, the results obtained by applying ANN methodologies can be considered as covering potential needs in these areas. Another very interesting outcome of the following presentation is the large variety of problem solving approaches adopted, regarding the instrumentation and databases involved, the selection of input and output variables, the choice of network architectures and the verification schemes adopted.

2. Solar Radiation

2.1. Estimating Solar Radiation: A Long Standing Issue

With regard to several applications of solar energy (e.g. engineering of solar systems, architecture, etc), a sufficient knowledge of the available solar resources is required. Indeed, in the design of any system that makes use of solar energy resources, it is absolutely necessary to know (sometimes in much detail) the solar energy regime of the particular place, in order to be able to simulate the operating conditions of the system and therefore to design it as efficiently as possible.

As a rule, solar radiation measurements are rather sparse. For this reason, traditionally, deterministic models have been developed for the estimation of solar radiation at places with no direct measurements. Such models have as inputs meteorological variables that are measured at synoptic or climatological stations. Such variables include sunshine, cloud cover, temperature and humidity.

The earliest such model was first proposed by Ångström (1924). Ångström's approach has traditionally been used in many parts of the world and in many respects, it was considered as a standard procedure where solar radiation measurements were scarce. For this reason, it is worth presenting briefly this regression procedure.

Several variations of Ångström's regression relationship have been developed since it was first introduced. The most widely used is the modified version proposed by Prescott (1940). In this modified linear regression model, the dependent variable comprises the ratio of the total solar radiation (G) to the available radiation at the top of the atmosphere (G_0); the independent variable comprises the ratio of the measured sunshine duration (S) to the theoretical available sunshine duration (S_0). The original formulation of Ångström-Prescott linear model, which has been adopted in many parts of the world, is:

$$\frac{G}{G_0} = \alpha + \beta \frac{S}{S_0} \quad (1)$$

where α and β are constants.

For several decades, this relationship was almost exclusively used for estimating total solar irradiance. However, more recently, different methods have been adopted in an effort to generate sufficiently long artificial time series, including Fourier analysis (e.g. Hittle and Pedersen, 1981), autoregressive moving average (e.g. Goh and Tan, 1977) and stochastic models (e.g. Richardson, 1981; Richardson and Wright, 1984; Hokoi *et al.*, 1990), to name but a few.

In another, theoretical methodology, complex radiative transfer models, replicating physical processes have been used. This approach takes into consideration physical interactions between solar radiation and the terrestrial atmosphere, such as Rayleigh scattering, radiative absorption by ozone and water vapor and aerosol extinction (Leckner, 1978; Bird and Riordan, 1986; Gueymard, 1989; 2001; La Casiniere *et al.*, 1997).

The introduction of ANN techniques opened new horizons in the effort of estimating solar irradiance by using as inputs recorded parameters which could yield sufficiently accurate approximations to the real solar irradiance on the earth's surface.

2.2. Estimation of Daily Solar Radiation with ANN

One of the earliest efforts to estimate the daily solar radiation by using ANNs is that reported by Elizondo *et al.* (1994). The aim of this research was the development of a neural network model which could predict daily solar radiation as a function of both observed weather and calculated environmental variables. The input variables for the neural network were, on the one hand, *observed* weather variables such as daily precipitation and maximum and minimum air temperatures; on the other hand, calculated *environmental* variables, such as daylength and daily total clear sky radiation, were also considered as inputs to the neural network. The backpropagation neural network algorithm was selected for this investigation.

Four sites in the southeastern USA with long-term daily weather datasets, which included solar radiation, were selected for model training and evaluation. A combined total of 23 complete years of weather datasets were available which were separated into two groups: 11 years were used as training dataset and 12 years as testing dataset.

The neural network model was evaluated by comparing the predicted daily solar radiation results with observations that were not used for model development. The neural network model gave reasonably good predictions of daily solar radiation, with a root mean square error as low as 2.92 MJ/m^2. Although the final neural network model had a tendency to overpredict solar radiation at low values and underpredict it at high values, average errors and coefficients of variation were low and the coefficients of determination were high.

At about the same time, Williams and Zazueta (1996) reported on a similar study which was carried out in order to determine whether a neural network could adequately estimate the solar radiation for a location that does not have suitable collection equipment. Their proposition was tested with data from Gainesville, FL in the USA and the results were contrasted to those from an analytical approach. This study used actual weather data collected at the above site to train a backpropagation neural network to estimating solar radiation. Results indicated that the neural network performed better than the analytical approach. The correlation coefficient for the neural network was 0.86 versus 0.77 for the analytical approach.

Mohandes *et al.* (1998a) presented a study that uses a neural network technique for modeling monthly mean daily values of global solar radiation on horizontal surfaces. The neural networks use as input the latitude, longitude, altitude and sunshine duration for the prediction of solar radiation values. In this study, Mohandes *et al.* (1998a) used the backpropagation algorithm for training several multi-layer feedforward neural networks. In their study, they concluded that the best network consists of 4 input neurons, 10 neurons in one hidden layer and one neuron in the output layer. Latitude and longitude were provided in degrees and the altitude in meters. The sunshine duration values were provided as a ratio of the actual values divided by the maximum possible values for each location. The output of the network is the ratio of monthly mean daily value of the global solar radiation divided by extraterrestrial radiation received at the top of the atmosphere. A total of 41 meteorological stations were employed in this research.

Data from 31 stations were used for training the neural network. The data from the remaining 10 stations were used for testing the models developed. The testing data were not used in the modeling phase in order to obtain an indication of the performance of the system at locations unknown to the network. The results from these 10 independent locations indicated a relatively good agreement between the observed and predicted values.

In a further expansion of the above idea, Mohandes *et al.* (2000), use a Radial Basis Functions (RBF) neural networks technique (see Wassereman, 1993; Bishop, 1996) for modeling monthly mean daily values of global solar radiation on horizontal surfaces and compares its performance with that of a Multilayer Perceptron (MLP) model and a classical regression model. The basic problem remains the same as in the above study. They have also adopted the same database separation procedure for training and testing.

The RBF network used here has five inputs: latitude, longitude, altitude, sunshine duration and the month number. In the MLP application, there is no indication of the month of the year other than that implicitly provided in the sunshine data. The model output is the ratio of monthly mean daily value of the global solar radiation divided by extraterrestrial radiation outside the atmosphere. The performance of both the RBF and MLP networks was tested against the independent set of data from 10 stations. The mean absolute percentage error was used as the testing statistic.

In the same study, Mohandes *et al.* (2000) compare the above RBF and MLP networks to the empirical Ångström's regression model, as it was later modified by Rietveld (1978). From the comparative results of the three approaches, namely the RBF, MLP and the Linear Regression (LR) model, it was suggested that no one exhibits a clear advantage when the results are looked upon station by station. Nevertheless, the mean absolute percentage error, averaged for all the stations in the testing dataset, for the RBF networks was 10.1, for the MLP network was 12.6 and for the LR model was 11.35. These results suggest that overall, the RBF performed better than the others.

Mellit *et al.* (2005) developed a hybrid model to predict the daily global solar radiation data using a combination of ANN and a library of Markov transition matrices (MTM). The database employed consisted of daily global solar radiation data from 60 meteorological stations in Algeria during 1991–2000.

The proposed architecture of the ANN model is a feedforward neural network with three inputs (latitude, longitude and altitude), a hidden layer and an output layer consisting of 12 neurons (monthly radiation). The size of the database and the ANN architecture employed demanded a fast modern backpropagation algorithm, since one of the problems when implementing the classical backpropagation is the slow convergence and the huge computing power needed. An adaptive method for fast backpropagation algorithm was chosen.

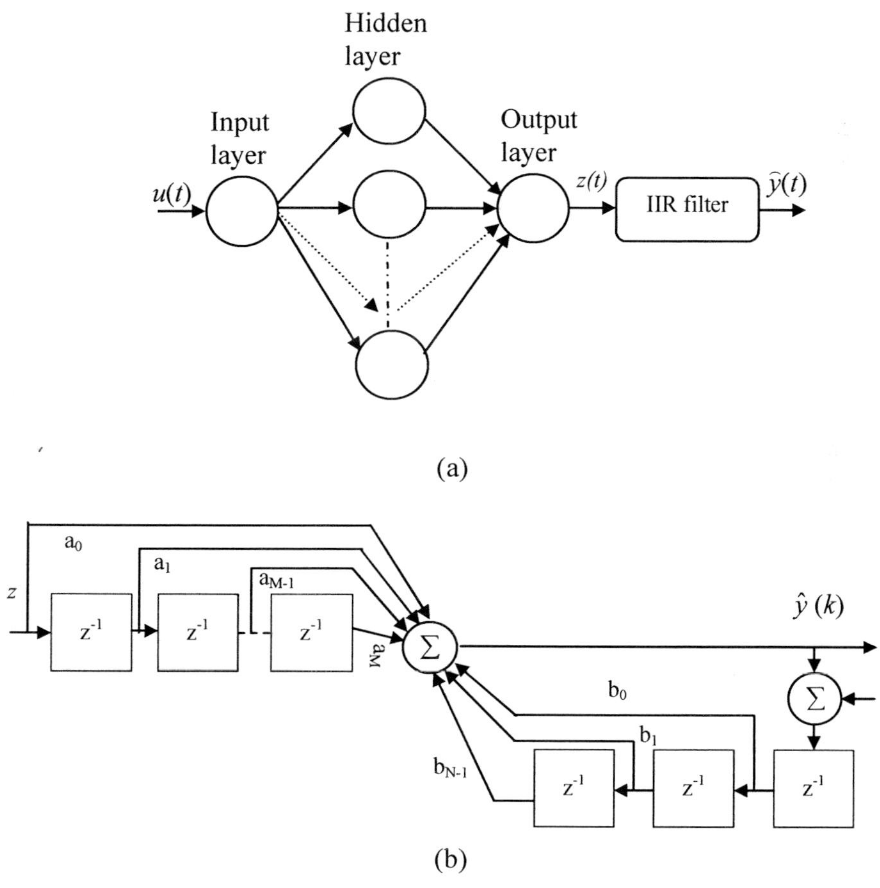

Fig. 1. Impulse Infinite Response adaptive wavelet network structure: (a) Neural network architecture, (b) Impulse Infinite Response model (from Mellit *et al.*, 2006)

Firstly, a neural network block was trained based on 56 known monthly solar radiation data from the database. In this way, the network was trained to accept and even handle a number of unusual cases. Secondly, these data were divided by the corresponding extraterrestrial value, in order to obtain the monthly clearness index values. Based on these monthly clearness indices and using a library of MTM block, the sequences of daily clearness indices were generated.

To check the accuracy of the ANN model proposed, the model was trained with the data from 56 stations leaving 4 stations aside for validation. The unknown validation data set produced very accurate predictions, with a RMSE not exceeding 8%; also, a correlation coefficient between predicted and observed data ranging from 90% and 92% was obtained. The results obtained indicate that the proposed model can successfully be used for the estimation of the daily solar radiation data for any locations in Algeria by using as input the altitude, the longitude, and the latitude.

In a more recent approach, Mellit *et al.* (2006) have investigated the suitability of the wavelet network architecture with Impulse Infinite Response filter for modelling and prediction of daily total solar radiation. This architecture provides a double local structure which results in an improved speed of learning. Wavelet networks are feedforward networks using wavelets as activation functions and have been used successfully in classification and identification problems. Figure 1 shows a local Impulse Infinite Response (IIR) block structure in cascade with the network.

The objective of this latter research is to predict the value of daily total solar radiation from the preceding values in order to solve the problem of missing data; the solar radiation for a day n, i.e., $\hat{G}_{(n)}$, is estimated by the function $f(G_{(n-1)}, G_{(n-2)},.....G_{(n-k)})$ which minimizes the mean square error, by using a wavelet-network algorithm. The block diagram of the approach followed in modelling the number of input and output neurons is shown in Fig. 2.

The meteorological data that have been used in this work are the recorded solar radiation values during the period extending from 1981 to 2001 from a meteorological station in Algeria. Two datasets have been used for the training of the network. The first set includes the data for 19 years of G and the second dataset is using data for one year (365 values) which is selected from the database. In both cases the data for the year 2001 are used for testing the network.

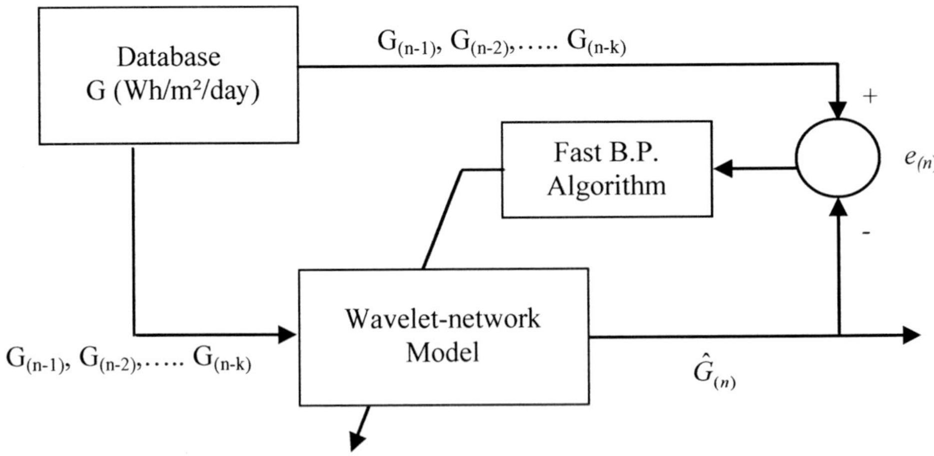

Fig. 2. Block diagram of modified wavelet network architecture (from Mellit *et al.*, 2006)

From the results obtained it was shown that the best model is the one used to predict the total solar radiation of a day from the preceded 5 values (5 days). This model gives more accurate results but it is limited to the prediction of one value from five preceded values of total daily solar radiation data.

However, total solar radiation data are not always available for a long period, (especially if there are missing data in the database). For this reason, one year from the 20 was set aside in order to investigate the prediction performance. Thus, one year of G selected from the 20-year is used for the training of the network and the year 2001 for testing the network. The wavelet-network model is validated by comparing the predicted results and actual measured total solar radiation data and was shown that there is almost a complete agreement between measured and predicted times series, with a correlation coefficient equal to 0.968 and mean square error equal to 5.6%.

2.3. Comparison with Traditional Techniques

Tymvios *et al.* (2002; 2005) also followed a comparative approach between the above Ångström-type linear model (Eq. 1) and several ANN models that are elaborated in the following. Multi-layer feedforward networks with one and two hidden layers trained with the backpropagation learning algorithm were investigated. The dataset used in these studies consists of measurements taken at Athalassa radiometric station in Cyprus. The raw radiation data consist of hourly total values of global solar irradiance, *measured* on a horizontal surface (in W/m^2 per hour) and sunshine duration (in hours). The daily totals and the monthly mean values of daily totals of the global irradiance and the daily and monthly mean values of daily duration of sunshine, were subsequently used for building the models, as explained below. The database was supplemented with the daily totals and monthly mean values of daily totals of the *calculated* extraterrestrial solar irradiance and the daily and monthly mean values of daily theoretical sunshine duration. The daily maximum temperature was also used in building some of the ANN models. The data cover the period 1986-1992. Global solar irradiance was measured using a Kipp-Zonen model CM-1, while another Kipp-Zonen model CM-6B6 was operated in parallel for error-checking purposes. Sunshine duration was measured with a Campell-Stokes heliograph.

The ANN models constructed by Tymvios *et al.* (2002; 2005) were trained by using the daily values of the measured sunshine duration, the theoretical sunshine duration, the month of the year and the daily maximum temperature. The predicted parameter is the daily value of total global radiation which was also used in the training process, accordingly. Because solar irradiance largely depends on season, all the ANN models that were build in this research consider at least one input variable that inherently refers to seasonality. Therefore, only models with at least two input variables were constructed and tested.

For the training of the ANN models, data for 1858 days (out of a total of 2090) and for the cross-validation (a procedure necessary for the models' convergence), data for the remaining 232 days were used. Seven ANN models were developed (see Table 1 for more details on the input parameters and number of hidden layers). The models trained can be separated into two groups.

The first group comprises three plain models with only two input variables, denoted by ANN-1, ANN-2 and ANN-3. For these plain models, only one hidden layer was used, since the use of more layers did not improve substantially these models' performance. The second group comprises ANN models which are more complicated, in the sense that they were trained by using three input variables and with more than one hidden layers. These more complicated networks, were trained with one hidden layer (ANN-4-i and ANN-5-i) and two hidden layers (ANN-4-ii and ANN-5ii) because it was realized that models with three hidden layers tend to memorize the training sets, thus giving excellent results for known inputs but behaving poorly against unknown data (see also Kalogirou *et al.*, 1997; 2002). An excessive number of hidden layers often prove to be unproductive. Generally speaking, a network with one hidden layer should be adequate as universal approximation of any nonlinear function (Barron, 1993; Funahashi, 1989; Hornik *et al.*, 1989). It has been further demonstrated that a three-layer network suffices to separate any (convex or nonconvex) polyhedral decision region from its background. In summary, one or two hidden layers should be adequate for most similar applications. (Principe *et al.*, 2000).

Table 1. Input variables and number of hidden layers in various ANN models (from Tymvios *et al.*, 2002; 2005)

Model	Input variables	Hidden layers
ANN-1	measured daily sunshine duration theoretical daily sunshine duration	1
ANN-2	month measured daily sunshine duration	1
ANN-3	month daily maximum temperature	1
ANN-4-i	measured daily sunshine duration month daily maximum temperature	1
ANN-4-ii	measured daily sunshine duration month daily maximum temperature	2
ANN-5-i	measured daily sunshine duration theoretical daily sunshine duration daily maximum temperature	1
ANN-5-ii	measured daily sunshine duration theoretical daily sunshine duration daily maximum temperature	2

Figure 3 presents the comparison between the predicted and measured radiation values for one of the seven Neural Network models developed, namely ANN-5-ii. For this particular model, three input parameters were used: measured daily sunshine duration, theoretical daily sunshine duration and daily maximum temperature. Comparable results were obtained for the other models too implying that they all have a good long-term representation of the physical problem.

The comparison of the performance of the above ANN models with various Ångström-type models has revealed that the widely used Ångström-Prescott formula can easily be substituted by better performing models based on more refined data (i.e., the raw dataset of daily values, rather than the derived mean monthly values). The above findings render the ANN methodology as a promising alternative to the traditional approach for estimating global solar radiation, especially in cases where radiation measurements are not readily available.

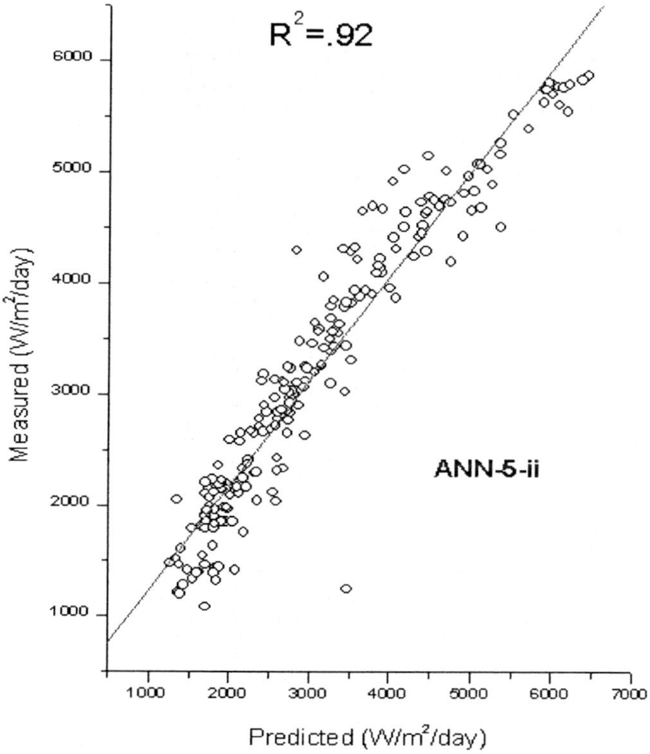

Fig. 3. Predicted versus measured radiation values for an ANN model (ANN-5-ii in Table 1) with three inputs: measured daily sunshine duration, theoretical daily sunshine duration and daily maximum temperature. The fitted linear regression is shown, together with the corresponding coefficient of determination, R^2 (from Tymvios *et al.*, 2002; 2005)

2.4. Solar Radiation Time Series

Mihalakakou *et al.* (2000) presented a simulation of time series for several years for the total solar radiation in Athens, Greece by using neural networks, in an effort to predict future values of total solar radiation time series. In this research, the ANN's ability to model and forecast the total solar radiation is investigated.

The time series generated in the present paper are total solar radiation measured on a horizontal surface at the Institute of Meteorology and Physics of the Atmospheric Environment of the National Observatory of Athens.

Integrated hourly, daily and monthly values of total solar radiation (in MJ/m^2) are measured at the Observatory with Kipp-Zonen and Eppley actinometers and pyranometers, respectively. Hourly values of total solar radiation for twelve years (1984-1995) and for various months of the year are used for training and testing the network. The data are divided into a training set that provides a fitting approximation function and a testing set which is used to validate the ability of prediction of the already trained network. Nine years (1984-1992) were used for training the neural network and three years (1993, 1994 and 1995) for testing the network. The nighttime values of total solar radiation, which are zero values, are omitted from the training and testing sets and therefore they are not used in the training and testing processes.

In their study, a multi-layer feedforward network based on backpropagation learning procedure is designed for predicting the total solar radiation time series. This type of neural network is extensively used in the time series prediction. The selected neural network architecture consists of one hidden layer of 16 log-sigmoid neurons followed by an output layer of one linear neuron. This scheme was selected after trying several ones as it gave the better convergence. The past eight values of the total solar radiation time series are used for predicting the next (ninth) value.

Two predictive schemes were considered:

- the one-lag scheme is used for predicting the next value of the time series given a number of past values of the series, thus generating a one-step-ahead prediction, and
- for the prediction of several steps ahead a multi-lag scheme is used in which the predicted output is fed back to the input for the next prediction thus allowing the generation of a new prediction two time steps ahead of the time series.

In their study, Sözen *et al.* (2005), used ANNs for the estimation of solar potential in Turkey, based on meteorological and geographical data. Such meteorological and geographical data, as utilized in this study are: the latitude, longitude, altitude, month, mean sunshine duration, and mean temperature. These variables were used as input to the network, whereas, the output was solar radiation. Twelve stations spread over the entire country were selected representing different regions at which the radiation data and sunshine duration measurements were collected since 2000.

In order to train and test the neural network, four years of meteorological data recorded in the period 2000–2003 from the above 12 stations were used. Data from nine of these stations were used for training, whereas, the data for the remaining three stations were used for testing the ANN models built.

The backpropagation learning algorithm has been used in a feedforward single hidden layer neural network. Variants of the algorithm used in the study are Scaled Conjugate Gradient, Pola-Ribiere Conjugate Gradient, and Levenberg–Marquardt (Hagan and Menhaj, 1994). Inputs and outputs are normalized in the (-1, 1) range; the logistic sigmoid transfer function was used in the network. In the training, they used an increased numbers of neurons (from 5 to 7) in a single hidden layer.

The performance of the neural network model was tested using various statistical measures and the results indicated that the ANN based estimation technique for solar radiation is more suitable to predict the solar radiation than the classical regression models proposed by other researchers. This study confirms the ability of the ANN to predict solar radiation values precisely and the results indicate that the ANN model seems promising for evaluating solar resource potential at the places where there are no monitoring stations.

Al-Alawi and Al-Hinai (1998) adopted ANNs to develop a model for analyzing the relationship between the global radiation and climatological variables, and to predict global radiation for locations not covered by the model's training data. The multilayer feedforward network used in this work is trained using the backpropagation paradigm (Rumelhart *et al.*, 1986).

The network architecture is composed of many simple processing elements that are organized into a sequence of layers. These are the input layer, the hidden layer and the output layer. The neurons in the input layer receive eight input signals representing the location, month, mean pressure, mean temperature, mean vapor pressure, mean relative humidity, mean wind speed and mean duration of sunshine; hence, eight neurons were used for input in the ANN architecture. The output layer, on the other hand, consists of one output neuron representing the Global Radiation (GR). Between the input and output layers one hidden layer has been used in this study. The data employed cover the years 1987 to 1992 and refer to six weather stations in the Sultanate of Oman.

Results indicate that the model predicted the GR values with a good accuracy of approximately 93% and a mean absolute percentage error of 7.30. In order to demonstrate the generalization capability of this approach over data unseen during the training effort and its ability to produce accurate estimates, the model was also tested to predict GR values for a location the archived data of which were not included in the training dataset. The monthly predicted values of the ANN model compared to the actual GR values for this independent dataset produced an accuracy of 93% and a mean absolute percentage error of 5.43 (i.e., comparable to the validation with the dataset used for training).

2.5. Maximum Solar Radiation

The objective of the research by Kalogirou *et al.* (2002) was to produce a neural network model for the prediction of maximum solar radiation based on simple data. The input data that are used in the present approach are those which influence mostly the availability and intensity of solar radiation, namely, the month, day of month, Julian day, season, mean ambient temperature and mean relative humidity.

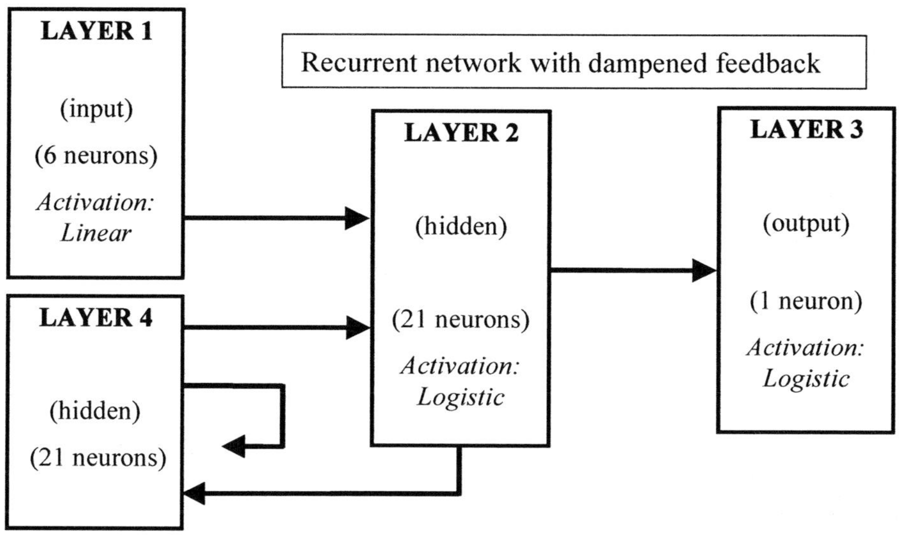

Fig. 4. The selected neural network architecture in Kalogirou *et al.* (2002)

Various network architectures, such as with three, four and five-layers, a number of recurrent types and a number of feedforward ones have been investigated aiming at finding the one that yields the best overall performance. In addition, a number of different network sizes and learning parameters have been tried. The architecture that was ultimately selected is shown in Fig. 4. It is of the recurrent type and is composed of four layers (slabs), one of which is hidden and one is used for dampened feedback. The extra layer is connected to the hidden layer. This architecture is commonly called "Jordan Elman recurrent network". It holds the contents of one of the layers as it existed when the previous pattern was trained. In this way, the network sees the previous knowledge it had about previous inputs. This extra layer is sometimes called the network's "long term" memory. The long-term memory remembers the hidden layer, which contains features detected in the raw data of previous patterns. Recurrent neural networks are particularly suitable for prediction of sequences so they are excellent for time series data, as in the present case. A backpropagation network with standard connections responds to a given input pattern with exactly the same output pattern, every time the input pattern is presented. A recurrent network may respond to the same input pattern differently at different times, depending upon the patterns that have been presented as inputs just previously.

Thus, the sequence of the patterns is as important as the input pattern itself. Recurrent networks are trained in the same manner as standard backpropagation networks, except that patterns must always be presented in the same order, i.e., random selection is not allowed. Six input elements have been used corresponding to the values of the input parameters listed above (i.e., the month, day of month, Julian day, season, mean ambient temperature and mean relative humidity). The learning procedure was implemented by using the backpropagation algorithm.

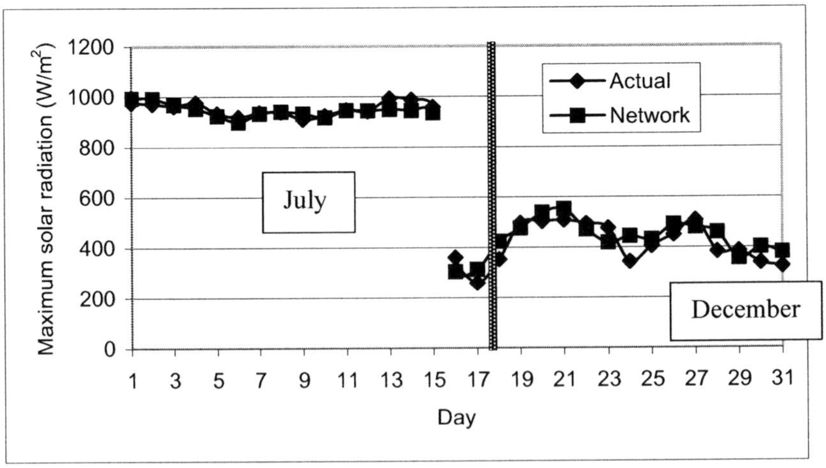

Fig. 5. Comparison of the actual and ANN predicted values of maximum solar radiation for the validation dataset (first two weeks of July and last two weeks of December) (from Kalogirou *et al.*, 2002)

Once a satisfactory degree of input-output mapping has been reached, the network training is frozen and a set of completely unknown test data was applied for validation. The trained network was used to predict the maximum solar radiation for the days contained in the validation dataset. It should be noted that the network has not seen these data before, i.e., these cases were completely unknown to the network. The correlation coefficient obtained for the validation dataset is 0.9867. In this respect, the closer to unity this value is the better the prediction accuracy. A graph of the actual and ANN predicted values of maximum solar radiation for the validation dataset is shown in Fig. 5. As expected, the prediction is much more accurate during the summer period when most of the days are almost always cloud free.

2.6. Daily Solar Radiation Time Series

Kemmoku *et al.* (1999) developed a multi-stage neural network to forecast the total insolation of the next day in MJ/m^2 per day. The concept of multi-stage is applied because of the following considerations. Insolation depends on the weather, which is closely related to the atmospheric pressure. The change of weather condition depends on the change of the atmospheric pressure, so there is a correlation between the insolation of the next day and the difference between average atmospheric pressures of the next day and previous day. Therefore, the average atmospheric pressure can be forecasted by the first stage neural network from meteorological data.

From this average atmospheric pressure and meteorological data, the insolation level of the next day is forecasted by the second stage neural network. A further stage refines the insolation prediction depending on three classes of the insolation level forecasted above. Therefore, this final stage considers high, middle and low levels, and proceeds with three different neural networks, accordingly.

The neural networks were trained with daily weather data of 6 years (1988–1993) for Omaezaki in Japan and forecast the daily insolation for 1994. The set of meteorological parameters used in this neural network modeling approach are:

- the difference between the first and the second local maxima of atmospheric pressure,
- the difference between the first and the second local minima of atmospheric pressure,
- the average atmospheric pressure,
- the clearness index,
- the temperature at 18:00 h,
- the atmospheric pressure at 18:00 h, and
- the insolation at a time before sunset (set at 17:00 h in April, 18:00 h in August and 16:00 h in October and December).

The performance of the multi-stage model outlined above was tested for four months in 1994, namely April, August, October and December. For this purpose, the percentage mean error statistic was used; the performance of the multistage neural network model was estimated to be around 20% for all the months in the evaluation dataset.

Santamouris *et al.* (1999) estimated hourly values of global solar radiation using the following three models : a neural network system based on backpropagation techniques, a deterministic atmospheric model and a fuzzy logic method based on fuzzy sets for modeling nonlinear functions. Training was performed using hourly values of the input climatic parameters for the estimation of integrated hourly global solar radiation values for 11 years (1984–94) and for various months of the year. The data used were taken on a hill at the center of Athens, Greece. The network was trained over a part of these data, and once training was completed, the network was tested over the remaining data. The input parameters of the neural network model were the air temperature, relative humidity, sunshine duration in hours and calculated extraterrestrial radiation. The output was the global solar radiation values.

A multiple network based on a backpropagation learning procedure was implemented. The selected neural network architecture consists of one hidden layer of 15 log-sigmoid neurons followed by an output layer of one linear neuron. Results from trial runs indicated that adding more hidden layers or nodes did not significantly improve the network's prediction capabilities; rather this only slowed the convergence.

The accuracy of the neural network estimations was tested by comparing the measurements of the testing set of data, which consists of the radiation values of the year 1995, with the estimated values reached with the neural network approach and the mean-square error was found to be quite acceptable (e.g. 0.22 MJ/m^2 for July and 0.20 MJ/m^2 for January). The present results are quite encouraging for developing a feedforward backpropagation neural network approach able to simulate and predict future values of global solar radiation time series by extracting knowledge from their past values.

In the same study, Santamouris et al (1999) performed also a comparison of the three methodologies studied, namely, the deterministic atmospheric model and the two intelligent techniques (i.e., the ANN method and the fuzzy logic method). The comparison of the three methods showed that the intelligent techniques put forward could be successfully used for the estimation of global solar radiation during the warm period of the year, while during the cold period the atmospheric deterministic model gives better estimations.

2.7. Hourly Solar Radiation Time Series

Hontoria *et al.* (1999; 2002) presented a methodology to solve a typical problem in solar energy by using a MLP neural network (Haykin, 1998; Kohonen, 1995; Lippmann, 1987). A feedforward – feedback architecture is proposed for the generation of hourly solar radiation series. In their approach, they have adopted the concept of atmospheric transmittance or clarity index (which is defined as the quotient between the global irradiation and the extra-atmospheric irradiation). They proceeded by decomposing this variable into two components: one is the ***trend*** (or mean) component (following the evolution of the clarity index during the day) and the other is the ***random*** component (which consists of random fluctuations in the trend component).

In the following, they adopted a MLP neural network model for the trend component. Regarding the random component, they used also a MLP for building models of this component's autocorrelation coefficient and standard deviation.

In summary, three different networks have been developed, one for generating the trend component and the other two for generating the random component. It came out that the former neural network model reproduced the trend component satisfactorily. Also, with the latter two networks, the random component has also been obtained satisfactorily.

For their calculations, they have used data from four to eight years for seven Spanish locations representing a wide spectrum of climatic characteristics of the country. All the data but those for one year were employed in the training of the neural network. The remaining one year data were used for validation and testing.

The work of Atsu *et al.* (2002) is along the same lines as the above research by Hontoria *et al.* (1999; 2002). They investigated the estimation the clearness index (defined as the ratio of the observed solar radiation to the maximum solar radiation) by using Radial Basis Functions and Multilayer Perceptron models.

Data from eight meteorological stations in Oman averaged over at least ten years (1986–1998) were used to train and validate the MLP neural networks and a RBF network. The input parameters were latitude, longitude, altitude, sunshine ratio, (ratio of sunshine hours to maximum sunshine hours) and month of the year. The output parameter is the clearness index. The estimated solar radiation was subsequently obtained by multiplying the estimated clearness index by the maximum solar radiation. Data from six of the above eight stations were used to train the networks, whereas, the data from the remaining two stations were employed in the validation of the models. The selection of the training and the validation stations were made based on all possible combinations (i.e., a total of twenty-eight training-validation station combinations were used).

Three MLP neural networks (with 1, 2 and 3 hidden layers) and one RBF network were implemented. Statistical testing of the models has shown that all the four networks performed very well. The performance of the various networks were similar but Atsu *et al.* (2002) recommend the use of the RBF network as it requires less computational effort compared to the MLP neural networks.

Sfetsos and Coonick (2000) focus on forecasting of hourly solar radiation by using two artificial intelligence based techniques: ANNs and Adaptive Neuro-Fuzzy Inference Systems (ANFIS). The proposed approaches make use and predict raw solar radiation values (univariate models). Additionally, their study investigates the use of additional meteorological time series as potential inputs to the models developed, such as temperature, pressure, wind speed and wind direction (multivariate models). They also compare different artificial intelligence models for the solar radiation data, which were considered as ideal for such a comparison, as they comprise time series combining strong periodical structure and discontinuities.

Different feedforward ANNs were trained and tested with two different learning algorithms; the first is the backpropagation algorithm with an adaptive learning rate and a momentum term and the second is the Levenberg-Marquardt algorithm (Hagan and Menhaj, 1994), which comprise a radial basis function network (Ranaweera *et al.*, 1995) and an Elman recurrent network (Elman, 1990).

The data used in the study were mean hourly solar radiation values on a horizontal level (in W/m^2) measured on the island of Corsica and they cover a period of 63 days during late spring and early summer of 1996. The data were split into three subsets: a training, an evaluation and a prediction subsets.

The training set is used for the training of the Feedforward, the Elman and the Neuro-fuzzy approaches. The evaluation set, which was unknown during the training phase, was used to check the progress of the network. The model that is used for the prediction set is the one whose parameters minimize the Root Mean Squared Error (RMSE) of the evaluation set. The performance of each forecasting method is evaluated with the prediction set.

On the basis of prediction error and training time, a comparison between the various models has indicated the Levenberg-Marquardt network as the optimum prediction model. Finally, the performance of the Levenberg-Marquardt forecasting method was improved when the wind direction is included as an input.

3. TEMPERATURE

3.1. Temperature Forecasting

Forecasting of temperature is one of the regular operational practices carried out by Meteorological Services worldwide. The great interest in developing methods for more accurate site specific predictions of the temperature regime has led to the invention of several such methods of which the most commonly used can be considered to fall into three groups.

The first group comprises physical methods utilizing the laws of radiation, (see Bagdonas *et al.*, 1978). However, the complexity of the process governing the temperature regime of a specific site for which a prediction is required renders such methods difficult to manipulate and their practical application is limited. The second group comprises statistical-empirical methods that are basically mathematical relationships fitted to archived meteorological data by means of some curve fitting procedure (e.g. Boyden, 1937; Craddock and Pritchard, 1951; Michaelides, 1991). Because such relationships are relatively easy to establish, the techniques in this group have been widely used for many decades. With the arrival of numerical weather modeling, a third group of numerical-statistical methods has evolved which have progressively been adopted by many Meteorological Services (e.g. Klein *et al.* 1967; Kelley *et al.*, 1988; Lemcke and Kruizinga, 1988). In this approach, the forecast temperature is statistically linked to archived observed or simulated meteorological parameters. Numerical model products are subsequently used as input to obtain a temperature forecast.

3.2. Forecasting Minimum Temperature with ANN

One of the first efforts to employ ANNs in meteorology refers to the forecasting of minimum temperature as a result of nocturnal cooling (Schizas *et al.*, 1991; 1994). The set of raw data involved in this work has been extracted horn the meteorological observations recorded at three-hour intervals at Larnaca Airport in Cyprus. From each of the eight sets of observations available daily, the following elements have been extracted, preprocessed and used as input to the ANN: total cloud amount (TC), visibility (VI), present weather condition (PW), atmospheric pressure (AP), dry-bulb temperature (DB), wet-bulb temperature (WB), and low cloud amount (LC). Using the wind direction and the wind speed, the eastward (W1) and northward components (W2) of the wind vector were calculated. Besides these records, the astronomical day length (DL) and the observed minimum temperature of the previous night (PM) were used as additional inputs. All of the above inputs were chosen after consultation with local weather forecasters and are considered as having an impact on nocturnal cooling.

A complete explanation of the above meteorological elements, their respective measuring units and coding practices is given by WMO (1995). However, it is worth outlining briefly here the use of one qualitative parameter that has been used, namely, the present weather. Although there exist general rules for reporting this weather element out of a total of 100 different weather situations, its final definition is largely subjective (more specifically, it is at the discretion of a weather observer to define it, although currently many automatic systems can simulate to a large extent this decision even in the absence of a human observer). To provide the reader with an insight regarding the use of such a qualitative element in the process of model training, a grouping of the 100 possible codes for present weather is presented in Table 2.

In the their study, Schizas *et al.* (1991; 1994) have used a total of 133 days in winter and spring of 1984 for training and evaluating the forecasting models. Of these days, 103 were randomly selected and used for training while the rest 30 days were used for testing the models constructed.

There are many possible ways to implement ANNs and determine the optimal network and this is not an easy task. In each case, the modeler has to consider the uniqueness of each problem and, in most cases, has no other choice but to resort to a systematic testing of some of the available possibilities based on experience. Some of the available choices refer to the input and output selection, the network architecture and the training algorithm. Table 3 displays how the various input parameters were used with different models.

The backpropagation training algorithm was used for building the investigated ANN models (Rumelhart *et al.*, 1986). Table 4 shows the set of ANN models that were trained and evaluated with the above mentioned data. For all the examined models, the learning rate and the momentum were chosen to be 0.01 and 0.9, respectively. For each model, the number of inputs (the first number shown by the architecture) is determined by the number of parameters that are used for training the model. The number of units in the first and the second hidden layers of the model are represented by the second and the third number of the architecture, respectively.

Table 2. Abstract of numerical coding of present weather (from Schizas *et al.*, 1991; 1994)

Range of Numerical codes	Description
0-19	Generic weather phenomena but mostly with no precipitation, fog, ice fog duststorm, drifting or blowing snow at the station at the time of observation.
20-29	Precipitation, fog. ice fog or thunderstorm at the station during preceding hour but not at the time of observation.
30-39	Duststorm, sandstorm, drifting or blowing snow.
40-49	Fog or ice fog.
50-59	Drizzle.
60-69	Rain.
70-79	Solid precipitation not in showers.
80-99	Showery precipitation or precipitation with current or recent thunderstorm.

Table 3. Input parameters to various models: total cloud amount (TC), eastward (W1) and northward components (W2) of the wind vector, visibility (VI), present weather condition (PW), atmospheric pressure (AP), dry-bulb temperature (DB), wet-bulb temperature (WB), low cloud amount (LC), astronomical day length (DL) and the observed minimum temperature of the previous night (PM) (from Schizas *et al.*, 1991; 1994)

Model	Input parameters										
	TC	W1	W2	VI	PW	AP	DB	WB	LC	DL	PM
1	✓	✓	✓	✓	✓	✓	✓	✓	✓	✓	✓
2	✓	✓	✓	✓	✓	✓	✓	✓	✓	✓	✓
3	✓	✓	✓	✓	✓	✓	✓	✓	✓	✓	✓
4					✓						
5					✓		✓	✓			✓
6		✓	✓				✓	✓			✓
7	✓	✓	✓				✓	✓	✓		
8		✓	✓				✓	✓			✓
9		✓	✓				✓	✓		✓	✓
10		✓	✓				✓	✓			✓
11		✓	✓		✓		✓	✓			✓
12		✓	✓				✓	✓			✓
13		✓	✓		✓		✓	✓			✓

These numbers were chosen empirically. The number of outputs (last number in the architecture) was chosen to be forty for all the models. The forty outputs represent the temperature range 0°C to +20°C with 0.5°C intervals that form the available classes of the models.

Most of the models learned the training set very well, as suggested by the Total Sum of Squares which is defined as the sum of the squared difference between the expected and the calculated output for all cases (days) in the training cycle (one epoch). The forecasting yield, also shown in Table 4, is a measure of the success of a model during the testing (evaluation) phase. The percentage of days in the testing set of data, for which the minimum temperature was successfully forecast, is defined as the forecasting yield of a model. This forecasting yield is given for temperature confidence ranges: ±0.5, ±1.0, ±2.0, and ±3.0°C, as shown in Table 4. It is obvious that, an increase of the temperature confidence range will increase the forecasting yield, at the expense of accuracy.

To cover a basic need regarding temperature forecasting, Tasadduq *et al.* (2002) have used ANNs for the prediction of hourly mean values of ambient temperature 24 hours in advance. One year hourly values of ambient temperature are used to train a neural network model for a coastal location in Saudi Arabia. The ANN has one input, one output and one hidden layer. The neural network was trained off-line using backpropagation and a batch learning scheme. The model developed requires only one temperature value as input to predict the temperature for the following day for the same hour. The trained neural network was tested with temperatures for three years other than the ones used for training and it was found that the model was able to forecast hourly temperatures 24 hours in advance with sufficient accuracy.

Table 4. Various models' architectures and forecasting yield (from Schizas *et al.*, 1991; 1994)

Model	Architecture	Epochs	Total sum of squares	Forecasting yield (%)			
				±0.5°C	±1.0°C	±2.0°C	±3.0°C
1	74-120-160-40	290	0.9	21	32	45	53
2	74-80-120-40	410	0.9	16	29	45	50
3	64-40-60-40	670	2.5	18	37	60	68
4	9-16-32-40	8060	30.6	24	332	39	53
5	25-40-60-40	2460	7.3	21	29	47	66
6	32-40-80-40	4070	5.0	24	29	47	50
7	48-60-100-40	431	0.9	18	26	37	42
8	33-40-80-40	4020	2.0	18	29	53	58
9	34-40-80-40	587	0.9	16	32	47	55
10	41-50-100-40	4180	3.0	18	29	50	60
11	41-50-100-40	453	0.9	24	26	39	55
12	29-40-80-40	1330	2.2	21	39	50	58
13	36-50-100-40	1100	0.9	21	32	42	53

Arca *et al.* (1998) also developed a model based on ANN for short-term prediction of the daily minimum temperature of air, analyzed its performance under different meteorological conditions and evaluated its capabilities as a tool for decision-making related to frost occurrence.

The database used consisted of measurements collected from December 1996 to March 1998 at an agrometeorological station in Florence, Italy. The measured variables were:

- air temperature,
- soil temperature,
- relative humidity of air,
- temperatures of a black body in the sun and in the shade,
- global solar radiation,
- diffuse radiation,
- net radiation,
- wind speed and direction,
- rain intensity, and
- atmospheric pressure.

The data were acquired every two minutes and a mean value of each variable was calculated and stored every twenty minutes. In addition, at sunset, the form and the total amount of cloud were recorded. For each quantity, the value at sunset and the daily minimum, maximum and mean value were calculated. The data collected were divided in two parts: the first, consisting of ninety days (patterns) refer to the period December 1996 - February 1997 and were used for training the ANN; the second, composed of twenty-eight records, collected in February 1998, were used for testing.

Seven different feedforward ANN with three layers were developed. In all of the seven ANN, both the input layer, which receives the values of selected quantities, and the hidden layer, which performs the data computations, have a variable number of neurons. The output layer has a unique neuron yielding the estimate of minimum air temperature. The input variables were chosen on the basis of physical laws describing the radiative cooling of soil and atmosphere.

The best performing architecture yields predictions of minimum air temperatures with a low standard error of estimate (1.63°C) and a good correlation with the observed values (0.91). The accuracy of estimates provided by the neural model developed in this study is high in comparison with other statistical or semi empirical methods (Brünt, 1941). In addition, the ANN models require few, easily measurable atmospheric parameters while other models prove to be more demanding.

3.3. Temperature Module in Energy Demand Controller

Argiriou *et al.* (2000) have demonstrated how ANNs can be usefully applied in optimizing the energy demand of buildings and systems that maximize the use of solar energy for space heating. An ANN model developed for this purposes plays the role of a controller which is able to forecast the energy demand leading to a simple On/Off decision of electrical heating system maintaining acceptable indoor conditions. The controller developed is of a modular structure. The relevance to meteorological ANN forecasting lies in the inclusion of two weather modules that make use and forecast the solar irradiance and the ambient temperature by using the values of these parameters at the previous time steps and parameters characterizing the specific time interval (hour and day). Other modules comprising the ANN controller predict the On/Off operation status of the heating system, the indoor temperature and the final output from the controller to the heating system, that is the On/Off switch position.

This review focuses on the two weather predictive modules for solar irradiance and ambient temperature. These modules were created by applying the feedforward neural network model and supervised training with the method of backpropagation with momentum term. The inputs to these modules were the measured solar irradiance and ambient temperature at regular time intervals of 15 min. The ANNs developed were trained using part of real meteorological data for 1993 in Athens, Greece; the remaining data were used for verification.

The module forecasting the ambient (outdoor) temperature uses a simple ANN with ten input neurons, one hidden layer of eight and another of four neurons, and one output neuron. The inputs to this ANN are the (daily normalized) time value for the next interval, the last and three previous values of the outdoor temperature, the (yearly normalized) day number for the next and the last and three previous values of the solar irradiance. Hourly values are normalized by dividing them by 24 and the day number values by 365. Also all other parameters are normalized to fit in the interval [0,1]. The output is the one-step-ahead prediction of the outdoor temperature difference.

The solar irradiance forecast module has twenty-eight input neurons, one hidden layer of sixteen and another of eight neurons, and one output neuron. The inputs are the last and six previous values of the following parameters: daily-normalized time, ambient temperature, day of the year and solar irradiance. The output is the one-step-ahead prediction of the difference of the global solar irradiance.

The ANN module for forecasting the one-step-ahead outdoor temperature performed satisfactorily with differences from the recorded values up to one degree in just a few cases. The prediction of solar irradiance was not so accurate; however, the difference exhibits a correlation with the hour of the day, being larger at times close to sunrise and sunset, therefore their effect on the controller are not so important.

3.4. Greenhouse Indoor Temperature

In another temperature-energy related problem, Ferreira *et al.* (2002) model the inside air temperature of a hydroponics vegetable production greenhouse. The greenhouse climate is affected by many factors (such as the outside weather, the actuators and the crop). Therefore, methods aimed at controlling the greenhouse climate environment must take these influences into account. Ferreira *et al.* (2002) addressed one aspect of this problem, namely the designing of models of the greenhouse inside air temperature as a function of the outside solar radiation and temperature, and the inside relative humidity. The modeled temperature is intended for use in a greenhouse adaptive predictive hierarchical control scheme.

In their endeavor, Ferreira *et al.* (2002) compared several off-line and on-line learning methods for RBF networks. Recently, this type of neural network has received growing attention because of its structural simplicity. The three different off-line methods used in this study are simply referenced here:

- a clustering procedure known as the optimal adaptive k-means algorithm;
- the orthogonal least squares learning algorithm which selects a suitable set of center locations from the set of input data in a constructive fashion, and
- the Levenberg–Marquardt algorithm that exploits the sum-of-the squares characteristic of the problem.

In addition, the three on-line learning methods used consist of:

- a constructive method (usually starting with an empty hidden layer and then, based on the novelty of the input training samples, start adding units to the network);
- the Strict Interpolation problem with generalized RBF with regularization (see Yee and Haykin, 1999);
- the Levenberg–Marquardt method.

The basic input dataset is composed of 4257 points acquired with a sample rate of 5 min. The results obtained led the authors to conclude that RBF Networks can adequately model the greenhouse internal temperature. All methods achieved good fittings and acceptable prediction errors.

Along the lines of greenhouse indoor temperature modeling with ANNs is the work by Seginer (1997) who presents three alternative architectures for predicting the greenhouse indoor temperature being the model output. In the first single-level model, indoor temperature depends on the outside temperature, solar radiation, heat and ventilation flux. Bearing in mind the difficulties in directly measuring the ventilation flux, a two-level model is subsequently put forward, in which the ventilation flux is indirectly estimated by using the wind and the area of the vents. The third model is obtained by collapsing the latter model into a single-level which makes direct use of the outside temperature, solar radiation, wind and heating flux as inputs. Single-hidden-layer feedforward neural modeling was used in this endeavor to model the climate of greenhouses.

3.5. Ground-Level Temperature

In another temperature application problem, Shao (1998) have used a Backpropagation Neural Network in order to improve nowcasts (i.e., very short term forecasts of the order of a few hours) of road surface temperature. In this respect, the applicability of neural networks was tested for a possible improvement of road ice prediction by using an already established forecasting model. In this approach, a three-layer and error-backpropagated neural network was provided with preliminary meteorological parameters and original model prediction errors in the previous 3 h. The network output is a set of temperature adjustments to the original model's forecasts for the next 3 or 6 h.

The input consists of a set of 20 variables. It is interesting to outline briefly these variables in order to get an idea of the spectrum of potential input to such a neural network:

- time of day,
- forecast errors 1, 2 and 3 h earlier,
- variations of surface temperature, air temperature and dewpoint temperature 1 and 2 h earlier,
- derivatives of surface temperature, air temperature and dewpoint variations,
- differences of surface and air temperatures, and air temperature and dewpoint 1 h earlier,
- mean wind speed for the last 3 h, and
- road surface state 1 h earlier.

The approach was tested in a wide range of localities in Europe and Japan. As a broad conclusion, this study has shown that a three-layer neural network was generally able to improve the accuracy of 3 and 6 h numerical model forecasts of road surface temperature by learning from historical data, especially at the sites with complex environments and unknown underlying mechanisms.

4. Wind

4.1. Wind Energy Potential and Wind Speed Forecasting

Wind is one of the earliest forms of energy exploited by humans for driving wind mills, grinding grains, pumping water and propelling ships, to name a few of such early applications. However, partly due the uncertainty of wind usage (e.g. diurnal variability, seasonal dependence, etc) but also due to the availability of plentiful and cheap fossil fuels, wind exploitation as a source of energy was diminished. Recently, in search for cheaper alternative energy sources the exploitation of wind's free source of energy has been brought to the surface again. The re-consideration of wind energy is also related to current environmental concerns, being a clean and inexhaustible source of energy.

With regard to the usage of wind data in energy related issues, several aspects can be considered. One aspect, which is usually addressed, is related to the establishment of the potential of a locality for generation of wind power; in this respect, mean monthly and daily wind speed is important for *planning* of wind power farms. Another important aspect refers to the *operational* usage of wind power and in this respect, accurate forecasting of the wind speed, normally with a few hours lead time can be crucial; indeed, such predictions can help producers take decisions for the optimization of energy production by calculating the wind power availability in advance.

Modeling the wind is a highly nonlinear issue due to the numerous factors that can influence both the wind speed and its direction (terrain configuration and characteristics, elevation, prevailing weather pattern, etc). Forecasting the wind is traditionally sought with dynamical modeling (i.e., with Numerical Weather Prediction models in which the wind at various levels is routinely predicted); also, stochastic analysis and modeling have been used extensively and related methodologies include time series modeling (e.g. Kamal and Jafri, 1997) and Kalman filtering techniques (Bossanyi, 1985), to name but a few.

ANNs have recently been employed in wind forecasting and, as it will be shown in the following, ANNs can be an alternative to conventional techniques and in many respects can be more advantageous for tackling such nonlinear problems.

4.2. Monthly Average Wind Speed

The potential of ANN modeling in estimating the wind regime in areas with no data by using wind data from adjacent stations has been investigated by Kalogirou *et al.* (1999). The geographical area in Cyprus involved in this study is shown in Fig. 6 where three meteorological stations are in operation. The effort in this study was to use data from two of these stations (called control stations) in order to estimate the wind at the third adjacent station (called target station). Wind speed is measured at two levels, namely, 2 and 7m and the observed data used in the study cover the years 1986 to 1997.

Two types of networks were trained, differing in the number of inputs. In the first type, the input data comprised the co-ordinates of all three stations, the month of the year and mean monthly values of the wind speed at two levels (2m and 7m) for each of the two control stations (i.e., 11 input neurons); in the second type, the co-ordinates of the stations were omitted (i.e., 5 input neurons). The output of both networks is the mean monthly values of wind speed of the target station. The learning procedure was implemented by using the backpropagation algorithm. Different network structures, sizes and learning parameters have been tried but the architecture that was ultimately selected is composed of five layers, three of which are hidden. This type of architecture has been used successfully for other applications of neural networks in estimating rainfall (Kalogirou *et al.*, 1997; 1998).

The wind database used was divided into two sets: a training dataset having all wind speed records for each month for the years from 1986 to 1996 (i.e., 11 years), and a verification dataset for all months of 1997. The training dataset has been used for the training and testing of the ANN, while the verification dataset has been used for validation of the network.

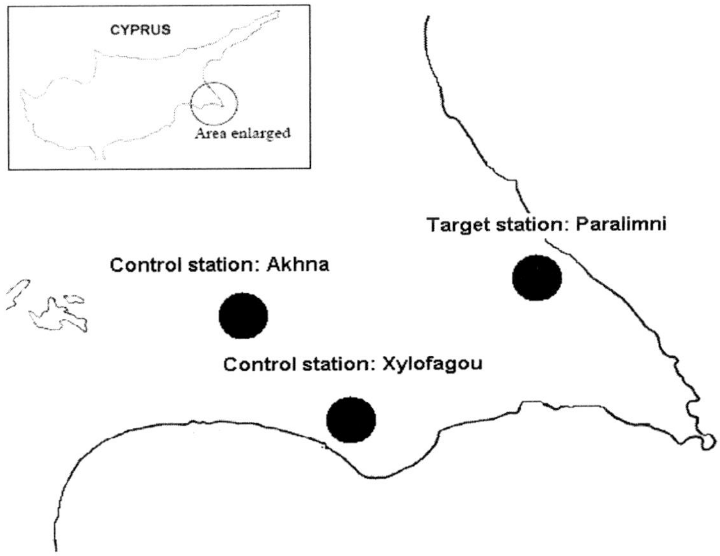

Fig. 6. Location of the meteorological stations used in Kalogirou *et al.* (1999)

The verification results for the 11- and 5-input networks are shown in Table 5. As can be seen from this table, the network with only five input parameters is more successful, giving a maximum percentage difference of only 1.8%. Kalogirou *et al.* (1999) recommend that the two networks can be used for two different types of assignments:

- The first 11-input network can be used more effectively in cases where the objective is to use the network to predict the wind speed in locations other than the three station locations. In this case, the coordinates must be used as an input parameter. The target station can be located either within the area marked by the three stations (interpolation) or outside (extrapolation).
- The second 5-input network can be used to predict mean wind speed in fixed locations more successfully, therefore, it can be used to fill missing data values from the meteorological databases.

Table 5. The verification results for the 11- and 5-input networks

Network architecture	Mean wind speed-Actual (m/s)		Mean wind speed-ANN predicted (m/s)		Percentage difference (%)	
	Height2m	Height 7m	Height 2m	Height 7m	Height2m	Height7m
11-input neurons	2.4	3.35	2.43	3.52	1.2	5
5-input neurons			2.4	3.41	0	1.8

4.3. Mean Daily Wind Speed

Mohandes *et al.* (1998b) describe the utilization of neural networks in predicting the monthly and daily mean wind speed for Jeddah in Saudi Arabia. These predictions are subsequently compared with those of autoregressive models. For the training of the multilayer neural network adopted in this study, the backpropagation algorithm has been used.

The observed data used in this study cover a period of 12 years, from 1970 to 1982. For the mean monthly wind speed prediction study, the data were first normalized and were subsequently split into two sets: one for training (120 months) and the other for testing (24 months). A similar procedure was adopted for the mean daily wind speed prediction study, in which case the data were also divided into two sets: a training set (730 days) and a testing set. The testing data were not used in either defining the coefficients of the autoregressive models or for adjusting of the weights of the neural networks. The same training and testing datasets were used for both the autoregressive and the neural networks approaches.

Two types of models were built and tested:

- single-step models (one month ahead for the monthly models and one day ahead for the daily models) by using one input and one output;
- multi-step models (several days ahead) by feeding the output of the system as input.

Several networks with different number of hidden neurons were trained. Based on the root mean square error over the training data, Mohandes *et al.* (1998b) conclude that, for the daily predictions, the best network was one with 24 hidden units. The results on a testing data indicate that for both the single-step and the multi-step models, the neural network approach outperforms the autoregressive approach.

4.4. Wind Classification and Prediction

For most applications, users of wind speed predictions are interested in ranges or classes of wind values rather than in an exact wind values. Kretzschmar *et al.* (2004) addressed the prediction of average hourly local ground wind speed and hourly local ground wind gust maximum values for prediction lead times between 1 and 24 h; in their effort the local wind prediction task is formulated as a classification task in order to meet the aforementioned need.

The wind speed and gust classes were defined on the basis of the widely used Beaufort scale. The neural networks employed were conventional feedforward neural networks with two variants: directly (as classifiers) and indirectly (as function approximators that predict the exact wind values). The classifiers presented exhibited a superior quality when compared with persistence with respect to performance as well as hit and false alarm rates for most of the prediction lead times for wind speed and wind gusts. For the prediction of strongest winds, satisfying results could only be obtained for the one-hour lead time.

The inputs selected for the classifiers were several lags of the local wind speed, wind gust and wind direction time series, time and date. To test the potential of additional input from Numerical Weather Prediction models, such model data from the European Centre for Medium-Range Weather Forecasts (ECMWF) analyses were also used. Except for the model features, all inputs were obtained from a single wind observation device at the site of interest.

The chosen model input led to improvements for wind speed prediction for 12 h and 24 h lead times but the improvement was smaller than expected.

4.5. Damaging Winds

Marzban and Stumpf (1999) developed a neural network to diagnose which circulations detected by the Mesocyclone Detection Algorithm of the National Severe Storms Laboratory (USA) yield damaging winds. Circulations detected in data from radar volume scans can be associated with a report of a tornado and/or damaging winds in excess of 25 m/s. If a circulation is detected within 20 min prior to a ground report of damaging wind or within 5 min after a report, the circulation is classified as a prediction of damaging wind.

Twenty three variables characterizing the circulations are selected to be used as the input nodes of a feedforward, supervised neural network. These variables were linearly scaled to lie in the range 0.1 to 0.9. It is interesting to list these variables for the reader to get an idea of the variety of input variables one can look for:

- height of the bottom of the circulation,
- depth of the circulation,
- a non-dimensional number related to the range-dependent strength parameters (rotational velocity and shear) of a circulation (termed as strength rank),
- diameter of the circulation at its lowest elevation detected,
- largest diameter of the circulation throughout its entire depth,
- height of the largest diameter of the 3-dimensional circulation; rotational velocity of the circulation at its lowest elevation detected,
- largest rotational velocity of the circulation throughout its entire depth,
- height of the largest rotational velocity of the 3-dimensional circulation,
- low-altitude shear,
- largest shear of the circulation throughout its entire depth,
- height of the largest shear of the 3-dimensional circulation,
- low-altitude gate-to-gate velocity difference,
- maximum gate-to-gate velocity difference,
- height of maximum gate-to-gate velocity difference,
- the height of the lowest 2-dimensional feature whose strength rank is greater than or equal to the strength rank above,
- continuous depth of the 2-dimensional features in the 3-dimensional detection whose strength rank is greater than or equal to the strength rank,
- age of the circulation; vertically integrated strength ranks for all 2D features within the 3D detection,
- rank derived using the vertically integrated rotational velocity and shear within the 3-dimensional detection,
- ratio between the depth of the circulation and the depth of the entire storm cell,
- low-altitude convergence, and
- mid-altitude convergence.

The outputs of the network represent the existence/nonexistence of damaging wind, based on ground observations.

Another very interesting component of this research is the large variety of performance measures that were employed in an attempt to assess the network; indeed, fourteen scalar, non-probabilistic measures (which include measures of accuracy, association, discrimination and skill) and two multidimensional probabilistic measures (reliability and refinement diagrams) are employed to assess the performance of the network. Two methods of classification were also considered.

It was found that a neural network with two hidden nodes is optimal for thirteen of the fourteen measures of performance and both methods of classification. In terms of reliability diagrams, it was shown that the process by which the outputs are converted to probabilities allows for the forecasts to be completely reliable.

4.6. Wind Nowcasting

Nowcasting refers to forecasting meteorological variables over a period of a few hours ahead. Accurate wind nowcasts can be crucial in several applications. Regarding wind power generation, accurate short-term wind power forecasts permit to operate optimally wind farms, maximize wind power penetration, plan efficiently for storage, reserves, maintenance, etc.

Storch (1999) used neural networks to forecast winds at the Kennedy Space Center and the Cape Canaveral Air Station launch pads (in the USA). Variables are developed from a network of tower observations, surface and buoy observations and an upper-air sounding. It is very interesting to note that initially, 153 experimental variables were developed from the different types of observations. Although it is not possible to present all of these variables in detail, it is worth discussing some of them in order to appreciate the large diversity of variables that can be incorporated in such a neural network approach.

Wind measurements dominated the variables chosen to train the network. The wind was transformed to eastward and northward components so it could be treated as two scalars; this circumvented the 0-360° discontinuity inherent in the wind direction measurements thus providing the network with a series of continuous wind measurements. Subtracting components of the wind at different heights developed measures of vertical wind shear; subtracting components of the wind at the same height, but different locations developed horizontal wind shear measurements. Sea level pressure observations were also included. Other variables were largely measures of differential heating, temperature, dewpoint, relative humidity and cloud ceiling heights that were included to aid the neural network in recognizing frontal weather; vertical and horizontal gradients of temperature and moisture were developed through component subtractions. The solar elevation angles for Kennedy Space center were calculated and included as well.

From the initial 153 variables, a smaller set of inputs was chosen, using a signal-to-noise variable screening method. Once the most salient set of variables was identified, the random variable was removed, and an Elman Recursive Neural Network was trained to forecast the fifteen target variables fifteen minutes into the future. These target variables are the eastward, northward wind components and peak wind speed at various tower locations and heights. The network forecasts are compared to persistence and peak wind predictions and found skillful compared to persistence.

Along a similar line was the research by Cloys (2000) who demonstrated the potential for using time-delay neural networks to provide advance warning of wintertime (November-March) peak wind speeds at the Atlas Space launch pad in the USA.

The following elements were used to compute inputs to the neural networks: peak wind speed, time of day, wind direction, directional deviation and 30-minute maximum peak wind. One-hour moving averages of observed peak wind speed were chosen. To capture the trend in the signal, previous one-hour averages were also included. Previous values of an input element are known as "delays," thus the network designator "time-delay neural network". The target sets (network output) were the 16 observed 30-minute maximum peak wind speeds that occurred starting 30 minutes after the current input time.

This work used data from March 1995 through March 1999. From the original database in these four years period, the 25 longest no-gap (i.e., without any missing data) periods were used in developing, training, and testing the neural network. Each of these 25 data periods was randomly assigned to one of three subsets:

- the first subset (containing nine periods) was used to develop and test a preliminary neural network to determine the best architecture to use in the network training,
- the second subset (eight periods) was used in the training phase in order to determine weights and biases for a new network, and
- the third subset (eight periods) was used to test the network developed.

The network performance was compared to three other forecasting options, namely, persistence, climatology and randomly selected wind speeds from a climatologically based distribution. For forecasts at the end of the forecast period, networks that were tested with data near in time to the networks' training data showed skill over the other three forecasting options.

The objective of a study by Sfetsos (2000) was to examine and compare various linear models and different artificial intelligence based approaches in forecasting mean hourly wind speed. The non-linear forecasting models that were examined were neural network based, i.e., an Adaptive Neuro-Fuzzy Inference System (ANFIS) and a Neural Logic Network. The author ascribes to the latter two a combination of the strength of neural networks with those of fuzzy logic and expert systems, respectively.

The experiments were carried out for a location on the Greek island of Crete and the data employed are mean hourly wind speed values measured at this location and covering a period of one month, that of March 1996; the dataset contained 744 values, representing the total number of hours in a month. This set of data was split into three sub-sets: a training, an evaluation and a prediction sub-sets. The training set was used for both the estimation of the parameters of the linear model and the training of the non-linear models. The evaluation set, which was not used in the training phase, was used to check the progress of the network. The performance of each model was subsequently measured by using the prediction set.

The comparison of the various time series forecasting approaches indicated that the Artificial Intelligence based models outperformed the respective linear ones. The model with the least error was identified as a Neural Logic Network that incorporated logic rules.

The final application, which will be presented in this review, refers to the use of a feedforward neural architecture adopted by Flores *et al.* (2005) in an effort to establish a predictive system for wind speed forecasting in locations in northern Spain. It is interesting to note that wind speed data at the level of the wind turbine were collected with high frequency, one sample per minute, and therefore the prediction system could be developed for very short lead times.

The ANN was developed for predicting the wind speed for the next hour by taking into account the wind speed measured in the last three hours. The method was verified and the results obtained were considered as quite acceptable. The real data used in the verification process were totally different from the data used in the training process. As the basic database used in this endeavor was too short (two months in 2002 for training and another two months for verification), the authors suggest a periodic training of the network as more data become available.

5. Neural Approach: An Effective Alternative

From the application areas discussed in this chapter, it is evident that neural networks have been used effectively in a large variety of meteorologically based applications. What is required for such neural network approaches are records of the specific meteorological parameters for a sufficiently long time period and the setting-up of a suitable ANN model, which can be trained with these data in order to make predictions of the parameters in study. The selection of the appropriate model is usually done empirically, following rigorous testing of several alternative configurations. The performance of the selected models needs to be tested with independent data, which are taken out from the available dataset and are not used for the network training. It is essential that the meteorological data used are of high quality; preferably, such records should be obtained with properly calibrated instruments of the highest possible standards.

The various ANN models presented here were able to predict solar radiation, temperature and wind speed with acceptable accuracy. In most of the cases, multilayer backpropagation neural networks are employed, while the other most popular architecture is the recurrent, which is considered suitable for time series prediction.

In most of the research efforts presented above, the use of neural network models were primarily investigated in order to prove the suitability of the neural approach in predicting future values of various meteorological parameters. Additionally, neural network models can be used in filling missing values in existing databases. In order to develop rigid tools which will be more useful to scientists and engineers working in the fields of meteorology and renewable energy, the models developed need to be "enriched" with more training cases that will cover a broader spectrum of weather conditions (which are quite often unavailable at the beginning of the task). As more data are collected, these can be used to retrain the networks so that more useful operational tools are produced. It can be anticipated that the neural prediction methodology will be further improved, as the training database expands, thus covering more diverse cases.

The site specificity of the ANN models that are normally developed is an issue which must be given serious consideration. Indeed, as the values of the meteorological parameters discussed in this chapter are largely controlled by local factors, the ANN models trained on such parameters are highly site specific too and different neural models need to be trained for different sites. However, it is possible to develop models which can include site characteristics as inputs to the neural network in an effort to create general purpose or universal tools.

Like all approximation techniques, ANNs have advantages and disadvantages. There are no rules as to when this particular technique is more or less suitable for an application. Based on the work presented here, it is believed that ANNs offer a promising alternative method which should be considered seriously when dealing with nonlinear energy related issues that make use of meteorological inputs.

In this survey, the original papers which were pioneer in this area were reviewed. The inventory of applications presented above is neither complete nor exhaustive; nevertheless, it comprises a quite large number of publications that demonstrate the usefulness of artificial neural networks in energy related applications based on the use of meteorological variables.

REFERENCES

Aguiar R., Collares-Pereira M., 1991. Statistical properties of hourly global radiation. *Solar Energy*, Vol. 48, pp. 157–167.

Al-Alawi S.M., Al-Hinai H.A., 1998. An ANN-based approach for predicting global radiation in locations with no direct measurement instrumentation. *Renewable Energy*, Vol. 14, pp. 199-204.

Ångström A., 1924. Solar and terrestrial radiation. *Quarterly Journal Royal Meteorological Society*, Vol. 4, pp. 121–126.

Arca B., Benincasa F., De Vincenzi M., Fasano G., 1998. A neural model to predict the daily minimum of air temperature. *7th CCTA - International Congress for Computer Technology in Agriculture, Computer technology in agricultural management and risk prevention*. Firenze, 15-18 November 1998. pp. 1-9.

Argiriou A.A., Bellas-Velidia I., Balaras, C.A., 2000. Development of a neural network heating controller for solar buildings. *Neural Networks*, Vol. 13, pp. 811-820.

Atsu S.S. Dorvlo, Jervase, J.A., Al-Lawati, A., 2002. Solar radiation estimation using artificial neural networks. *Applied Energy*, Vol. 71, pp. 307–319.

Bagdonas A., Georg J.C., Gerber J.F., 1978. Techniques of frost prediction and methods of frost and cold protection, Technical note No. 157. *World Meteorological Organization*, Geneva.

Barron A.R., 1993. Universal approximation bounds for superpositions of a sigmoidal function. *IEEE Transactions on Information Theory,* Vol. 39, pp. 930-945.

Bird R.E., Riordan C., 1986. Simple solar spectral model for direct and diffuse irradiance on horizontal and titled planes at the earth's surface for cloudless atmospheres. *Journal of Climate and Applied Meteorology,* Vol. 25, pp. 87-97.

Bishop C. M., 1996. *Neural networks for pattern recognition*, Oxford University Press.

Bossanyi E.A., 1985. Short-term wind prediction using Kalman filters. *Wind Engineering*, Vol. 9, pp. 1-8.

Boyden C.J., 1937. A method of predicting minimum temperatures. *Quarterly Journal Royal Meteorological Society*, Vol. 63, pp. 383-392.

Brünt D., 1941. *Physical and Dynamical Meteorology*, 2nd ed., Cambridge University Press, New York.

Cloys K.P., 2000. A neural network solution to predicting wind speed at cape Canaveral's atlas launch pad., *M.S. thesis*, Air Force Institute of Technology, Wright-Patterson Air Force Base, Dayton, OH, 133 pp.

Craddock J.M., Pritchard D., 1951. Forecasting the formation of radiation fog- a preliminary approach, Meteorological Research Paper No. 624. British Meteorological Office.

Elizondo D., Hoogenboom G., McClendon R., 1994. Development of a neural network to predict daily solar radiation. *Agricultural and Forest Meteorology*, Vol. 71, pp. 115–132.

Elman J., 1990. Finding structure in time. *Cognitive Science*, Vol. 14, pp. 179–221.

Ferreira P.M., Faria E.A., Ruano A.E., 2002. Neural network models in greenhouse air temperature prediction. *Neurocomputing*, Vol. 43, pp. 51–75.

Flores P., Tapia A., Tapia G., 2005. Application of a control algorithm for wind speed prediction and active power generation. *Renewable Energy*, Vol. 30, pp. 523-536.

Funahashi K., 1989. On the approximate realisation of continuous mappings by neural networks. *Neural Networks*, Vol. 2, pp. 183-192.

Gardner M.W., Dorling, S.R., 1998. Artificial neural networks (the multilayer perceptron)- a review of applications in the atmospheric sciences. *Atmospheric Environment*, Vol. 32, No. 14/15, pp. 2627-2636.

Goh T., Tan K., 1977. Stochastic modelling and forecasting of solar radiation data. *Solar Energy*, Vol. 19, pp. 755–757.

Gueymard C., 1989. A two-band model for the calculation of clear sky solar irradiance, illuminance and photosynthetically active radiation at the earth's surface. *Solar Energy*, Vol. 43, pp. 253-265.

Gueymard C., 2001. Parameterized transmittance model for direct beam and circumsolar spectral irradiance. *Solar Energy*, Vol. 71, pp. 325-346.

Hagan M., Menhaj M., 1994. Training feed-forward networks with the Marquardt algorithm. IEEE Transactions. *Neural Networks*, Vol. 5, pp. 989–993.

Haykin S., 1998. *Neural networks: A comprehensive foundation*, 2nd edition. Macmillan College Publishing.

Hittle D.C., Pedersen C.O., 1981. Periodic and stochastic behaviour of weather data. *ASHRAE Transactions*, Vol. 87, pp. 173-183.

Hokoi S., Matsumoto M., Kagawa M., 1990. Stochastic models of solar radiation and outdoor temperature. *ASHRAE Transactions*, Vol. 2, pp. 245–252.

Hontoria L., Riesco J., Zufiria P., Aguilera J., 1999. Improved generation of hourly solar irradiation artificial series using neural networks. *Proceedings of Engineering Applications of Neural Networks, EANN99 Conference.* Warsaw, Poland, 13-15 September 1999, pp. 87-92.

Hontoria, L., Aguilera, J., Zufiria, P., 2002. Generation of hourly irradiation synthetic series using the neural network multilayer perceptron. *Solar Energy*, Vol. 72, pp. 441–446.

Hornik K., Stinchcombe M., White H., 1989. Multilayer feedforward networks are universal approximators. *Neural Networks*, Vol. 2, No. 5, pp. 359-366.

Hsieh W.W., Tang B., 1998. Applying neural network models to prediction and data analysis in meteorology and oceanography. *Bulletin of the American Meteorological Society*, Vol. 79, pp. 1855–1870.

Jain A.K., Mao J., Mohiuddin K., 1996. Artificial neural networks: A tutorial. *IEEE Computer,* Vol. 29, pp. 31-44.

Kalogirou S., Neocleous C., Michaelides S., Schizas C., 1997. A time series reconstruction of precipitation records using artificial neural networks. *Proceedings of EUFIT'97, Aachen, Germany*, pp. 2409–2413.

Kalogirou S., Neocleous C., Michaelides S., Schizas C., 1998. Artificial neural networks for the generation of isohyets by considering land configuration, *Proceedings of the Engineering Applications of Neural Networks (EANN'98) Conference*, Gibraltar, pp. 383-389.

Kalogirou S., Neocleous C., Pashiardis S., Schizas C., 1999. Wind speed prediction using artificial neural networks, *Proceedings of the European Symposium on Intelligent Techniques ESIT'99 on CD-ROM*, Crete, Greece.

Kalogirou S., Michaelides S., Tymvios, F., 2002. Prediction of maximum solar radiation using artificial neural networks. *Proceedings of the World Renewable Energy Congress VII (WREC 2002) on CD-ROM*. Cologne, Germany.

Kamal L., Jafri Y.Z., 1997. Time series models to simulate and forecast hourly averaged wind speed in Quetta, Pakistan. *Solar Energy*, Vol. 61, pp. 23-32.

Kelley J.G.W., Kousso J.M., Eyton J.R., Carlson T.N., 1988. Mesoscale forecasts generated from numerical weather-prediction model output. *Bulletin American Meteorological Society*, Vol. 69, pp. 7-15.

Kemmoku Y., Orita S., Nakagawa S., Sakakibara T., 1999. Daily insolation forecasting using a multi-stage neural network. *Solar Energy*, Vol. 66, pp. 193-199.

Klein W.H., Lewis B.M., Caseley G.P., 1967. Automated nationwide forecasts of maximum and minimum temperature. *Journal of Applied Meteorology*, Vol. 6, pp. 216-228.

Kohonen T., 1995. *Self-organising maps*, Springer Verlag.

Kretzschmar R., Eckert P., Cattani D., 2004. Neural network classifiers for local wind prediction. *Journal of Applied Meteorology*, Vol. 43, pp. 727-738.

La Casiniere A. de, Bokoye A.I., Cabot T., 1997. Direct solar spectral irradiance measurements and updated simple transmittance models. *Journal of Applied Meteorology,* Vol. 36, pp. 509-520.

Leckner B., 1978. The spectral distribution of solar radiation at the earth's surface- elements of a model. *Solar Energy*, Vol. 20, pp. 143-150.

Lemcke C., Kruizinga S., 1988. Model output statistics forecasts: Three years of operational experience in the Netherlands. *Monthly Weather Review*, Vol. 116, pp. 1077-1090.

Lippmann R.P., 1987. An introduction to computing with neural nets. IEEE ASSP Magazine, pp. 4–22, April.

Marzban C., Stumpf G.J., 1999. A neural network for damaging wind prediction. *Weather and Forecasting*, Vol. 13, pp. 151-163.

Mellit A., Benghanem M., Hadj Arab A., Guessoum A., 2005 A simplified model for generating sequences of global solar radiation data for isolated sites: Using artificial neural network and a library of Markov transition matrices approach. *Solar Energy,* Vol. 79, pp. 469–482.

Mellit A., Benghanem M., Kalogirou S.A., 2006. An adaptive wavelet-network model for forecasting daily total solar radiation, *Applied Energy*, Vol. 83, No. 7, pp. 705-722.

Michaelides S.C., 1991. Comparative regression analysis coupled with rational verification for forecasting night cooling. *Applied Stochastic Models and Data Analysis*, Vol. 7, pp. 237-255.

Mihalakakou G., Santamouris M., Asimakopoulos D.N., 2000. The total solar radiation time series simulation in Athens, using neural networks. *Theoretical and Applied Climatology*, Vol. 66, pp. 185-197.

Mohandes A.M., Rehman S., Halawani T.O., 1998a. Estimation of global solar radiation using artificial neural networks. *Renewable Energy*, Vol. 14, pp. 179-184.

Mohandes, A.M., Rehman, S., Halawani, T.O., 1998b. A neural network approach for wind speed prediction. *Renewable Energy*, Vol. 13, No. 3, pp. 345-354.

Mohandes M., Balghonaim A., Kassas M., Rehman S., Halawani T.O., 2000. Use of radial basis functions for estimating monthly mean daily solar radiation. *Solar Energy*, Vol. 68, pp. 161–168.

Prescott J.A., 1940. Evaporation from a water surface in relation to solar radiation. *Transactions of the Royal Society of South Australia*. Vol. 64, pp. 114-125.

Principe J., Euliano N., Lefebvre W., 2000. *Neural and adaptive systems*, J. Wiley.

Ranaweera D., Hubele N., Papalexopoulos A., 1995. Application of radial basis function network model for short term load forecasting. *IEE Proc. Gen. Trans. Dist.*, Vol. 142, pp. 45–50.

Richardson C.W., 1981. Stochastic simulation of daily precipitation, temperature, and solar radiation. *Water Resources Research*, Vol. 17, pp. 182-190.

Richardson C.W., Wright D.A., 1984. WGEN: a model for generating daily weather variables. US Department of Agriculture, Agricultural Research Service, ARS-8, 83 pp.

Rietveld M.R., 1978. A new method for estimating the regression coefficients in the formula relating solar radiation to sunshine. *Agricultural Meteorology*, Vol. 19, pp. 243–352.

Rumelhart D.E., Hinton G.E., Williams R.J., 1986. Learning internal representation by error propagation. In: Rumelhart D.E., McClelland J.L., editors. Parallel distributed processing: explorations in the microstructure of cognition, Vol. 1, MIT Press, Cambridge, MA, pp. 318-362.

Santamouris, M., Mihalakakou, G., Psiloglou, B., Eftaxias, G., Asimakopoulos, D.N., 1999. Modelling the global solar radiation on the earth surface using atmospheric deterministic and intelligent data driven techniques. *Journal of Climate*, Vol. 12, pp. 3105-3116.

Schizas C.N., Michaelides S.C., Pattichis C.S, Livesay R.R., 1991. Artificial neural networks in forecasting minimum temperature. *Proceedings of Second International Conference on Artificial Neural Networks*, Bournemouth, UK. Institution of Electrical Engineers, Publ. No. 349, pp. 112-114.

Schizas C.N., Pattichis C.S, Michaelides S.C., 1994. Forecasting minimum temperature with short time-length data using artificial neural networks. *Neural Networks World*, Vol. 2, pp. 219-230.

Seginer I., 1997. Some artificial neural network applications to greenhouse environmental control, *Computers and Electronics in Agriculture*, Vol. 18, pp. 167–186.

Sfetsos A., 2000. A comparison of various forecasting techniques applied to mean hourly wind speed time series. *Renewable Energy*, Vol. 21, pp. 23-35.

Sfetsos A., Coonick A.H., 2000. Univariate and multivariate forecasting of hourly solar radiation with artificial intelligence techniques. *Solar Energy*, Vol. 68, pp. 169–178.

Shao J., 1998. Improving nowcasts of road surface temperature by a backpropagation neural network. *Weather and Forecasting*, Vol. 13, pp. 164-171.

Sözen A., Arcaklıoğlu E., Özalp M., Çağlar N., 2005. Forecasting based on neural network approach of solar potential in Turkey. *Renewable Energy*, Vol. 30, pp. 1075-1090.

Storch S.J., 1999. Predicting launch pad winds at the Kennedy space center with a neural network model. *M.S. thesis*, Air Force Institute of Technology, Wright-Patterson Air Force Base, Dayton, OH, 61 pp.

Tasadduq I., Shafiqur Rehman S., Bubshait K., 2002. Application of neural networks for the prediction of hourly mean surface temperatures in Saudi Arabia. *Renewable Energy,* Vol. 25, pp. 545–554.

Tymvios F.S., Jacovides C.P., Michaelides S.C., 2002. The total solar energy on a horizontal level with the use of artificial neural networks. *Proceedings of the 6th Hellenic Conference of Meteorology, Climatology and Atmospheric Physics*, Ioannina, 26-28 Sep., Greece. pp. 468-475.

Tymvios F.S., Jacovides C.P., Michaelides S.C., Scouteli C., 2005. Comparative study of Angstrom's and artificial neural networks' methodologies in estimating global solar radiation. *Solar Energy*, Vol. 70, pp. 752-762.

Wassereman P.D., 1993. *Advanced methods in neural networks*, Van Nostrand Reinhold.

Williams D., Zazueta F., 1996. Solar radiation estimation via neural network. *Proceedings of the 6th International Conference on Computers in Agriculture*, Cancun, Mexico, pp. 1143–1149 (see http://wcca.ifas.ufl.edu/archive/6thProc/williams.html).

WMO, 1995. Manual on Codes – International Codes - Vol. I.1, Part A - Alphanumeric Codes, World Meteorological Organization Publication No. 306. Geneva.

Yee P., Haykin S., 1999. A dynamic regularized radial basis function network for nonlinear, nonstationary time series prediction, *IEEE Transactions on Signal Process*, Vol. 47, pp. 2503–2521.

In: Artificial Intelligence in Energy and Renewable Energy Systems ISBN 1-60021-261-1
Editor: Soteris Kalogirou, pp. 83-130

Chapter 3

ARTIFICIAL NEURAL NETWORKS IN SOLAR THERMAL ENERGY SYSTEMS

Sylvain Lalot
Laboratory of Mechanics and Energetics (LME)
University of Valenciennes and Hainaut-Cambresis
Valenciennes, France

ABSTRACT

This chapter begins by demonstrating the power of Artificial Neural Networks (ANNs) in modelling complex thermal systems. For each of the studied systems, a simplified sketch of the system is shown, the neural network architecture is given, the accuracy of the model is assessed and potential applications are suggested.
Two applications of ANNs in PV systems optical design and sizing are presented followed by models dealing with solar thermal collectors. After the presentation of the identification of collector parameters at steady state, transient states are analyzed. Initially, techniques are presented using the traditional multilayer ANNs, one trying to find directly a model of the collector and one trying to find parameters of second order systems. The latter is classically based on the step response of the collector. Secondly, recurrent networks specifically developed for identification are presented. The models embedded in these networks are described along with the training algorithm. Subsequently, the on-line techniques, using these specialized networks, are presented. The approach is validated by the comparison of a Single Input-Single Output (SISO) neural model with a model obtained using the Laplace transform. Then Multiple Inputs-Single Output (MISO) models are studied and it is shown that, in some cases, it is possible to get a Multiple Inputs-Multiple Outputs (MIMO) model as MISO models coupled in parallel. Finally, validation of the models based on experimental data is presented.
To conclude the chapter, it is shown that neural models obtained using on-line techniques can help to determine that a drift (such as coating degradation, dusty cover, fouling, etc.) has occurred and that a maintenance operation should be carried out to keep the collectors in a good working order.

Keywords: Thermal collectors, PV, drift, identification, SISO, MIMO, MISO, MLP, RBF

NOMENCLATURE

$\text{avg}\left[Y\left(t_{init}:t_{final}\right)\right]$	Average value of Y between initial time t_{init} and final time t_{final}
A	Constant in equation 11 (W/m^2)
A_a	Frontal absorber area (m^2)
A_c	Convective area for a given frontal absorber area (m^2)
B	Constant in equation 14 (expression given by equation 15)
c_f	Specific heat of the fluid (J/kg K)
C	Constant in equation 11 (W/m^2 K)
C_a	Thermal capacity of the absorber (J/K m^2)
CA	Capacity of the generator in a photovoltaic system (dimensionless)
CS	Capacity of the accumulator in a photovoltaic system (dimensionless)
$dA_a = w\ dx$	Elementary frontal area (m^2)
dt	Distance based test function
dt_{max}	Threshold for the dt function
D	Constant in equation 11 (J/K m^2)
E	Constant in equation 11 (J/K m^2)
F	Constant in equation 11 (J s/K m^2)
h	Local convection coefficient (W/m^2K)
h_r	Reduced convection coefficient (W/m^2K)
H	Transfer function
I_S	Solar irradiation (W/m^2)
j	Day number
k	Time step (s)
K	Heat loss conductance (W/K)
K_B	Size of the accumulator in a photovoltaic system
K_{PV}	Size of the PV-array in a photovoltaic system
K_r	Reduced heat loss conductance (W/m^2K)
K_{TDM}	Yearly averaged clearness index (dimensionless)
L	Length of the collector (m)
$\mathcal{L}$	Laplace operator
$\dot{m}$	Mass flow rate (kg/s)
$\dot{m}_r$	Reduced mass flow rate (kg/sm^2)
n_1, n_2, n_3	Order of the first, second, and third inputs

n_y	Number of past values of the output taken into account
R^2	Coefficient of multiple determination
s	Laplace variable
s_1	First root of the denominator of a second order system
s_2	Second root of the denominator of a second order system
t_0	Length of the initial observation window (s)
t_1	Length of the current observation window (s)
T_a	Temperature of the absorber (K)
T_f	Temperature of the fluid (K)
T_i	Temperature of the fluid at the inlet (K)
T_o	Temperature of the fluid at the outlet (K)
T_∞	Ambient temperature (K)
tanh(x)	Hyperbolic tangent of x
u_1, u_2, u_3	First, second and third inputs
V_r	Volume of fluid per unit frontal absorber area (m^3/m^2)
w	Width of the collector (m)
$W_{k,i}^j$	k^{th} connection weight at date j, between the input layer and the ith neuron of the hidden layer
x	Abscissa (m)
y	Output variable
$\overline{y}$	Mean value of y
$\hat{y}$	Estimated value of y
z	Variable used in the z-transform
z^{-1}	Time delay operator

Greek symbols

α	Absorptivity of the coating (dimensionless)
E	Efficiency of the fins (dimensionless)
η	Efficiency of a collector or of a system (dimensionless)
$\eta_0 = \alpha E \tau$	Global "optical" efficiency (dimensionless)
θ	Temperature difference (K)
θ_∞	Temperature difference in steady state (K)
ω_n	Natural pulsation
τ	Transmissivity of the glass cover (dimensionless)

ς	Damping factor
ξ	Threshold in the definition of the distance based test function

ABBREVIATIONS

ANN(s)	Artificial Neural Network(s)
ARX	AutoRegressive model with eXogeneous inputs
CFD	Computational Fluid Dynamics
GMDH	Group Method of Data Handling
HVAC	Heating, Ventilating, and Air Conditioning
LOLP (or LLP)	Loss of load probability
MIMO	Multiple Inputs Multiple Outputs
MISO	Multiple Inputs Single Output
MLP	MultiLayer Perceptrons
MSE	Mean Square Error
PID	Proportional Integral Derivative
PV	PhotoVoltaic
RBF	Radial Basis Function
RMSE	Root Mean Square Error
SISO	Single Input Single Output
SSE	Sum of Square Error

1. INTRODUCTION

The use of Artificial Neural Networks (ANNs) in the analysis of solar energy systems has begun in the early 90s. By performing a literature survey in the global field of "solar energy", it comes out that only a limited number of papers deal with "thermal systems", such as collector-storage or collector-steam generator and an even smaller number of papers deal with photovoltaic (PV) systems. Irrespective of the ANNs capabilities, very little attention has been paid to the application of this technology to solar energy collectors.

It is well known that ANNs are very versatile tools. The main characteristics of ANNs are:

- They can represent very small to very large systems in a single "black-box",
- They can handle non-linear systems even when data are noisy and/or missing,
- Once trained, ANNs are very fast.

This chapter deals with applications of ANNs in solar energy systems. The systems examined include solar water heating, which comprise thermosiphon and forced convection systems, solar power plants and photovoltaics. With respect to the application of ANNs in solar collectors both steady state and transient models are considered. Finally the use of ANNs for drift detection of solar systems is presented.

2. THERMAL SYSTEMS

It is possible to consider three levels of applications of ANNs. The first level is identification in which modelling of a system is considered. At the second level, the models obtained from the first level were used extensively by changing the inputs representing space or the inputs representing time. In the first case, a size optimization can be achieved, whereas in the second case a long term performance assessment can be carried out. The last level is control and the ANNs are used to tune classical PID controllers, or to predict at least one-step ahead values for advanced control.

Concerning the first level, it can be considered that one of the pioneering works is presented in (Mager, 1996) where a comparison with the F-Chart method (University of Wisconsin, 1985) is carried out. Unfortunately the data were quite incomplete and the ANNs available at that time were not as powerful as those available only a few years ago. Afterwards, ANNs have been used for identification of water heating systems (Kalogirou *et al.*, 1999a; 1999b; Cetiner *et al.*, 2005), of a steam generator during the start-up phase (Kalogirou *et al.*, 1998) and of an ejector-absorption refrigeration system (Sözen and Akçayol, 2004). Reviews are given in (Kalogirou, 1999c; 2000a; 2001).

Concerning the second level and regarding space, it is shown by Kalogirou (2004), that the optimum size of a system is determined in a short time. Regarding time, it is shown in (Kalogirou, 2000b) and (Kalogirou and Panteliou, 2000c) that long term performances can be obtained once the neural model of a system is valid.

Combining the second and the third level, it is shown in (Berenguel *et al.*, 1998; Arahal *et al.*, 1998; Henriques *et al.*, 1999; Gil *et al.*, 2002; Henriques *et al.*, 2002), that a solar plant can be efficiently controlled once a neural model is obtained.

The common characteristic of all these systems is the fact that they involve a large variety of components. For example, a solar water heating system (the simplest system studied) is typically composed of a collector, some tubing, and storage. To take precise account of all the components in an analytical model would require a lot of time to get a satisfactory accuracy. In a very recent paper, (Wongsuwan and Kumar, 2005) shows that ANNs are as powerful as TRNSYS® (transient systems simulation program developed at the University of Wisconsin, Madison, USA). It is shown that, after the determination in a detailed analysis of the number of input data that are necessary, the ANN model is faster than the TRNSYS® model.

It is worth noting that, although neural networks are seen as "black box models", the relationships between the inputs and the outputs are mathematically computed; cf. (Sözen and Akçayol, 2004) for a detailed description of the computations required.

2.1. Water Heating

In this section the use of ANNs for the modelling of thermosiphon and forced convection water heating systems is presented.

2.1.1. Thermosiphon Systems

The simplest system, in terms of technology, is the thermosiphon water heating system. The system is using the difference in density between the cold and hot water to create a small flow, therefore there is no need for external mechanical power to transfer solar energy from the collector to storage as shown in Fig. 1.

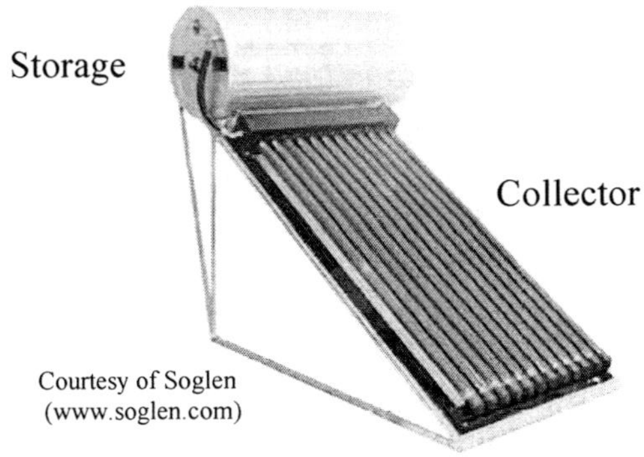

Fig. 1. A typical "all in one" thermosiphon solar water heating system

Four configurations are studied by Kalogirou *et al.* (1999a), for which the storage volume varies from 125 to 200 litres. The collector area is about 2 m^2. Seventy one experiments have been carried out under different weather conditions. Forty six are used for learning, eight for tests and seventeen are kept for validation. The 7 inputs and the 2 outputs of the ANN are listed in Table 1.

Table 1. Inputs and outputs considered for the modelling of solar domestic water heating systems (Kalogirou *et al.*, 1999b)

Input number	Parameter
1	System type (open (1) or close (2) thermosiphon)
2	Mean storage tank heat loss coefficient (W/m^2K)
3	Storage volume (litres)
4	Total radiation (MJ/m^2)
5	Diffuse radiation (MJ/m^2)
6	Mean air temperature (°C)
7	Initial cold water temperature (°C)
Output number	**Parameter**
1	Maximum water temperature rise
2	Useful energy (actually extracted from the system)

The neural architecture is a standard 3-layer one as shown in Fig. 2.

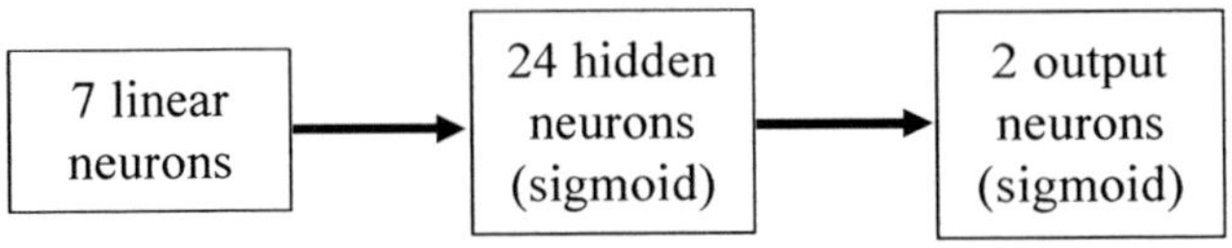

Fig. 2. Schematic diagram of the standard 3-layer neural network

The training algorithm is the standard back-propagation. Early stopping was used; the criterion being a stable error for more than 4,000 epochs. Then the validation set is used to assess the quality of the predictions on data unseen during the training. Concerning the useful energy, the mean and median relative errors are slightly positive (1.5 % and 1.6 %); the maximum deviation is limited to 9.8 % for two cases and to 5 % in all other cases. It is interesting to note that there is no link between the storage volume and the relative error as shown in Fig. 3.

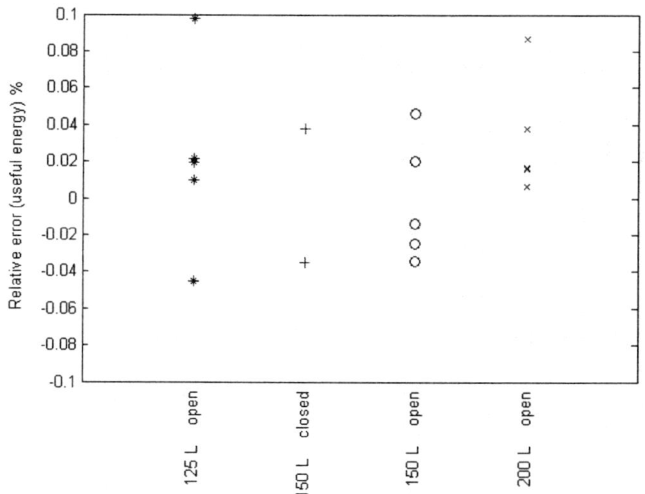

Fig. 3. Relative error in useful energy prediction for thermosiphon heating systems versus storage volume and type of system

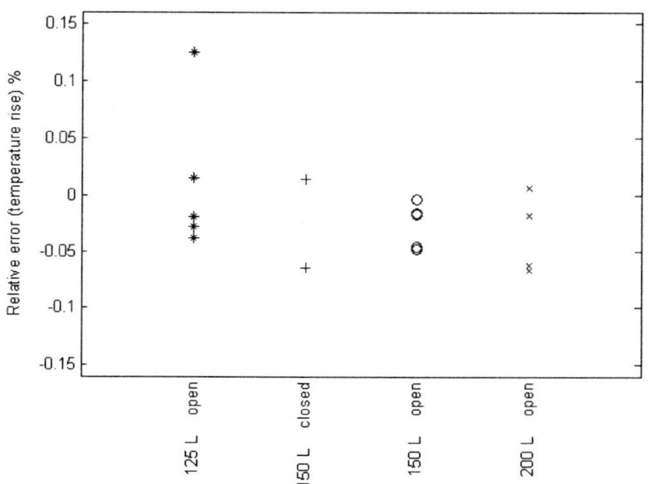

Fig. 4. Relative error in temperature rise prediction for thermosiphon heating systems versus storage volume and type of system

Concerning the temperature rise, the mean and median relative errors are slightly negative (-1.5 % and -1.8 %); the maximum deviation is limited to 12.5 % (for the case corresponding to the lowest temperature rise), and to 6 % in all other cases. Again, it is interesting to note that there is no link between the storage volume and the relative error as shown in Fig. 4.

A generalization of this study is presented in (Kalogirou and Panteliou, 2000c). The water tank is either vertical or horizontal. The collector area varies from about 1.8 m^2 to 4.4 m^2. Three locations (in Greece) are studied. To be able to get accurate results a large number of inputs are necessary as shown in Table 2.

Table 2. Inputs considered for the long-term performance prediction of thermosiphon units (Kalogirou and Panteliou, 2000c)

Input number	Parameter
1	Month
2	Mean ambient air temperature during the day
3	Average ambient temperature during night-time
4	Maximum temperature of the draw-off water
5	Heat loss coefficient of the storage tank
6	Daily solar irradiation
7	Mean cold water temperature
8	Tank position (vertical or horizontal)
9	Draw-off normalized temperature profile
10	Water demand temperature
11	System type (open or close)
12	Storage volume per collector unit area
13	Three constants computed for the standard performance curve of the system
14	Mixing draw-off normalized profile

The outputs given in Table 3 make possible a long term performance analysis.

Table 3. Outputs considered for the long-term performance prediction of thermosiphon units (Kalogirou and Panteliou, 2000c)

Output number	Parameter
1	Solar energy output of the system for a draw-off equal to the volume of the storage tank
2	Solar energy output of the system for a demand temperature of 35°C
3	Average quantity of hot water per month for a demand temperature of 35°C

To study such diverse cases, a specific architecture has been used. It is composed of a hidden layer which is divided in three sub-layers as shown in Fig. 5. The 8 neurons of the first sub-layer use a Gaussian activation function, the 8 neurons of the second sub-layer use a Gaussian complement activation function and the 8 neurons of the third sub-layer use the classical hyperbolic tangent activation function. For the output layer the activation is the logistic function.

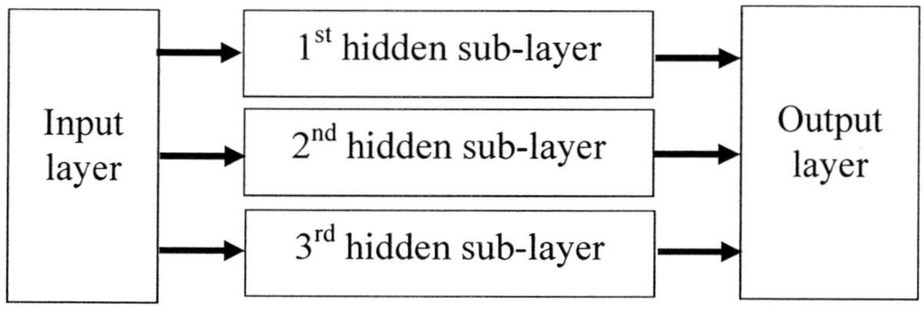

Fig. 5. Flow chart of the neural architecture developed for the study of a broad spectrum of heating systems

Very good results are obtained, as can be concluded from the coefficient of multiple determination which is equal to 0.97 for the first output and higher than 0.99 for the other two outputs (using the validation set). The latter is defined as follows:

$$R^2 = \frac{\sum(y-\bar{y})^2 - \sum(\hat{y}-y)^2}{\sum(y-\bar{y})^2} \tag{1}$$

where $\bar{y}$ and $\hat{y}$ are the mean value and the estimated value of the output y.

By comparing this work with the previous ones of the same research group it can be seen that in (Kalogirou *et al.*, 1999a) and (Kalogirou *et al.*, 1999b) less inputs are necessary, while the number of neurons in each sub-layer are increased to 18. It is shown in (Kalogirou *et al.*, 1999b) that very accurate results are obtained when looking for a model of the useful energy extracted and for a model of the temperature rise of the water in the storage tank; the two latter quantities being the outputs of the ANN. In fact more neurons (18 instead of 8) are needed in the hidden layer not only because the number of inputs is decreased, but also because the ranges of the characteristics of the systems are slightly broader. The collector area varies from about 1.8 m^2 to about 4.4 m^2; the tank is either horizontal or vertical; the mean storage tank heat loss coefficient varies from 0.84 to 5.1 W/m^2K; the storage volume varies from 100 to 200 litres; and the system is either closed or open. Apart from these five characteristics, the inputs are limited to the solar radiation (MJ/m^2), mean ambient air temperature and the mean cold water temperature.

The training set is composed of thirty systems (ten readings from experiments per system), while the validation set is composed of three randomly selected systems. Although the size of the validation set could seem relatively small, it is in accordance with what is given in (Haykin, 1999).

Good results are obtained for the energy extracted, as can be seen in Fig. 6; the solid line is the identity function and the crosses are the values computed by the neural network. Concerning the temperature rise, the results are a little less satisfactory (see Fig. 7). The neural network underestimates quite all values, but as this is conservative, it is not a problem for the end user.

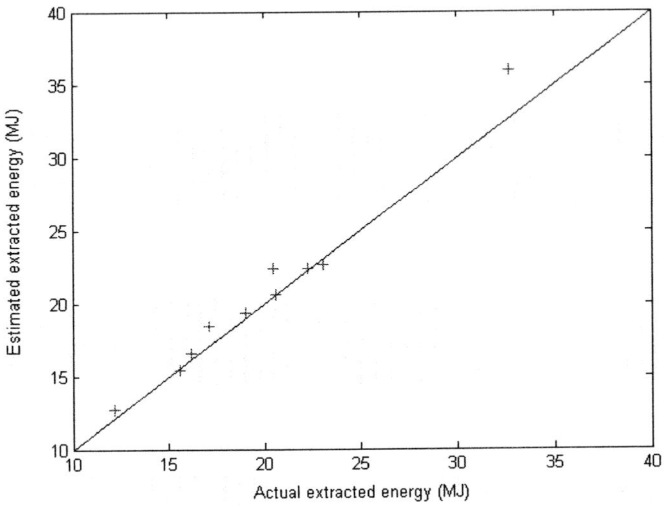

Fig. 6. Comparison between the actual and ANN estimated extracted energy, for a thermosiphon water heating system

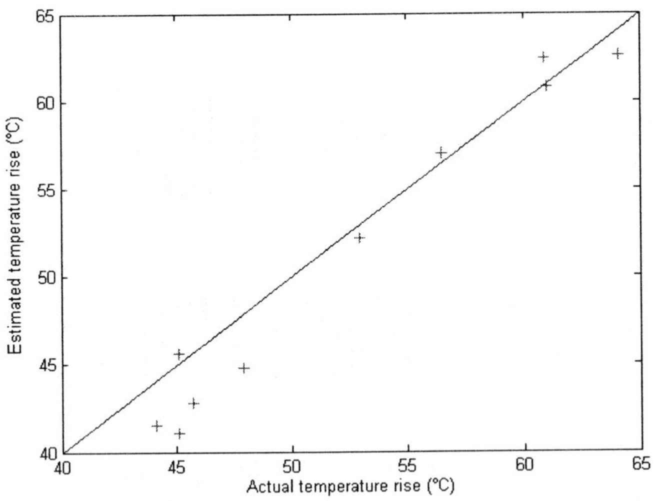

Fig. 7. Comparison between the actual and ANN estimated temperature rise, for a thermosiphon water heating system

It has to be noted that such good results could only be obtained with the use of a CFD (Computational Fluid Dynamics) code however such an attempt would have needed a very long computational time. A possible extension of this work would be to consider ground heat storage using packed rock beds or phase change materials.

2.1.2. Forced Convection Water Heating

The difference of this type of system with the previous one is that the circulation of the fluid is achieved by using mechanical power. Two types of systems fall in this category; open and closed, as shown schematically in Figs. 8 and 9 respectively.

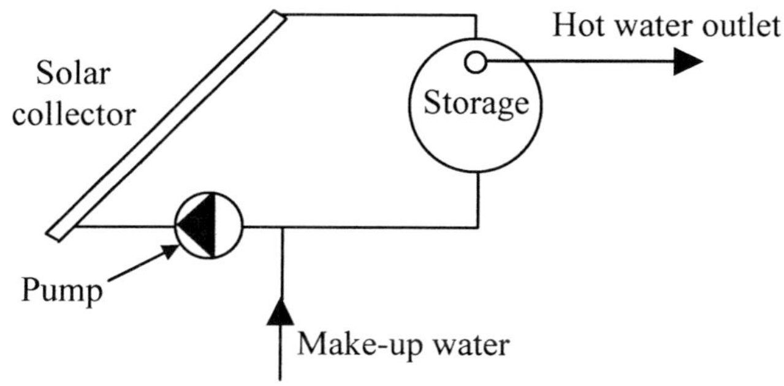

Fig. 8. Schematic of an open loop forced circulation water heating system

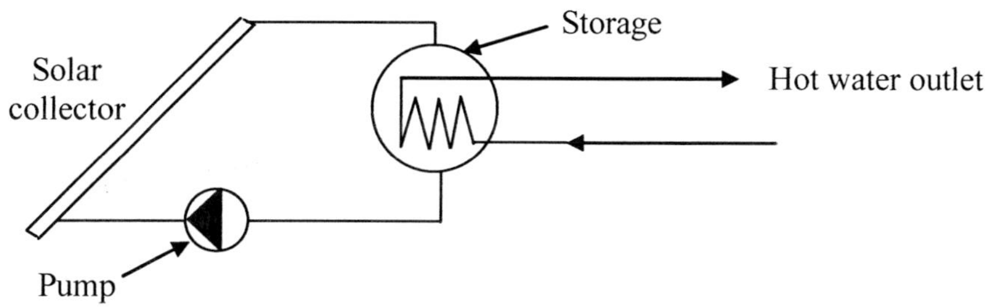

Fig. 9. Schematic of a closed loop forced circulation water heating system

Two kinds of collectors have been used in these studies; non-concentrating or stationary and concentrating, depending on the temperature level required in the storage tank.

A. Low Temperature Systems

Similarly to the thermosiphon systems a long-term study for the forced-convection systems has also been carried out by Kalogirou (2000b). Only vertical tank systems have been studied, leading to a reduced number of hidden neurons (7 in each sub-layer). The comparison between the actual energy and the predicted energy for the whole year shows that the neural network is well trained (see Fig. 10).

A very interesting study is presented by Wongsuwan and Kumar (2005) who showed that neural networks can produce more accurate results than TRNSYS® software. The system is more complicated as it is a two-tank system as shown in Fig. 11.

A standard three layer network is used. For the hourly-based predictions, the input layer is composed of thirty neurons; ten for the hour of the day, ten for the diffuse horizontal radiation and ten for the global radiation (at the tilt angle); the transfer function for this layer is the log-sigmoid function. The hidden layer is composed of 16 neurons using the hyperbolic tangent transfer function.

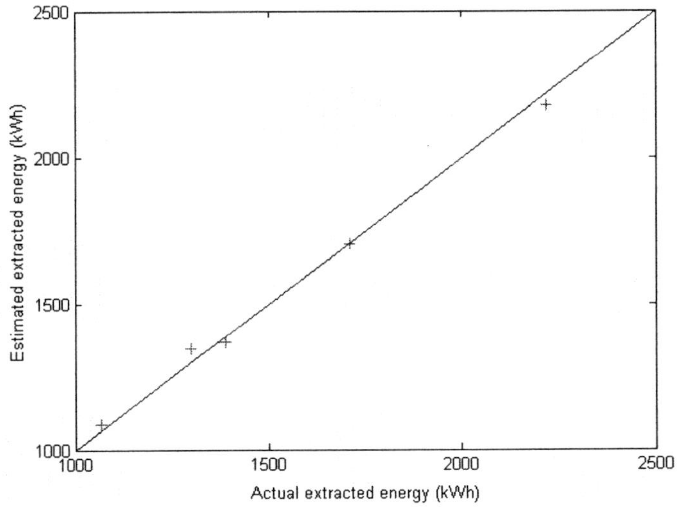

Fig. 10. Comparison between the actual and the ANN predicted extracted energy, for a forced-circulation water heating system

Two applications have been considered. In the first, the output layer is composed of three linear neurons; one for the water temperature at the outlet of the collectors, one for the water temperature in the first storage tank and one for the water temperature in the second storage tank.

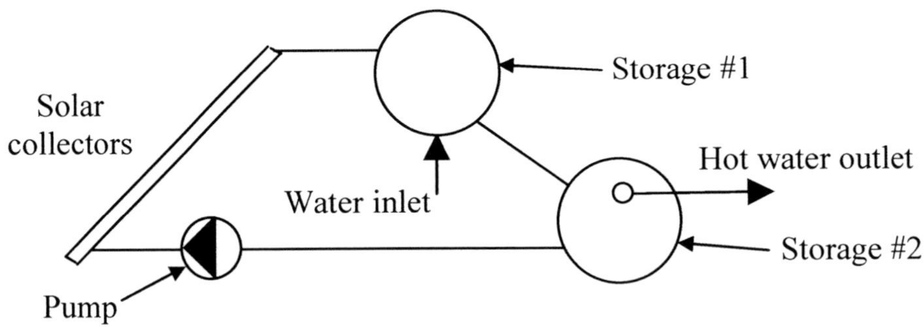

Fig. 11. Two-tank water heating system used to compare TRNSYS® and neural networks

The quality of the results concerning the water temperature at the outlet of the collectors is assessed using the coefficient of multiple determination. The results are given in Table 4. It should be noted that in both cases the comparison is done against actual experimental results.

Table 4. Comparison of the coefficient of multiple determination between TRNSYS and ANN (Wongsuwan and Kumar, 2005)

Day	TRNSYS®	ANN
Clear	0.987	0.990
Partly cloudy	0.976	0.963
Cloudy	0.978	0.953

It can be concluded from the results presented in Table 4 that both techniques give good results.

In the second application, a unique neuron is used in the output layer representing the useful energy. The results for this model are shown in Fig. 12. As can be seen TRNSYS® quite systematically overestimates the useful energy. It should be noted that again actual results were used for the comparison. The average value of the relative error is about 12%. The median value of the relative error is about 7%. On the contrary, ANNs slightly underestimate the useful energy; the average value of the relative error is about -1.7% and the median value of the relative error is about -0.4%. It can be concluded that, concerning energy savings, ANNs perform better than TRNSYS®. It has to be noted that, once trained, ANNs are much faster than TRNSYS® and this is quite important for a long-term performance prediction.

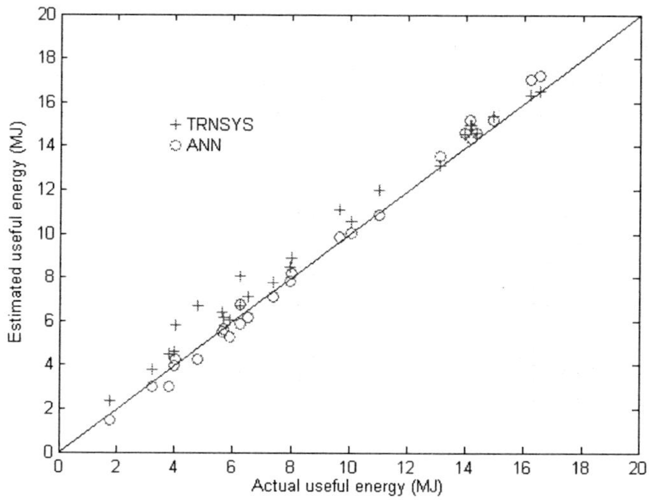

Fig. 12. Comparison between the actual and estimated extracted energy from TRNSYS® and neural network -ANN, for the two tank forced convection water heating system

TRNSYS® is considered as a program giving accurate results which can be used to train a neural network. In an optimization study, twenty configurations of a solar system (storage volume ranging from 5 to 30 m^3, and collector area ranging from 100 to 500 m^2) are modelled using TRNSYS® (Kalogirou, 2004). Then a neural version of the GMDH (Group Method of Data Handling) algorithm is used to model the system in any configuration. Subsequently, a genetic algorithm is used to find the optimum system.

B. High Temperature Systems

In a first attempt to model systems involving concentrating collectors, a study of a closed-type water heater is presented (Cetiner *et al.*, 2005). The collector mirror area is about 54 m^2; the focal length is 12 m and the system is employing a polar-mode east-west tracking. A standard three layer network is used with seven neurons in the hidden layer. The inputs are the cold water inlet temperature, the direct solar radiation, the ambient temperature and the wind velocity. The experiment lasts eight hours and data are collected every quarter of an hour, so thirty two sets of experimental results are available. Among them, twenty six randomly selected sets compose the training database and the remaining six sets compose the validation set.

One of the outputs is the useful heat rate (kW) and as it is shown in Fig. 13 for the training and validation set, the predicted values are globally within 25% of the actual values.

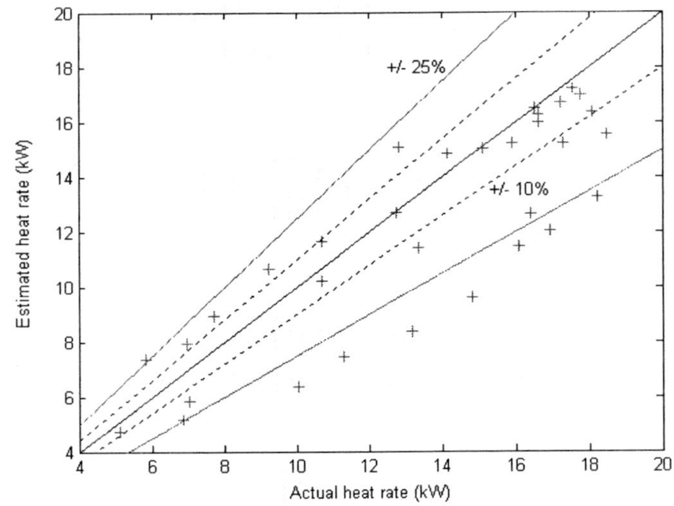

Fig. 13. Comparison between actual and ANN estimated heat rate, for a concentrating system

Another study involving a parabolic trough collector is presented by Kalogirou *et al.* (1998). Apart from the parabolic collector, the installation is mainly composed of a pump and a flash vessel as shown in Fig. 14. The collector area is 3.5 m^2, the concentrating ratio is about 21, the collector axis is horizontal, the mode of tracking is east-west and the mass flow rate is constant at about 0.04 kg/s. Three different flash vessels were used in order to find the one which has the quickest warming-up time.

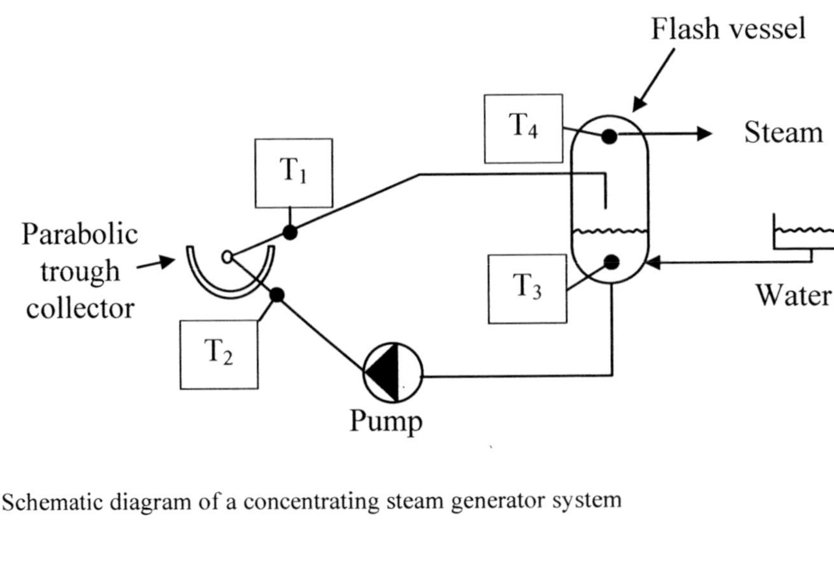

Fig. 14. Schematic diagram of a concentrating steam generator system

The aim of the study is to determine the time needed for the system to produce steam and this occurs when the temperature is uniform in the flash vessel, i.e., top temperature is equal to bottom temperature of vessel ($T_3=T_4$). The output of the neural network comprises these two temperatures together with the collector inlet and outlet temperatures, T_1 and T_2 (see Fig. 14). It is shown that a complex architecture is necessary to get accurate predictions. This architecture is the one composed of three hidden sub-layers (see Fig. 5). In this study, 8 neurons per sub-layer were used. The inputs are given in Table 5.

Table 5. The inputs considered for the parabolic trough steam generation system ANN model (Kalogirou *et al.*, 1998)

Input number	Parameter
1	Set of values of time
2	Ambient temperature
3	Solar radiation
4	Effective aperture-area
5	Flash-vessel content
6	Flash-vessel dimensions (height, diameter, and thickness)

The learning database consists of 349 experimentally obtained results (about 130 readings per temperature sensor and per vessel). The duration of each experiment is about three quarters of an hour. So, a broad range of weather conditions is covered in the learning database. The test database consists of 47 experimentally obtained results having the same characteristics. Figure 15 gives the estimated temperatures (T_1 to T_4) during the start-up phase (one out of ten for clarity reasons) for the three flash vessels.

It can be seen that the agreement is excellent. This is confirmed by the narrowness of the histogram of the relative error of predicted temperature (see Fig. 16). The mean value is about 0.03%, the median value is about 0.06% and the standard deviation is about 1.27%.

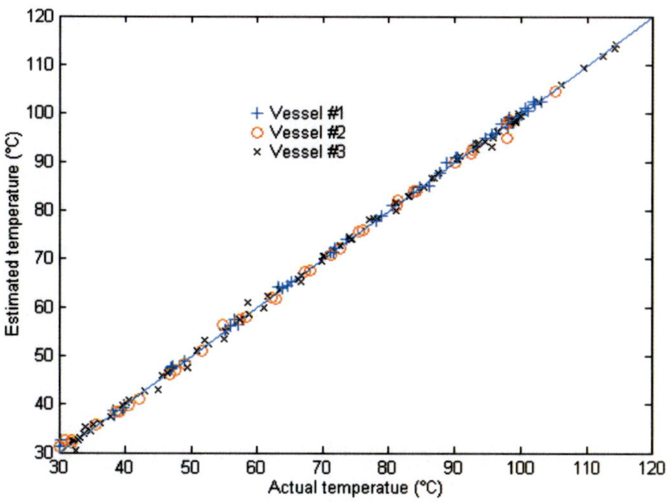

Fig. 15. Comparison between actual and ANN estimated temperature during the starting phase of a solar steam generator

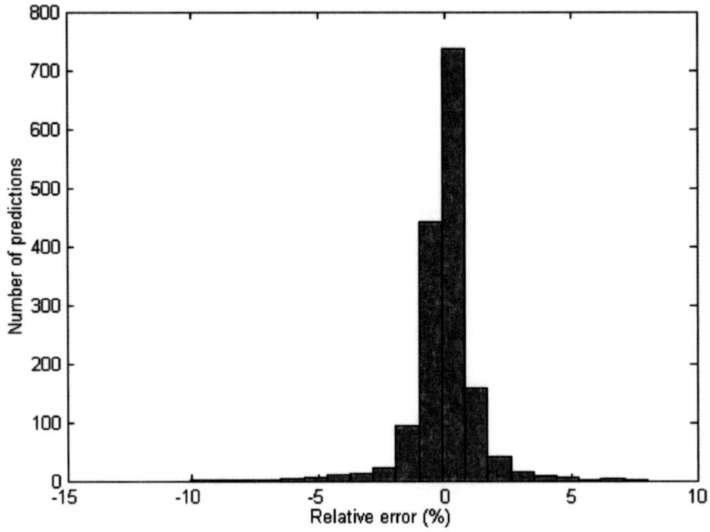

Fig. 16. Histogram of the relative error of predicted temperature during the starting phase of a solar steam generator

As an extension of this work, it could be considered that the steam generator is the evaporator of a high temperature heat pump. It has also been shown by Lecoeuche *et al.* (2005) that a condenser can be modelled by coupled neural networks. So, for some components of a heat pump the neural models are already available. It has also been shown that a whole heat pump (Bechtler *et al.*, 2001) and HVAC (Heating, Ventilating, and Air Conditioning) systems (Fock *et al.*, 2003) can be modelled using ANNs. So, it is quite sure that by coupling the two approaches, neural networks could be of a real benefit when trying to model solar boosted/assisted heat pumps, such as those presented in (Chaichana *et al.*, 2004; Wongsuwan *et al.*, 2004; Ozgener and Hepbasli, 2005).

2.2. Solar Power Plants

More complex systems are also studied using neural networks. One of these systems is the ACUREX solar power plant which is part of the "Plataforma Solar de Almería", located in south-eastern Spain in the Desert of Tabernas (Berenguel *et al.*, 1998; Arahal *et al.*, 1998; Henriques *et al.*, 1999; Gil *et al.*, 2002; Henriques *et al.*, 2002). The power plant produces electricity and desalinated seawater as shown in Fig. 17. The distributed collector field consists of 480 solar collectors arranged in 20 rows which form 10 parallel loops [see for example (Zarza *et al.*, 2004) for a photograph of the system].

The most detailed study, concerning the identification of the whole solar plant, is done in (Arahal *et al.*, 1998). A specific neural version (linear or non-linear) of the ARX model – **A**uto**R**egressive model with e**X**ogeneous inputs - is used in which the input layer is directly linked to the output layer for the linear part of the model (see Fig. 18). In fact the hidden layers are not used for the linear configuration. Using this kind of representation, once the inputs are selected, it is necessary to find the right sample period and the right order for each input. The latter is the number of past values that are to be taken into account.

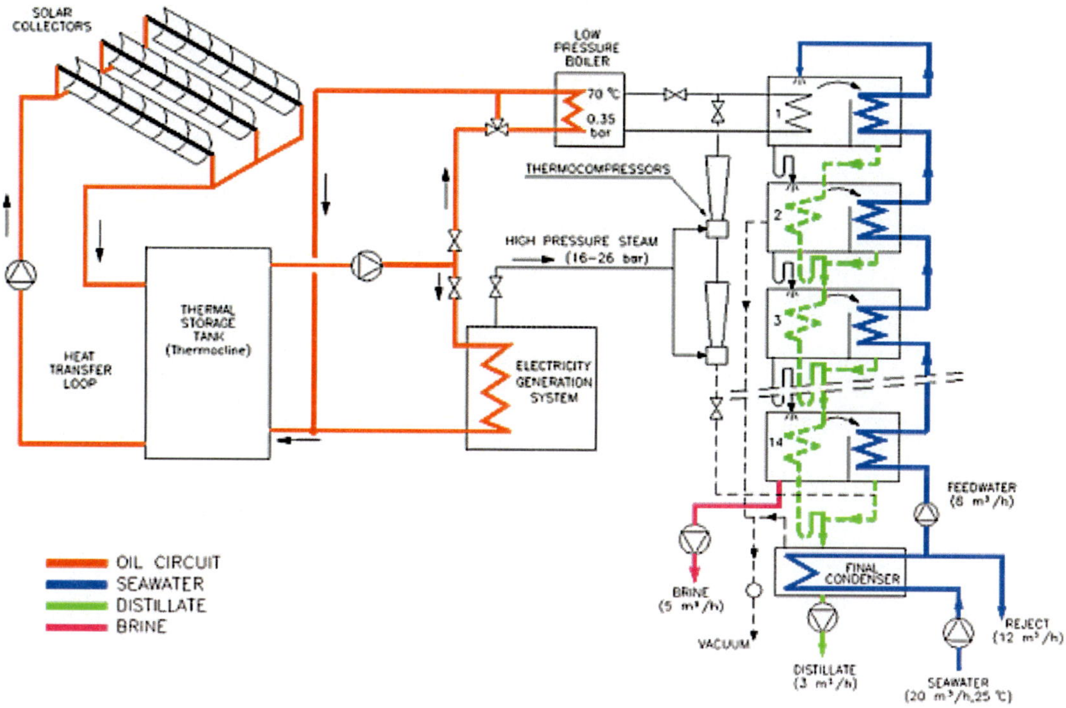

Fig. 17. Schematic of the solar plant used to produce electricity and desalinated water in the desert of Tabernas

The available relevant inputs are the temperature of the working fluid at the outlet of the collector field y(k), i.e., at the inlet of the storage tank, the flow rate of the working fluid $u_1(k)$, the solar irradiation $u_2(k)$ and the inlet temperature of the working fluid $u_3(k)$. The output is the estimated temperature of the working fluid at the outlet of the collector field at the next time step $\hat{y}(k+1)$. All data are normalized in the [-1,1] range. The sampling period has to be chosen in accordance with the dynamics of the system studied. For example, the number of samples should be sufficient when considering the response to a step function of one of the inputs to capture the dynamics of the system. For this study, the sampling period is 39 seconds.

Looking at the variation of the sum of squared errors (SSE) versus the number of past inputs $[n_y n_1 n_2 n_3]$ is a good way to begin the optimization of the orders of the inputs. For example, as can be seen from Fig. 19 from the case [2111] onwards, n_1 should probably be at least 2 and that increasing n_3 from 1 to 2 does not bring a significant improvement.

The main drawback of the ARX model is the fact that the quality of the x-step ahead prediction (using the estimated outputs instead of the actual values in the model-Fig. 18) gets worse as x increases. This is true for MLP (MultiLayer Perceptrons) networks, as shown by the results presented in Fig. 20. This is also true for RBF (Radial-Basis Function) networks. It is shown in (Arahal *et al.*, 1998) that the SSE (sum of square error) is about 2 for a 20-step ahead prediction for a number of RBF neurons ranging from five to twenty. It can be concluded that this kind of model can only be used when very few step-ahead predictions are required as in control.

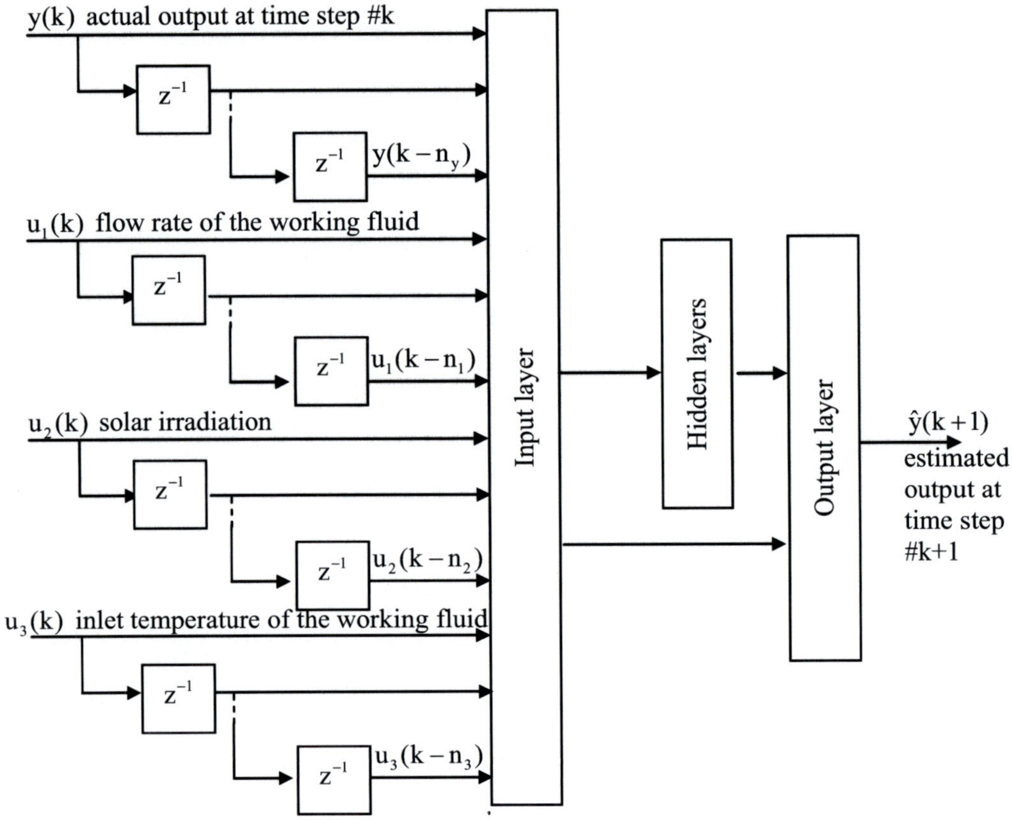

Fig. 18. Specific ARX neural model used in the study of the solar plant

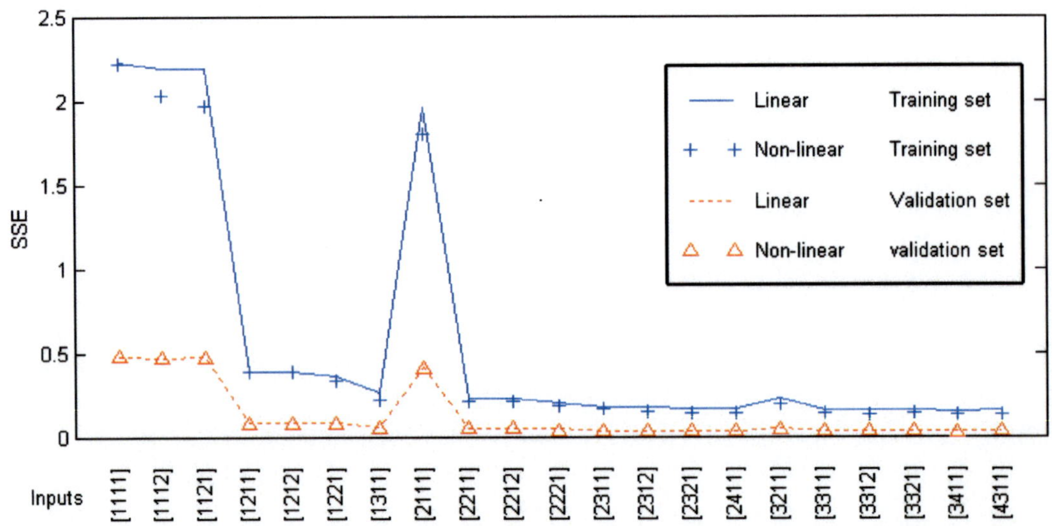

Fig. 19. Overview of the effect of the neural architecture on the sum of squared errors for the solar plant

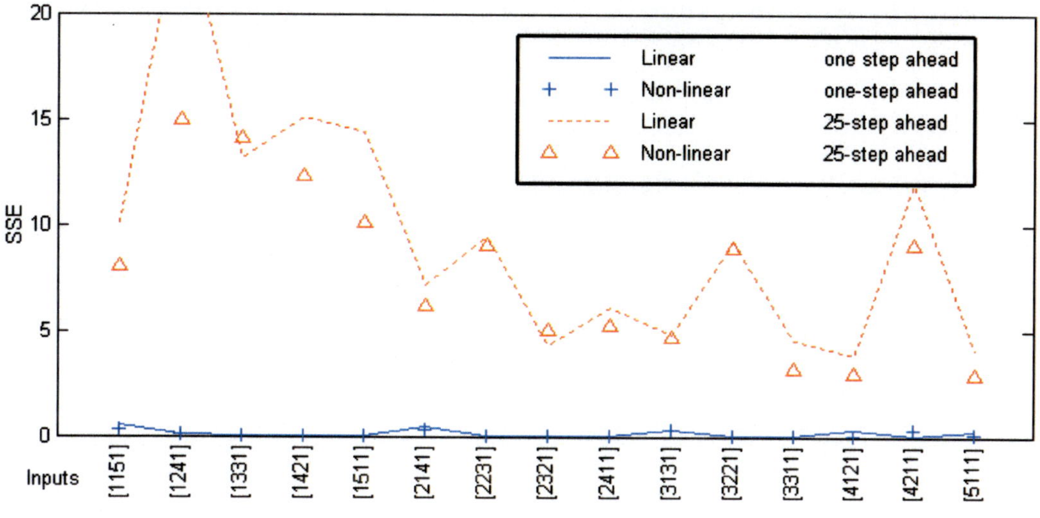

Fig. 20. Overview of the effect of the number of past inputs on the sum of square errors during one- and twenty five-step ahead prediction for the solar plant

3. Photovoltaic Systems

In this section it is indicated how ANNs can be used for the optimal operating point selection and for sizing stand alone systems.

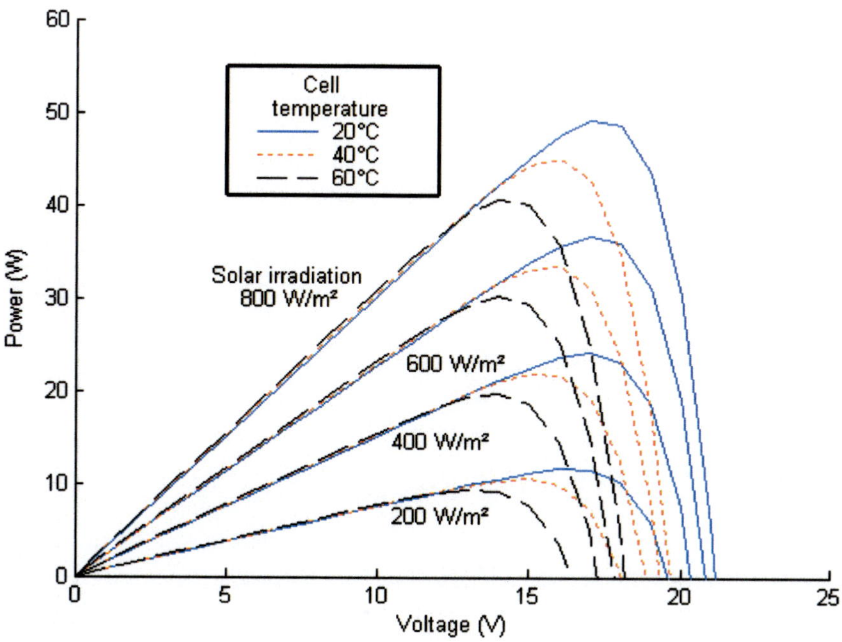

Fig. 21. Typical power-voltage curves for a PV system

3.1. Optimal Operating Point

When exploiting PV systems, the user can choose between a simple "constant voltage" type output and a "maximum power" type control (Bahgat *et al.*, 2004; Veerachary and Yadaiah, 2000). The latter makes possible a significant increase, up to 30%, of the energy savings when operating the PV at the maximum point which depends on the cell temperatures and solar radiation as shown in Fig. 21. This is due to the characteristic curve of a PV module (Walker, 2001).

This is comparable to the net power a solar collector can obtain by taking into account the mass flow rate. An example for an air collector is given in Fig. 22.

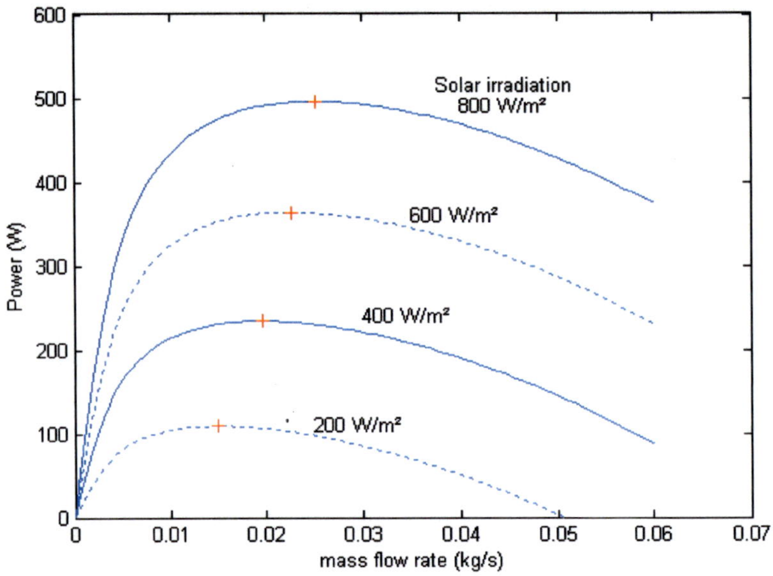

Fig. 22. Typical curves of useful energy for an air collector taking account of the pressure losses

Bahgat *et al.* (2004) used a neural network to determine the maximum power of a PV module. The ANN consists of one hidden layer with 4 neurons. The inputs are the solar irradiation and the temperature of the PV module. The output is the optimal operating point (voltage). The transfer function of all neurons (in the hidden layer and for the output neuron) is the sigmoid function. It is shown that this quite simple architecture leads to accurate predictions, both for a sunny and for a cloudy day.

It is shown by Zhang and Yun Fei Bai (2005) that a well trained RBF network can also give satisfactory results. The inputs are the solar radiation, ambient temperature and load voltage. In this case the number of hidden RBF neurons is determined using a genetic algorithm. For the practical applications of PV arrays located in Saudi Arabia and England, the authors found that 26 and 18 RBF neurons are the optimum sizes respectively.

By considering solar radiation and inlet temperature of a thermal collector as inputs and the mass flow rate as output, these techniques could potentially be applied to determine the optimal operating point (optimal mass flow rate) in solar thermal collectors.

3.2. Sizing Standalone PV Installations

The classical method for sizing stand alone PV systems is based on the loss of load probability (LOLP or LLP), which is the complement of reliability. Typical values are 0.1 for home appliances, 0.01 for rural home lighting, and 0.0001 for telecommunication devices. The goal of the study presented by Hontoria *et al.* (2005) is to find the link between the LOLP, the capacity of the generator (CA), the capacity of the accumulator (CS) and the yearly averaged clearness index (K_{TDM}). The latter is considered as being representative of the location where the standalone PV system is to be installed. This is performed by using a standard MLP network with nine neurons in the hidden layer. The ANN has three inputs (CA, CS and K_{TDM}) and one output (LOLP).

A comparison between MLP and RBF networks is given in (Mellit *et al.*, 2005). The goal is to find two optimal coefficients, one for the size of the PV-array, K_{PV}, and one for the size of the accumulator, K_B. The neural architecture is similar to the one shown in Fig. 18. Among the 200 available data for three locations, 180 were used for training and 20 served for the validation of the ANN. The comparison with an analytical approach shows that the MLP network has better generalization abilities than the RBF network as shown by the results presented in Table 6.

Table 6. Comparison of the generalization abilities of MLP and RBF networks (Mellit *et al.*, 2005)

	Test location #1		Test location #2		Test location #3	
	K_{PV}	K_B	K_{PV}	K_B	K_{PV}	K_B
Analytical model	2.453	2.781	1.435	0.812	1.893	1.196
MLP network	2.456	2.890	1.439	0.847	1.945	1.230
RBF network	2.473	2.894	1.440	0.851	1.951	1.234

These two approaches could also be used for the sizing of water heating systems. As the long term performance of such systems is available [e.g. (Kalogirou, 2000b)], a training database could easily be built and then the optimal size of the water tank could be linked to the size of the collectors for a given location.

4. Collectors

In this section it is indicated how ANNs can be used for the steady state and transient state modelling of solar systems. The transient case includes single input-single output (SISO) and multiple inputs-single output (MISO) models.

4.1. Steady State

The main component of a solar thermal system is the collector. Only flat-type collectors are considered here. So, the basic data that are available are the efficiency curve (linear or second order degree equation) and the influence of the incidence angle. A typical linear efficiency curve is shown in Fig. 23.

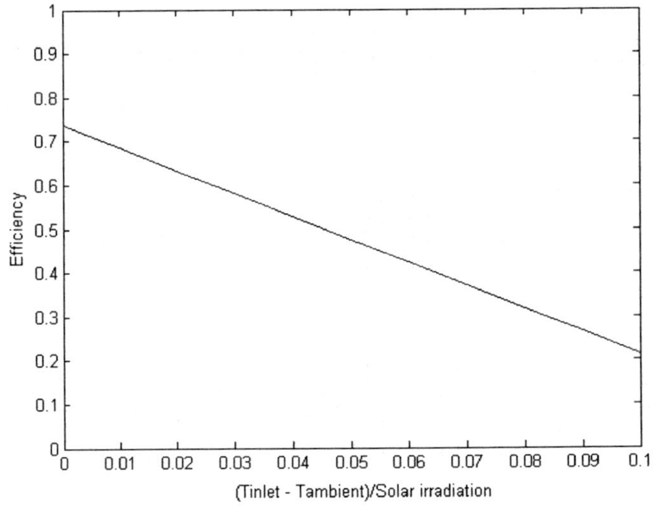

Fig. 23. Example of the standard efficiency curve of a flat type solar collector

The efficiency curve is valid for a unique flow rate, a unique ambient temperature and a unique wind velocity. This can be shown by the governing equation in steady state, when radiation from the collector is included in the thermal heat loss conductance. Considering a one-pass collector the energy balances can be written as follows:

$$w\ dx\left[\eta_0\ I_S - \frac{K}{w\ L}\ \left(T_a(x) - T_\infty\right)\right] = \dot{m}\ c_f\ dT_f(x) \tag{2}$$

and

$$\dot{m}\ c_f\ dT_f(x) = h\ \frac{dA_a}{A_c}\ \left(T_a(x) - T_f(x)\right) \tag{3}$$

Introducing the reduced mass flow rate $\dot{m}_r = \dfrac{\dot{m}}{w\ L}$, the reduced convection coefficient $h_r = h\dfrac{A_a}{A_c}$ and the reduced loss conductance $K_r = \dfrac{K}{w\ L}$ in Equations (2) and (3) the governing differential equation of the evolution of the fluid temperature can be obtained as:

$$\frac{dT_f(x)}{\eta_0\ I_S - K_r\left(T_f(x) - T_\infty\right)} = \frac{h_r}{(h_r + K_r)\ \dot{m}_r\ c_f} d\left(\frac{x}{L}\right) \tag{4}$$

The solution of this equation is:

$$T_f(x) = T_i + \frac{\eta_0 \, I_S - K_r (T_i - T_\infty)}{K_r}\left(1 - \exp\left[-\frac{h_r K_r}{(h_r + K_r)\, \dot{m}_r \, c_f}\frac{x}{L}\right]\right) \tag{5}$$

Hence the outlet temperature of the fluid is:

$$T_o = T_i + \frac{\eta_0 \, I_S - K_r (T_i - T_\infty)}{K_r}\left(1 - \exp\left[-\frac{h_r K_r}{(h_r + K_r)\, \dot{m}_r \, c_f}\right]\right) \tag{6}$$

and the efficiency is given by:

$$\eta = 2\frac{\dot{m}_r \, c_f}{K_r} \tanh\left[\frac{h_r \, K_r}{2(h_r + K_r)\dot{m}_r \, c_f}\right]\left(\eta_0 - \frac{K_r\left(\frac{T_i + T_o}{2} - T_\infty\right)}{I_S}\right) \tag{7}$$

This equation shows the relation between the inlet fluid temperature, mass flow rate, irradiation, ambient air temperature and outlet fluid temperature. The dependence on the wind velocity is embedded in the reduced loss conductance. When the solar radiation varies very slowly, this equation can be used for all day long calculations (Farkas and Géczy-Víg, 2003). A neural model of the collector is used to determine the outlet temperature by knowing the solar radiation, ambient air and inlet fluid temperature. A standard one-hidden layer network with 7 neurons is used. As no scaling is carried out, the output neuron is linear. The results are shown in Fig. 24.

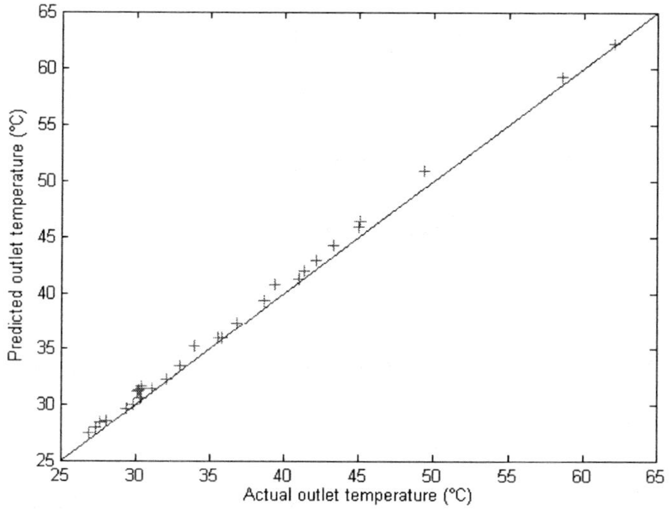

Fig. 24. Comparison between actual and ANN predicted outlet temperatures for a water collector during steady state

Another steady state model is determined by Zárate *et al.* (2004) in which the same inputs, the same number of hidden neurons and the same output are used. It is concluded that the larger the learning database the better the prediction; although the learning time is longer.

To be able to handle a broad range of collectors, Kalogirou (2005) proposes to use the architecture developed for the solar systems (see Fig. 5). By using 25 inputs such as the collector material, type of coating and the operating pressure and using 7 hidden neurons for each sub-hidden layer, good results are obtained. For example, by training the network with 61 patterns and testing it on 6 patterns, the maximal error of the efficiency is lower than 2% (for a temperature difference ranging from 10°C to 50°C and a solar radiation of $800 W/m^2$).

The much more complicated hybrid heat pipe collectors are considered by Facão *et al.* (2004) where a hot gas flows between the heat pipe and the back insulation as shown in Fig. 25.

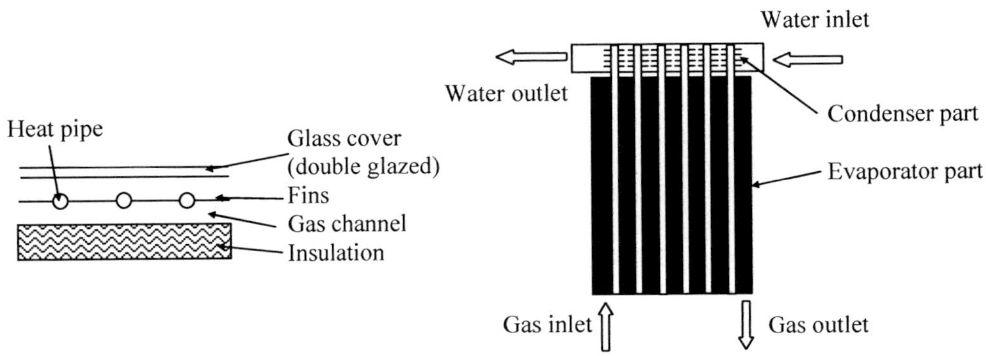

Fig. 25. Schematic of the hybrid solar collector

In this case the governing equations are very complex. The operating temperature is generally much higher, which requires precise consideration of the radiation of the absorber. The efficiency of the collector is then defined as the ratio of the energy transferred to the flowing water to the total input energy (solar energy transferred from the flowing gas):

$$\eta = \frac{\dot{m}_{water} c_{water} \left(T_{water,outlet} - T_{water,inlet}\right)}{I_S A_a + \dot{m}_{gas} c_{gas} \left(T_{gas,inlet} - T_{gas,outlet}\right)} \tag{8}$$

The outputs of the neural model are this efficiency and the heat output, which is given by the numerator of Eq. (8), i.e., $\dot{m}_{water} c_{water} \left(T_{water,outlet} - T_{water,inlet}\right)$. The inputs are the solar radiation, ambient temperature, evaporator length, condenser length, gas mass flow rate and the water mass flow rate. Two types of networks are tested; MLP and RBF. About seven hundreds cases are available; fifty were used for validation, fifty for testing whereas the remaining ones were used for the training of the network. Early stopping is used classically corresponding to the increase of the validation error. The evolution of the MSE (Mean Square error) shown in Fig. 26 for the two outputs indicates that the MLP network having nine neurons in the hidden layer is the optimal architecture.

The quality of the prediction is assessed through a comparison between the actual normalized heat output and its predicted value. The normalized heat output is defined as:

$$\eta = \frac{2}{\left[\dot{m}_{water} c_{water}\left(T_{water,outlet} - T_{water,inlet}\right)\right]_{max}} \dot{m}_{water} c_{water}\left(T_{water,outlet} - T_{water,inlet}\right) - 1 \qquad (9)$$

and so varies from –1 to +1.

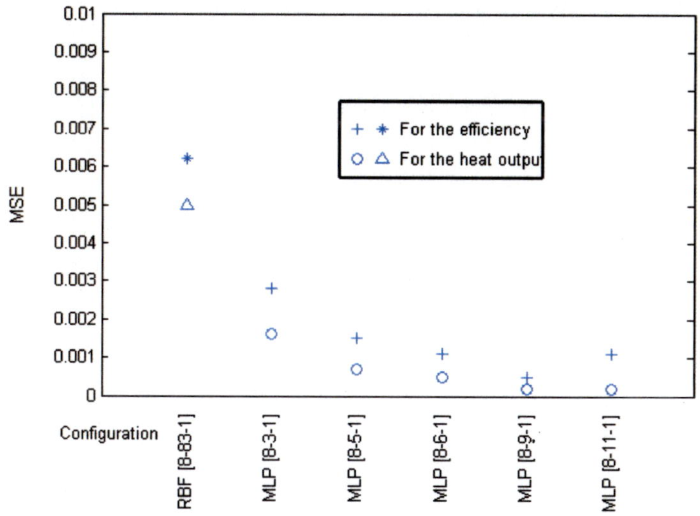

Fig. 26. Evolution of the mean squared error for various neural configurations for the hybrid collector model

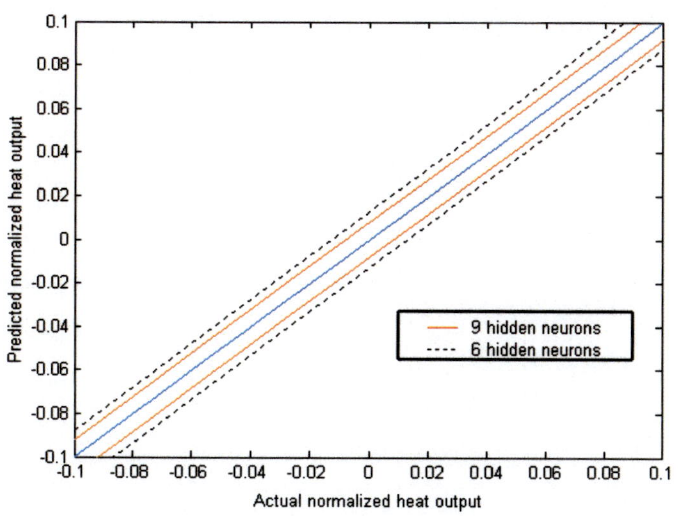

Fig. 27. Comparison between the actual and ANN predicted normalised heat output for a 6 and 9 hidden neurons network

The comparison of the actual and ANN predicted normalized output for the range [-0.1 ; 0.1] shown in Fig. 27 proves the excellent agreement between the two values (only the upper and lower limits are plotted). It has to be noted that this zone is representative of the whole range [-1 ; +1].

Therefore, it can be deduced that, if the results are needed in the shortest possible time, six neurons on the hidden layer are sufficient.

However, when the solar radiation varies quickly and frequently, the steady state model is no longer valid and it is necessary to find new models as outlined in the next section.

4.2. Transient State

For the transient state, it is necessary to introduce the thermal capacity of the absorber (C_a) in the governing equation. Therefore:

$$\eta_0\, I_S = K_r\left(T_f - T_\infty\right) + \frac{h_r + K_r}{h_r}\, L\, \dot{m}_r\, c_f \frac{\partial T_f}{\partial x} + \left(C_a + \frac{h_r + K_r}{h_r}\, \rho_f\, c_f\, V_r\right)\frac{\partial T_f}{\partial t}$$
$$+ \frac{C_a\, \dot{m}_r\, c_f\, L}{h_r}\, \frac{\partial^2 T_f}{\partial x\, \partial t} + \frac{\rho_f\, c_f\, V_r\, C_a}{h_r}\, \frac{\partial^2 T_f}{\partial t^2} \tag{10}$$

Equation (10) can be written as:

$$A = K_r\, \theta + C\, L\, \frac{\partial \theta}{\partial x} + D\, \frac{\partial \theta}{\partial t} + E\, L\, \frac{\partial^2 \theta}{\partial x\, \partial t} + F\, \frac{\partial^2 \theta}{\partial t^2} \tag{11}$$

where $\theta = T_f(x,t) - T_\infty$

This equation has no exploitable analytical solution. Numerical solutions can be obtained only if the characteristics of the collectors are known. When using this option only trial and error can be used to determine the collector characteristics and it is quite impossible to introduce the collector model in a larger system and get fast simulations.

Table 7. Characteristics of the air collector considered

Number	Parameter
1	K_r=3 W/m²K
2	$\dot{m}_r = 0.015$ kg/s m²
3	$h_r = 60$ W/m² K
4	$C_A = 10$ kJ/kg m²
5	$\eta_0 = 0.87$
6	The inlet temperature is the ambient temperature

Hence, it is necessary to search other ways to find a model for the collector. Nevertheless, the solutions of this equation will serve as references in the subsequent applications. A finite difference based software is used and the values obtained are referred to as actual values. The solutions are obtained using the characteristics given in Table 7 which represent an air collector which is well insulated, quite light and having a large convective area compared to the frontal area.

4.2.1. Single Input Single Output Models

In this subsection both off-line and on-line identification models are examined.

A. Off-Line Identification

To get a first approximation of the neural model and mainly of the order of the model, it is interesting to use the Laplace transform. In this case the response to a step function of the solar radiation can easily be found. The transformed governing equation is:

$$\frac{A}{s} = K_r\,\mathcal{L}(\theta) + C\,L\,\frac{d\mathcal{L}(\theta)}{dx} + D\,s\mathcal{L}(\theta) + E\,Ls\,\frac{d\mathcal{L}(\theta)}{dx} + Fs^2\,\mathcal{L}(\theta) \tag{12}$$

or

$$\frac{A}{s\left(K_r + Ds + Fs^2\right)} = \mathcal{L}(\theta) + \frac{(C+Es)}{\left(K_r + Ds + Fs^2\right)}\,L\,\frac{d\mathcal{L}(\theta)}{dx} \tag{13}$$

The solution, in the Laplace domain is:

$$\mathcal{L}(\theta) = \frac{A}{s\left(K_r + Ds + Fs^2\right)} + B\exp\left(\frac{-\left(K_r + Ds + Fs^2\right)}{(C+Es)}\frac{x}{L}\right) \tag{14}$$

B is determined using two of the properties of the Laplace transform:

$$\left.\begin{aligned} &\lim_{t\to+\infty}(\theta) = \frac{A}{K_r}\left[1-\exp\left(-\frac{K_r}{C}\frac{x}{L}\right)\right] = \lim_{s\to 0}\left(s\,\mathcal{L}(\theta)\right) \\ &\lim_{t\to 0}(\theta) = 0 = \lim_{s\to+\infty}\left(s\,\mathcal{L}(\theta)\right) \end{aligned}\right\} \Rightarrow B = -\frac{A}{s(K_r + Ds)} \tag{15}$$

So, the solution is:

$$\mathcal{L}(\theta) = \frac{A}{s\left(K_r + Ds + Fs^2\right)} - \frac{A}{s\left(K_r + Ds\right)}\exp\left(\frac{-\left(K_r + Ds + Fs^2\right)}{(C+Es)}\frac{x}{L}\right) \tag{16}$$

and the transfer function of the collector is:

$$H = \frac{1}{\left(K_r + Ds + Fs^2\right)} - \frac{1}{\left(K_r + Ds\right)} \exp\left(\frac{-\left(K_r + Ds + Fs^2\right)}{\left(C + Es\right)} \right) \quad (17)$$

Considering that the characteristics of the collector are constant, it is possible to plot the evolution of the gain and of the phase of the transfer function as shown in Figs. 28 and 29 respectively.

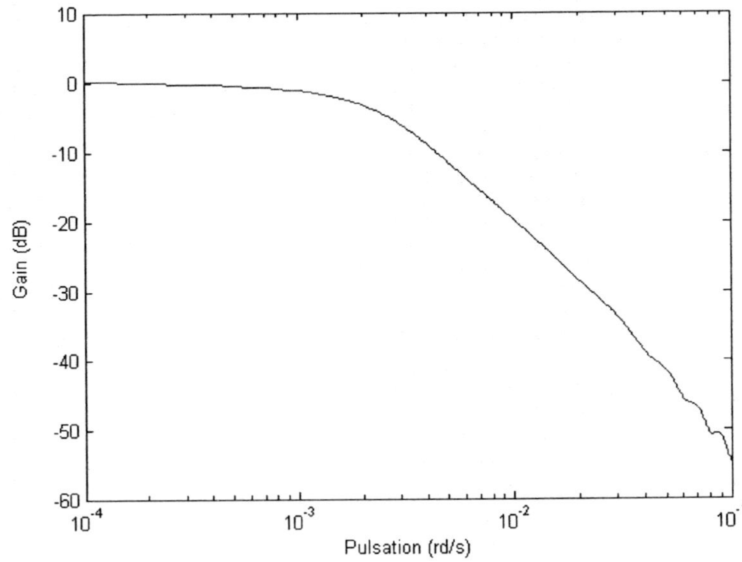

Fig. 28. Gain of the transfer function

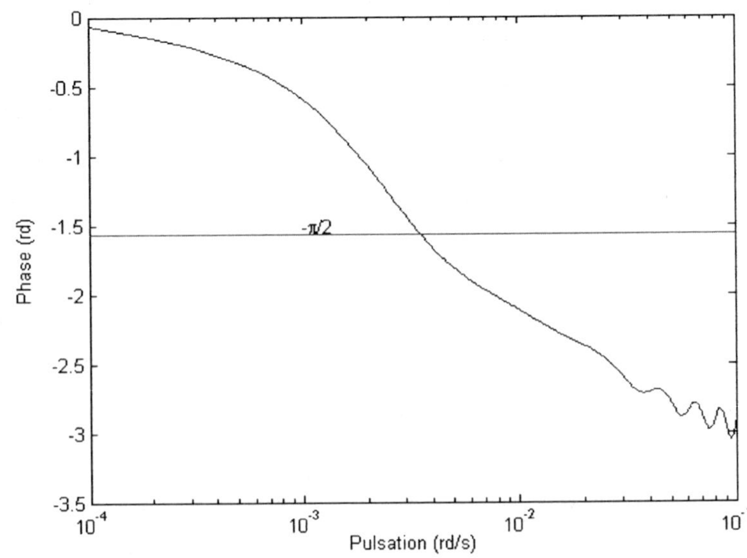

Fig. 29. Phase of the transfer function

It can be seen that both characteristics (gain and phase) are close to the characteristics of a second order system. Using classical identification criteria (Ljung, 1999), it is possible to determine the natural pulsation and the damping factor of the equivalent system:

$\omega_n = 0.003552$, $\tau = 1.2045$ and

finally the transfer function of the model (including the static gain) is given by:

$$H(\text{approximated system}) = \frac{39.87}{79{,}260\ s^2 + 678.2\ s + 1} = \frac{39.87}{79{,}260\ (s + 0.006663)(s + 0.001894)}$$

It should be noted that, if the transfer function is developed using power series, at least 35 terms have to be taken into account to get a good agreement, over the whole range of pulsations.

It is now possible to compare the response of the collector and the response of the approximated model to a step change in irradiation. This comparison shows a relatively good agreement and the RMSE (Root Mean Square Error) is about 1.1 as shown in two different ways in Fig. 30.

It can therefore be deduced from Fig. 30 that the collector is close to a second order system. But in practice, as already mentioned, the characteristics are unknown. The procedure to solve this problem with ANNs is then as follows:

- measure the final value of the response curve θ_∞,
- write the transfer function as: $H(\text{approximated system}) = \dfrac{\theta_\infty}{s^2/\omega_n^2 + 2\varsigma\, s/\omega_n + 1}$,
- estimate the intervals for $s_1 = -\omega_n\left(\varsigma + \sqrt{\varsigma^2 - 1}\right)$ and $s_2 = -\omega_n\left(\varsigma - \sqrt{\varsigma^2 - 1}\right)$ so that the step response of the collector stands within the identification domain as shown in Fig. 31,
- compute the step responses for at least 10 values of s_1 and s_2 in their respective interval,
- train and validate a standard back propagation network with 4 neurons (hyperbolic tangent transfer function), taking as inputs the values of the computed step responses at 400 seconds and at 1,000 seconds and as outputs the s_1 and s_2 values,
- take as inputs the values of the actual step responses at 400 seconds and at 1,000 seconds and get the estimated s_1 and s_2 values.

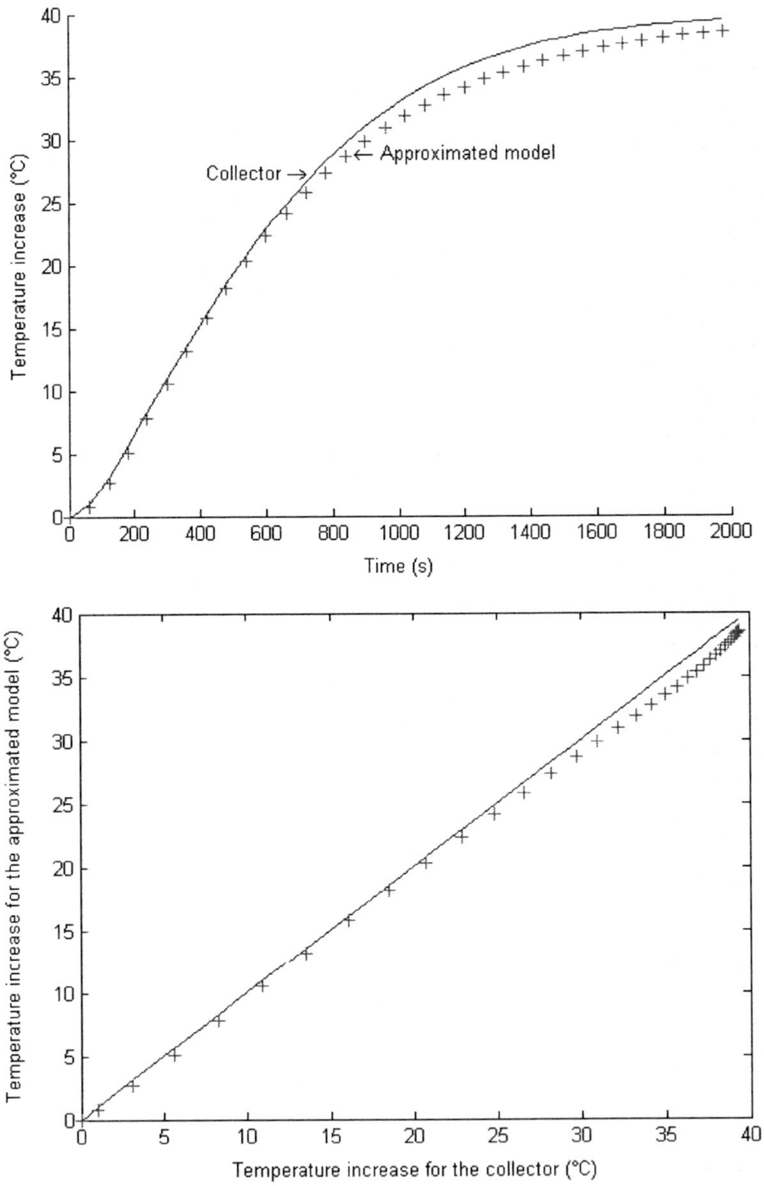

Fig. 30. Temperature increase of the approximated model compared to the actual values

Applying this procedure to the specific collector, the following transfer function is obtained:

$$H(\text{estimated system}) = \frac{39.87}{87{,}647\ s^2 + 633.5\ s + 1} = \frac{39.87}{87{,}647\ (s+0.00489867)(s+0.00232909)}$$

Figure 32 shows the response to a step function of the solar radiation, the temperature differential between the estimated values and the actual values for both models. It can be concluded that the step response of the ANN model is closer to the actual one than the step response of the model identified using the Laplace transform (referred to as the Laplace model). This is confirmed by the RMSE errors, which are equal to 1.1042 for the Laplace model and 0.3871 for the ANN model.

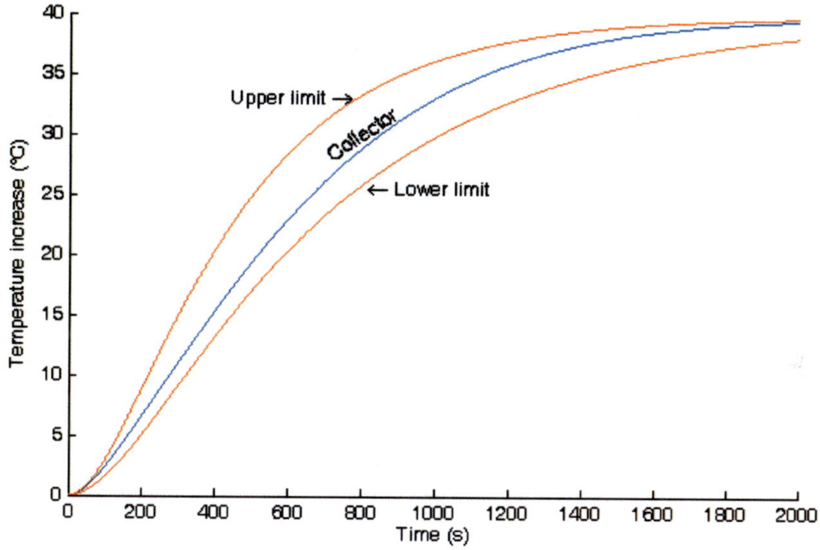

Fig. 31. Lower and upper limits of the identification domain

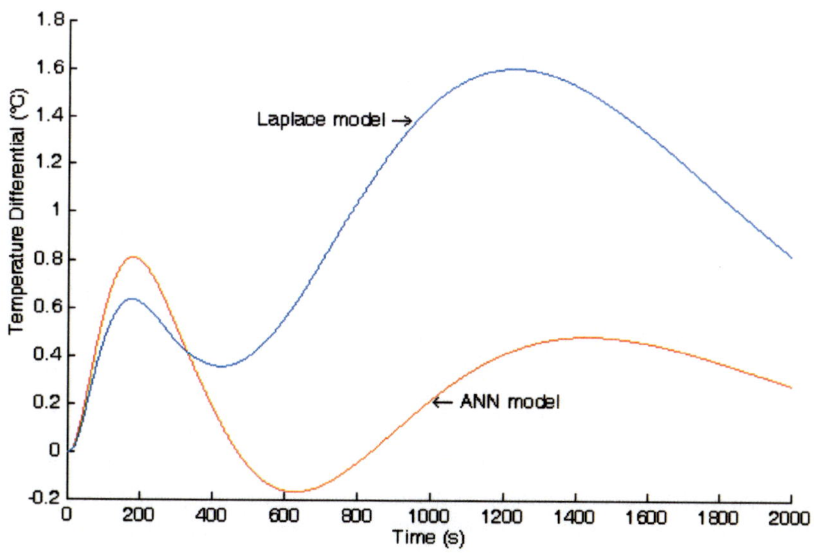

Fig. 32. Temperature differential for a step change in solar irradiation between the values obtained by the Laplace models and ANN and the actual values

It has been shown by Lalot (2000a) that a slight improvement is obtained when considering the solar collector as a third order model.

It should be noted however that it is quite difficult to get very accurate MISO (Multiple Inputs Single Output) and MIMO (Multiple Inputs Multiple Outputs) models using off-line identification techniques. This is due to the fact that it is not easy to reproduce natural phenomena (wind, rain, fog) and to reproduce the exact installation conditions (tilt angle, back insulation, etc.); see for example (Bosanac and Jensen, 1997).

B. On-Line Identification

On-line identification techniques are preferred to off-line identification techniques as they consider the data obtained in actual exploitation conditions. One of these techniques is based on the specialized neural networks developed by Nørgaard and presented in (Sjöberg, 1996; Nørgaard *et al.*, 2000; Nørgaard, 1997; 2003). These networks have been extensively used in identification of thermal devices (Lalot and Lecoeuche, 2000b; 2001a; 2001b; 2001c; Lecoeuche *et al.* 2005a). Among them, the architecture which has been selected (see Fig. 33) for the present study is close to the architecture presented in section 2.2 for the study of the solar plant (see Fig. 18) and the two architectures can be considered as mathematically equivalent. As the estimated output (not the actual one) is taken as an input, it represents an output error model. The training is based on the Levenberg-Marquard algorithm and the derivatives are analytically computed.

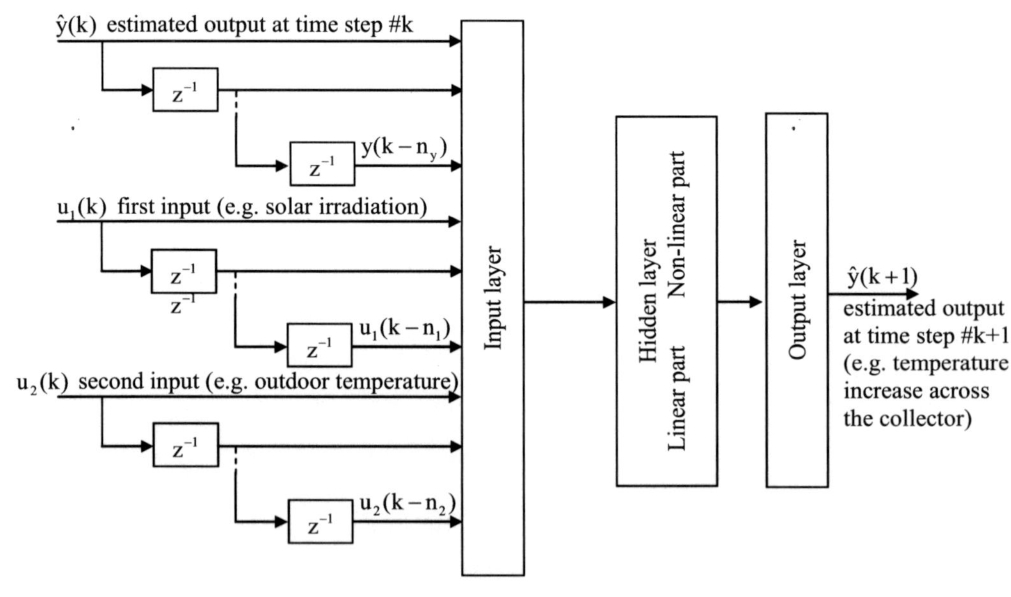

Fig. 33. Architecture dedicated to on-line identification

In its linear version and for a SISO model, it is possible to determine the initial values for the training process of the connection weights corresponding to the denominator of the transfer function. These are deduced from the discrete version of the transfer function. For example, the bilinear approximation leads to the following representation of the transfer function determined in the previous section:

$$H = \frac{0.0001133z^2 + 0.0002267z + 0.0001133}{z^2 - 1.993z + 0.9928} = \frac{0.0001133 + 0.0002267z^{-1} + 0.0001133z^{-2}}{1 - 1.993z^{-1} + 0.9928z^{-2}}$$

The actual temperature increase across the collector y in response to a random pattern of solar radiation u_1 during one day (see first plot of Fig. 34 in the next section for an example) is split in two equal parts. The first part serves as the training database. The second half of the day is used for validation. The values of the connection weights obtained by this process are 1.9468 and 0.9476. These values have to be compared to the values of the denominator of the transfer function, i.e., 1.993 and 0.9928. Still considering a SISO model, it is possible to include a non-linear part. In this case, it is found that introducing two non-linear neurons on the hidden layer and keeping a linear hidden neuron, leads to accurate results as shown below.

To compare the accuracy of the model obtained from the analysis of the gain and phase curves of the linear neural model and the non-linear neural model, the RMSE has been computed for one hundred validation days, which have a different random solar irradiation pattern for each day. Table 8 shows the mean values, the median values and the standard deviations of the RMSE.

Table 8. Comparison of the accuracies of a Laplace model and ANN models

Parameter	Laplace model	Linear neural model	Non-linear neural model
Mean RMSE	0.2682	0.0772	0.0716
Median RMSE	0.2694	0.0773	0.0711
Standard deviation	0.0116	0.0015	0.0035

It can be concluded that the neural models are more accurate than the model deduced using standard identification tools and that a non-linear model is slightly more accurate than the linear model.

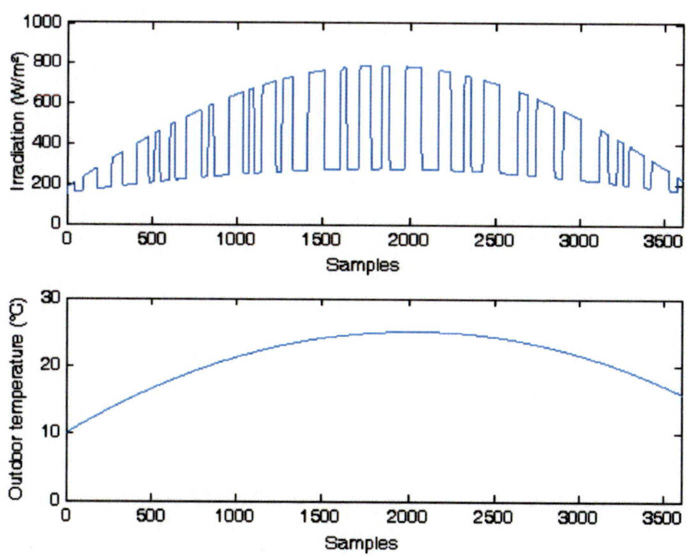

Fig. 34. Inputs considered for the MISO model of the air collector

4.2.2. Multiple Inputs Single Output Models

In this case, the inputs are the solar irradiation u_1 and the outdoor temperature u_2 shown in Fig. 34. The output y is still the temperature increase through the solar collector.

It has been shown in the previous section that the collector is very close to a second order system. So, it is considered that for the MISO version, the order is two for all the inputs. Keeping the same number of hidden neurons (one linear, two non-linear), good results are obtained when training the network using the first half day and validating it on the whole day as shown in Fig. 35 (RMSE= 0.0927). In this case the prediction of the daily energy gained by the collector is very accurate, as indicated by Lalot and Lecoeuche (2003a).

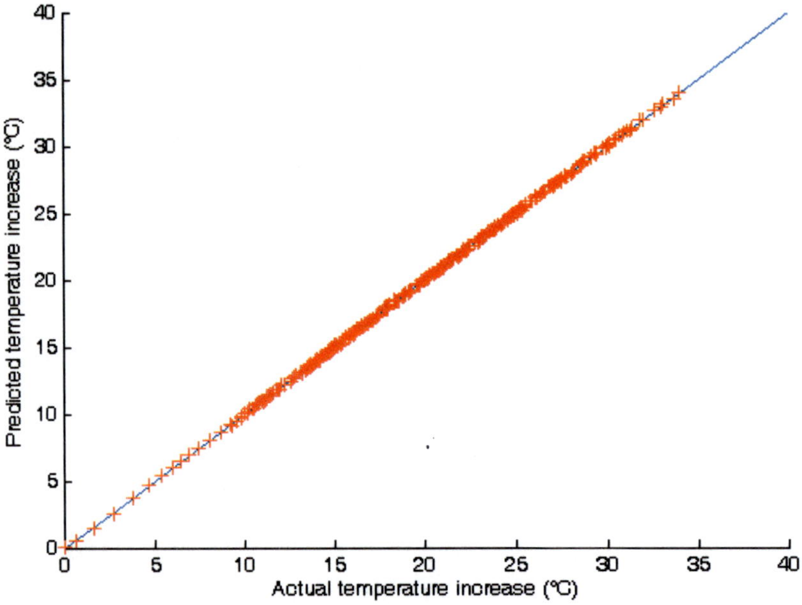

Fig. 35. Comparison between the actual and ANN predicted temperature increase for a MISO model

Other types of MISO models could be considered. For example, the inputs could be the wind velocity and the mass flow rate of the heated fluid.

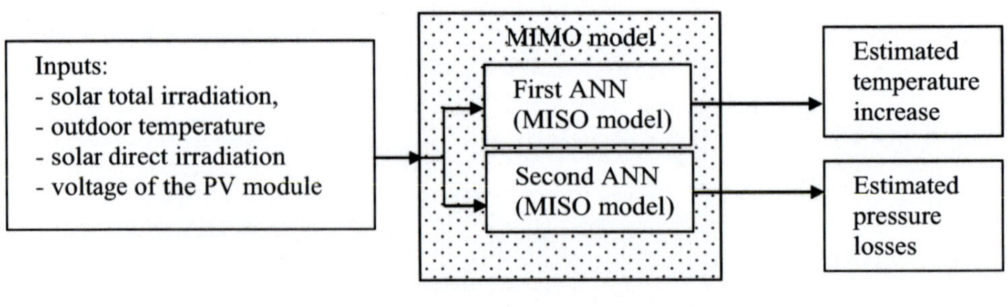

Fig. 36. Example of a MIMO model based on multiple MISO models

To get Multiple Inputs Multiple Outputs (MIMO) models and as far the outputs depend on the same inputs, it is possible to couple several MISO models. For example, if the pressure losses are considered for the second output, the architecture shown in Fig. 36 could be used for a solar thermal system which employs a PV module to supply electrical energy to the fan or the pump.

4.2.3. Validation of the On-Line Identification Procedure Using Experimental Data

In this section it is shown that experimental data can be used to assess the capability of ANNs for modelling of solar collectors. The data come from experiments carried out by Amer *et al.* (1998) and used by permission. The response of a solar system to a step change in solar irradiation (provoked artificially by a shield) clearly shows that the collector is not a first order system. Hence the neural models that have to be tested are at least second order systems. In fact, considering orders higher than two does not bring significant improvements. So, the results shown below have been obtained with second order models. Due to the fact that the collector is quite small (2.11 m^2), a small number of non linearities have to be taken into account and the neural model is also small. It is composed of one non-linear hidden neuron and one linear output neuron. It has to be noted that the solar radiation and the heat gain rate (the output of the neural models) are divided by 900 and the ambient temperature is divided by 40, so that input and output values are in the interval [0 ; 1]. The error on the heat gain rate due to measurement errors on temperature, flow rate and solar irradiation are ±0.5°C, ±0.05 litre per minute and ±2% respectively and are estimated at ±10 W/m².

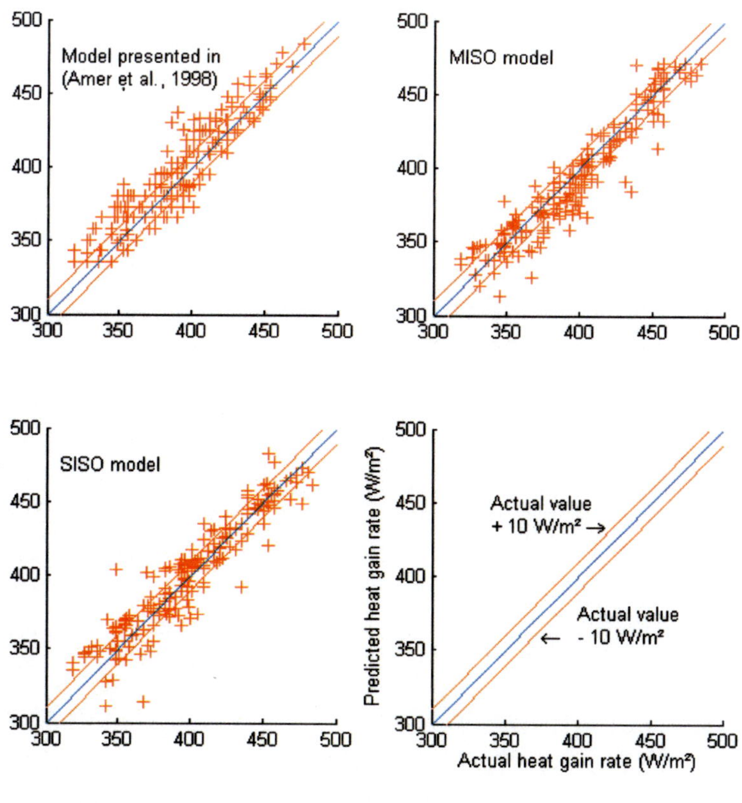

Fig. 37. Assessment of neural models using experimental data

The variations of the ambient temperature are quite small. So, two models are considered; a simple SISO model which considers the solar irradiation as the unique input and a MISO model in which the solar irradiation and the ambient temperature are used as inputs. Both are trained on one half of the dataset. In fact the whole dataset is evenly divided in eight parts and the training dataset is composed of the second, third, fifth and sixth parts. The output is the heat gain rate. Figure 37 shows the comparison between the three models. In these figures the experimental measurement error is also plotted as the two solid lines off the identity line. It can be concluded that by considering the ambient temperature as an input, slightly improves the accuracy of the predictions.

To get even better results, at least another two quantities should be taken into account; the wind velocity and the incidence angle. The influence of the wind velocity has not been evaluated, but the influence of the incidence angle is clearly shown by Amer *et al.* (1998). The coefficients of the standard steady state efficiency curve, shown in Fig. 23, are evolving along with the position of the sun and both show a maximum at midday. So, taking time as an explicit input should lead to better results.

5. Drift Detection

It has been shown that when coupled with an autoadaptive neural classifier (Lurette *et al.*, 2001; Lecoeuche *et al.*, 2004), neural models can be used to detect fouling in heat exchangers (Riverol and Napolitano, 2002; 2005), electrical heaters (Lalot and Lecoeuche, 2003b) and in a water circulating temperature controller (Lecoeuche and Lalot, 2005b).

In the case of solar collectors and to be able to operate them always at their maximum efficiency, it is necessary to get a tool which is able to detect any drift (fouling, dusty cover, etc.). Two examples dealing with the detection of the deterioration of the transmissivity of the collector cover and of the insulation are presented here.

5.1. Deterioration of the Transmissivity of the Cover

In this case, it is assumed that a drift is occurring during the whole month of May. It is supposed that the global "optical" efficiency evolves exponentially with time according to the following equation:

$$\eta_0 = 0.87 - \frac{\left(\exp\left(-\frac{31-j}{10}\right) - \exp(-3)\right)}{1-\exp(-3)} 0.27 \tag{18}$$

where j is the day number.

Hence, its value decreases from 0.87 down to 0.6 in 31 days as shown in Fig. 38.

The first method which can be used to detect this evolution is to keep the model constant and determine the differential between the predicted value and the actual temperature increase. Figure 39 clearly shows that the RMSE is a very good indicator as it increases when the "optical" efficiency decreases.

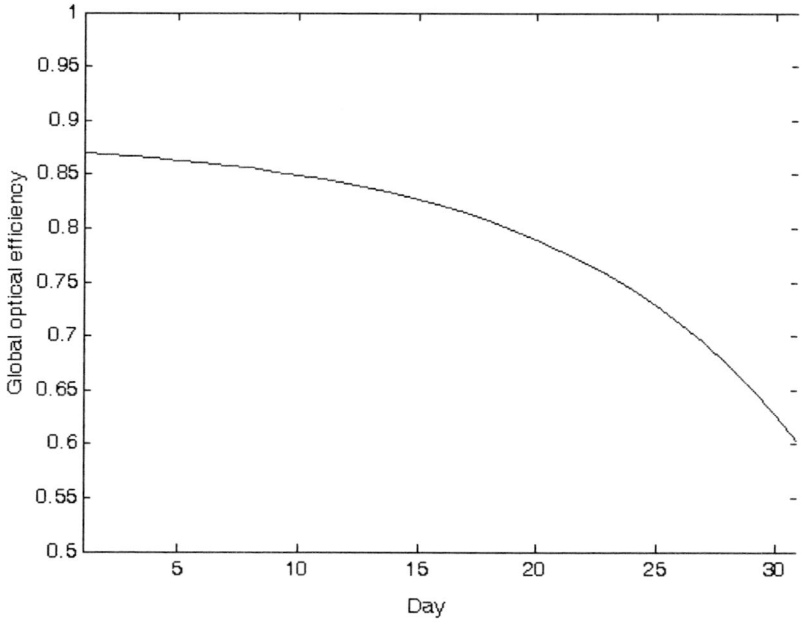

Fig. 38. Evolution of the global "optical" efficiency taken into account during the drift detection

The main drawback of this technique is the fact that it is impossible to discriminate the origin of the drift. To be able to do so, it is suggested to study the evolutions of the connection weights, as done in (Lalot and Lecoeuche, 2003b). To do so, the neural network is trained every day, taking the values of the connection weights obtained for the preceding day as their initial values (instead of random values). If there is no drift, the connection weights do not evolve, on the contrary, their evolutions form a pattern linked to the drift.

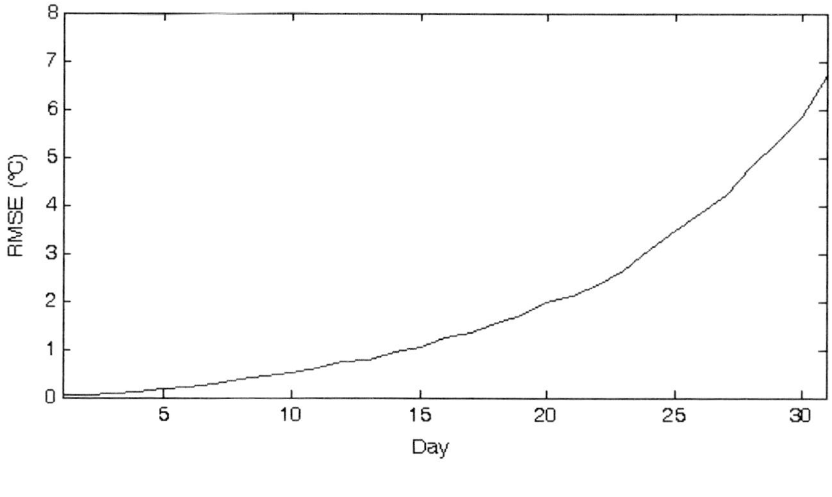

Fig. 39. Evolution of the RMSE when a constant model is kept throughout the drift

In the MISO neural model developed for on-line identification, two past values of the solar irradiation are considered as inputs. Having three neurons on the hidden layer, the network has six connection weights which affect the relation between the solar irradiation and the temperature increase, without being influenced by another input as shown in Fig. 40.

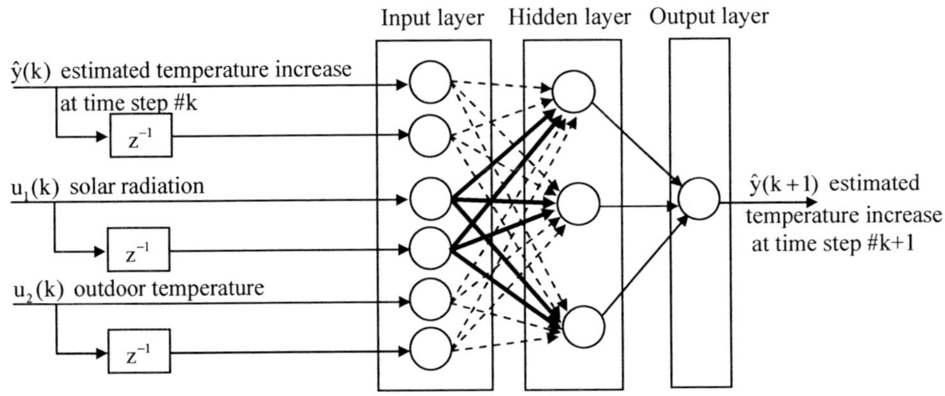

Fig. 40. The six connection weights linking the solar radiation and the temperature increase

The value of the global optical efficiency is embedded in these connection weights. Hence, at least one of them should evolve if, for example, dust accumulates on the collector. This is clearly shown in Fig. 41, which plots the evolution of the connection weights when the global optical efficiency decreases over the days examined.

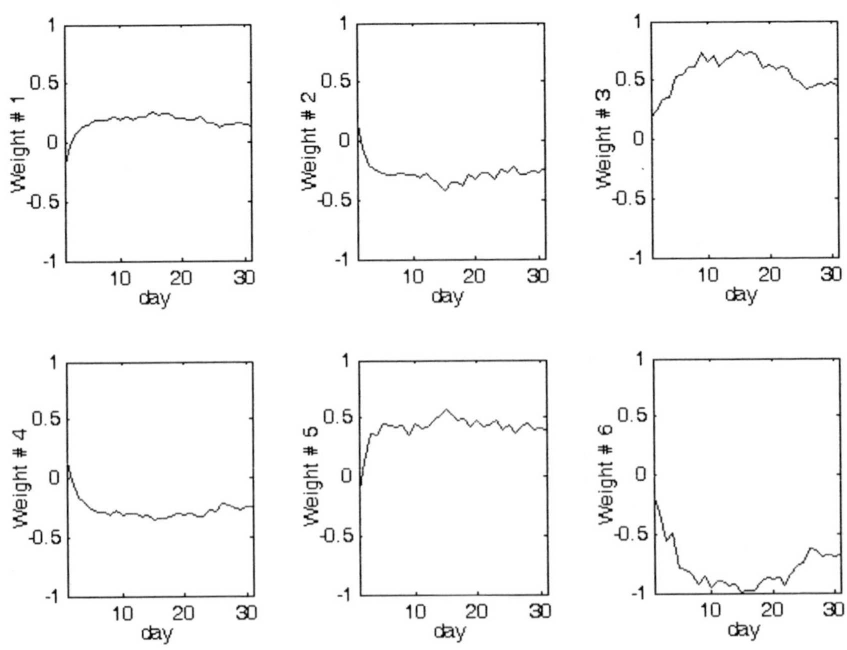

Fig. 41. Evolution of the connection weights when the global "optical" efficiency decreases

Then a simple distance-based detection tool is used. In a first step, the following quantities are computed (one per connection weight):

$$\begin{cases} dt(i,0)=0 \\ dt(i,j,k)=\max\left(0,\left|\text{avg}\left[W_{k,i}^{j}(j-t_1:j)\right]-\text{avg}\left[W_{k,i}^{0}(1:t_0)\right]\right|-\xi\right) \end{cases} \tag{19}$$

where:

- $W_{k,i}^{j}$ is the kth connection weight at date j, between the input layer and the ith neuron of the hidden layer,
- ξ is an adjustable threshold

Then a flag is set to one if $dt(i,j) > dt_{max}$, and finally the user is alerted when all the connection weights have their flag set to one. The threshold dt_{max} has to be adjusted to the drift and is up to now determined heuristically.

Applying this tool to the connection weight sequences with a threshold dt_{max} of 0.1, leads to detect the drift during the seventh day as shown in Fig. 42, which shows only the part of the curves from day 1 to day 10. It has to be noted that after the tenth day the values still increase.

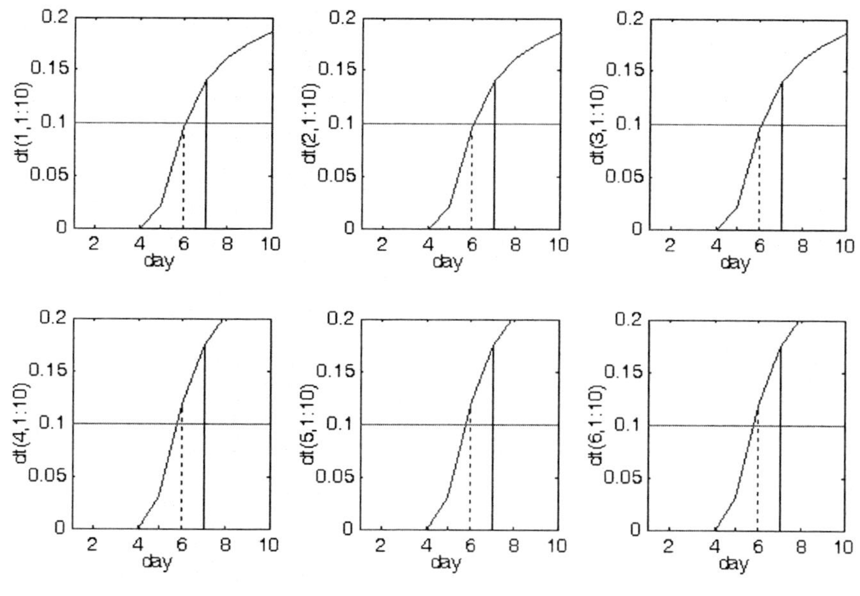

Fig. 42. Detection of the drift in the seventh day when applying the distance based test to the six connection weights

In fact, it can be seen on Fig. 42 that all flags are set to one during the seventh day (the dt values being higher than the threshold). It can be concluded that, for this drift, it is only necessary to monitor one connection weight. In that case, depending on which connection weight is chosen, the drift would be detected on the sixth day or on the seventh day.

5.2. Deterioration of the Insulation

If the drift is due to another factor, such as the degradation of the quality of the insulation, the evolutions are different. It is supposed here that the reduced loss conductance varies as follows:

$$K_r = 3 + \frac{\left(\exp\left(-\frac{31-j}{10}\right) - \exp(-3)\right)}{1-\exp(-3)} 3 \tag{20}$$

where j is the day number.

Hence, its value increases from 3 up to 6 in 31 days.

In this case, the connection weights which are concerned are the ones linking the estimated temperature increase and the outdoor temperature, as is shown on Fig. 43. Figure 44 shows that the distance-based test leads to thc detection of the drift during the fourteenth day if the same threshold is applied.

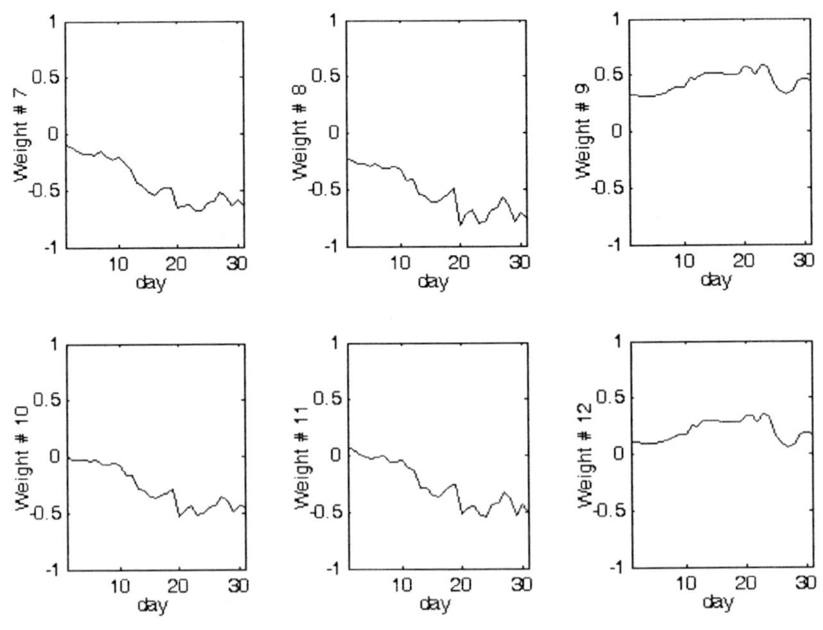

Fig. 43. Evolution of the connection weights when the quality of the insulation decreases

In fact, it can be seen on Fig. 44 that all flags are set to one during the fourteenth day (the dt values being higher than the threshold). It can be concluded that, as in the previous case, it is only necessary to monitor one connection weight. In that case, depending on which connection weight is chosen, the drift would be detected on the thirteenth day or on the fourteenth day.

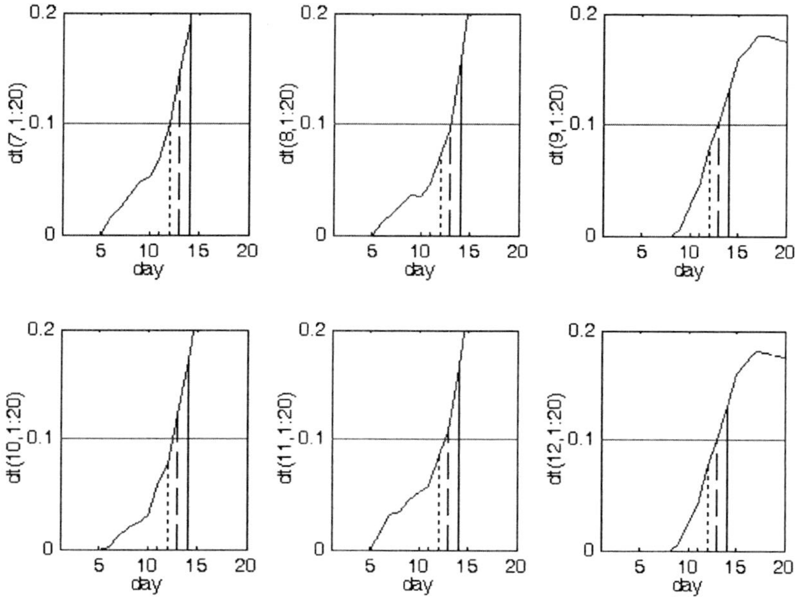

Fig. 44. Detection of the drift in the fourteenth day when applying the distance based test to the six connection weights

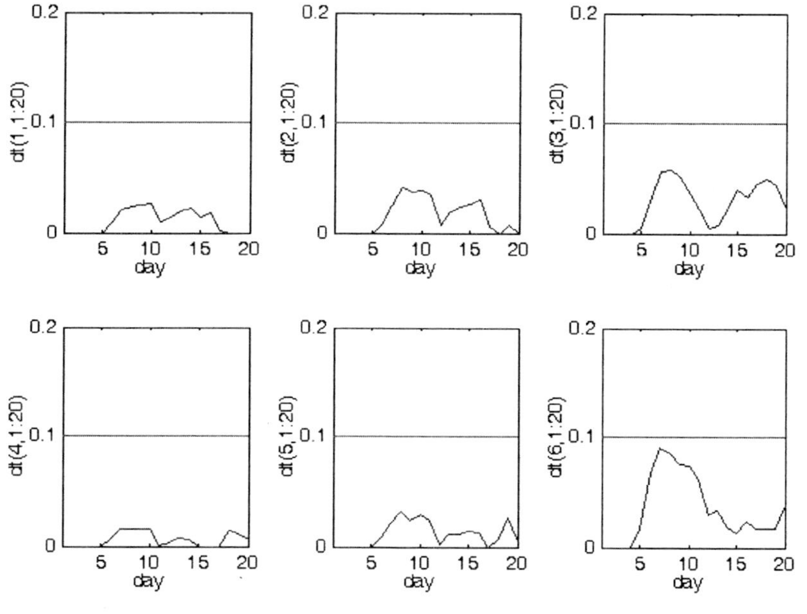

Fig. 45. The decrease of the insulation does not lead to a significant modification of the connection weights linking the irradiation and the temperature increase

It is important to note that in this case the connection weights linking the solar irradiation and the estimated temperature rise do not vary significantly. In fact, as they should not depend on the outdoor temperature, they should be constant, and the dt values should be zero during the whole drift. In practice small variations are observed, but the threshold is never attained as shown in Fig. 45.

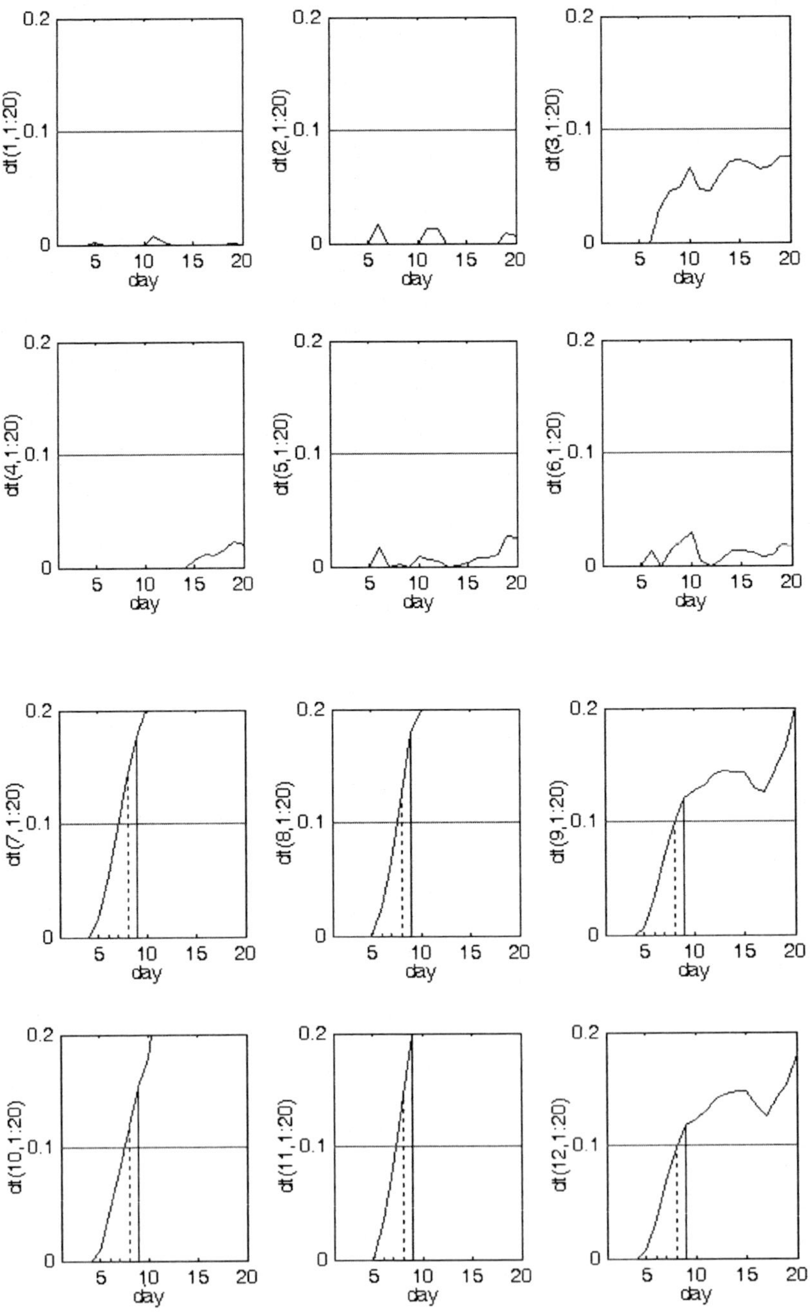

Fig. 46. Evolution of the connection weights when the quality of the insulation decreases along with the transmissivity of the cover

5.3. Deterioration of the Insulation and of the Transmissivity of the Cover

If both previous drifts are occurring simultaneously, the connection weights evolve differently as shown in Fig. 46. The analysis of the patterns in this case is more difficult and further research is required in this domain. It can be seen that the drift which is detected is the one coming from the deterioration of the insulation (weights number 7 to 12).

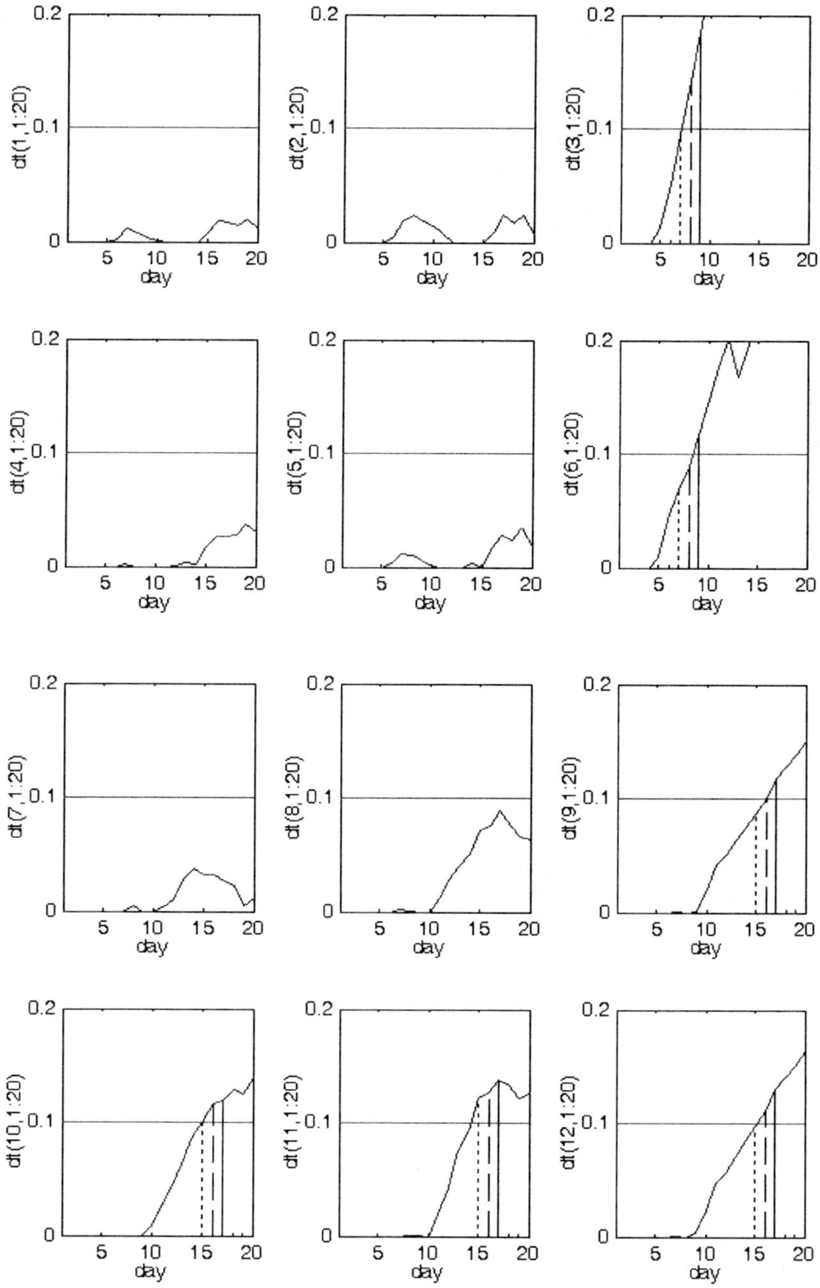

Fig. 47. Evolution of the connection weights when the quality of the insulation decreases along with the transmissivity of the cover (second case)

If the deterioration of the insulation is not as severe as simulated in the previous case (K_r varying from 3 up to 5), the evolutions of the connection weights present another pattern as shown in Fig. 47.

It can be seen that the distance-based test makes possible the detection of both drifts if one considers the connection weights number three and number six for the detection of the drift due to the deterioration of the transmissivity (on the eighth or ninth day) and weights number nine to twelve for the deterioration of the insulation (on the fifteenth, sixteenth or seventeenth day, depending on which weight is monitored).

These preliminary results show that a drift can be detected by the analysis of the connection weights. More inputs could surely lead to more accurate models. But this would surely require the use of automated classification tools such as those used in (Lurette et al, 2001; Lecoeuche *et al.*, 2004; Moreno *et al.*, 2003; Pöyhönen *et al.*, 2005; Lv *et al.*, 2005).

6. Conclusions

It has been shown that, when the model is well tuned, very accurate results can be obtained, either for short term or long term estimations. This applies both for the collector and for the whole solar plant.

The procedures used during the analysis of PV systems could be applied to determine either the size of a collector field or the optimal mass flow rate of the fluid to be heated.

Although neural networks are seen as black boxes, it has been shown that the connection weights, determined during an on-line identification, can be linked to a specific characteristic of a collector.

Further investigations are needed to get more general neural models that would be able to take account of the tilt angle, latitude, longitude, cover material, absorber material and so on. This would surely lead to more complex models but it would lead to an easier determination of the origin of a drift.

References

Amer E. H., Nayak J. K., Sharma G. K., 1998. Transient method for testing flat-plate solar collectors, *Energy Conversion Management*, Vol. 39, No. 7, pp. 549-558.

Arahal M. R., Berenguel M., Camacho E. F., 1998. Neural identification applied to predictive control of a solar plant, *Control Engineering Practice*, Vol. 6, No. 3, pp. 333-344.

Bahgat A. B. G., Helwa N. H., Ahamd G. E., El Shenawy E. T., 2004. Estimation of the maximum power and normal operating power of a photovoltaic module by neural networks, *Renewable Energy*, Vol. 29, No. 3, pp. 443–457.

Bechtler H., Browne M. W., Bansal P. K., Kecman V., 2001. Neural networks - a new approach to model vapour-compression heat pumps, *International Journal of Energy Research*, Vol. 25, No. 7, pp. 591-599.

Berenguel M., Arahal M. R., Camacho E. F., 1998. Modelling the free response of a solar plant for predictive control, *Control Engineering Practice*, Vol. 6, No. 10, pp. 1257-1266.

Bosanac M, Jensen S. Ø., 1997. *In-situ solar air collector array test*, Test report, Solar Energy Laboratory, Danish Technical Institute, Taastrup, Denmark.

Cetiner C., Halici F., Cacur H., Taymaz I., 2005. Generating hot water by solar energy and application of neural network, *Applied Thermal Engineering*, Vol. 25, No. 8-9, pp. 1337–1348.

Chaichana C., Wongsuwan W., Lu Aye, Charters W.W.S., 2004. An experimental study on natural working fluids in solar-boosted heat pump water heater, *proceedings of the 6th IIR Gustav Lorentzen Natural Working Fluids Conference*, Glasgow, UK, 29 August - 1 September, pp.3/C/10.10.

Facão J., Varga S., Oliveira A. C., 2004. Evaluation of the use of artificial neural networks for the simulation of hybrid solar collectors, *International Journal of Green Energy*, Vol. 1, No. 3, pp. 337-352.

Farkas I., Géczy-Víg P., 2003. Neural network modelling of flat-plate solar collectors, *Computers and Electronics in Agriculture*, Vol. 40, No. 1-3, pp. 87-102.

Fock E., Mara T. A., Lauret P., Boyer H., 2003. Black-box modelling of HVAC system: improving the performances of neural networks, *Proceedings of the eighth International IBPSA Conference*, Eindhoven, The Netherlands, August 11-14, pp. 347-354.

Gil P., Henriques J., Carvalho P., Duarte-Ramos H., Dourado A., 2002. Constrained adaptive non-linear neural model-based predictive control of a distributed solar collector field, available at http://www.psa.es/webeng/projects/web_ihp/papers_year.htm#2002.

Haykin, S., 1999. *Neural networks - a comprehensive foundation*, second edition, Prentice Hall, Upper Saddle River.

Henriques J., Cardoso A., Dourado A., 1999. Supervision and c-means clustering of PID controllers for a solar power plant, *International Journal of Approximate Reasoning*, Vol. 22, No. 1-2, pp. 73-91.

Henriques J., Gil P., Dourado A., 2002. Neural output regulation for a solar power plant, available at http://www.psa.es/webeng/projects/web_ihp/papers_year.htm#2002.

Hontoria L., Aguilera J., Zufiria P., 2005. A new approach for sizing stand alone photovoltaic systems based in neural networks, *Solar Energy*, Vol. 78, No. 2, pp. 313-319.

Kalogirou S. A., Neocleous C. C., Schizaz C. N., 1998. Artificial neural networks for modelling the starting-up of a solar steam-generator, *Applied Energy*, Vol. 60, No. 2, pp. 89-100.

Kalogirou S., Panteliou S., Dentsoras A., 1999a. Artificial neural networks used for the performance prediction of a thermosiphon solar water heater, *Renewable Energy*, Vol. 18, No. 1, pp. 87-99.

Kalogirou S. A., Panteliou S., Dentsoras A., 1999b. Modelling of solar domestic water heating systems using artificial neural networks, *Solar Energy*, Vol. 65, No. 6, pp. 335-342.

Kalogirou S. A., 1999c. Applications of artificial neural networks in energy systems: A review, *Energy Conversion and Management*, Vol. 40, No. 10, pp. 1073-1087.

Kalogirou S. A., 2000a. Applications of artificial neural-networks for energy systems, *Applied Energy*, Vol. 67, No. 1-2, pp. 17-35.

Kalogirou S. A., 2000b. Long-term performance prediction of forced circulation solar domestic water heating systems using artificial neural networks, *Applied Energy*, Vol. 66, No. 1, pp. 63-74.

Kalogirou S. A., Panteliou S., 2000c. Thermosiphon solar domestic water heating systems: long-term performance prediction using artificial neural networks, *Solar Energy*, Vol. 69, No. 2, pp. 163-174.

Kalogirou S. A., 2001. Artificial neural networks in renewable energy systems applications: a review, *Renewable and Sustainable Energy Reviews*, Vol. 5, No. 4, pp. 373-401.

Kalogirou S. A., 2004. Optimization of solar systems using artificial neural-networks and genetic algorithms, *Applied Energy*, Vol. 77, No. 4, pp. 383-405.

Kalogirou S. A., 2006. Prediction of flat-plate collector performance parameters using artificial neural networks, *Solar Energy*, Vol. 80, No. 3, pp. 248-259.

Lalot S., 2000a. Identification of the time parameters of solar collectors using artificial neural networks, *Proceedings of the third ISES Europe Solar Congress - Eurosun2000*, Copenhagen, Denmark, June 19-22, on CD-ROM.

Lalot S. and Lecoeuche S., 2000b. Identification of the time parameters of electrical heaters using artificial neural networks, *Proceedings of the 34th ASME national heat transfer conference on CD-ROM*, Pittsburgh (PA), August 20-22, 2000, paper NHTC2000-12153.

Lalot S.and Lecoeuche S., 2001. On-line identification of heat dissipaters using artificial neural networks, *Journal of Mechanical Engineering*, Vol. 47, No. 8, pp. 411-416.

Lalot S. and Lecoeuche S., 2001. On-line and off-line identification of electrical heaters using neural networks, *Transactions of the CSME*, Vol. 25, No 3-4, pp. 277-296.

Lalot S. and Lecoeuche S., 2001. On-line identification of circulation electrical heaters as MISO systems using artificial neural networks, *Proceedings of the 35th ASME National Heat Transfer Conference*, Anaheim (CA), June 10–12, 2001, Paper NHTC01-1811 on CD-ROM.

Lalot S. and Lecoeuche S., 2003a. Neural models of solar collectors for prediction of daily performance, *International Journal of Sustainable Energy*, Vol. 23, No 1-2, pp. 39-49.

Lalot S. and Lecoeuche S., 2003b. On-line fouling detection in electrical circulation heaters using neural networks, *International Journal of Heat and Mass Transfer*, Vol. 46, No. 13, pp. 2445-2457.

Lecoeuche S., Lurette C., Lalot S., 2004. New supervision architecture based on on-line modelling of non-stationary data, *Neural Computing and Applications*, Vol. 13, No. 4, pp. 323-338.

Lecoeuche S., Lalot S., and Desmet B., 2005a. Modelling a non-stationary single tube heat exchanger using multiple coupled local neural networks, *International Communications in Heat and Mass Transfer*, Vol. 32, No. 7, pp. 913-922.

Lecoeuche S. and Lalot S., 2005b. Neural network based on-line detection of fouling in a water circulating temperature controller (WCTC), *Oral presentation at the international conference: Heat Exchanger Fouling and Cleaning: Challenges and Opportunities*, Irsee, Germany, June 5-10.

Ljung L., 1999. *System identification: theory for the user*, 2nd edition, Prentice Hall, Upper Saddle River.

Lurette C., Lalot S., Lecoeuche S., 2001. Detection and diagnostic of drifts by unsupervised and autoadaptative neural network, *Proceedings of the International conference on Engineering Applications of Neural Networks - EANN2001*, Cagliari, Italy, July 16-18, pp. 239-246.

Lv G., Cheng H., Zhai H., Dong L., 2005. Fault diagnosis of power transformer based on multi-layer SVM classifier, *Electric Power Systems Research*, Vol. 75, No. 1, pp. 9-15.

Mager J. S., 1996. Prediction of performance of solar domestic hot water systems using artificial neural networks, *Master of Science Thesis*, Graduate School of the University of Florida.

Mellit A., Benghanem M., Hadj Arab A., Guessoum A., 2005. An adaptive artificial neural network model for sizing stand-alone photovoltaic systems: application for isolated sites in Algeria, *Renewable Energy*, Vol. 30, No. 10, pp. 1501-1524.

Moreno P. J., Ho P. P., Vasconcelos N., 2003. A Kullback-Leibler divergence based kernel for SVM classification in multimedia applications, *On-line paper of NIPS 2003, available at http://books.nips.cc/nips16.html.*

Nørgaard M., 1997.*Neural network based system identification*, Technical report 97-E-851, Department of Automation, Technical University of Denmark, Lyngby, Denmark.

Nørgaard M., Ravn O., Poulsen N. K., Hansen L. K., 2000. *Neural Networks for Modelling and Control of Dynamic Systems*, Springer-Verlag, London.

Nørgaard M., 2003. *The NNSYSID toolbox for use with Matlab®*, available at http://www.iau.dtu.dk/research/control/nnsysid.html.

Ozgener O., Hepbasli A., 2005. Exergoeconomic analysis of a solar assisted ground-source heat pump greenhouse heating system, *Applied Thermal Engineering*, Vol. 25, No. 10, pp. 1459-1471.

Pöyhönen S., Arkkio A., Jover P., Hyötyniemi H., 2005. Coupling pairwise support vector machines for fault classification, *Control Engineering Practice*, Vol. 13, No 6, pp. 759-769.

Riverol C. and Napolitano V., 2002. Estimation of the overall heat transfer coefficient in a tubular heat exchanger under fouling using neural networks. Application in a flash pasteurizer, *International Communications in Heat and Mass Transfer*, Vol. 29, No. 4, pp 453-457.

Riverol C. and Napolitano V., 2005. Estimation of fouling in plate heat exchanger through the application of neural networks, *Journal of Chemical Technology and Biotechnology*, Vol. 80, No.5, pp. 594-600.

Sjöberg J., 1996. *Non-linear system identification with neural networks*, Ph.D. thesis, Department of Electrical Engineering, Linköping University, Sweden.

Sözen A., Ali Akçayol M., 2004. Modelling (using artificial neural-networks) the performance parameters of a solar-driven ejector-absorption cycle, *Applied Energy*, Vol. 79, No. 3, pp. 309-325.

University of Wisconsin, 1985. F-Chart software -User's manual.

Veerachary M., Yadaiah N., 2000. ANN based peak power tracking for PV supplied DC motors, *Solar Energy*, Vol. 69, No. 4, pp. 343-350.

Walker G., 2001. Evaluating MPPT converter topologies using a MATLAB PV model, *Journal of Electrical and Electronics Engineering, Australia*, Vol. 21, No. 1, pp.49-56.

Wongsuwan W., Chatchawan C., Kumar S., 2004. Performance study of the coupled solar thermal-chemical heat pump system, *Proceedings of the 6th IIR Gustav Lorentzen Natural Working Fluids Conference*, Glasgow, UK, 29 August - 1 September, pp.7/B/2.00.

Wongsuwan W., Kumar S., 2005. Forced circulation solar water heater performance prediction by TRNSYS® and ANN, *International Journal of Sustainable Energy*, Vol. 24, No. 2, pp. 69-86.

Zárate L. E., Duarte Pereira E. M., Silva J. P. D., Vimeiro R., Cardoso Diniz A. S., 2004. Neural representation of a solar collector with statistical optimization of the training set, in Bob Orchard B., Yang C., Moonis A. editors, *Lecture Notes in Computer Science*, Volume 3029 - *Innovations in Applied Artificial Intelligence: 17th International*

Conference on Industrial and Engineering Applications of Artificial Intelligence and Expert Systems, IEA/AIE 2004, Ottawa, Canada, May 17-20, Springer, pp. 87-96.

Zarza E., Valenzuela L., León J., Hennecke K., Eck M., Weyers H.-D., and Eickhoff M., 2004. Direct steam generation in parabolic troughs: final results and conclusions of the DISS project, *Energy*, Vol. 29, pp. 635-644.

Zhang L., and Yun Fei Bai, 2005. Genetic algorithm-trained radial basis function neural networks for modelling photovoltaic panels, *Engineering Applications of Artificial Intelligence*, Vol. 18, pp. 833–844.

In: Artificial Intelligence in Energy and Renewable Energy Systems ISBN 1-60021-261-1
Editor: Soteris Kalogirou, pp. 131-162

Chapter 4

Artificial Neural Networks and Genetic Algorithms for the Optimisation of Solar Thermal Systems

Soteris A. Kalogirou
Higher Technical Institute, Nicosia, Cyprus

Abstract

In this chapter, it is shown how artificial neural networks (ANNs) and genetic algorithms (GA) have been used for the optimisation of solar energy systems. The ultimate purpose when designing a solar system is to maximise its economic benefits. This can be achieved either by maximising the life cycle savings of the system or by minimising its net solar energy price. The method suggested in this chapter employs an ANN to learn the correlation of some variable parameters and a GA is used subsequently to find the optimum solution. Two cases are presented in this chapter. In the first, the collector area and storage tank size are correlated with the auxiliary energy required and thus on the net solar energy price. Subsequently a GA is employed to estimate the optimum size of the two parameters, which minimise the net solar energy price. In the second, the variable parameters are the collector area, collector slope, mass flow rate through the collectors and volume of the storage tank. In this case, the genetic algorithm is used to estimate the optimum size of the variable parameters, which maximise life cycle savings. By following this procedure in both cases, the design time is reduced substantially. The system is initially modelled with TRNSYS computer program. An artificial neural network is trained using the results of a small number of TRNSYS simulations, to learn the correlation of the above variable parameters on the auxiliary energy required by the system from which either the net solar energy price or the life cycle savings can be estimated. As examples of the two variations of the method, the optimization of an industrial process heat (IPH) system and a large hot water system, both employing flat-plate collectors are presented. The optimum solution obtained from the present methodology is achieved very quickly in both cases as compared to the time required to obtain the same solution by the traditional trial and error method, which would require thousands of runs of TRNSYS to cover all possible combinations considered by the genetic algorithm. Therefore, the present method greatly reduces the time required by design engineers to find the optimum characteristics of a solar system.

Keywords: Artificial neural networks, genetic algorithms, optimization, industrial process heat, solar water heating.

NOMENCLATURE

A	Collector area (m^2)
b_o	Incidence angle modifier equation constant
c_0, c_1, c_2	Coefficients used in standard collector equation
C	Total annual cost (€/year)
C_a	Area dependent cost (€/m^2)
C_{aux}	Cost of auxiliary (€)
C_f	Area independent or fixed cost (€)
C_{FA}	Cost rate of auxiliary energy (€/kJ)
C_{FL}	Cost rate of conventional fuel (€/kJ)
C_{load}	Cost of fuel to cover load (€)
C_o	Operation cost (€)
C_m	Rate of electivity (€/kWh)
C_s	Investment cost of solar system (€)
C_{sl}	Investment cost for the increased slope above the 25° minimum (€/°)
C_v	Storage volume dependent cost (€/m^3)
d	Market discount rate (%)
F_R	Heat removal factor
G_t	Total global solar radiation (kJ/m^2)
h_s	Hours of operation (hr)
i	Interest rate (%)
$k_{\alpha\tau}$	Incidence angle modifier
m	Mass flow rate per unit collector area (m^3/min-m^2)
N	Number of years
n	Collector efficiency
P_m	Pump power (kW)
PW	Present worth (€)
Q_d	Flow rate (m^3/min)
Q_u	Useful energy extracted (kJ)
$Q_{u,a}$	Total annual Q_u (kJ)
Q_{aux}	Auxiliary energy (kJ)
Q_{load}	Energy required to cover load (kJ)
Q_{loss}	Energy lost from storage tank and pipes (kJ)
Q_{net}	Net energy collected
Q_{rel}	Energy relieved from relief valve (kJ)
T_a	Ambient temperature (°C)
T_i	Inlet temperature to the collector (°C)
U_L	Overall heat loss coefficient of the collector (kJ/m^2°C)
V	Storage volume (m^3)
x_i	Input value

x_{max}	Maximum value of a parameter in a set
x_{min}	Minimum value of a parameter in a set
y_i	Output value

Greek

α	Annuity
ΔP	Differential pressure (bar)
ΔS	Change of slope from the minimum of 25°
η_m	Pump mechanical efficiency
θ	Angle of incidence (deg.)
$\tau\alpha$	Transmittance-absorptance product

ABBREVIATIONS

ANN	Artificial Neural Network
FYFS	First Year Fuel Savings
GA	Genetic Algorithm
GMDH	Group Method of Data Handling
LCS	Life Cycle Savings (€)
LFO	Light Fuel Oil
NSEP	Net Solar Energy Price (€/kWh)
TMY	Typical Meteorological Year

1. INTRODUCTION

During the design of a solar system, the engineer tries to find a solution, which maximises the economic benefits of the installation. This can be achieved either by maximising the life cycle savings (LCS) of the system or by minimising its net solar energy price (NSEP). LCS represent the money that the user/owner will save because of the use of a solar energy system instead of buying fuel. NSEP is the final cost of each kWh produced by the system during its life by considering both the initial cost of the system and the amount of auxiliary (conventional fuel) required to cover the demand. An optimal configuration of a system is a combination of the optimum size of various characteristics of the system, which give the maximum performance at a lower cost. In this chapter, two optimisation studies are presented. In one the most critical characteristics of a solar system which need to be optimized are the collector area and storage tank size whereas in the second the collector area, slope, flow rate and storage tank size are considered. The traditional optimisation process is carried out by running a system simulation and varying the desired parameters until the optimum solution, i.e., the one that maximises the economic benefits of the system, is reached by trial and error. Usually the optimum solution cannot be easily identified as sometimes is not easy to foresee the economic implication that a modification has on other parameters. For example, by increasing the flow rate up to a point the solar system gives more energy but more money would be required to be spend on parasitic energy (electricity) to run the solar pump. Therefore, the extra money spent because of the use of the increased parasitic energy has to be compensated by the extra energy obtained by the solar system. Often the optimum system cannot be easily found and a lot of simulations and/or calculations are required in order to decide which combinations give the best financial benefits.

As an example, a graph of life cycle savings against the solar collector area is shown in Fig. 1. For this graph, all other variable parameters except the solar collector area are kept constant. The solar collector area however, is one of the most crucial parameters as it affects strongly the energy collected and the initial expenditure of the system and thus its economic benefits. As can be seen the life cycle savings start at a negative value for a collector area equal to zero. This represents the total value of money required for fuel for a non-solar system. As solar collector area is added to the system, the life cycle savings reach a maximum. An increase in collector area from the maximum point onwards gives lower life cycle savings (than the maximum value), as the bigger initial expenditure required for the solar system cannot replace the cost of the fuel saved. For large areas in multiplication of the optimum value negative life cycle savings can be obtained representing the money lost by the owner in erecting and operating the solar system instead of buying the fuel. It should be noted that a large number of simulations is required in order to obtain the data plotted in Fig. 1. Therefore, it is very crucial to be able to select the appropriate optimum area of solar systems accurately and quickly.

The optimization process becomes even more complicated when more than one parameter is optimized because these are interrelated. For example, an increase in the storage tank size also affects the performance of the system, as more energy can be stored which affects concurrently the optimum collector area and flow rate and thus the initial and running costs of the system. In such an optimisation exercise as the number of parameters investigated increase, the number of simulations increase, with a consequent raise in the time and effort required.

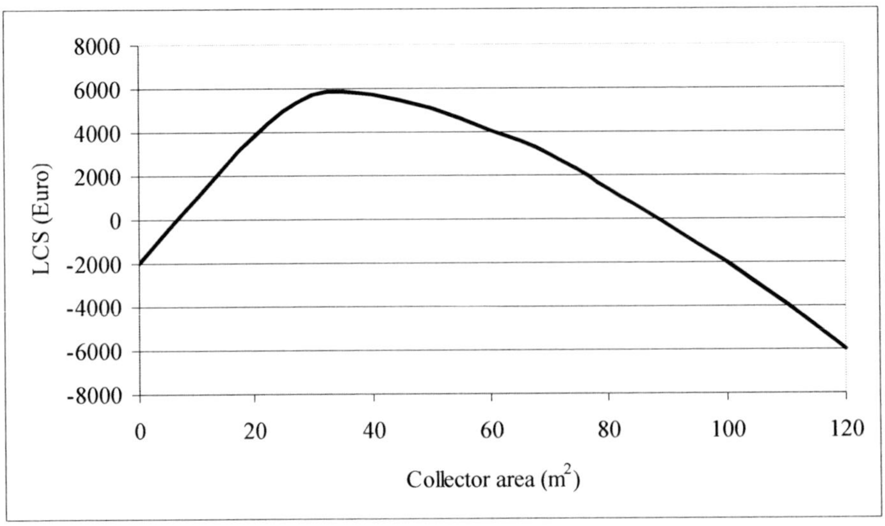

Fig. 1. Variation of collector area with life cycle savings

It is therefore very important for solar system engineers to be able to select the optimum system characteristics quickly and accurately.

A number of optimization methods and software are available in the market, which can be used for the design of solar systems. These range from simple to very detailed and complicated ones, based on detailed system simulations. The first one, which is in the category of the simple methods, is called the Φ-method. This was initially developed by Hottel and Whillier (1958) who developed the concept of "utilizability" and the Φ-curves, which are a set of performance curves that dependent only on the location of the installation. From these curves, the monthly thermal performance of a system of any size and construction can easily be estimated. The Φ-method was later extended by Liu and Jordan (1963), Klein (1978) and Collares-Pereira and Rabl (1979). Other simple methods include the f-chart (Klein *et al.*, 1976), developed from various runs of the program TRNSYS, the P_1, P_2 method, which is an economic evaluation and optimization method (Brandemuehl and Beckman, 1979) and design nomograms (Hobson and Norton, 1989). All these simple methods however suffer from various limitations. For example the basic limitation of the Φ-method is that an estimation of the average critical radiation intensity is required, which is not easily obtained unless long-term data are available. The f-chart although can be used very quickly to estimate the long-term performance of solar heating systems can be applied only to standard system configurations and the results suffer from the fact that storage tanks are assumed to be fully mixed, which leads to overestimation of the collector inlet temperature. In addition, in f-chart, systems are considered to be well built and flow distribution in collectors is considered uniform. Additionally, in most of the methods the results suffer from uncertainties arising from the use of average meteorological data, which are also considered symmetrical about solar noon, and from the consideration of average daily load utilization.

In the market there are also a number of user-friendly software tools, which can be used to simulate solar systems. These include the f-chart program (Klein and Beckman, 1981), Watsun (1992), Polysun (2000) and many others. These can be used mostly for modelling standard types of systems and include also routines to perform economic analysis. There is also TRNSYS program (2005), which is not very user-friendly but offers detailed simulation capabilities (see section 2).

In this chapter, a new method to optimize solar energy systems in order to maximise their economic benefits is introduced. For this purpose, an artificial neural network (ANN) is trained to learn the correlation of various system parameters such as the collector area, collector slope, flow rate through the collectors and storage tank size on the auxiliary energy required. From the auxiliary energy, the net solar energy price or the life cycle savings of the system can be estimated. Subsequently a genetic algorithm (GA) is employed as an optimizing tool in order to determine the optimum size of the variable parameters, which would minimise the net solar energy price or maximise the life cycle savings of the installation. Thus, the design time could be reduced substantially and the optimum solution eventually reached could be more correct than a solution reached by trial and error used traditionally for this purpose (see section 3.2) or by simple modelling methods. To illustrate the method two examples are presented as follows:

(1) An industrial process heat system in which the variable parameters are the collector area and storage tank size and the function to minimise is the net solar energy price.

(2) A large hot water system in which the collector area, slope, flow rate and storage tank size are the variable parameters and the function to maximise is the life cycle savings.

The importance of the optimization method suggested here is that, apart from reducing the overall time required, it is possible to find solutions that make solar energy applications more viable and thus more attractive to potential investors, which will lead to move wide applications of solar energy systems.

Artificial neural networks differ from the traditional modelling approaches in that they are trained to learn solutions rather than being programmed to model a specific problem in the normal way. They are usually used to address problems that are intractable or cumbersome to solve with traditional methods. ANNs are systems of weight vectors, whose component values are established through various machine-learning algorithms in which a set of pattern inputs is used and produce as output a numerical pattern representing the actual output. ANNs mimic somewhat the learning process of a human brain. Instead of complex rules and mathematical routines, ANNs are able to learn key information patterns within a multi-information domain.

Neural networks are widely accepted as a technology offering an alternative way to tackle complex or ill-defined problems. They can learn from examples, are fault tolerant in the sense that they are able to handle noisy and incomplete data, are able to deal with non-linear problems and once trained can perform predictions at very high speed. ANNs have been used in many engineering applications such as in control systems, in classification and in modelling complex process transformations (Kah *et al.*, 1995; Kreider and Wang, 1995; Kalogirou and Bojic, 2000, Curtiss *et al.*, 1995). Some applications in the solar energy field can be found in (Kalogirou *et al.*, 1998; 1999a; 1999b; Kalogirou, 2000; Kalogirou and Panteliou, 2000).

A Genetic algorithm (GA) as an optimisation technique is widely used for optimisation of engineering problems. Many engineering design problems are very complex and therefore difficult to solve with conventional optimization techniques (Gen and Cheng, 1997). GAs have also been used by the author for the optimal design of flat plate solar collectors (Kalogirou, 2003a). In particular, the optimum number of riser tubes has been selected based on their cost and the increase of the collector effectiveness.

The combination of artificial neural networks and genetic algorithms for the optimization of coal combustion performance is presented by Zhou *et al.* (2003). In this work, the relationships between the unburned carbon in fly ash and the coal properties as well as the operating parameters are obtained by using an artificial neural network. Then a genetic algorithm is employed as an optimizing tool to determine the optimal operating conditions for various types of coals and conditions in order to obtain the highest boiler heat efficiency.

2. System Model

Simplified analysis methods have advantages of computational speed, low cost, rapid turnaround, which is especially important during iterative design phases and ease of use by persons with little technical experience. Disadvantages include limited flexibility for design optimization, lack of control over assumptions and a limited selection of systems and system configurations that can be analyzed. Thus, if the system application, configuration or load characteristics under consideration are significantly non-standard, a detailed computer simulation may be required to achieve accurate results.

The proper sizing of the components of a solar system is a complex problem, which includes both predictable (collector and other performance characteristics) and unpredictable (weather data) components. For the modelling and simulation of the systems presented in this chapter, the well-known TRNSYS program, which has detailed modelling capabilities, is employed. This program is considered as the most accurate for the modelling of solar systems.

2.1. TRNSYS Program Description

TRNSYS is an acronym for a "transient simulation program" and is a quasi-steady simulation model (TRNSYS, 2005). The program consists of many subroutines that model subsystem components. The mathematical models for the subsystem components are given in terms of their ordinary differential or algebraic equations. With a program such as TRNSYS, which has the capability of interconnecting system components in any desired manner, solving differential equations and facilitating information output, the entire problem of system simulation reduces to a problem of identifying all the components that comprise the particular system and connecting them together to form the complete system model.

The latest version of TRNSYS (version 16) works in a graphic interface environment called the design studio. In this environment, icons of readymade components are dragged and dropped from a list and connected together according to the real system configuration. Each icon represents the detailed program of each component of the system and requires a set of inputs (from other components or data files) and a set of constant parameters, which are specified by the user. Each component has its own set of output parameters, which can be saved in a file, plotted, or used as input in other components. Thus, once all the components of the system have been identified and as the mathematical description of each component is readily available, the user all he has to do is to construct an information flow diagram for the system, the purpose of which is to facilitate identification of the components and the flow of information between them. The flow diagram also contains information on the weather data file and the output format. An information flow diagram of a solar system, similar to the ones considered here, is shown in Fig. 2.

TRNSYS is employing the standard second-order collector performance equation to model the collector, given by:

$$n = c_0 k_{\alpha\tau} - c_1 \frac{(T_i - T_a)}{G_t} - c_2 \frac{(T_i - T_a)^2}{G_t} \tag{1}$$

where $k_{\alpha\tau}$ is the incidence angle modifier given by:

$$k_{\alpha\tau} = 1 - b_o\left(\frac{1}{\cos(\theta)} - 1\right) \tag{2}$$

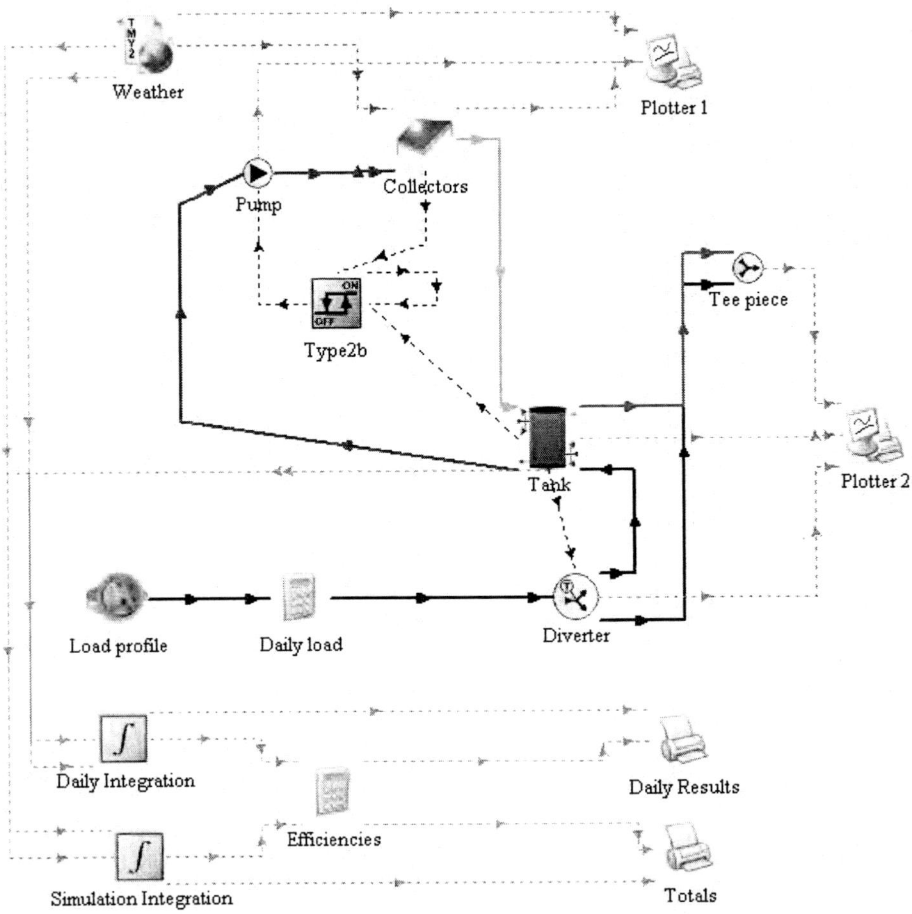

Fig. 2. Information flow diagram for a solar system model

The values of c_0, c_1, c_2 and b_o are obtained by experimental testing of collectors in accredited laboratories. The useful energy extracted from the collectors is given by:

$$Q_u = F_R A\left[k_{\alpha\tau}(\tau\alpha)G_t - U_L(T_i - T_a)\right] \tag{3}$$

The total useful energy for the whole year is obtained from:

$$Q_{u,a} = \sum_{d=1}^{365}\sum_{h=1}^{24} Q_u \tag{4}$$

and the auxiliary energy required, Q_{aux} is:

$$Q_{aux} = Q_{load} - \lfloor Q_{u,a} - Q_{loss} - Q_{rel} \rfloor \quad (5)$$

where Q_{load} is the energy required by the load, Q_{loss} is the energy lost from the storage tank and pipes and Q_{rel} is the energy relieved from the storage tank relief valve.

As can be seen from the above equations the energy obtained from the solar collector field depends on the collector area (A), collector slope (affects cosθ), flow rate (affects F_R) and the storage tank size (affects T_i). The collector inlet temperature depends also on the load pattern, make-up water temperature, the losses from the storage tank and pipes and the energy relieved from the relief valve. The storage tank losses and energy relieved from relief valve depend on the temperature of the stored water, i.e., it depends on the energy collected and storage tank size.

The systems presented in this chapter are simulated with TRNSYS using Typical Meteorological Year (TMY) data for Nicosia, Cyprus. TMY is defined as a year, which sums up all the climatic information characterizing a period as long as the mean life of the system. The selection of typical weather conditions for a given location is very crucial in computer simulations for performance predictions and has led various investigators either to run long periods of observational data or to select a particular year, which appears to be typical from several years of data. The TMY for Nicosia, Cyprus, was generated from hourly measurements, of solar irradiance (global and diffuse on horizontal surface), ambient temperature, wind speed and direction and humidity ratio, for a seven-year period, from 1986 to 1992 using the Filkenstein – Schafer statistical method (Kalogirou, 2003b). The measurements were recorded by the Cyprus Meteorological Service at the Athalassa region, an area at the suburbs of the town of Nicosia. The TMY is considered as a representative year for the Cypriot environment.

2.2. Economic Analysis

In the two different optimisation examples presented, two different economic analysis methods are used; a simplified and a detailed.

2.2.1. Simplified Method

In this method, the economic analysis is performed in order to obtain the total annual cost of the system, the net energy collected and thus the net solar energy price (NSEP). The investment cost of the solar system is obtained from:

$$C_s = C_f + C_a A + C_v V \quad (6)$$

where C_f is the area independent cost, C_a is the area dependent cost, both applied to the solar collectors, and C_v is the cost of storage per m^3 of storage volume.

For the operation cost (C_o), maintenance and parasitic cost are considered. The former are estimated to be 1% of the initial investment and are assumed to increase at a rate of 1% per year of the system operation. The latter accounts for the energy required (electricity) to drive the solar pump.

The total annual cost is given by:

$$C = C_s \alpha + C_o \tag{7}$$

where α is the annuity given by:

$$\alpha = \frac{i(1+i)^N}{(1+i)^N - 1} \tag{8}$$

where i is the inflation rate and N is the number of years the system is operational, i.e., the life of the system (20 years in the present work).

The net energy collected from the collector (Q_{net}) is obtained by subtracting the annual amount of auxiliary energy from the annual amount of conventional fuel required to cover the load for a fuel only system, i.e.,

$$Q_{net} = Q_{load} - Q_{aux} \tag{9}$$

Finally, the solar energy price, in €/kWh, is obtained by:

$$NSEP = \frac{C}{Q_{net}} \tag{10}$$

2.2.2. Detailed Method

In the detailed method, a life cycle analysis, in which the various costs are estimated annually, is considered. A life cycle analysis is performed in order to obtain the total cost (or life cycle cost) and the life cycle savings (LCS) of the system. This kind of economic analysis can be performed either within the TRNSYS environment or in a spreadsheet program. The spreadsheet application is preferred here, because the genetic algorithm employed can be readily used within the same environment. A detailed description of the method of economic analysis of solar systems using spreadsheets is given in (Kalogirou, 1996).

In general, the present worth (or discounted cost) of an investment or cost C at the end of year N, at a discount rate of d and interest rate of i is obtained by:

$$PW_N = \frac{C(1+i)^{N-1}}{(1+d)^N} \tag{11}$$

In this method, the various costs and savings are estimated annually. From the addition of fuel savings incurred because of the use of the system and the tax savings, the mortgage, maintenance and parasitic costs are subtracted and thus the annual solar savings of the system are estimated which are converted into present worth values of the system. These are added up to obtain the life cycle savings according to the equation:

$$PW_{LCS} = \sum_{N=1}^{N} \frac{Solar\ Savings}{(1+d)^N} \tag{12}$$

The fuel savings are obtained by subtracting the annual cost of the conventional fuel used for the auxiliary energy from the fuel needs of a fuel only system. The integrated cost of the auxiliary energy use for the first year, i.e., solar back up, is given by the formula:

$$C_{aux} = \int_0^t C_{FA} Q_{aux}\, dt \tag{13}$$

The integrated cost of the total load for the first year, i.e., cost of conventional fuel without solar, is:

$$C_{load} = \int_0^t C_{FL} Q_{load}\, dt \tag{14}$$

where C_{FA} and C_{FL} are the cost rates for auxiliary energy and conventional fuel respectively. In case that the same fuel is used for both $C_{FA}=C_{FL}$.

The investment cost of the solar system in this case is estimated from the following equation:

$$C_s = C_f + (C_a + \Delta S\, C_{sl})A + C_v V + C_m h_s P_m \tag{15}$$

The pump power, P_m, is estimated from:

$$P_m = Q_d \frac{\Delta P}{60} \eta_m \tag{16}$$

where Q_d is the mass flow rate in m^3/min (=m*A).

The maintenance and parasitic cost (C_o) considered are the same as in the simplified method (see section 2.1.1). In the present example, the pumping power was considered to change proportionally with the solar collector area. The total annual cost in this case is given by:

$$C = C_s + C_o \tag{17}$$

3. Industrial Process Heat System

3.1. System Description

The industrial process heat (IPH) system considered for the first case is shown schematically in Fig. 3.

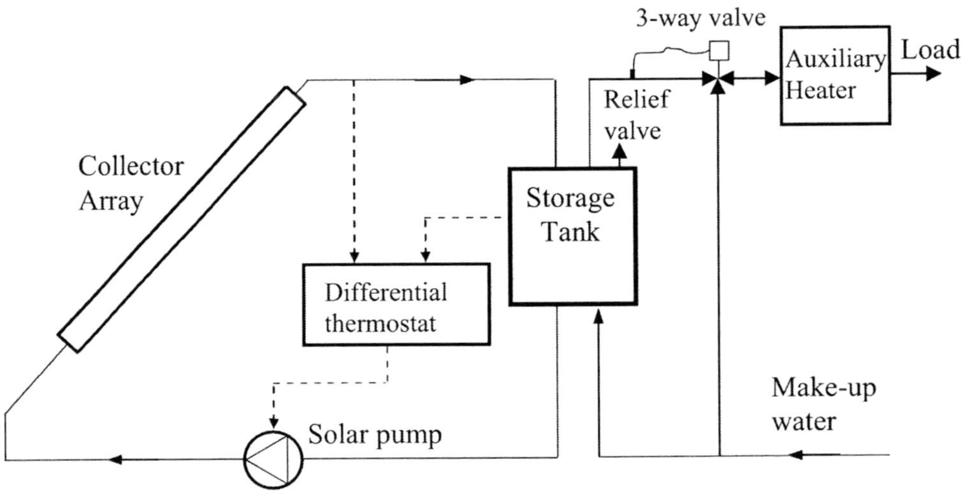

Fig. 3. Schematic diagram of the IPH solar system

The system consists of an array of collectors, a circulating pump and a storage tank. It includes also the necessary controls and thermal relief valve, which relieves energy when storage tank temperature is above a preset value. The system is once through, i.e., there is no hot water return to storage, which is what usually happens in many industrial food applications. The used hot water is replaced by mains water and mean monthly ground temperature values are used in simulations. When the temperature of the stored water is above the required process temperature, this is mixed with mains water to obtain the required temperature. If no water of adequate temperature is available in the storage tank its temperature is topped-up with an auxiliary heater before use.

The system considered is one where 2000 litres/hr of hot water are used at a temperature of 85°C (load). The load is required for the first three quarters of each hour. The industry is assumed to work on a single shift basis from 8.00 to 16.00. The system characteristics are shown in Table 1.

Flat plate collectors are employed for the system, which are by far the most used type of solar collector. Flat plate collectors are usually permanently fixed in position and require no tracking of the sun. The collectors should be oriented directly towards the equator, facing south in the northern hemisphere and north in the southern. Due to the introduction of selective coatings, flat plate collectors can reach stagnation temperatures of more than 200°C. With these collectors however, good efficiency can be obtained up to temperatures of about 80-90°C. The characteristics of the collector considered are shown in Table 2.

Table 1. Characteristics of the IPH system

Parameter	Value/Type
Load temperature	85°C
Load flow rate	2000 lt/hr
Collector to storage distance	30 m
Piping UA value	20 W/°C
Piping diameter	76 mm
Relief valve set temperature	100°C
Use pattern	5 days a week, 8.00-16.00 hours each day, load used for the first ¾ of each hour

Table 2. Characteristics of the flat-plate collectors

Parameter	Value
Fixing of risers on absorber plate	Embedded
Absorber coating	Black mat paint
Glazing	Low-iron glass
Flow rate per unit area at test conditions	54 lt/hr-m^2
Intercept efficiency (c_0)	0.792
Negative of first-order coefficient of the efficiency (c_1)	6.67 W/m^2 °C
Negative of second-order coefficient of the efficiency (c_2)	0.06 W/m^2 °C^2
Incidence angle modifier constant (b_o)	0.1

3.2. System Simulation

With the aid of TRNSYS and the TMY for Nicosia, Cyprus, a number of simulations were performed. The optimisation parameter used in the net solar energy price (NSEP), i.e., the optimum system is the one, which gives minimum NSEP. As the load pattern is the same for all the cases considered, the parameters on which the performance of the system depends are the collector area and the storage tank volume. Also all other parameters that might influence the performance of the system are considered constant. This is a reasonable assumption, as the two parameters considered are the most influential in the determination of the system performance. By increasing the collector area more solar energy is collected but the solar system costs more, whereas by increasing the storage tank volume more energy can be stored but the losses from the storage tank to the environment are increased. Small storage tank volumes exhibit lower environmental losses but increase the losses of energy though the relief valve as the stored water can easily reach the relief valve setting temperature.

The system was simulated for a number of collector areas and a range of storage tank volumes. In this way, a database is created which gives a correlation of the collector area and storage volume on auxiliary energy required by the system, for various cases. The objective is to create a small database of combinations of values and use these values to create a fitness function for the NSEP. The data collected from the present system from 7 runs of TRNSYS are shown in Table 3.

Table 3. IPH system-training data set

Area (A) [m^2]	Storage tank volume (V) [m^3]	Auxiliary energy (Q_{aux}) [GJ]
100	10	644.7
100	30	628.8
200	30	462.3
300	20	360.6
400	10	305.8
500	10	250.8
500	30	212.4

As can be seen from Table 3 as the collector area and the storage tank are increased the auxiliary energy required is reduced because the solar system can collect and store more solar energy, but the system obviously costs more. It is important to cover the low and upper extreme ends of the cases to be investigated plus some intermediate values. This is important for two reasons; first, the neural network learns all the range of possible values and thus it will not need to extrapolate and second the scaling functions for the input and output data need to be determined only once, based on the minimum and maximum values of the parameters in the data set. For those extreme conditions, rules of thump can be used such as storage tank must be between 40 and 300 litres per square meter of collector area.

The traditional method for finding the optimum solution is to perform many runs of TRNSYS and decide by trial and error the characteristics of the system, which gives the optimum solution. This method may lead to solutions far away from the optimum, as the method strongly depends on the intuition of the engineer and the peculiarities of the system and of the site weather conditions. Additionally, depending on the computer system frequency and the complexity of the system, each run might need several seconds to several minutes to be performed. Thus, it is required for the present system not only to be able to find the optimum solution but also to reduce the time required for such task to be performed.

The costs of the various system components considered in this work are:

Area independent cost (C_f) = 850 €
Area dependent cost (C_a) = 135 €/m^2
Storage volume dependent cost (C_v) = 510 €/m^3

3.3. Optimisation Method Description

A different approach to optimise the system, based on artificial intelligence systems is suggested here (Kalogirou, 2002; 2004). The idea is to use genetic algorithms to find the optimum values of collector area and storage tank volume, which will minimise the net solar energy price of the system. It should be noted that all other parameters that might influence the performance of the system are considered constant. For this purpose, an accurate correlation of collector area and storage tank volume on the auxiliary energy required to cover the load is necessary and a neural network is used which when trained gives a complex polynomial equation correlating these parameters. The neural network and the genetic algorithm are described in this section.

3.3.1. Group Method Data Handling (GMDH) Neural Network

There are various methods that can be used to model the data, i.e., correlate the collector area and storage tank volume with the auxiliary energy required. These could be based on simple regression analysis, multiple regression analysis, pace regression, sequential minimal optimization (SMO), decision tables, neural networks and many others. Here the neural network approach is selected because of its accuracy. One type of neural networks, which is very suitable for the present application, is the group method of data handling (GMDH). This is used to model the data of the IPH system. GMDH works by building successive layers with links that are simple polynomial terms. These polynomial terms are created by using linear and non-linear regression. The initial layer is simply the input layer. The first layer created is made by computing regressions of the input variables and then choosing the best ones. The second layer is created by computing regressions of the values in the first layer along with the input variables. Again, only the best are chosen by the algorithm. These are called survivors. This process continues until the network stops getting better (according to a pre-specified selection criterion).

The resulting network can be represented as a complex polynomial description of the model. The complexity of the resulting polynomial depends on the variability of the training data. In some respects GMDH, it is very much like using regression analysis, but it is far more powerful than the latter. GMDH can build very complex models while avoiding over fitting problems. An advantage of GMDH is that it recognizes the best variables as it trains.

The GMDH network is not like regular multilayer perceptron (MLP) network and was not originally represented as a network. GMDH networks are usually implemented with polynomial terms in the links and a genetic component to decide how many layers to build. The result of training, which is obtained at the output layer, is represented as a polynomial function of all or some of the inputs.

The main idea behind GMDH is that it is trying to build a function (called a polynomial model) which would behave in such a way that the predicted value of the output would be as close as possible to the actual value of output. For many end users it may be more convenient to have such a model, which is able to make predictions using familiar polynomial formulas that are widely understood. Many more details of this type of networks are given in chapter 1 of this book.

3.3.2. Genetic Algorithm

Details of GAs are given in chapter 1 and may not be repeated here. Genetic algorithms have been widely applied to optimization problems where methods that are more traditional fail. In particular a GA (Kalogirou, 2005):

- Do not require the knowledge of the gradient of the fitness functions, which is very suitable for optimization problems where an analytical expression for the fitness function is unknown.
- Have the ability to depart from the local optimum.
- Search for the optimum solution by using a progressive and black box method, which is suitable for complex problems including the processes with very complicated or even unknown mechanism.

What is required in this example is to find the optimum collector area and storage tank volume, which minimise NSEP. However, setting up the problem in the form of an optimisation analysis, the search space can be defined in terms of the input parameters under investigation, which are subject to constraints, and an optimisation function can be utilised.

Genetic algorithms (GA) are suitable for finding the optimum solution in problems were a fitness function is present. Genetic algorithms use a "fitness" measure to determine which of the individuals in the population survive and reproduce. Thus, survival of the fittest causes good solutions to progress. A genetic algorithm works by selective breeding of a population of "individuals", each of which could be a potential solution to the problem. In this case, the genetic algorithm is seeking to breed an individual that in this case minimises the net solar energy price of the IPH system.

During the setting up of the GA, the user has to specify the adjustable chromosomes, i.e. the parameters than would be modified during evolution to obtain the minimum value of the fitness function. In this work, these are the collector area and the storage tank volume. Additionally the user has to specify the ranges of these values. It is important that the ranges specified are the same as the extreme cases used when setting up the neural network. In the present work, these are equal to 100-500m^2 for the collector area and 10-30m^3 for the storage tank volume. The step sizes adopted in this case are 1m^2 for the collector area and 0.1m^3 for the storage tank.

The genetic algorithm parameters used in the present work are:

- Population size=50
- Crossover rate=90%
- Mutation rate=1%
- Generation gap=98%
- Chromosome type=continuous

3.4. Optimum IPH System

Initially the GMDH network needs to be trained with the training data set (Table 3). The final equation obtained from GMDH is:

$$M = -0.29565 - 0.93105\, N_1 + 0.32545\, N_1^2 - 0.07628\, N_2^3 \tag{18}$$

All the data required for the GMDH need to be scaled from –1 to 1. The data used with Eq. (18), for the prediction of Q_{aux}, needs also to be scaled in the same interval. For this purpose, the following equation was used:

$$y_i = \frac{2(x_i - x_{\min})}{x_{\max} - x_{\min}} - 1 \tag{19}$$

Consequently, for the present problem the following equations apply:

$$N_1 = \frac{2(A_i - 100)}{400} - 1 \tag{20}$$

$$N_2 = \frac{2(V_i - 10)}{20} - 1 \tag{21}$$

Similarly, for the network output:

$$M = \frac{2(Q_{aux(i)} - 2.124)}{4.323} - 1 \tag{22}$$

Therefore, the way to estimate the auxiliary energy is as follows; for each set of collector area (A_i) and storage volume (V_i), the scaling factors N_1 and N_2 are evaluated using Eqs. (20) and (21) respectively. Subsequently, the auxiliary energy scaling factor (M) is estimated with Eq. (18) and finally Eq. (22) is used in order to obtain the auxiliary energy ($Q_{aux(i)}$) corresponding to the two input parameters A_i and V_i.

The training data set was learned with very good accuracy with R^2-value equal to 0.9986. A plot of the actual (modelled) and network predicted data is shown in Fig. 4. It should be noted that a multiple linear regression method could only produce correlation with R^2=0.9653 which is not acceptable for the kind of predictions required in this type of problems. A similar figure for data, which are completely unknown to the network and used for validation of the ability of the network to produce accurate results is shown in Fig. 5. It should be noted that in this case also the network provided good predictions with R^2=0.9906.

Relations (18) and (20) to (22) were used together with other characteristics and economic parameters of the system in combination with a genetic algorithm program to find the optimum collector area and storage tank size (adjustable chromosomes) that minimise the NSEP of the system. The whole model was set in a spreadsheet program in which the various parameters and equations (like Eq. 18 and 20 to 22) are entered into different cells.

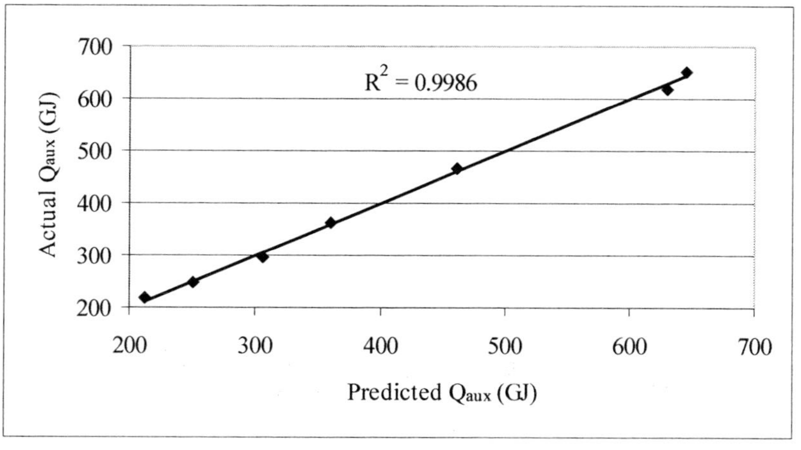

Fig. 4. Comparison of actual (modelled) and ANN predicted data for the IPH system (training data set)

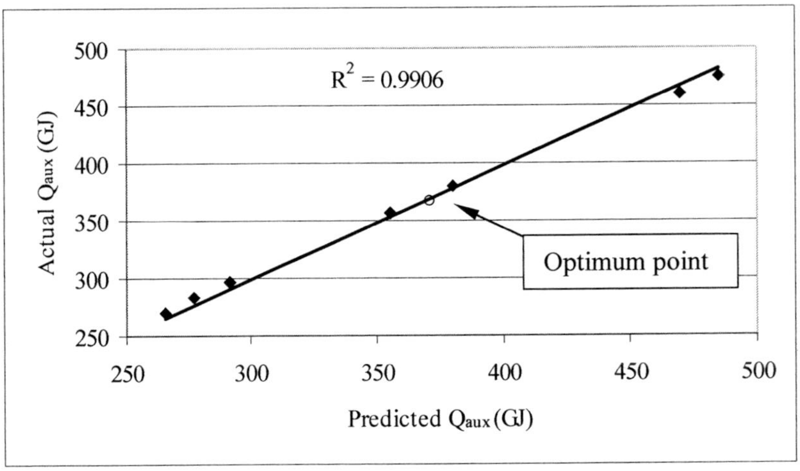

Fig. 5. Comparison of actual (modelled) and ANN predicted data for the IPH system (validation data set)

The adjustable chromosomes are set in different cells and the fitness function is the cell that contains an equation giving the NSEP. It should be noted that, the spreadsheet file described above need to be constructed once. The only changes required for different problems would be to modify the cells with minimum and maximum values of the input parameters and the cell containing Eq. (18). It should be noted however that for each problem a new neural network need to be set-up and trained.

Figure 6 shows the fitness function against the number of generations during the running of the genetic algorithm program. As can be seen the best fitness is found at the 59th generation, which is very fast. It should be noted however that the shape of the graph depends on the initial conditions.

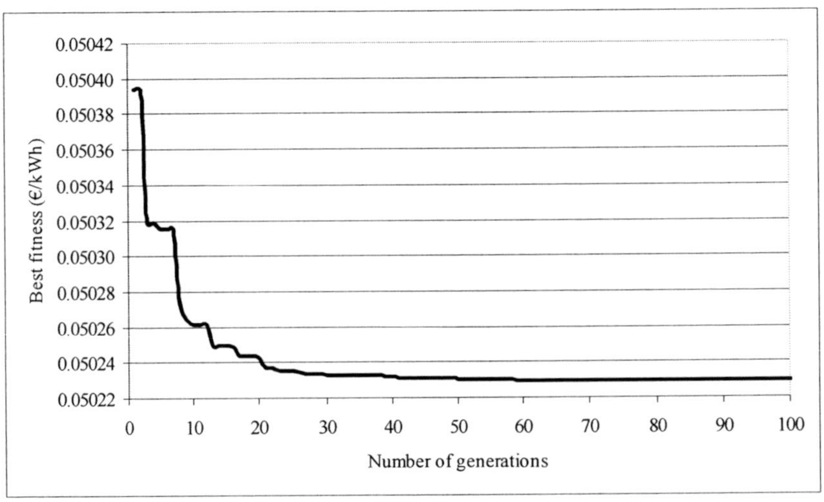

Fig. 6. Plot of best fitness against the number of generations-IPH system

The particular case presented is for initial collector area equal to 400 m^2 and initial storage tank volume equal to 30 m^3. As can be seen after about 30 generations values near the optimum ones have been obtained.

The optimum solution reached is; collector area equal to 245.5 m^2 and storage tank size equal to 12.23 m^3. The net solar energy price for this solution is 0.05023 €/kWh, whereas the net solar energy price of a more practical solution with A=250 m^2 and V=12 m^3 are 0.05024 €/kWh. Both values obtained are very close to round values, which could have been obtained by the trial and error method (depending on the step size used), but this just happened in this example. The performance of the GMDH network to predict Q_{aux} for this case is very good as it is shown by the point marked on Fig. 5 (optimum point), which compares the auxiliary energy predicted with the neural network with that obtained from TRNSYS. This case is checked on purpose to evaluate the prediction accuracy for the optimum case.

It should be noted that both the training of the neural network and the execution of the genetic algorithm program required just a few seconds to be performed whereas each run of TRNSYS requires 2.5 minutes, all on a Pentium 400 MHz machine. Thus, by reducing the number of runs of TRNSYS the time required to find the optimum solution is greatly reduced and the solution reached is more correct (depending on the step size selected) than by using the traditional trial and error method which most of the times relies on the intuition of the user to find a good solution.

4. Large Hot Water System

4.1. System Description

Another example that illustrates the optimization method suggested in this chapter deals with a large solar hot water system, which is optimized with respect to selecting the collector area, slope, flow rate and the storage tank size that maximise LCS.

A schematic diagram of the system is shown in Fig. 7. The system consists of a collector array, a storage tank, solar pump and auxiliary heater. A differential thermostat is used which compares the temperature at the exit of the collectors and the storage tank and gives a signal to switch on the pump. Again, in this application flat plate collectors are used. Their characteristics are the same as the ones used in the IPH system shown in Table 2.

The economic scenario used here is that 30% of the initial cost of the solar system is paid at the beginning and the rest is paid in equal instalments in 10 years. The period of economic analysis is taken as 20 years (life of the system), whereas the inflation rates of fuel and electricity, used for pumps, are mean values of the last 10 years. Light fuel oil (LFO) is assumed to be used for a fuel-only system.

The solar system examined can satisfy the hot water needs of 10 houses or flats or any other similar application of same hot water requirement. For this application, a hot water consumption (load) profile is required. This load is subject to a high degree of variation from day to day and from consumer to consumer, however, it is impractical to use anything but a repetitive load profile. This is not quite correct for the summer period, where the consumption pattern is somewhat higher due to frequent bathing.

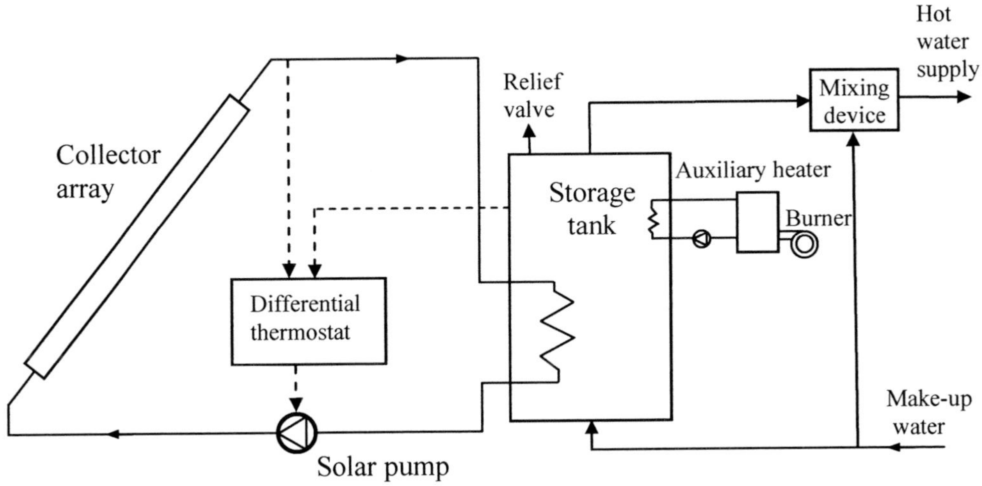

Fig. 7. Schematic diagram of the large solar hot water system

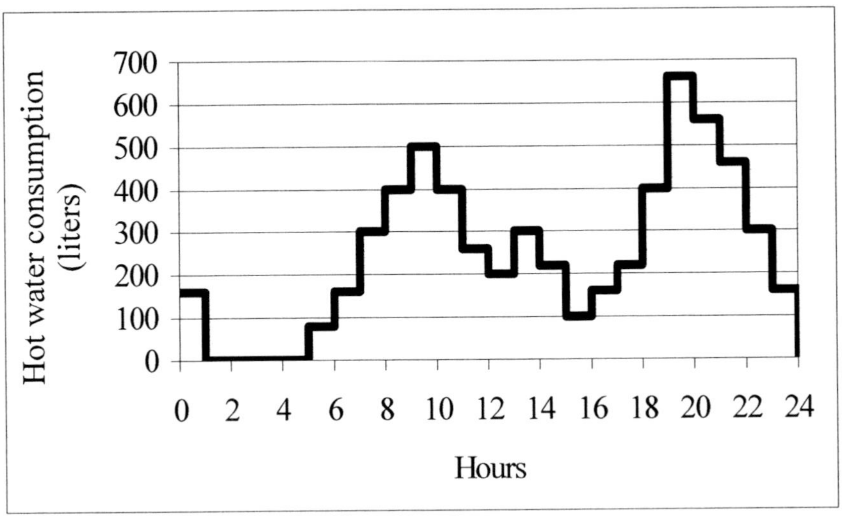

Fig. 8. Hot water daily consumption profile

However, during this period, the temperature requirement for hot water is not as high as during winter. Consequently, the total thermal energy requirement is reasonably constant throughout the year. For the present study, the hot water consumption profile, illustrated in Fig. 8 is used. The same daily hot water consumption profile, of 120 litres at 50°C, is assumed for each house (families of four persons, 30 litres/person).

4.2. System Simulation

The system simulation is again carried out with the aid of TRNSYS and the TMY for Nicosia, Cyprus. It should be noted that the load pattern is the same for all the cases considered as the same daily hot water consumption profile (shown in Fig. 8) is used. The parameters on which the performance of the system depends are:

- the collector area,
- collector slope,
- flow rate in the collector circuit, and
- the storage tank volume.

All these parameters influence the performance of the system differently. For example, by increasing the collector area more solar energy is collected but the solar system costs more, whereas by increasing the storage tank volume more energy can be stored but the losses from the storage tank to the environment are increased. Small storage tank volumes exhibit lower environmental losses but increase the losses of energy though the relief valve as, depending on the collector area used, the relief valve setting temperature could easily be reached. Additionally, by increasing the slope from a minimum of 25° up to 45° more energy is collected (local latitude=35°). For the flow rate values from 30 to 60 lt/hr-m^2 were considered. Note that the specific flow rate is used, i.e., flow rate per m^2 of collector area. Usually the higher the value of flow rate, up to 60 lt/hr-m^2, the better the solar system performance is, but more parasitic energy is required. Parasitic cost is the money paid for electricity for the solar pump. The usual value of flow rate used by solar engineers is 54 lt/hr-m^2. This parameter is included in the present analysis to investigate if a lower flow rate with smaller parasitic cost gives higher LCS.

For the present optimization method, which is described in the next section, the results of a small number of simulations are required. The optimization parameter used in this case is the life cycle savings (LCS), i.e., the optimum system is the one, which gives maximum LCS. The LCS are estimated from the auxiliary energy required by the system to cover the load demand. Various cases are simulated for a combination of the variable system parameters outlined above. In this way, a database is created with the different combinations of the variable parameters and their effect on auxiliary energy. The objective again is to create a small database of combinations of values of the variable parameters, i.e., collector area, slope, flow rate and storage volume, and use these values to create a fitness function for the LCS. The total number of cases simulated is 48. From these, 42 were used for the training and testing of the neural network and the remaining six were used for the network validation. A sample of the training data set is shown in Table 4 and the validation data set is shown in Table 5. An increased number of patterns is required in this case, compared to the previous system (see section 3) because here the variable parameters are four, compared to two used before, and these are required in order to allow the ANN to learn the required interactions correctly. It is important to cover both low and upper extreme ends of the cases investigated plus some intermediate values. This is important for the reasons outlined in section 3.2 and apply here as well. For the selection of the extreme conditions, again rules of thump can be used such as storage tank can be between 40 and 300 litres per square meter of collector area.

The traditional method for finding the optimum solution in this case also is to perform many runs of TRNSYS and determine by trial and error the characteristics of the system, which gives the optimum solution. This method may lead to solutions far away from the optimum as the method strongly depends on the intuition of the engineer, the peculiarities of the system, the site weather conditions and the selected step size (large step size is usually chosen to minimise the required effort). Additionally, depending on the computer hardware speed and the complexity of the system, each run might need several seconds to several minutes to be performed (30 seconds for the present example).

Table 4. Large hot water system-training data set (sample)

Area (m^2)	Slope (°)	Flow rate (lt/hr-m^2)	Storage tank volume (m^3)	Auxiliary energy (GJ)
20	30	60	1.0	28.91
40	35	54	1.5	14.69
70	35	55	3	5.78
60	30	30	1.0	10.63
40	45	50	1.5	13.80
70	40	60	3	5.99
...	...	...	...	...

Table 5. Large hot water system-validation data set

Area (m^2)	Slope (°)	Flow rate (lt/hr-m^2)	Storage tank volume (m^3)	Auxiliary energy (GJ)
10	25	54	0.5	41.39
20	45	30	1.7	23.87
25	30	50	2.0	20.34
35	35	40	1.5	14.75
15	25	35	1.8	29.57
17	30	30	1.2	27.94

Thus, it is required for the present method not only to be able to find the optimum solution but also to reduce the required time. The range and step size of values considered for the four variable parameters are shown in Table 6. By considering all the cases that could be investigated, according to the step size of each parameter, it can easily be found that 900,000 runs of TRNSYS would be required in order to cover all combinations. This would required about 7500 hours of computer time just to run the program, without considering the time required to input data and of course analyze the output.

Table 6. Range and step size of variable parameters considered in large hot water system

Parameter	Range	Step size
Collector area (m^2)	10-70	1
Collector slope (°)	25-45	1
Mass flow rate (lt/hr-m^2)	30-60	1
Storage tank capacity (m^3)	0.5-3	0.1

The values of the various cost rates used in the present work are shown in Table 7.

Table 7. Cost rates used in the large hot water system

Item	Value
Area independent or fixed cost (C_f)	850 €
Area dependent cost (C_a)	135 €/m^2
Cost due to change of slope from a minimum of 25° (C_{sl})	0.5 €/°
Storage volume dependent cost (C_v)	510 €/m^3
Rate of electricity cost (C_m)	0.12 €/kWh

4.3. Optimization Method Description

In this case a genetic algorithm is used to find the optimum values of collector area, slope, flow rate and storage tank volume, which will maximise the life cycle savings of the system. For this purpose, an accurate correlation of the above variable parameters on the auxiliary energy, which is necessary to cover the load, is required. For this job a feedforward multilayer artificial neural network is used (Kalogirou, 2005).

Using the neural network to model the solar system, the function between the input parameters and the auxiliary energy can be determined. From the auxiliary energy, the life cycle savings can be estimated by applying the detailed economic method described in section 2.2.2. Then if one considers the life cycle savings estimated from the auxiliary energy (calculated by the neural network) as the fitness function, the optimum size of the input parameters yielding the maximum savings can be determined with the aid of a GA.

4.3.1. Feedforward Neural Network

There are various methods that can be used to find the functional relationship between the inputs and outputs, i.e., correlate the collector area, slope, flow rate and storage tank volume with the auxiliary energy. The most popular method to do this is by using multiple regression analysis. However in this work neural networks are used and as will be proved later in this section, neural networks are superior to multiple regression analysis.

The ability to approximate a complicated function and information processing of the neural network completely depends on the weight of the links between the neurons. These weights cannot be pre-determined but are adjusted during the learning or training phase. In this phase, all patterns of a training data set are presented to the artificial neural network. A lot more information on how ANNs operate is shown in chapter 1.

To obtain an accurate ANN model for the whole range of input parameters, a sufficient number of training patterns covering the whole range must be used. In addition, the distribution of the input parameters over the range is decisive for the prediction performance of the ANN model. Additionally, a sufficient number of training patterns on the borders of these ranges must also be included in the training data set. After training, the prediction of the performance of the model is checked by using a validation set which includes patterns, which were not used for the training of the ANN. For the present method, the data were obtained by running TRNSYS for the required range and random combination of input parameters. The amount of data included in the training and testing sets is indicated in the previous section (see Tables 4 and 5).

Various network architectures have been investigated to find the one that could provide the best overall performance. The architecture, that gave the best results and adopted for the present work, is a feedforward neural network with three hidden slabs as shown in Fig. 9. This architecture has been used in a number of engineering problems for modelling and prediction, with very good results (Kalogirou *et al.* 1998; 1999a; 1999b; Kalogirou, 2000; Kalogirou and Panteliou, 2000).

Four neurons were used in the input slab, corresponding to the values of the input parameters, which are the following:

- Collector area (m^2),
- Collector slope (degrees),
- Mass flow rate (lt/hr-m^2), and
- Storage tank volume (m^3).

The output is a single-element vector corresponding to the values of the auxiliary energy required by the system (Q_{aux}) to cover the load (GJ).

The accuracy of the ANN model depends on the choice of the hidden neurons and the activation function. In ANN modelling, typically the activation function is chosen by the designer and then the weighting factors and biases will be adjusted by some learning rule so that the neuron input/output relationship meets some specific goal.

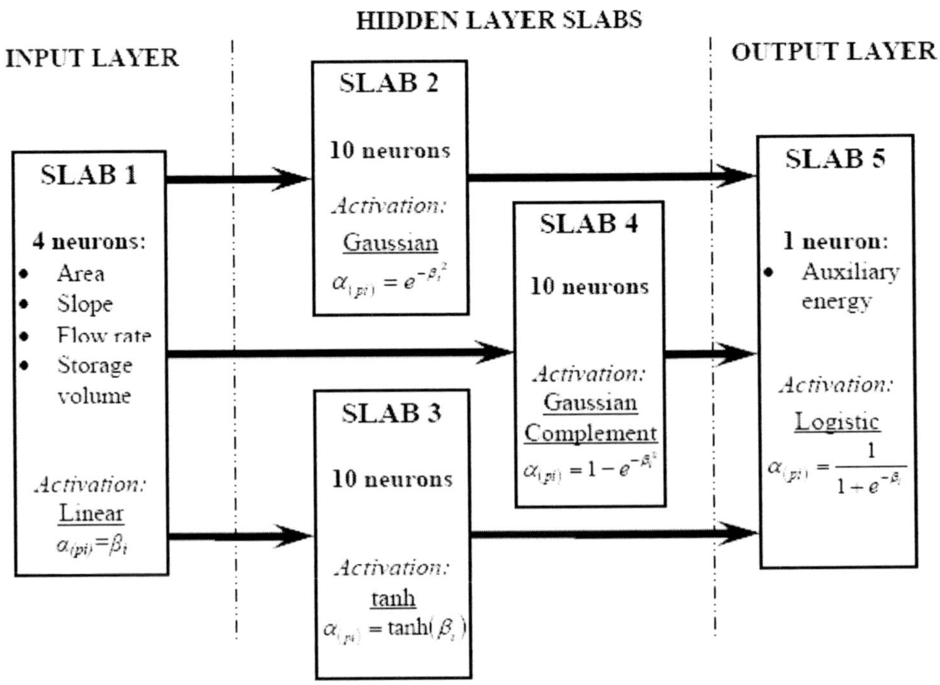

Note: In above slabs α(pi) is the activation for each node and βi is the weighted average obtained by combining all input numerical information from upstream nodes

Fig. 9. Neural network architecture employed for the large hot water system

Different activation functions were applied to the hidden layer slabs in order to detect different features in a pattern processed through the network. The activation functions used in each slab are shown in Fig. 9. The network consists of ten neurons in each hidden slab. All the data required for the neural network need to be scaled from –1 to 1. For this purpose, Eq. (19) was used. More details of this architecture are included in chapter 1.

The training of the ANN was done with the back-propagation learning algorithm, described in chapter 1. The network gain was set to 0.1 and the momentum factor to 0.3. The training was stopped when the average error, obtained by comparing the actual and the ANN modelled data, remained constant for 200,000 events. This is considered a good value enabling the network to learn the input patterns satisfactorily and to give good predictions, while avoiding overtraining. The coefficient of multiple determination (R^2-value) obtained from the training phase is 0.9999, indicating that the mapping was performed at a very satisfactory level. A scattered plot of the actual (modelled) and network predicted data is shown in Fig. 10.

It should be noted that a multiple linear regression method could only produce a correlation with R^2=0.9032 which is not acceptable for the kind of predictions required in this type of problems as will give false values of auxiliary energy which will lead to wrongly estimated LCS that will drive the genetic algorithm to wrong solutions.

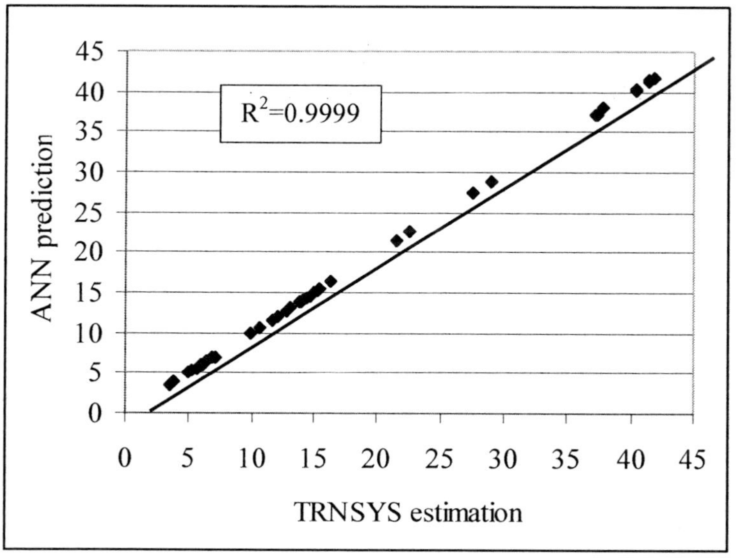

Fig. 10. Comparison of actual (modelled) and ANN predicted data for the large hot water system (training data set)

For the validation of the network the validation data set (see Table 5) is used which is completely unknown to the network. These data are not required for the present optimization method but were produced only in order to validate the neural network. The results are shown graphically in Fig. 11, in which the auxiliary energy predicted by the neural network is compared to that obtained from TRNSYS. As can be seen the performance of the neural network to predict auxiliary energy for the cases examined is excellent with R^2=0.9885. The closer this value is to unity the better is the accuracy of the network and its ability to produce accurate results. It should be noted that the training of the neural network required about 5 minutes on a Pentium 400 MHz machine whereas the prediction for unknown cases is instant.

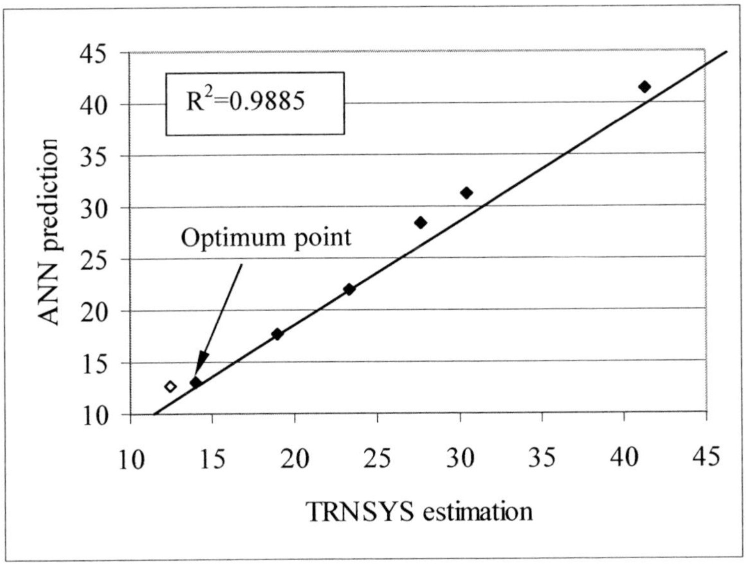

Fig. 11. Comparison of actual (modelled) and ANN predicted data for the large hot water system (validation data set)

4.3.2 Genetic Algorithm

During each step (called a generation) in the reproduction process, the individuals are evaluated by a fitness function value, which is a measure of how well the individual solves the problem. Genetic algorithms are suitable for finding the optimum solution in problems were a fitness function is present. In this case, the genetic algorithm is seeking to breed an individual that maximises the life cycle savings of the solar system.

For this example, the user specified adjustable chromosomes, i.e. the parameters that would be modified during evolution to obtain the maximum value of the fitness function, are; the collector area, slope, flow rate and the storage tank volume. Additionally the user has to specify the ranges of these values called constraints. It is important that the ranges specified to be the same as the extreme cases used when setting up the neural network. In the present work, these constraints or ranges are shown in Table 6. It should be noted that for each combination of collector area, slope, flow rate and storage tank volume, the auxiliary energy and costs are estimated from which LCS are calculated. As the method concentrates very fast towards the optimum solution very fine step sizes could be adopted in this case (steps of 1 m^2 for the collector area, 1° for slope, 1 lt/hr-m^2 for flow rate and 0.1 m^3 for the storage tank) , as shown in Table 6.

The genetic algorithm parameters used in the present work are:

- Population size=100
- Crossover rate=90%

- Mutation rate=1%
- Generation gap=96%
- Chromosome type=continuous

The genetic algorithm was stopped after best fitness remained unchanged for 75 generations.

PARAMETERS:

Collector area	[m^2]	(Adjustable chromosomes)
Collector slope	[°]	(Adjustable chromosomes)
Flow rate	[lt/hr-m^2]	(Adjustable chromosomes)
Storage tank	[m^3]	(Adjustable chromosomes)
Auxiliary energy (Q_{aux})	[GJ] (estimated by ANN)	
Area dependent cost	850 [€]	
Area independent cost	(estimated with Eq. 15 without C_f and values shown in Table 7) [€]	
Price of electricity	0.12 [€/kWh]	
Annual increase in electricity price	4.5 [%]	
Period of economic analysis	20 [years]	
Annual market discount rate	6.5 [%]	
Inflation rate	5.2 [%]	
Maintenance in year 1	1.0 (of total system cost) [%]	
Annual increase in maintenance	1.0 [%]	
Pumping requirement	(estimated with Eq. 16) [kW]	
Operation hours	3500 [hours]	
Initial payment	30 [%]	
Load (Q_{load})	55.3 [GJ]	
First year fuel savings (FYFS)	(Fuel savings are estimated by subtracting Q_{aux} from Q_{load} and estimating savings from the fuel price.) [€]	

DETAILED ECONOMIC ANALYSIS:

1 Year	2 Mortgage payment	3 Maintenance cost	4 Pumping cost	5 Total M&P cost	6 Tax savings	7 Fuel savings	8 Solar savings	9 PW of solar savings
0								
1	*to cover*	*inflated*	*inflated*	*columns*	*estimated*	*FYFS*	*columns*	*estimated*
2	*loan*			*2+3*	*according*	*inflated*	*6+7-2-5*	*with Eq. 12*
....					*to tax law*			
19								
20								
							Sum:	

(Fitness function)

Fig. 12. Representation of the spreadsheet used for the optimization exercise for the large hot water system

4.4. Optimization Method

The relations of the economic analysis presented in section 2.2.2 are used together with other characteristics and economic parameters of the system in combination with a genetic algorithm program to find the optimum variable parameters that maximise the LCS of the system. The variable parameters, i.e., collector area, slope, flow rate and storage tank size are the adjustable chromosomes. The whole model was set in a spreadsheet program in which the various parameters and equations are entered into different cells (see Fig. 12).

The trained neural network can also be called and evaluated within the spreadsheet. Evaluation of a trained network is generally the process of feeding an array of inputs to the trained network and receiving back the appropriate output, (Q_{aux}). As can be seen from Fig. 12, the adjustable chromosomes are entered in different cells and the fitness function is the cell that contains the equation giving the LCS, i.e., the sum of the present worth of the solar savings over all years. During the execution of the program, the genetic algorithm tries to find a combination of the adjustable chromosomes, i.e., a combination of the variable parameters that maximise the fitness function. For each combination of chromosomes the spreadsheet is recalculated automatically, i.e., when a new set of parameters is used the ANN is evaluated to give a new value of Q_{aux} from which all the economic parameters are calculated as shown in Fig. 12. It should be noted that, as in the previous example, the spreadsheet model required needs to be constructed only once. The only changes required for different problems would be to modify the cell with the ANN call function to refer to the trained network of a particular problem and any new cost figures of a particular problem. It should be noted however, that for each problem a new neural network need to be set-up and trained.

4.5. Optimum Large Hot Water System

The results of the optimization of the system considered are presented in this section. Figure 13 shows the fitness function against the number of generations during the running of the genetic algorithm program. The shape of the graph and the speed of reaching the optimum solution strongly depend on the initial conditions specified by the user at the beginning. Two such cases are plotted in Fig. 13.

The results plotted with the thick curve are for an initial collector area of 30 m^2, slope of 25°, flow rate of 50 lt/hr-m^2 and initial storage tank volume of 2 m^3 whereas for the thin curve are for an initial collector area of 70 m^2, slope of 25°, flow rate of 54 lt/hr-m^2 and initial storage tank volume of 3 m^3. As can be seen irrespective of the initial conditions used the same result is obtained (€5918). However, in the first case values close to the best fitness are obtained at about the 120th generation while in the second at the 240th generation. The optimum solution is reached in about 130 seconds in the first case and 250 seconds in the second, which are very fast. As can be seen however, after about 25 generations, values near the optimum one have been obtained in both cases.

The results of the optimization exercise are shown in Table 8 where the actual results as obtained from the genetic algorithm program are presented together with the results of a more practical solution (rounding of numbers) and the traditional trial and error method. As can be seen, higher LCS were obtained by the genetic algorithm method than the ones obtained from the traditional method. The percentage difference between the LCS obtained from the practical optimum solution and those obtained by the traditional method is equal to 15.8%. Of course, the magnitude of the difference between the obtained solution and the one reached from the traditional trial and error method and thus the magnitude of the percentage difference depends on how close an optimum solution obtained from the genetic algorithm is to a step size value chosen for the traditional method. It should be noted that large step sizes are usually employed to reduce the effort required. The present difference is obtained by considering a solution, which is very near to the optimum one except for the case of flow rate, for which the standard value of 54 lt/hr-m^2 that is usually used from solar engineers is considered. The optimum solution obtained for this parameter is for a flow rate of 30 lt/hr-m^2. This low value is due to the cost of electricity required to drive the solar pump (parasitic cost) which is generally expensive, therefore it is proved that it is preferable to use a smaller flow rate, which consumes less electricity, as apparently the extra solar energy collected at the increased flow rates does not pay off for the electricity consumed.

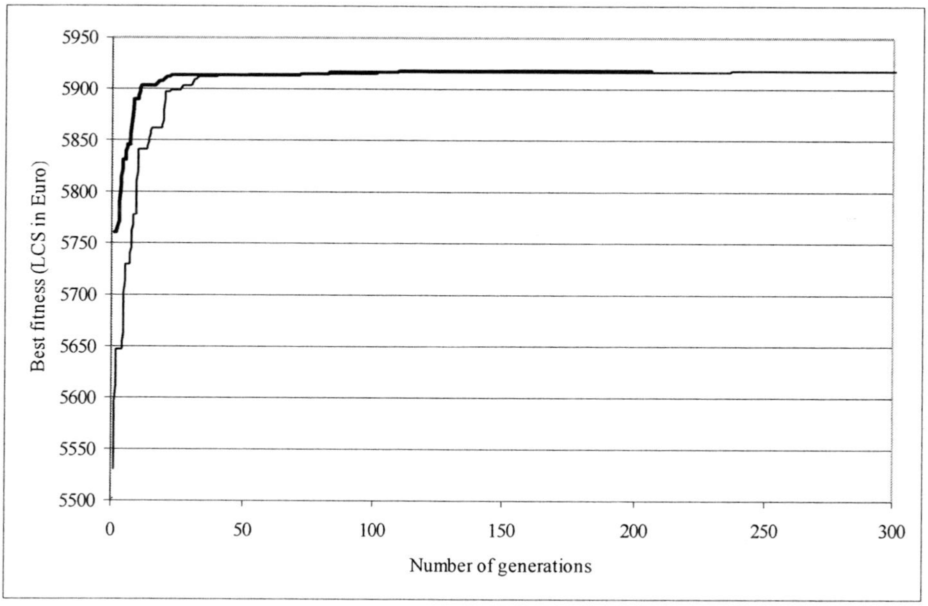

Fig. 13. Plot of best fitness against the number of generations for the large hot water system for two initial conditions

The ability of the ANN to predict auxiliary energy accurately, compared to the value obtained from TRNSYS for the optimum case, is also shown on Fig. 11. This was investigated in order to evaluate the prediction accuracy of the neural network for the optimum case. The percentage difference between the ANN predicted value and that of TRNSYS for the optimum value of Q_{aux} is 1.8%, which is very satisfactory. The run time of the genetic algorithm program required 130 or 250 seconds (depending on the initial conditions used), to be performed whereas each run of TRNSYS requires about 30 seconds, both on a Pentium 400 MHz machine.

Table 8. Large hot water system-results of the solar system optimization

Parameter	Optimum system obtained from GA	Practical GA solution	Traditional method
Area (m^2)	36	36	35
Slope (°)	39	40	40
Flow rate (lt/hr-m^2)	30	30	54
Volume (m^3)	1.79	1.8	2.0
LCS (€)	5918	5915	4980

As was seen before the training of the ANN required 5 minutes on the same machine. It should be noted however, that the ANN and GA need to run only once whereas TRNSYS need to run many thousands of times in order to try all possible combinations. Therefore, the overall time required is reduced substantially. Additionally, the time specified for the run of TRNSYS program is actual computer time, without considering the time required to modify the input data and to analyze the results in order to find the optimum solution. TRNSYS has a function to run many simulations, one after the other, for values of variable parameters included in a "parametric analysis table". However, even if this function is used in TRNSYS to run groups of variations of the input data automatically, the analysis of the results obtained in order to find the optimum solution still needs to be performed manually. Thus, by reducing the number of runs of TRNSYS (48 runs in total) the time required to find the optimum solution is greatly reduced. Additionally, the solution reached is more correct than the one obtained by using the traditional trial and error method, which in most of the times relies on the intuition of the engineer to find a good solution.

5. CONCLUSIONS

The method suggested in this chapter is using two artificial intelligence techniques, i.e., artificial neural networks and genetic algorithms, to find the optimum combination of various collector system variable parameters, which either minimise the NSEP or maximise the LCS of a solar installation. Two systems were considered here to demonstrate the method; an industrial process heat and a large hot water system. Initially the solar system needs to be modelled with TRNSYS in order to obtain a small number of simulation results, which are used to train the neural network. Two types of neural networks, have been used here (GMDH and a feedforward one) to model the systems examined. However, any other suitable ANN can be used for this purpose. In both cases, the systems were modelled with high accuracy, and the R^2-value obtained equal to 0.9986 and 0.9999 for the two systems respectively. The prediction accuracy for unknown data sets was also very high for both systems. Subsequently a genetic algorithm was used to select the combination of the system parameters, which in the first case minimise the net solar energy price (NSEP) and in the second maximise the life cycle savings (LCS) of the system. Both the ANN (when trained) and the GA require just few seconds to provide the optimum results. It is thus believed that the present method would decrease the time required by design engineers to find the optimum solution and in many cases to obtain a solution that could not be easily obtained by the traditional trial and error method, or by simple modelling programs, which by definition cannot be very accurate. Additionally, the trial and error method in most of the cases depends on the intuition of the engineer. Moreover, the method presented here can be applied to any solar energy system irrespective of its size or configuration.

The advantages of the present method are the following (Kalogirou, 2005):

- The method does not suffer from limitations of other simple programs, such as the use of standard system configurations only, assumption that systems are well built and use of average meteorological data.

- Other programs rely on the accuracy of the model. Here a proven model is used (TRNSYS) which is considered as one of the most accurate and if experimental data were available, which can easily be obtained for some types of systems where a range of applications already exists, then the model accuracy is completely eliminated.
- The weather conditions are embedded into the input data used to train the ANN model. Therefore, no pre-processing or further analysis is required.
- The optimum system is given directly, i.e., there is no need to analyze TRNSYS output data in order to find the optimum system characteristics.
- The method can be used for the design and optimization of new, novel or more complicated systems, which until now have not been investigated in detail.

The latter one is the biggest advantage of the present method.

Finally, it is believed that other parameters of a solar or any other energy system could be optimized using the method suggested here. This could only be possible however, if a relationship (fitness function) between the optimization parameter/s and the system characteristics can be derived (from a successfully trained ANN) which can be minimised or maximised using a genetic algorithm.

References

Brandemuehl M.J., Beckman W.A., 1979. Economic evaluation and optimization of solar heating systems. *Solar Energy*, Vol. 23, pp. 1-10.

Collares-Pereira M., Rabl A., 1979. Derivation of method for predicting long term average energy delivery of solar collectors. *Solar Energy*, Vol. 23, pp. 223.

Curtiss P.S., Brandemuehl M.J., Kreider J.F., 1995. Energy management in central HVAC plants using neural networks', In Haberl J.S., Nelson R.M. and Culp C.C. (Eds.). *The use of artificial intelligence in building systems.* ASHRAE, pp. 199 – 216.

Gen M., Cheng R., 1997. *Genetic Algorithms and Engineering Design*. London, Wiley.

Hobson P.A., Norton B., 1989. A design nomogram for direct thermosyphon solar-energy water heaters. *Solar Energy*, Vol. 43, No. 2, pp. 85-95.

Hottel H.C., Whillier A., 1958. Evaluation of flat-plate solar collector performance, *Transactions of the conference on the use of solar energy,* II. Thermal processes, pp. 74-104, University of Arizona, Tempe, Arizona.

Kah A.H., San Q.Y., Guan S.C., Kiat W.C., Koh Y.C., 1995. Smart air-conditioning system using multilayer perceptron neural network with a modular approach. *Proceedings of the IEEE International Conference ICNN'95*, Perth, Western Australia, Vol. 5, pp. 2314-2319.

Kalogirou S.A., 1996. Economic analysis of solar energy systems using spreadsheets. *Proceedings of the World Renewable Energy Congress IV,* Denver, Colorado, USA, Vol. 2, pp. 1303-1307.

Kalogirou S.A., Neocleous C., Schizas C., 1998. Artificial neural networks for modelling the starting-up of a solar steam generator. *Applied Energy,* Vol. 60, No. 2, pp. 89-100.

Kalogirou S.A., Panteliou S., Dentsoras A., 1999a. Artificial neural networks used for the performance prediction of a thermosyphon solar water heater. *Renewable Energy,* Vol. 18, No. 1, pp. 87-99.

Kalogirou S.A., Panteliou A.D., Dentsoras A., 1999b. Modelling of solar domestic water heating systems using artificial neural networks. *Solar Energy,* Vol. 65, No. 6, pp. 335-342.

Kalogirou S.A., 2000. Long-term performance prediction of forced circulation solar domestic water heating systems using artificial neural networks. *Applied Energy,* Vol. 66, No. 1, pp. 63-74.

Kalogirou S.A., Bojic M., 2000. Artificial neural networks for the prediction of the energy consumption of a passive solar building. *Energy-The International Journal*, Vol. 25, No. 5, pp. 479-491.

Kalogirou S.A., Panteliou S., 2000. Thermosyphon solar domestic water heating systems long-term performance prediction using artificial neural networks. *Solar Energy*, Vol. 69, No. 2, pp. 163-174.

Kalogirou S.A., 2002. Use of artificial intelligence for the optimisation of solar systems, *International Journal of Renewable Energy Engineering*, Vol. 4, No. 3, pp. 499-505.

Kalogirou S.A., 2003a. Use of genetic algorithms for the optimal design of flat plate solar collectors. *Proceedings of the ISES 2003 Solar World Congress on CD-ROM*, Goteborg, Sweden.

Kalogirou S.A, 2003b. Generation of typical meteorological year (TMY-2) for Nicosia, Cyprus, *Renewable Energy,* Vol. 28, No. 15, pp. 2317-2334.

Kalogirou S.A., 2004. Optimisation of solar systems using artificial neural networks and genetic algorithms, *Applied Energy*, Vol. 77, No. 4, pp. 383-405.

Kalogirou S.A., 2005. Use of artificial intelligence for the optimal design of solar systems, *International Journal of Computer Applications in Technology,* Special Issue on *Intelligent systems for intelligent energy in the new millennium*, Vol. 22, Nos. 2/3, pp. 90-103.

Klein S.A., Beckman W.A., Duffie J.A., 1976. A design procedure for solar heating systems. *Solar Energy*, Vol. 18, pp. 113-127.

Klein S.A., 1978. Calculation of flat-plate collector utilizability. *Solar Energy*, Vol. 21, pp. 393.

Klein S.A., Beckman W.A., 1981. *F-Chart User's Manual*, University of Wisconsin.

Kreider J.F., Wang X.A., 1995. Artificial neural network demonstration for automated generation of energy use predictors for commercial buildings. In Haberl J.S., Nelson R.M. and Culp C.C. (Eds.). *The use of artificial intelligence in building systems.* ASHRAE. pp. 193-198.

Liu B.Y.H., Jordan R.C., 1963. The long-term average performance of flat-plate solar energy collectors. *Solar Energy*, Vol. 7, No. 2, pp. 53.

Polysun, 2000. *User's Manual for Polysun 3.3*, SPF, Switzerland.

TRNSYS version 16 program manual, 2005. *Solar Energy Laboratory*, University of Wisconsin, Madison, USA.

Watsun users manual and program documentation, 1992. *Watsun Simulation Laboratory*, University of Waterloo, Canada.

Zhou H., Qian X., Cen K., Fan J., 2003. Optimizing pulverized coal combustion performance based on ANN and GA. *Fuel Processing Technology*, Vol. 85, No. 2-3, pp. 113-124.

In: Artificial Intelligence in Energy and Renewable Energy Systems ISBN 1-60021-261-1
Editor: Soteris Kalogirou, pp. 163-200

Chapter 5

ARTIFICIAL NEURAL NETWORKS APPLIED IN PV SYSTEMS AND SOLAR RADIATION

Leocadio Hontoria*, Jorge Aguilera*, Florencia Almonacid*, Gustavo Nofuentes* and Pedro Zufiria*

*Research Group "IDEA"
Department of Electronics Engineering, Polytechnics School.
Jaén. University of Jaén, Jaén, Spain
**Research Group "Neural Networks"
Department of Applied Mathematics to Information Technologies
Telecommunications School. Polytechnics University of Madrid
Madrid, Spain

ABSTRACT

An artificial neural network (ANN) is an information-processing system that has certain performance characteristics in common with biological neural networks. Artificial neural networks have been developed as generalisations of mathematical models, of human cognition or neural biology. During the last three decades, artificial neural networks have been extensively employed in numerous fields of science and technology. For instance they have been used in signal processing, medicine, pattern recognition, robotics, control, forecasting, speech production, speech recognition, business, manufacturing, power systems and also in the renewable energy and solar energy fields. As a computation and learning paradigm, they are proposed as an alternative approach for addressing complex problems.

This chapter consists of two parts. In the first part a bibliographic review of how artificial neural networks have been used in the renewable energy field in general and more specifically in the solar energy field is presented. This review includes also applications of artificial neural networks in the photovoltaic (PV) and in the solar radiation fields. Following this review, the chapter is centred on the research results obtained by the authors on using artificial neural networks, and more particularly the Multilayer Perceptron (MLP), in the solar radiation and in the photovoltaic fields. Among the many types of networks, supervised models have consolidated as the most robust and easy to employ when applicable. Usually, these models are implemented via feedforward architectures such as, the Multi-Layer Perceptron. The Multi-Layer Perceptron is the most widely used type of supervised neural network employed for approximation tasks. One of the most appealing properties of these neural network architectures is their potential use for function approximation, which is due to their universal approximation capabilities. The use of the MLP in the photovoltaic and solar radiation fields has been carried out by the authors and is giving very satisfactory results. The authors have used the MLP for generating the long-term solar radiation series needed in the design of photovoltaic systems. The MLP was also used as a very useful tool for generating Loss of Load Probability (LOLP) curves for stand-alone photovoltaic systems. Finally, the MLP was used to extrapolate V-I curves of PV modules recorded at non-Standard Test Conditions to Standard Test Conditions V-I curves. In all the above applications, the problems were solved satisfactorily with a very simple structure for the MLP.

Keywords: Solar radiation, photovoltaic systems, artificial neural networks, multilayer perceptron

NOMENCLATURE

CA	Generator Capacity
CS	Accumulator Capacity
G	Irradiance (W/m^2)
H_d	Daily solar irradiation (Wh/m^2)
H_{dm}	Monthly mean daily solar irradiation (Wh/m^2)
I	Current (A)
K_T	Daily clarity index
k_t	Hourly clarity index
k_{tm}	The trend component of the hourly clarity index
k_{t-p}	Predicted value of k_t
k_{t-ave}	Average value of k_t
N	Number of samples
P_{Oi}	Measured power of PV module (W)
P_{Ti}	Theoretical power of PV module (W)
R	Gaussian random number
T_{am}	Ambient temperature (°C)
V	Voltage (v)

Greek

α	The random component of the hourly clarity index
σ_α	Standard deviation of α for a daily clarity index K_T
ϕ	Autocorrelation coefficient

ABBREVIATIONS

AC	Aguiar and Collares-Pereira
ANN	Artificial Neural Network
ARMA	Autoregressive Moving Average
GH	Graham and Hollands
IES	Instituto Energía Solar (Solar Energy Research Centre)
LOLP	Loss of Load Probability
MBE	Mean Bias Error
MLP	MultiLayer Perceptron
MMV	Mean Monthly Value
MRV	Mean Relative Variance
NN	Neural Network
PV	PhotoVoltaics
QE	Quadratic Error
RMSE	Root Mean Square Error
STC	Standard Test Conditions
TSP	Time Series Prediction

1. INTRODUCTION

A literature review on the application of artificial neural networks (ANN) (Agarwal, 1997; Fausett, 1994; Haykin, 1994; Hush and Horne, 1993; Kohonen, 1995; Lippmann, 1987) in renewable energy, solar radiation and photovoltaics is presented in this section.

1.1. Renewable Energy

Artificial neural networks (ANN) have been used for modelling and prediction of renewable energy systems. A comprehensive review of how artificial neural networks have been applied in the renewable energy field was carried out by Soteris A. Kalogirou (Kalogirou, 2000; 2001). There are a lot of examples of applications employing neural networks in solving different problems in this field. It is not the intention of this chapter to make a literature review of these applications. The literature review is centred on the use of artificial neural networks in the solar radiation and photovoltaic fields. Nevertheless some examples of how artificial neural networks have been used to solve different problems in the renewable energy field are presented. For instance, ANNs have been used for modelling a solar steam generator (Kalogirou, 2001). In this case ANNs were employed to model different aspects of the solar steam generator like the estimation of the collector intercept factor or the modelling of the starting – up of the generator. Also, in the same research field, ANNs have been applied for solar water heating systems (Kalogirou *et al.*, 1996), in several aspects like, the modelling of solar domestic water heating systems or the prediction of the long-term performance of a thermosiphon solar domestic water heating system. The identification of the time parameters of solar collectors using ANNs with radial basis function was carried out by

Lalot (2000). Kreider and Wang (1995) used ANNs in commercial buildings to predict energy use of chillers. Curtiss *et al.* (1995) did a similar work for the energy consumption optimisation in a commercial building and Kajl *et al.* (1997) using fuzzy-neural assistant. The estimation of heating loads of buildings (Kalogirou *et al.*, 1997) and the prediction of airflow in a naturally ventilated test room (Kalogirou *et al.*, 1999) are two other applications of ANNs in the building energy field.

1.2. Solar Radiation

The knowledge of the magnitude and strength of solar radiation for photovoltaic applications is evident. One of the preliminary data necessary for the design of photovoltaic systems is the knowledge of long sequences of solar radiation, mainly *hourly* values. Finding records of long sequences of hourly solar radiation is rather difficult. There are two main approaches which can be used for this purpose. One may be called the classical approach that is based on traditional methods, which apply statistical properties of solar radiation series. The other one, called the neural network approach, is based on the use of artificial neural networks.

In the last decades and as part of the classical approach, many radiation models and several generators of synthetic solar radiation series have been proposed. The first attempts for generating solar radiation series were based on the use of radiation of a specific site as the variable to model. For instance Goh and Tan (1977) used ARMA (Autoregressive Moving Average) models for Singapore, Hittle and Pedersen (1981) using Fourier analysis, and Baloutktsis and Tsaides (1986) used the same analysis for Greece. Additional work in this area has also been performed in the 80's, by some research groups, for instance, the Solar Energy and Building Physics Laboratory (LESO-PB/EPFL) (Bochud *et al.*, 1991). Nevertheless using solar radiation as the variable to model is problematic as it is specific for each location so those models are only applicable locally.

A second attempt was made in which the variable to model was the atmospheric transmittance or clarity index. Liu and Jordan (1960) demonstrated that the probability features of this new variable are quasi-universal. The atmospheric transmittance or clarity index is defined as the quotient between the global irradiation and the extraterrestrial irradiation. Using the clarity index Dagleman (1976) made a first study about hourly series using distribution clarity index curves. Using this approach Excell (1981) produced a model of Thailand data. Mustachi *et al.* (1979) used an ARMA model. But the models more frequently used are those proposed by Aguiar and Collares-Pereira (1992), and Graham and Hollands (Graham *et al.*, 1988; Graham and Hollands, 1990).

Alternative approaches for generating solar radiation series are developed using artificial neural networks. It is worth mentioning the works of Negnevitsky and Le (1995) who applied an ANN to predict hourly solar radiation data which where applied for the evaluation of current rating of overhead lines. Alawi and Hinai (1998) applied an ANN approach for the prediction of global radiation. Mohandes *et al.* (1998) made a similar work with solar radiation data of Saudi Arabia. Also Kemouku *et al.* (1999) and Sfetsos and Coonick (2001) made similar applications of ANN in the solar radiation field. Additionally, Sözen *et al.* (2005) used ANNs to forecast the solar potential of Turkey.

1.3. Photovoltaic Systems

In the photovoltaic field artificial neural networks have been applied in several cases. In order to identify the optimal operating point of a photovoltaic system Hiyama (1995) and Hiyama *et al.* (1995) applied ANNs. Takashi and Imakubo (1995) also published a similar work and Veerachary and Yadaiah (2000) have carried out similar applications. For the estimation of the maximum power and normal operating power of a photovoltaic module, Baghat *et al.* (2004; 2005) have used successfully ANNs. Also very recently Mellit *et al.* (2005) used ANNs for sizing stand-alone PV systems in Algeria.

The authors of this chapter have also use the MLP as a very useful tool for generating Loss of Load Probability (LOLP) curves for stand-alone photovoltaic systems (Hontoria *et al.*, 2003; 2005a). They have also used the MLP in order to extrapolate V-I curves of PV modules recorded at non-Standard Test Conditions to Standard Test Conditions V-I curves (Hontoria *et al.*, 2005b).

2. Artificial Neural Networks for Solar Radiation Prediction

2.1. Introduction

In this section the use artificial neural networks (the multilayer perceptron in particular) to generate solar radiation series and to draw solar radiation maps is described. One of the first problems to solve in the design of solar systems (photovoltaic or solar thermal ones) is the search of long sequences of solar radiation, mainly *hourly* values. In the design of any solar system it is necessary to know the solar radiation series in order to simulate the environmental conditions in which the system will work. Nevertheless, in most locations these series are non-existent or even where available, their quality is very poor (Gansler *et al.*, 1994). Finding records of long sequences of hourly solar radiation is rather difficult.

In order to solve this problem, many researchers have demonstrated that it is possible to replace real series of solar radiation by synthetic ones. These synthetic series are generated by various methods. The statistical properties of these new series have to be the same as the real ones so both series must be indistinguishable (Aguiar and Collares-Pereira., 1991).

The models proposed by Aguiar and Collares-Pereira (1992), and Graham and Hollands (Graham *et al.*, 1988; Graham and Hollands, 1990), referred from now on as AC and GH respectively for short, may be considered as paradigms in the field of hourly radiation modelling. They are auto-regressive first-order models, based on a stochastic disaggregation methodology that generates hourly irradiation series by making use of daily values. These values can be obtained from historic daily records (which are more common than hourly records) or by using daily generation methods in turn (which are more validated than hourly methods). Complex empirical expressions are proposed to relate the hourly and daily values, obtaining the parameters in the formulae via regression analysis over historical data.

In both models the construction of hourly solar radiation series values is based on the decomposition of the hourly clarity index into two components; one is called the trend or mean component and the other is called the random component. The trend component follows the evolution of the clarity index along the day and the random component consists of random fluctuations over the trend component. This can be expressed by this equation:

$$k_t = k_{tm} + \alpha \tag{1}$$

where k_{tm} is the trend component and α the random component of the hourly clarity index k_t.

The clarity index is the ratio between the global radiation, i.e., the solar radiation received over a surface on earth and the extraterrestrial solar radiation, the solar radiation received outside the atmosphere.

When the day is clear the trend component is more influential. But in cloudy days the influence of the random component is greater. For generating the trend component both models are very similar. Both AC and GH proposed some equations which depend on the daily clarity index. For generating the random component the two models are quite different. GH used an ARMA model and Gaussian inversion. AC used a time dependent autoregressive Gaussian model. This is what is called the *classical approach* or *classical methodology*. Nevertheless, a new approach is proposed to generate solar radiation series with the *MultiLayer Perceptron* (MLP), in feedforward-feedback architecture and its results are compared to those obtained by the AC and GH classical methods. This new approach will be called *neural network (NN) approach* or *neural network methodology* in contrast with the classical approach. The neural network approach is based on the MLP capability for approximating any continuous function defined on a compact set within a prescribed error margin (Barron, 1993; Funahashi, 1989; Cybenko, 1989; Hornik *et al.* 1989). It was proved that it is enough to employ a MLP with a hidden layer, a required number of neurons and an appropriate training procedure. In practice, selection of appropriate topology as well as training algorithms may become a big challenge. It was proved that the NN approach improves previous results where an *a priori* knowledge of the initial values of the series to be generated was assumed.

Another important aspect addressed with the NN approach is the possibility of employing the presented architecture with a reduced knowledge of the problem to be considered. In that sense the work presented here defines a simple design methodology with quite general applicability. In consequence we have applied the neural network methodology not only for the generation of hourly solar radiation series but also to draw solar radiation maps which is a very important application of the present methodology, especially in selecting appropriate locations for PV installations. In the following section the MLP which was designed to obtain the trend component of the clarity index is described and the MLP used in order to get the random component is shown together with some results. Finally explanations are given on how the NN approach has been used to draw solar radiation maps.

2.2. Study of the Trend Component

For the generation of the trend component in the classical approach, AC, GH and some other authors proposed equations which are dependent on the daily clarity index. This idea was used in the present approach combined with the properties of the MLP.

The MLP, in static contexts, is usually trained via a supervised procedure, i.e., the backpropagation algorithm (Rumelhart and MacClelland, 1986). This is one of its greatest advantages as backpropagation is a very efficient training method. Also, in dynamic contexts, the Time Series Prediction (TSP) has been successfully addressed via feedback architectures of supervised neural models (Narendra and Parthasarathy, 1991; Weigend *et al.*, 1990; Zufiria, 1995). The problem of TSP via MLP makes use of the time series $\{s_n\}$ for obtaining the function G (assuming that such function exists) which relates each series value with the previous p values:

$$s_{n+1} = G(s_{n-p+1},\ldots,s_n) \cong MLP(s_{n-p+1},\ldots,s_n) \qquad (2)$$

By training a MLP with p inputs and 1 output with a representative training set, the MLP will be able to find the desired relationship (in case that it exists) by just approximating the function G (Zufiria, 1995).

2.2.1. Structure of the Multilayer Perceptron (MLP) Proposed

The methodology presented for Times Series Prediction and system identification via MLP defines the framework of the method developed for the generation of hourly clarity index series $\{k_t\}$. For the computational experiments, a set of hourly clarity index k_t values measured in the locations shown in Table 1 and its corresponding daily K_T values have been used. As a first approach, in order to evaluate the quality of a generated series, the first years were considered as a *training set* and the last year was employed for testing the validity of the generated series, in each location.

The proposed method has been developed via a step by step inclusion of the available associated information. The greatest advantage of the MLP-based methodology is that explicit knowledge of the relationship among all the information sources is not needed. Such information sources can be progressively incorporated in different steps upon the proposed method. The details of this step-by-step procedure can be found in (Zufiria *et al.*, 1999). The final procedure carried out by employing a MLP in a mixed feedback-feedforward configuration, is shown in Fig. 1.

The seven inputs required are (Hontoria *et al.*, 1999; Zufiria *et al.*, 1999):

- s: this is an input taking values $\{0,1\}$ that indicates whether an hour is between sunrise and sunset (k_t should be different from 0) or not (k_t has to be 0),
- d_n: distance (days) between the value to be generated and the day with maximum value in the $\{k_t\}$ annual distribution. This is used in order to keep the hourly global irradiation monthly periodicity and seasonal nature,
- h_n: indicates the hour order number of the k_t value (ensuring the hourly global irradiation periodicity and seasonal nature). Both inputs (d_n and h_n) are normalised to the range [0,1],
- K_T : daily clarity index,
- $k_t(h\text{-}3)$, $k_t(h\text{-}2)$, $k_t(h\text{-}1)$: hourly clarity index of the three previous hours

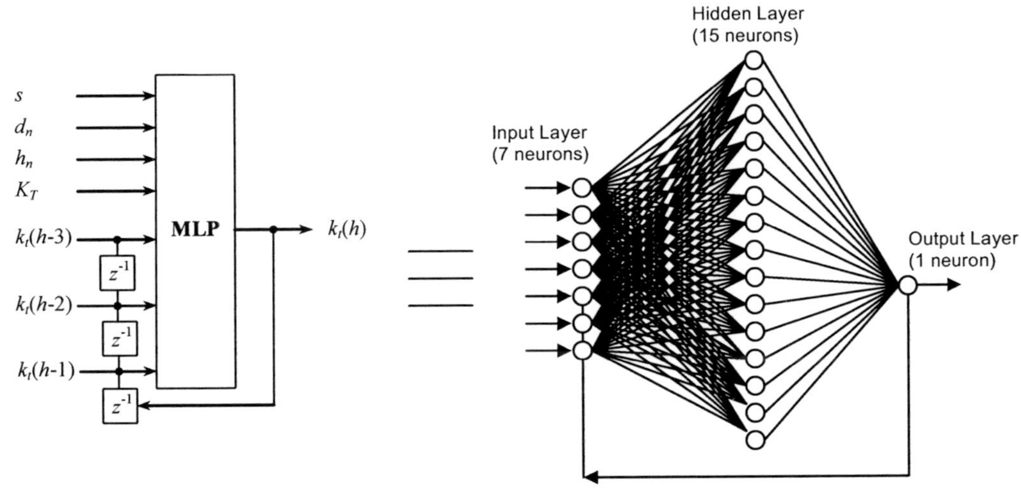

Fig. 1. MLP Architecture for predicting k_{tm} (trend component)

The training was done using the backpropagation algorithm (with momentum and random presentations), which is eventually combined with second order optimisation methods during the last few epochs. The hidden layer has fifteen nodes. For the training of the MLP and for the subsequent validation it is necessary to use real solar radiation series. The locations that have been chosen for this purpose are: Oviedo, Logroño, Madrid, Mallorca, Badajoz, Murcia and Málaga. These are Spanish locations with different latitudes and climates as shown in Table 1, together with the record years. The number in parenthesis in the last column represents the total number of years that data are available.

These locations are spread all over Spain as is shown in Fig. 2.

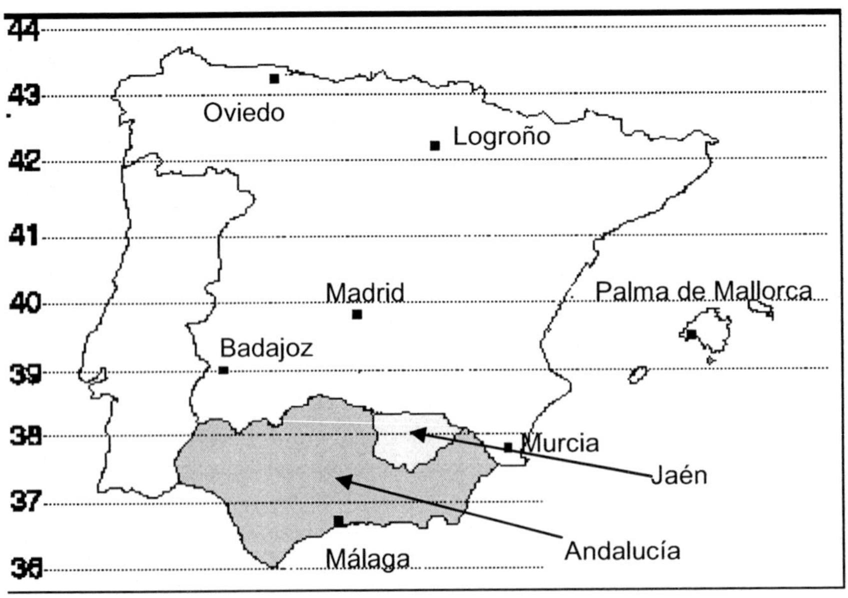

Fig. 2. Map of Spain showing the locations considered in this study and the regions considered in the section 2.5 to draw solar maps

Table 1. Locations considered in the study

Location	Climate	Latitude	Record Years
Oviedo	Atlantic	43.35 °	1977 - 1984 (8)
Logroño	Continental	42.46 °	1981 - 1984 (4)
Madrid	Extreme Continental	40.45 °	1977 - 1984 (8)
Palma de Mallorca	Island Mediterranean	39.33 °	1977 - 1984 (8)
Badajoz	Attenuated Continental	38.89 °	1976 - 1983 (8)
Murcia	Arid Mediterranean	38.00 °	1977 - 1984 (8)
Málaga	Mediterranean	36.66 °	1977 - 1984 (8)

2.2.2. Results for the Trend Component

As has been mentioned before the methodology proposed has been applied by using a set of hourly irradiation values (k_t) measured in the locations considered in this study and its corresponding daily values (K_T). The first years were considered as a *training set*, and the last year was employed for testing the validity of the generated series, in each location and also for comparisons with other methods.

Figures 3 and 4 present a comparison for some days of actual or real data with those generated by the MLP. As can be seen, the trend component follows perfectly the real data. Also, as it will be shown in the comparative study with other methods the results are quite satisfactory.

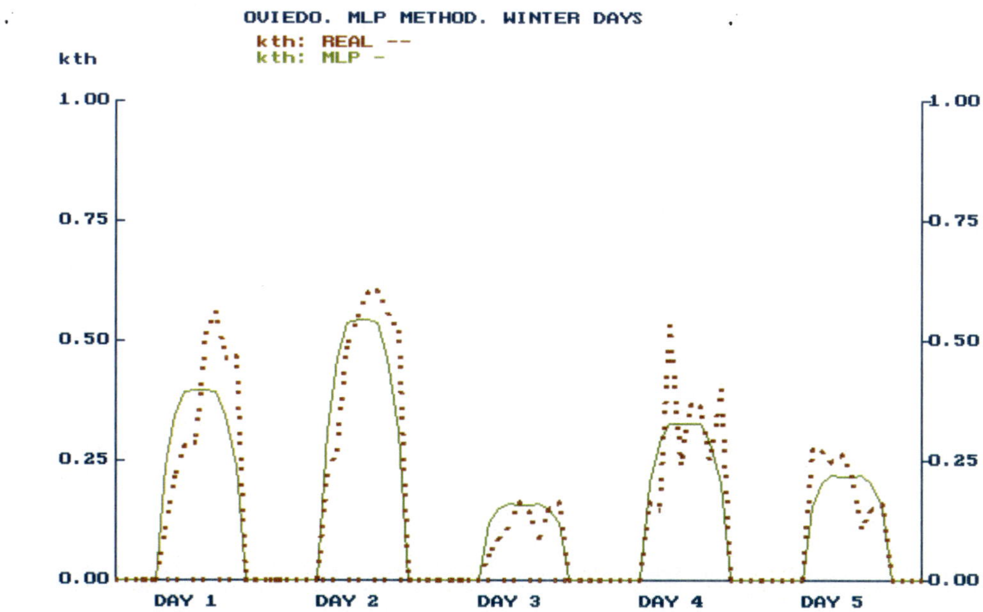

Fig. 3. Comparison between the real and MLP predicted clarity index values for Oviedo (Days 1-5, winter)

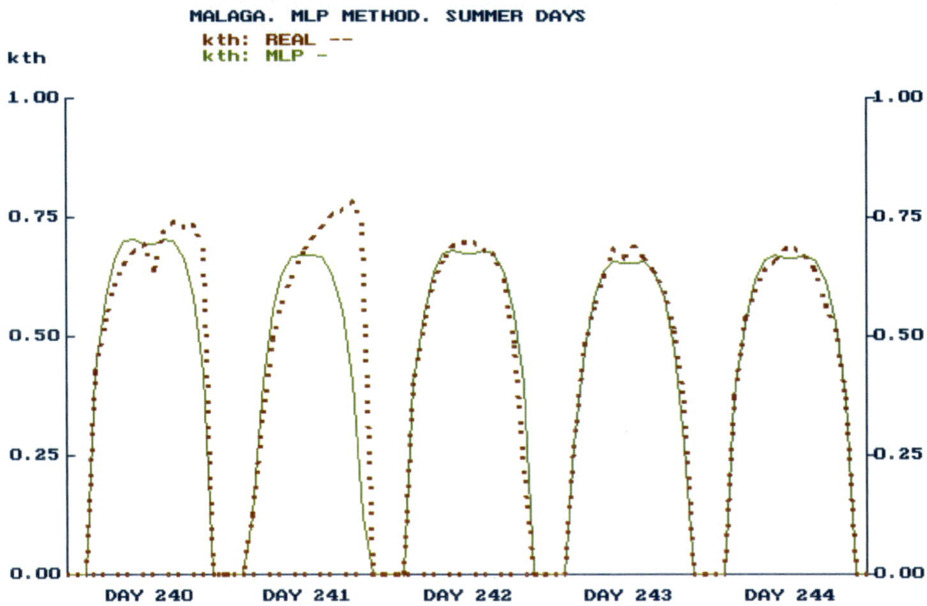

Fig. 4. Comparison between the real and MLP predicted clarity index values for Málaga (Days 240-244, summer)

2.3. Study of the Random Component

As has been mentioned before, the hourly clarity index can be expressed by Eq. (1), where the two components of k_t can be observed. One is the trend component, k_{tm}, which has just been studied and presented in section 2.2. The other is the random component, α. According to AC, the random component, α, can be generated following an ARMA model:

$$y(h) = \phi \cdot y(h-1) + r \tag{3}$$

where :

ϕ is the autocorrelation coefficient. This coefficient is dependent of the daily clarity index K_T,

R is a random number, which follows a Gaussian distribution, with null average and standard deviation $\sigma' = (1-\phi^2)^{0.5}$.

y(h) is the variable k_t (h) normalised at hour h. The normalisation is done by using the following equation:

$$y(h) = \frac{\alpha}{\sigma_\alpha(K_t, h)} \tag{4}$$

It should be remembered that $\alpha = [k_t(h) - k_{tm}(K_t,h)]$.

In Eq. (4), σ_α is the standard deviation of α for a daily clarity index K_T. σ_α is function of the hour and K_T. The parameter y(h-1) is the same as y(h) but for the previous hour. Therefore, once the series {y} is calculated it is easy to obtain the series $\{k_t\}$.

Again, using the properties of the MLP, a similar way for obtaining the random component as that of AC just described is used. Nevertheless there are some differences. For calculating both the autocorrelation coefficient (ϕ) and the standard deviation of α (σ_α), MLPs are used. The structures of those new MLPs are shown in Figs 5 and 6.

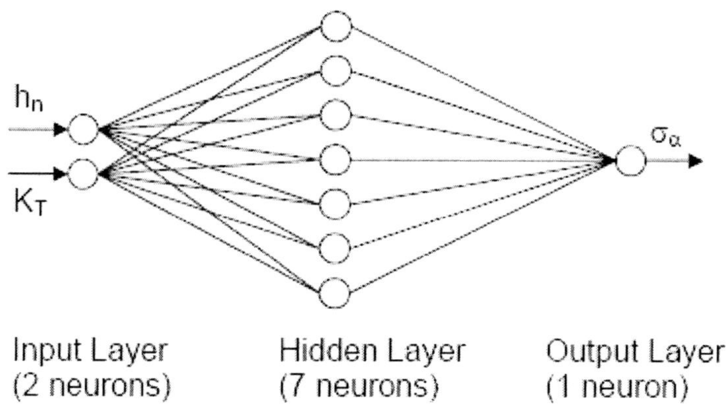

Fig. 5. MLP for the standard deviation σ_α

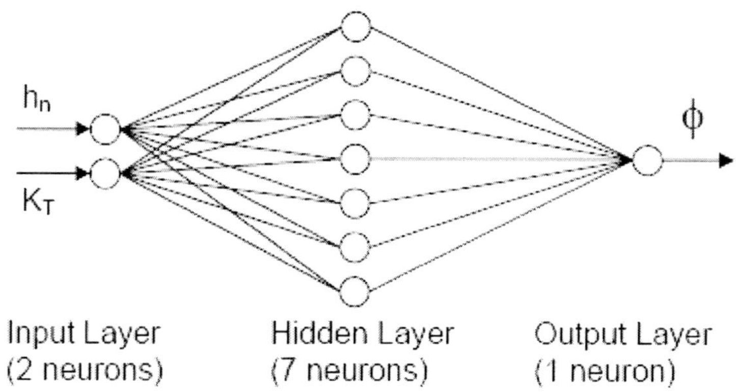

Fig. 6. MLP for autocorrelation coefficient ϕ

Both MLPs have a hidden layer with seven neurons. In this case the training of the MLP is the same as that of the MLP for the trend component.

2.4. Comparative Study with Classical Methods

2.4.1. Variables Considered

In order to validate the methodology proposed and test the synthetic data obtained a comparative study between actual and ANN predicted data has been performed. The variables taken into account are the Monthly Mean Value (MMV) and the Mean Relative Variance (MRV).

The first statistical feature studied to characterise solar radiation is the Monthly Mean Value which is defined as the mean of all the values in a *month*, using daily values (H_{dm}). The expression employed for this purpose is:

$$H_{dm} = \sum_d \frac{H_d}{n} \tag{5}$$

where n the number of days of the month.

The other selected parameter taken into account is the Mean Relative Variance (MRV). As a distance measure, this parameter commonly employed in the digital signal processing community, quantifies the relative error and it is defined as the quotient between the prediction error signal power and the power of the signal to be predicted (a value closer to zero indicates a very good prediction). The expression employed for this purpose is:

$$MRV = \frac{\sum_h \left(k_t - k_{t-p}\right)^2}{\sum_h \left(k_t - k_{t-ave}\right)^2} \tag{6}$$

where:

k_{t-p} is the predicted k_t value, and
k_{t-ave} is the average of k_t.

The parameter MRV is proportional to the standard deviation.

A synthetic year has been generated by the MLP method, and by the AC and GH methods for each location. The three synthetic years have been compared with the actual year. The results can be seen in Table 2.

As can be seen in Table 2, the values of the annual irradiation of the synthetic year obtained by the MLP method are closer to the actual data than the values of the years obtained by the other two methods. Also the neural method proposed gives the best (lower) MRV parameter.

Table 2. Comparisons between test year and the synthetic years generated by the three methods for the various cities

Location	Annual H_{dm} (kJm^{-2})				MRV		
	Actual	MLP	GH	AC	MLP	AC	GH
Oviedo	11.26	11.30	11.25	11.49	0.3476	0.4128	0.4527
Logroño	14.60	14.65	14.71	14.96	0.3082	0.3420	0.3913
Madrid	16.32	16.34	16.46	16.70	0.2307	0.3024	0.3274
Palma de Mallorca	15.58	15.60	15.76	16.03	0.2398	0.3261	0.3352
Badajoz	16.21	16.25	16.46	16.77	0.2539	0.3081	0.3162
Murcia	16.38	16.40	16.62	16.91	0.2519	0.3086	0.3083
Málaga	16.53	16.57	16.88	17.14	0.2571	0.3524	0.3717

2.5. Solar Radiation Maps

A very important application of the methodology proposed using the MLP in the solar radiation field is the drawing of solar radiation maps. As was mentioned in the introduction of section 2, the availability of information about the solar radiation in the location where a solar system is going to be installed is necessary for the designer of solar systems. This information, in case that exists, can be available, in several ways. The most common one is by means of several tables with a lot of very useful information (usually large solar radiation sequences), but extremely difficult to handle. Nevertheless, another way to present this information is by means of different solar radiation maps (one for each month) of the area where the installation is going to be made. This second way is usually more efficient, easy to handle and preferable by the designers and can be used at the initial stage of the solar system design and sizing.

Many procedures have been developed to draw solar radiation maps. In most of them, to obtain a solar radiation map is necessary to know the solar radiation of many points spread wide across the real zone of the map where it is going to be drawn. These solar data can be available in several time scales. Usually there are two time scales in which the data are available; daily and hourly.

Nevertheless, for most of the locations all over the world records of solar radiation data are non existent. Only very few locations have the privilege of having meteorological stations with proper registered records of solar radiation. Even in the locations with historical records of solar data, the quality of the solar series is usually not so good.

So the problem of obtaining a solar radiation map in a certain zone is difficult to solve, due to the fact that it is necessary to know the solar radiation of a large number of sites. In this section an application of the neural network methodology to obtain solar radiation maps is presented. In this case, an improvement of the neural system presented section 2.2 has been made, with the objective of the application of the neural supervised model to draw solar radiation maps.

2.5.1. Proposed Methodology to Draw Solar Radiation Maps

The preparation of solar radiation maps requires the previous knowledge of a lot of information, which can be summarised in the following three steps:

Step 1. Construct a grid of the zone that is going to be represented in the map. Each point of the grid corresponds to a certain real site that will have, besides the solar radiation data, some particular features as its latitude, longitude and altitude above sea level.

Step 2. For each point of that grid, knowledge of long sequences of hourly solar radiation is needed.

Step 3. Once the series of hourly solar radiation is available for the entire grid, it is necessary to obtain some particular statistical values to draw the maps. For instance, it is interesting to summarise all that information in 12 maps, one for each month of the year. So, in this case, it is necessary to obtain the mean monthly values of the solar radiation data. In order to do this, it is necessary to calculate daily solar radiation data from the hourly solar radiation data. These new data are obtained for each day by summing up all hourly values. Then with the daily solar radiation data, average monthly values must be calculated.

From the three steps the one that is more difficult to achieve is step number two because usually, in most sites there are no hourly solar radiation data available. Initially this can be solved by using an MLP as it was explained in the section 2.2. By training a MLP (with the architecture explained previously) it is possible to generate long sequences of hourly solar radiation data which are statistically equivalent to the real series, as has been applied previously (Hontoria *et al.*, 1999; 2000; 2001; 2002; Zufiria *et al.*, 1999). It must be remembered that in order to do this, the MLP needs as one important input the values of K_T (daily solar data). The problem arises because this information (daily solar data) is known in very few points of the grid.

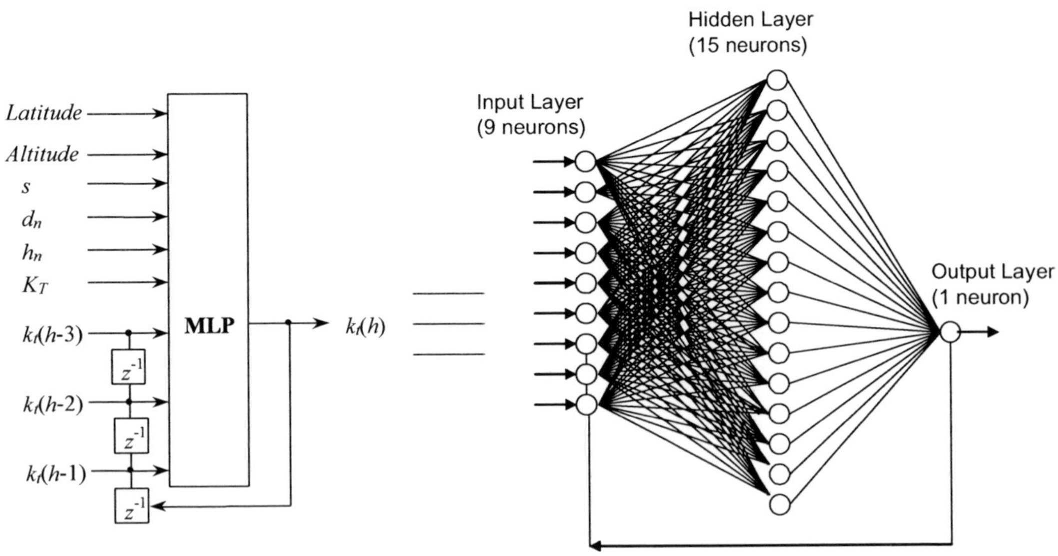

Fig. 7. MLP Architecture for clarity indexes prediction, including latitude and altitude

Particularly in Andalucía, the most southern region of Spain, where solar radiation maps are being prepared, this problem exists. Only in some cities of the region such as Almería, Cádiz, Córdoba, Huelva, Jaén, Málaga and Sevilla, records of hourly solar radiation (and so daily solar radiation data) are available. Also, for Jaén province, a small region of southern Spain, only in Jaén city hourly solar radiation data have been recorded. But for the rest of the sites of the grid, where the map is going to be drawn, there are no daily solar radiation data available that could be employed by the MLP to generate the corresponding hourly solar sequences.

To solve this problem, the idea of the authors (Hontoria *et al.*, 2000) was to include in the MLP other pieces of information available, such as latitude and altitude above sea level, which are available in all points of the grid. These two items were included in the MLP as new inputs. The final procedure employing a MLP in a mixed feedback-feedforward configuration is shown in Fig. 7.

A unique MLP was trained with the all the solar data available in Andalucía and the two new inputs corresponding to the latitudes and altitudes of these sites. This MLP is used for the generation of hourly solar radiation data at each point of the grid which are subsequently used to produce the monthly solar radiation maps of the region.

2.5.2. Examples o Solar Radiation Maps

Initially we are going to show the solar radiation maps for small regions. Particularly we are going to draw a map for Jaén (southern province in Spain). To obtain solar radiation maps the steps mentioned in section 2.5.1 are followed. For instance, to obtain solar radiation maps for Jaén province the following procedure is carried out. First, a grid with more than 250 points has been prepared. The grid points vary from parallels 38° 25' N to 37° 25' N and meridians 4° 10' W to 2° 10' W and have been chosen in intervals of 5'. For all the grid points solar hourly values must be generated.

Secondly, as mentioned previously, a unique MLP has been trained. To train this MLP, hourly solar radiation data from Madrid, Oviedo, Málaga and Jaén have been used. In this case, the solar radiation data for the first three locations mentioned are the same as the ones described before, and for Jaén we have ten years of solar radiation data recorded by our research group. The selection of the first three locations is due to the very different climates and latitudes they have. The choice of the location of Jaén is due to the fact that the maps are going to be made in its province. The MLP is the tool used for the generation of solar hourly radiation values in all the sites of the grid of the map. The solar information used for the generation is the Jaén solar data and the particular information of each site is the latitude and altitude (above sea level). Therefore the idea is, once the MLP is trained with solar radiation data of several sites (in the case of this example with data from Madrid, Oviedo, Málaga and Jaén) with different altitudes and latitudes, for the generation of solar data of each point of the grid it is only necessary to use the solar data of one place which belongs to the map (in this example is Jaén) and varying the values of the inputs "latitude" and "altitude" and the MLP would be able generate solar radiation data for every point on the grid.

Finally the solar global monthly mean values (Iqbal, 1983) are obtained for each point on the grid, so twelve solar radiation maps are possible to be drawn (one for each month). The solar radiation maps obtained by following this methodology for the Jaén region for the month of January (winter) and July (summer) are shown in Figs 8 and 9 respectively.

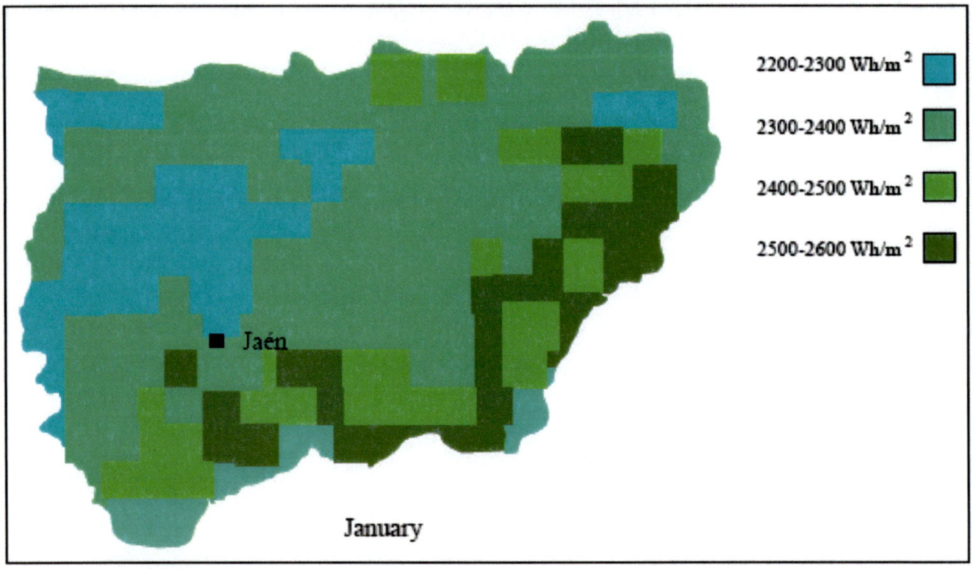

Fig. 8. Solar radiation map of Jaén province in January

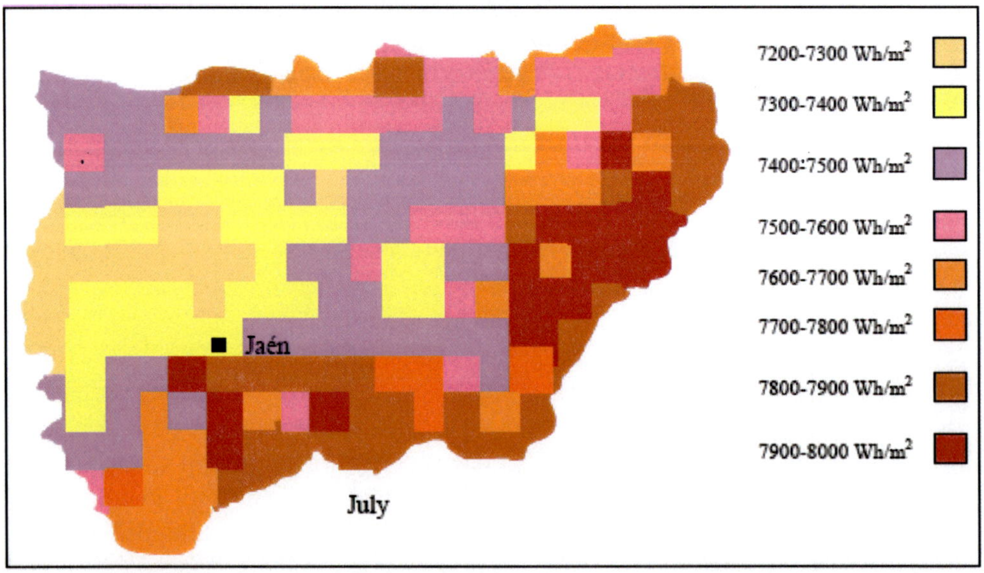

Fig. 9. Solar radiation map of Jaén province in July

The same procedure, but with some data variation has been carried out in order to obtain solar radiation maps of a larger region. In this case the region in study is Andalucía. This is the most southern region of Spain. A grid with more than 500 points has been prepared in this case. These points vary from parallels 38º 40' N and 36º 00' N, and meridians 7º 30' W and 1º 40' W and have been chosen in intervals of 10'. For all these grid points solar hourly values must be generated.

A MLP has been trained with hourly solar radiation data of sites in Andalucía. In this case, the solar radiation data available correspond to the following locations: Jaén, Córdoba,

Sevilla, Cádiz, Huelva, Málaga, Granada and Almería. For those eight cities solar radiation data are available from the National Meteorological Institute for eight years. As mentioned previously the solar global monthly mean values are obtained for each point on the grid, so twelve solar radiation maps were drawn (one for each month). The solar radiation maps obtained by this method for the Andalucía region for autumn and spring are shown in Figs 10 and 11 respectively.

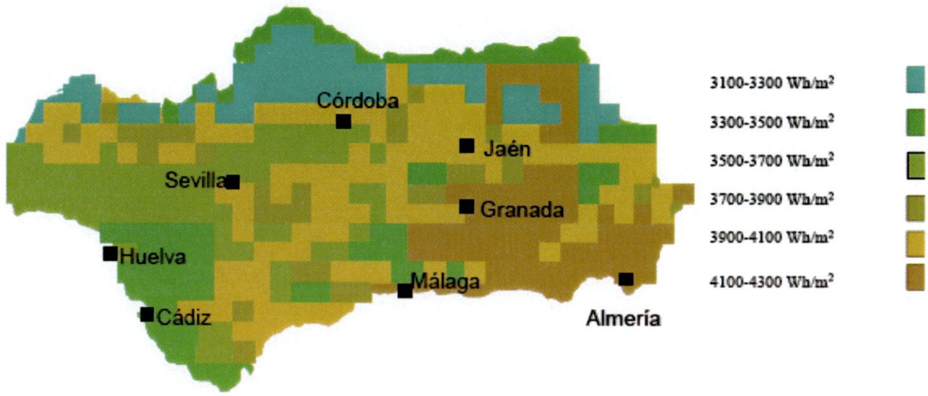

Fig. 10. Solar radiation map of Andalucía in autumn

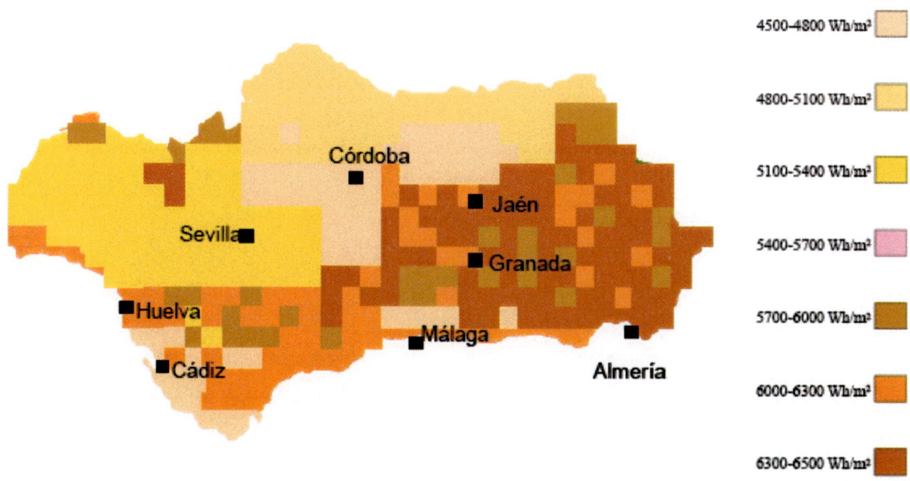

Fig. 11. Solar radiation map of Andalucía in spring

3. Artificial Neural Networks and Sizing Stand Alone Photovoltaic Systems

3.1. Introduction

There are many methods available for sizing stand alone photovoltaic (PV) systems. Several authors have classified them by paying attention mainly to their simplicity and accuracy. In a first group there are the more simplistic methods which usually are called intuitive methods. They are used as a first approach, but they are very inaccurate. Typically they use one or two intuitive parameters for sizing the generator and the storage system of the PV system and the results are not very reliable. Another group of methods is the numerical methods. These methods use system simulations and are more accurate than the previous ones but present the problem that they need to have long time series of solar radiation for the simulations. They work very well in sites where all the data needed for the sizing of a PV system such as radiation data, load data and so on are available. Nevertheless these simulations are very complicated to be performed. In short, the numerical methods are the more accurate but also the more difficult to put into practice. Finally in a third group there are methods which use equations to describe the PV system size as a function of reliability. These are called analytical methods.

The advantage of the latter methods is that they combine accuracy and simplicity. They allow the designer to optimise the energy and economic cost of the PV system. Many of the analytical methods employ the concept of reliability of the system or the complementary term, i.e., Loss of Load Probability (LOLP). The LOLP represents the percentage of time that the PV/storage system will not be able to satisfy the load. Several authors use this type of methodology for sizing stand alone PV systems (Chapman, 1987; Egido and Lorenzo, 1992; Lorenzo *et al.*, 1994; Sidrach-de-Cardona and Lopez, 1998; 1999; Barra *et al.*, 1984; Bartoli *et al.*, 1984). With these methods the relations between the generator capacity, the accumulator capacity and the LOLP lead to curves called isoreliability curves or LOLP curves.

Nevertheless, the analytical methods also present some disadvantages. For instance they usually need long term radiation data for the locations where the equations must be deduced or the LOLP curves must be drawn. So for locations with no radiation data those equations may not be valid or the LOLP curves may not be able to be obtained.

In order to overcome this disadvantage, a new approach for obtaining LOLP curves is presented here. It is based on a Multilayer Perceptron (MLP) neural network. Particularly, the method is based on the MLP's ability to extract, from a sufficiently general training data set, the existing relationships between variables whose interdependence is not known *a priori*. In the previous section we have used the MLP for generating long-term solar radiation series with very satisfactory results (Hontoria *et al.*, 2001; 2002). As was seen with a very simple structure of the MLP, the solar radiation series in locations where data are not available can be generated. For the case of LOLP curves prediction, a new architecture for the MLP is used. The MLP is trained with LOLP data (different LOLP values i.e. 0.1, 0.001, 0.05,...) and radiation data from different Spanish locations (first stage). After training, the MLP is able to generate as many LOLP curves as needed, for a particular site (second stage). The main advantage of this tool is that the effort required in the first stage is similar with that required in other methods, however in the second stage, when the MLP is trained, it is extremely simple to obtain new LOLP curves.

3.2. PV Stand Alone Systems

3.2.1. Definitions

As a first approach, but without loosing truthfulness, a very simple PV stand-alone system will be used. This system consists of two subsystems to be designed; the PV generator and the battery storage system. Let CA be the generator capacity, which is defined as the ratio of the average energy output of the generator in the month with worst solar radiation input divided by the average consumption of the load (assuming a constant consumption of load for every month). The battery storage system is related to the term CS, the accumulator capacity, which is defined as the maximum energy that can be extracted from the accumulator divided by the average daily consumption of the load.

There are two other concepts that must be defined; the reliability of a PV system and the loss of load probability. The system reliability is defined as the percentage of load satisfied by the PV system whereas the loss of load probability (LOLP) as the percentage of the mean load (in large periods of time) not supplied by the PV system, i.e., it is the opposite of reliability.

3.2.2. LOLP Curves

With the aim to establish relations between CA, CS and LOLP the numerical methods use system simulations while the analytical methods use equations. For instance, Egido and Lorenzo (1992) presented a method consisting in creating reliability maps for each LOLP value considered, (isoreliability lines). They proposed very simple equations to determine the isoreliability lines for many Spanish locations.

Once the LOLP curves are obtained, it is very simple to design both the capacity of the generator (CA) and the accumulator capacity (CS). Depending on the reliability needed for the PV system design, a specific value of the LOLP will be considered. For instance, in Table 3 some usual values for typical PV systems are shown.

Table 3.Recommended LOLP values for different applications

Application	LOLP
Domestic appliances	10^{-1}
Rural home lighting	10^{-2}
Telecommunications	10^{-4}

In Fig. 12 an example of LOLP curve is represented. In this example the value for LOLP is 0.01.

The LOLP curve represents pairs of CS and CA values that lead to the same value of LOLP. This means for example, that for the pair (CS, CA) = (2, 1.1) the reliability proposed will be achieved by having a “big” size of the generator and a “small” size of the storage system. Similarly, for the same reliability, the pair (CS, CA) = (8, 0.61) leads to a "small" generator and a "big" battery.

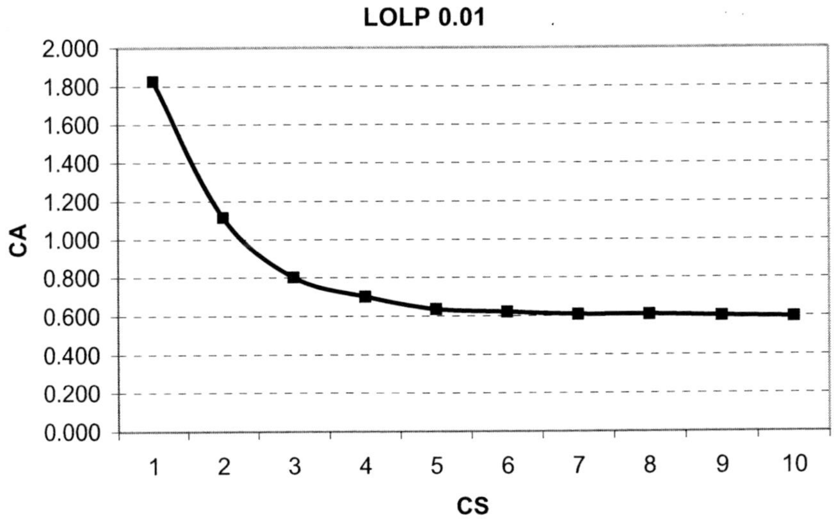

Fig. 12.LOLP curve for the value of LOLP = 0.01

3.3. Application of the MLP

In a first approach the case that the LOLP curves, treated as functions between the three parameters that are needed for the sizing of PV stand alone system, i.e., CA, CS and LOLP, is examined (Hontoria *et al.*, 2003). Let G be a general function which relates the three mentioned parameters. Hence, the problem reduces to the computation of function G. The MLP-based approach approximates such function via a MLP, making use of the available data (pairs of CS-CA values for each LOLP value). Therefore, the prediction procedure relies on the possibility of training the network such that:

$$CA_n = G\,(CS_n, LOLP_m) \tag{7}$$

wth $1 < n < N$ for N pairs (CS, CA) different values for each $1 < m < M$ different values for LOLP.

By training a MLP with 2 inputs and 1 output, with a representative training set, the MLP was able to find the desired relationships between those values just by approximating the function G. Once the approximation is performed, future values can be computed whenever they are needed. One of the greatest advantages of employing a MLP-based methodology for generating LOLP curves is that most of the computational resources are required during the training procedure, as opposed to the generation procedure.

Summing up, in the previous work mentioned, we trained several MLPs (one for each location considered) with many pairs of (CS, CA) data but for very few LOLP curves (just the values 0.01, 0.05 and 0.1) and after the training process each MLP was able to generate the

LOLP for other values (for instance the value 0.02 or 0.25). The results obtained were very satisfactory.

Nevertheless, working in that way, the problem of dependency on the location arises. Thus, for each location a new MLP must be implemented and trained and for sites where radiation data are not available the proposed methodology cannot be applied.

With the aim of solving this problem and with the idea of avoiding the local dependency of the method an improvement of that first approach is proposed. In this a unique neural network (MLP) was implemented for a number of Spanish locations considered in this study. The hypothesis applied is that the LOLP curves may be treated as functions between four parameters; CA, CS and LOLP as in the previous work and also K_{TDY} (the annual average clarity index). The inclusion of this new parameter is useful for making different LOLP curves in different locations. Let *H* be a general function that relates the four mentioned parameters. Again, the problem reduces to the computation of function *H*. The MLP-based approach approximates such function via a MLP making use of the available data (pairs of CS-CA values for each LOLP value and for each location considered with the parameter K_{TDY}). Consequently, the methodology applied relies on the possibility of training a MLP such that:

$$CA_n = H(CS_n, LOLP_m, K_{TDYu}) \quad (8)$$

With $1 < n < N$ for N pairs (CS, CA) different values for each $1 < m < M$ different values for LOLP and for each $1 < u < U$ different locations.

The MLP used in this case has 3 inputs and 1 output. The proposed methodology is described in next section.

3.4. Proposed Methodology

3.4.1. Architecture of the MLP

The structure of the Multi Layer Perceptron (MLP) proposed for the generation of LOLP curves is shown in Fig. 13. It consists of three layers. The first one or input layer has three inputs as follows:

CS: Accumulator Capacity.
LOLP: Loss of Load Probability.
K_{TDY}: Yearly Clarity Index.

The second layer, also called the hidden layer, has nine neurons or nodes. Different tests have been done in order to choose the number of neurons and the actual number selected produced the best results. Finally the last layer has only one node, the value of the generator capacity (CA) to be estimated.

The training is done with the backpropagation algorithm (Rumelhart and MacClelland, 1986; Weigend *et al.*, 1990).

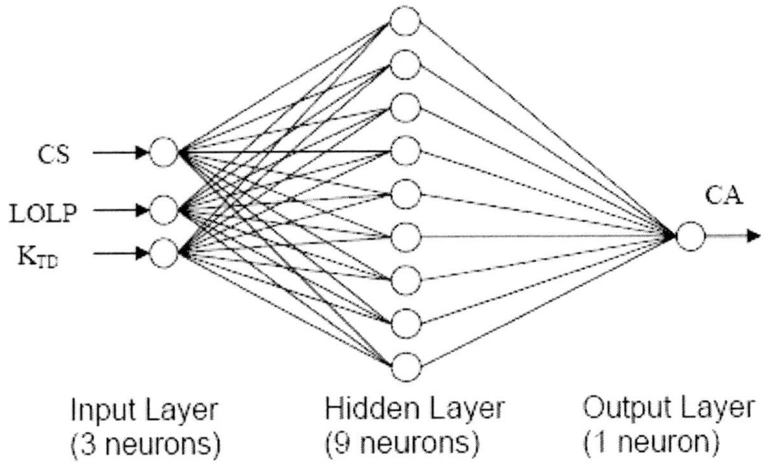

Fig. 13.MLP Architecture for the obtaining of the LOLP curves

3.4.2. Data Availability

As it was mentioned before, the aim of this work is to design a new MLP with the structure described in the previous subsection. This new MLP should generate LOLP curves for any location in Spain. So it must be trained with data from many Spanish sites and validated with data from other places not used in the training. The places considered in this study are the same as those used for the prediction of solar radiation presented in section 2. In this case two new localities were added; Santander and Valladolid (cities of Spain) which are used to validate the results.

The proposed methodology and the process of validation can be summarised as follows:

Step 1. Obtaining Real LOLP Curves

Initially, with the actual solar radiation data of the locations considered, simulations of stand alone PV systems were performed. These simulations are carried out over periods of more than ten years and LOLP curves are calculated. In these simulations a software which includes many variables such as the tilt of the PV modules and several models of batteries with different adjustable parameters have been employed. These LOLP, as they are obtained with real radiation data of each location, are called *"real LOLP curves"*. Different values of LOLPs are used varying from LOLP = 0.01 to LOLP = 0.1.

Step 2. Training the MLP

Subsequently, the training of the MLP is performed. For the training process, the MLP receives information of the real LOLP curves obtained before, but only from the first seven locations shown in Table 2, section 2.4.1. These locations are called the training locations ("T").

Step 3. Generation o LOLP Curves ad Validation of the MLP

Finally, once the MLP is trained, the generation of LOLP curves is performed for all locations considered (the seven locations used for the training and the two new locations used for validations "V"). The LOLP curves obtained are compared with real LOLP curves achieved in step 1.

In the following section some of the results obtained are presented.

3.5. Results

Applying the proposed methodology many different LOLP curves have been calculated. As it was mentioned before the generation of the LOLP curves has been carried out not only for the locations used for the training, but also for those used for the validation. As a quality measure the quadratic error between the real curves and those obtained via the MLP has been calculated.

The LOLP curves obtained by the MLP have been compared not only with the real ones but also with those obtained by other methods. In this case a very simple method called, the Egido and Lorenzo method (developed in the Instituto de Energía Solar Madrid, called the "IES method") has been applied. This method makes use of the following equation:

$$CA = f\, CS^{-u} \tag{9}$$

Where f and u are two parameters, which are different for each location and related with LOLP.

In Figs 14, 15 and 16 some examples of curves generated by the MLP are shown.

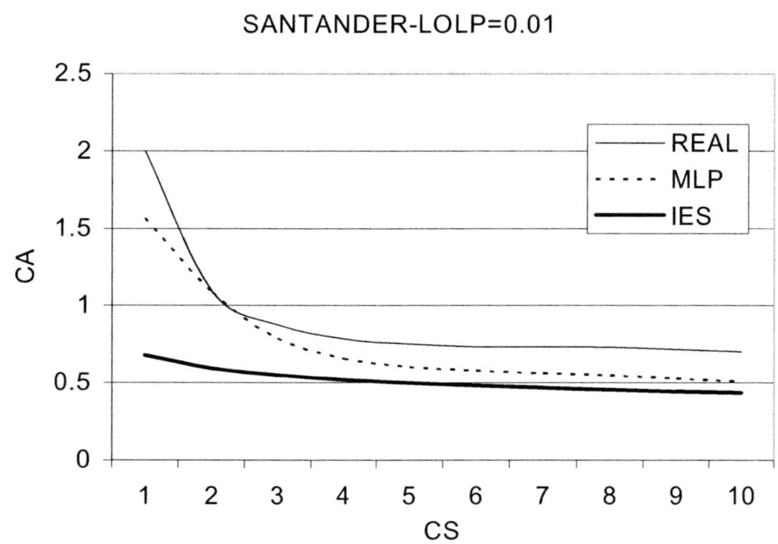

Fig. 14. LOLP curves obtained by different methods. LOLP = 0.01. Location Santander

As can be seen the LOLP curves obtained by the MLP are very close and fit better to the real ones than the LOLP curves obtained by the IES method. It is important to point out the ability of the MLP to generate a curve with values for which it has not been trained and for locations with data not used in the training. This indicates that any LOLP curve can be generated by the MLP, ensuring that the results will be very close to the real ones.

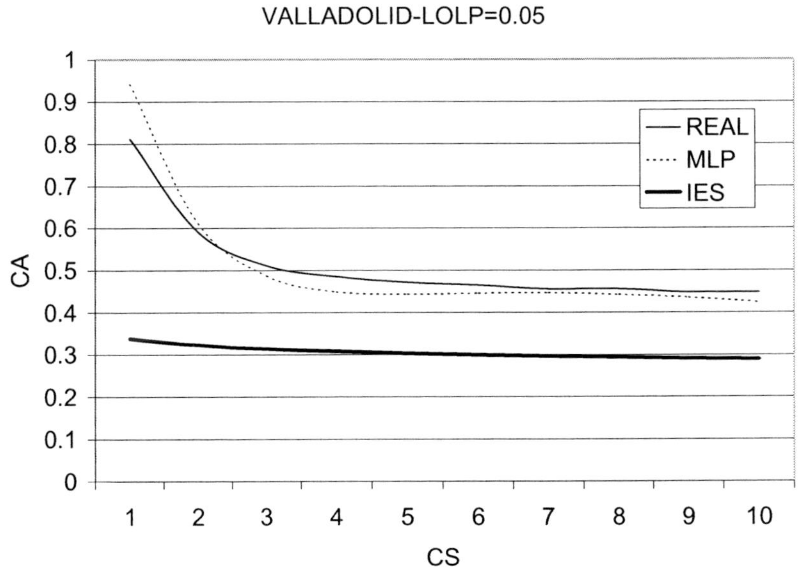

Fig. 15. LOLP curves obtained by different methods. LOLP = 0.05. Location Valladolid

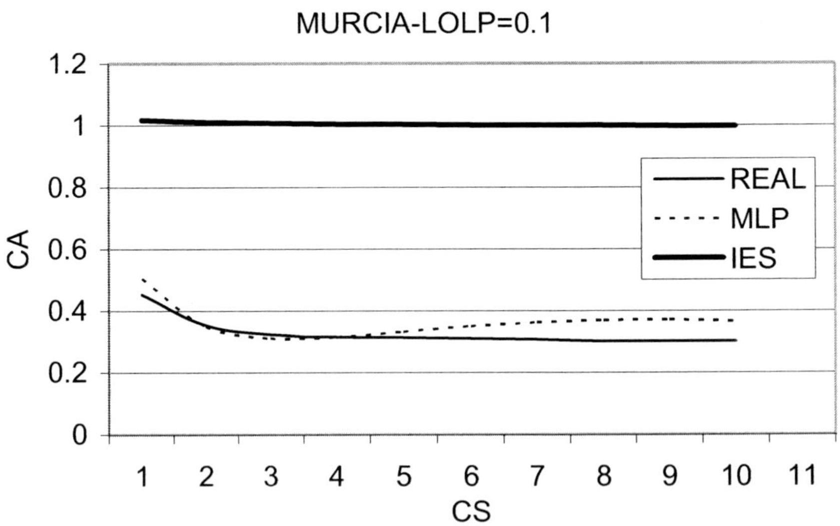

Fig. 16. LOLP curves obtained by different methods. LOLP = 0.1. Location Murcia

In order to prove the importance of the methodology the Quadratic Error (QE) between the real curves and those obtained by the MLP is calculated and presented in Table 4.

All locations considered in this work are included in Table 4. It must be remembered that the data of the seven first locations have been used for training the MLP, but the last two locations have been used for the validation process.

The first conclusion that can be drawn from the results presented in Table 4 is that the Quadratic Error is very low, which indicates a very good fit between the real and the MLP curves for each location. Only for the value LOLP 0.1 and for the locations of Santander and

Valladolid, the QE is a little bit higher. It has to be pointed out however, that the real data of those locations have not been employed for the training of the MLP. Nevertheless the match between real values and MLP values is considered as very adequate.

Table 4. Quadratic Error (QE) between real curves and MLP curves. LOLP 0.01. LOLP 0.1

Location	Type	LOLP 0.01	LOLP 0.1
Oviedo	T	0.050	0.048
Logroño	T	0.157	0.073
Madrid	T	0.089	0.050
P. Mallorca	T	0.101	0.039
Badajoz	T	0.092	0.044
Murcia	T	0.133	0.046
Málaga	T	0.077	0.038
Santander	V	0.091	0.073
Valladolid	V	0.201	0.035

The quadratic error between the real curves and those obtained by the Egido and Lorenzo method (IES method) and the MLP are shown in Table 5. In this case, only the locations used for the validation of the MLP method have been included, in order to confirm the good prediction capability of the MLP. As can be seen, the QE values between the MLP values and the real ones are lower than those between the real ones and the IES method.

Table 5. Comparison of the QE between the real curves and IES curves, and QE between the real curves and MLP curves for LOLP=0.01 and LOLP=0.1

Location	LOLP 0.01		LOLP 0.1	
	IES	MLP	IES	MLP
Santander	0.144	0.091	0.102	0.073
Valladolid	0.509	0.201	0.125	0.035

4. Quality Control of PV Modules Using the Multilayer Perceptron

4.1. Introduction

In this section another application of the MLP in the photovoltaic field is presented. In this case a new method for converting V-I curves measured at whatever outdoors conditions to the Standard Test Conditions (STC) is presented. STC correspond to Irradiance of 1000 W/m^2 and Cell Temperature of 25°C. For the quality control of PV modules a number of methods can be used in order to obtain the standard test conditions V-I curves of the PV modules (Caamaño *et al.*, 1999; Wagner, 2003). Once the STC V-I curve of the PV module is obtained the maximum power in STC of that module can easily be calculated and a comparison between the measured power and the one provided by the manufacturer can be performed.

The IDEA Research Group from Jaén University is implementing a methodology for quality control of PV modules. For this purpose the IDEA Research Group in the Solar Laboratory at the Polytechnic school, at Jaén has all the equipment required and so an outdoor measurement system was built based on a marketed capacitive load. Particularly the equipment needed is the following:

a) A *Photovoltaic Engineering* PVPM 2540C measuring instrument (connected to the module under test using a four-wire connection). The PVPM 2540C allows the measurement of V-I curves of PV modules and generators. With these results the device is able to calculate the peak power of the PV module directly on the place of installation.
b) *A meteorological station* (for measurements of solar radiation and temperature). The PVPM 2540C can also record incident spectral irradiance *(G)* and ambient temperature (T_{am}) through a Phox100/Pt100 sensor mounted near the module to be measured.

In Fig. 17 the outdoor measurement system is shown.

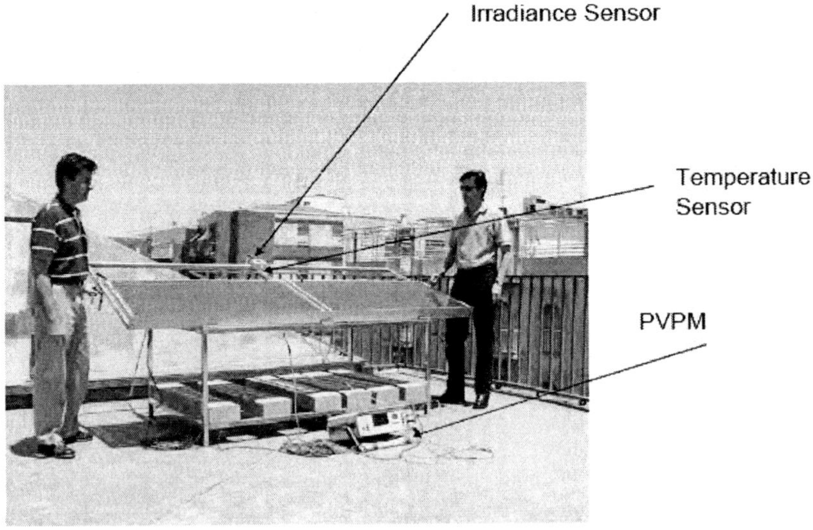

Fig. 17. Outdoor measurement system

The use of the MLP in the photovoltaic field is providing very satisfactory results. As was seen from the previous section the MLP was used for generating Loss of Load Probability (LOLP) curves for stand alone photovoltaic systems (Hontoria *et al.*, 2003). For this case a new architecture for the MLP is used in order to extrapolate V-I curves of PV modules recorded at non STC to STC V-I curves. The advantage of using the MLP is that the conversion method is simplified.

4.2. Hypothesis

The hypothesis assumed here is to consider the V-I curve of any PV module as a function of the form:

$$V = f(I, T_{am}, G) \tag{10}$$

Where:

V is the voltage
I is the current
T_{am} is the ambient temperature, and
G is the Irradiance

Typical V-I curves at different weather conditions are shown in Fig. 18. As can be observed for different test conditions of ambient temperature and irradiance there are different V-I curves. So it can be assumed that there is a function that relates the voltage of the PV module with the current, the ambient temperature and the irradiance.

Hence, the problem reduces to the computation of function *f*. The MLP-based approach approximates such function via a MLP, making use of the available data. Therefore, the prediction procedure relies on the possibility of training the artificial neural network so as to adjust function *f*. For this new investigation it is possible again to train a MLP with the appropriate architecture and with a representative training set.

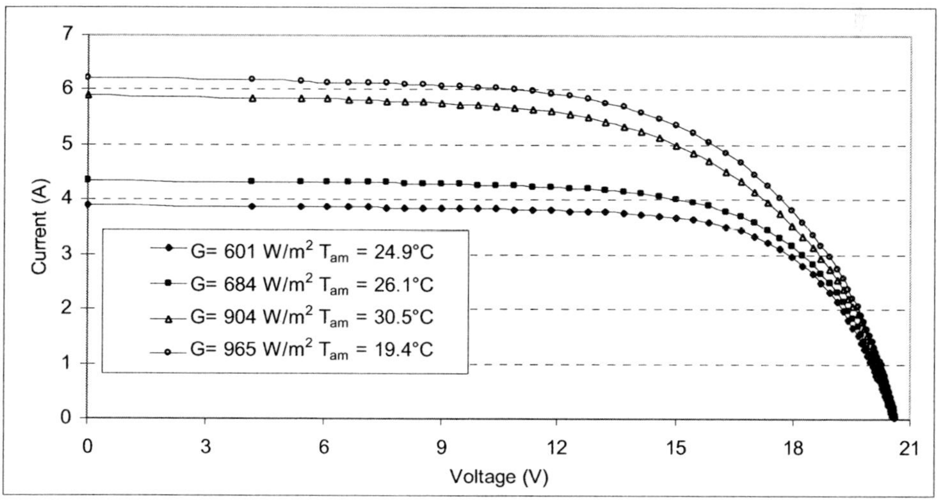

Fig. 18. Different V-I curves at different weather conditions

Thus, the MLP will be able to find the desired relationships between those values (provided that they exists) just by approximating the function *f*. Once the approximation is performed, future values can be computed whenever they are available. One of the great advantages of employing a MLP-based methodology for conversion of V-I curves to STC is that most of the computational resources are required during the training procedure, whereas when the trained network is used for new unknown patterns the prediction is simple and very quick.

4.3. Architecture of the Artificial Neural Network

The structure of the Multi Layer Perceptron (MLP) proposed for converting the V-I curves tested at non STC to STC V-I curves is shown in Fig. 19. It consists of three layers. The first one or *input layer* has two inputs, i.e., the irradiance (G) [W/m^2] and ambient temperature (T_{am}) [ºC].

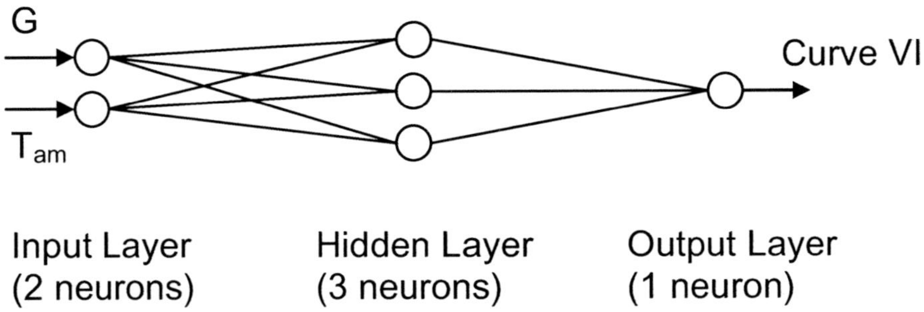

Fig. 19. Proposed architecture of the neural network Multilayer Perceptron for converting non STC V-I curves to STC V-I curves

The second layer, also called the *hidden layer*, has three neurons or nodes. Different tests have been performed in order to find the optimum number of neurons in this layer. Finally the last layer, *output layer*, has only one node; the V-I curve. Really this node is a vector with one hundred pairs of points (100 voltage points and 100 current points) which form the V-I curve. The training is done by the Levenberg-Marquart backpropagation algorithm.

4.4. Proposed Methodology for Converting Non STC V-I Curves to STC V-I Curves

4.4.1. Method Description

In this section, the method based in the MLP to obtain V-I curves at STC is described. Two stages are required for this purpose. In a first stage many data must be shown to the neural network (MLP). This first process is called the *training of the network*. For this process the information that the MLP needs is:

a) *Input information:* Different values of irradiance and ambient temperature,
b) *Output information:* Different V-I curves (pairs of V and I of many curves; 100 pairs of points for each V-I curve) at different outdoor conditions.

In a second stage, once the MLP is trained, the MLP can be used with the appropriate STC conditions as input to produce the V-I curve at STC.

4.4.2. PV Module Features

The PV module used in this study is the I-106 panel from Isofotón. Table 6 shows the main features of the PV module.

This PV module was calibrated by CIEMAT (Energy and Environmental Investigation Centre). Its maximum power at STC is 96.6 W ± 2.2 %. The PV module is located on the roof of Polytechnics building at the University of Jaén, facing south with a tilt from horizontal equal to 35° and is operated at its maximum-power point.

Table 6. Typical values at 100mW/cm^2, AM 1.5 spectrum, 25° C cell temperature

Specifications	Value
Maximum Power (P_{max})	106 W
Short Circuit Current (I_{sc})	6.54 A
Open Circuit Voltage (V_{oc})	21.6 V
Maximum power current (I_{max})	6.10 A
Maximum power voltage (V_{max})	17.4 V

4.4.3. Data Available

Initially the design of the MLP with the structure described previously needs to be performed. The MLP should generate the V-I curves for any irradiance (*G*) and ambient temperature (T_{am}) for the module considered. So it must be trained with data from many radiation and ambient temperature combinations and afterwards to be validated at other radiation and ambient temperature conditions. In Table 7 the irradiance and ambient temperature used for this study are tabulated.

As can be observed from the data presented in Table 7, a number of different combinations of irradiance and temperature have been used. The use of the data is indicated in the last column. Letter "T" indicates data used for the training of the MLP, and "V" designates data used for validation as it will be explained in next subsection.

4.4.4. Training of the MLP

For this purpose we have used the PV module mentioned before, and we have measured its V-I curve at different outdoors conditions, indicated in Table 7.

Table 7. Irradiance and Temperature values considered in this study

Irradiation (W/m^2)	Ambient Temperature (°C)	Use
716	29.1	T
601	24.9	T
1001	19.8	T
1018	28.6	T
728	27.9	T
1108	34.1	T
979	21.5	V
1001	19.8	V
717	29.0	V
1114	37.0	V

Once the V-I curves are measured and recorded the conversion to STC method must be applied. We have used the Blaesser and Munro method (Blaesser and Munro, 1995). This method is the best for the outdoor determination of maximum power of a module at STC.

With all of this information (the V-I curves at non STC and the curve at STC) the MLP is trained and this first stage is finished.

4.4.5. Obtaining V-I curves at STC with the MLP

Once the MLP is trained as it has been explained previously, it can be used for converting any other V-I curve measured at non STC to STC.

4.5. Results

Applying the methodology explained previously many different V-I curves have been generated. The generation of the V-I curves has been done not only for the radiation and ambient temperature values used for the training, but also for those used in the validation. As a quality measure the root mean square error (RMSE) and the mean bias error (MBE) between the real curves and those obtained via the MLP have been calculated.

The Root Mean Square Error (RMSE) which is also known as the square root of the medium quadratic error, has been used in this work to compare the values generated by the MLP with those values observed (what we call the real ones).

$$RMSE_P = \sqrt{\frac{\sum_{i=1}^{N} (P_{Ti} - P_{Oi})^2}{N}} \quad (W) \tag{11}$$

where:

P_{Ti} is the theoretical power,
P_{Oi} is the measured power, and
N is the number of samples.

The Mean Bias Error (MBE) is known as slant or deviation. It has been used in this work along with the RMSE to characterise the accuracy of each one of the methods studied.

$$MBE_P = \frac{\sum_{i=1}^{N} (P_{Ti} - P_{Oi})}{N} \quad (W) \tag{12}$$

In Fig. 20 some examples of curves generated by the MLP are shown.

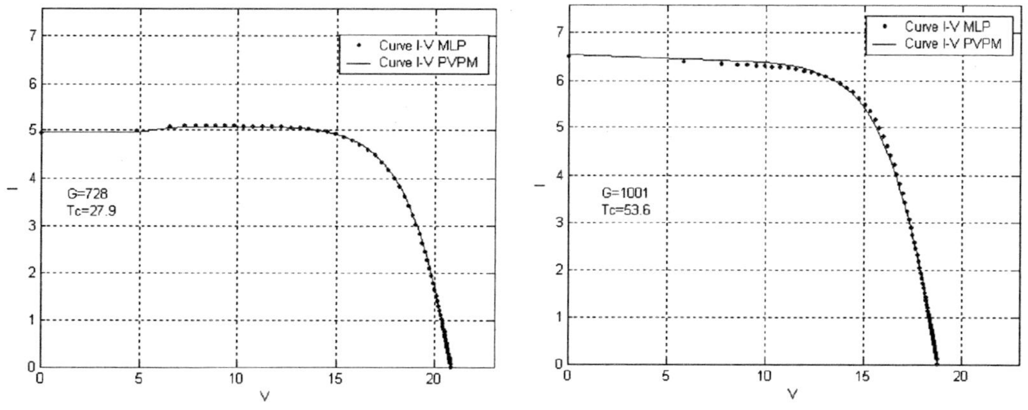

Fig. 20. V-I curves obtained by different radiation and module temperature

As can be seen the V-I curves obtained by the MLP are very close to the real ones. It is important to point out the ability of the MLP to generate a V-I curve with values for which it has not been trained. This indicates that any V-I curve can be generated by the MLP, ensuring that the results will be very similar to the real ones.

The Root Mean Square Error (RMSE) and Mean Bias Error (MBE) between the real curves and those obtained by the MLP are shown in Table 8. It must be remembered that the data of the six first patterns have been used for training the MLP, but the last four have not been used in the training process.

Table 8. RMSE and MBE between real curves and MLP curves

Irradiation (W/m^2)	Ambient Temperature (ºC)	RMSE	MBE
716	29.1	0.6281	-0.2165
601	24.9	0.3755	0.1522
1001	19.8	0.7187	0.2354
1018	28.6	0.1103	0.0194
728	27.9	0.1251	-0.0722
1108	34.1	0.2954	-0.1193
979	21.5	1.0515	0.4449
1001	19.8	0.7187	0.2354
717	29.0	0.6308	-0.1699
1114	37.0	1.0340	0.6588

As can be seen the Root Mean Square Error is low, which indicates a very good fitness between the real and the MLP produced curve for each radiation and module temperature. Thus, the validity of the method proposed is verified and this methodology can be used to obtain V-I curve of the module in standard test conditions. A comparative study has also been performed in order to compare the V-I curve extrapolated to STC conditions by the MLP and the following ones:

a) V-I curve in STC provided by the manufacturer,
b) V-I curve obtained with the PVPM 2540C (called the real one),

c) V-I curve calculated by the Blaesser method, and
d) V-I curve provided by an accredited Spanish laboratory called "Ciemat".

Figure 21 shows the V-I curve of the PV module in standard test conditions obtained by the different ways mentioned above and Table 9 shows the maximum power of the photovoltaic module, $V_{oc,STC}$ and $I_{sc,STC}$, obtained with each method.

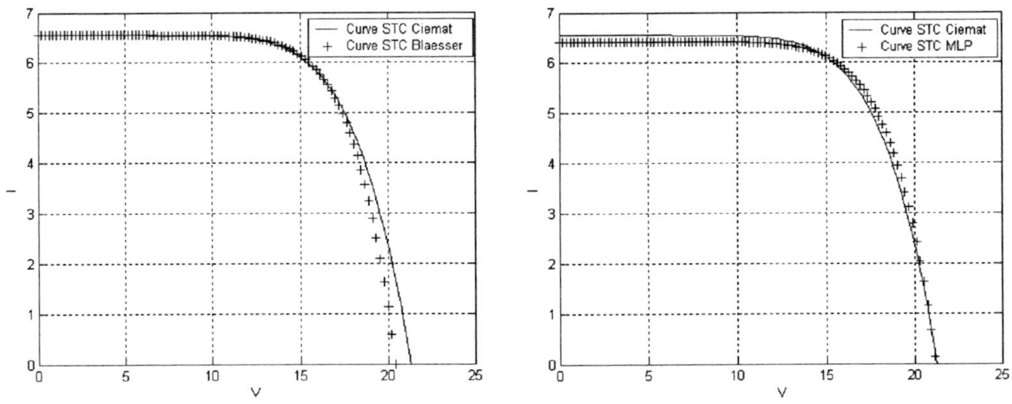

Fig. 21. V-I Curves obtained by different methods

As can be seen from Table 9 the data obtained by the MLP (maximum power of the module and open circuit voltage), are very similar to those obtained by the CIEMAT method.

If we consider now the curve provided by the CIEMAT as the real one (due to its accuracy) and compare it with the remainder of the methods the results presented in Table 10 can be obtained. As can be seen both the root mean square error and mean bias error are satisfactory.

Table 9. P_{max}, $V_{oc,STC}$ and $I_{sc,STC}$ obtained by each method

Method	P_{max}(W)	$V_{oc,STC}$ (V)	$I_{sc,STC}$ (A)
MLP	94.63	21.23	6.43
Blaesser	93.59	20.42	6.55
PVPM	92.77	21.5	6.54
Manufacturer	106.13	21.6	6.10
CIEMAT	96.6	21.39	6.42

Table 10. RMSE and MBE of each method compared to CIEMAT method

Method	RMSE	MSE
MLP	1.4293	-0.1027
Blaesser	13.2025	-4.2815
Manufacturer	10.4141	5.8614
PVPM	1.2760	-0.3377

5. CONCLUSIONS

In this chapter a literature review on applications of artificial neural networks in the renewable energy field has been presented. As was seen ANNs have been applied in many renewable energy applications. They have been used for instance for many solar thermal energy applications. Also the authors of this chapter have applied a particular supervised ANN called the Multilayer Perceptron (MLP) for the solar radiation and for two applications in photovoltaic system prediction.

In the solar radiation field a methodology based in ANN for generating solar irradiation hourly data has been presented. The MLP methodology is based on the possibility of implicitly employing information associated with the problem, without knowing the existing relationships between different variables and sources of information, i.e., the proposed method does not assume any *a priori* model, as opposed to the standard approximation techniques where polynomial regression techniques are employed. Furthermore, the test results have shown that the neural network approach outperforms those methods. A comparative study with the Aguiar and Collares-Pereira (AC) and Graham and Hollands, (GH) methods has been performed and the results obtained by the methodology proposed showed that improved accuracy has been obtained compared to the classical methods. The proposed methodology could be easily extended to the generation of minute-scale radiation series, a field that still lacks a reference theoretical frame and for which data are becoming available nowadays.

Also this methodology has been used for the generation of solar radiation maps. To obtain a solar radiation map of a certain zone it is necessary to know the solar radiation of a large number of sites spread wide across the zone. This information is usually non-existent. As it was shown with a MLP is quite easy to obtain this huge number of solar data to solve this problem. This generation is extremely simple and takes less time than the same generation produced with the classical methods. Besides, the classical methods are unable to generate solar radiation series in places where no solar information is available. Nevertheless, the methodology proposed is able to do this satisfactorily; it is more versatile than the classical methods and thus can be used to draw the solar radiation maps of a zone. This methodology can easily be extended and applied to other places. The only requirement is the knowledge of the hourly solar radiation from only one site of the zone where the map is going to be drawn. Therefore, maps of different provinces, regions or even countries could be prepared. The extension of this methodology to other areas is, as has been demonstrated, quite easy. So, drawing solar radiation maps of more extended areas could be carried out.

In the photovoltaic field we have presented two applications. One is the application of the MLP for generating LOLP curves for sizing stand-alone PV systems. A supervised network can be successfully trained with the set of available data. The proposed methodology exploits the possibility of implicitly employing information associated with the problem, without knowing the existing relationships between different variables and sources of information, i.e., the proposed method does not assume any *a priori* model, as opposed to the standard approximation techniques where polynomial regression techniques are employed. Simulations have shown that the LOLP curves generated by the proposed method fit the real curves well. In the comparison with other analytical methods, the MLP-based method shows an improved performance with respect to the Quadratic Error. In addition, the fact that once the MLP trained for some LOLP values and in different locations works satisfactorily for other diverse values of LOLP and other locations, suggests that the new method has very good generalisation capabilities. Further studies with data from geographically more distant sites, not only for Spain, but for other countries, could be used to confirm this tendency.

The last application presented is for using the MLP for converting non STC V-I curves to STC ones. Again the supervised MLP can be successfully trained with the set of available data. Simulations have shown that the STC V-I curves generated by the proposed method fit the reference curves well. It is very important to point out that this work is only a first version of the method based in the MLP to extrapolate V-I curves. So in these first results we have only made comparison with only one module. Further work in this area is expected in which this methodology will be applied for many other modules with different features. Also the comparison of the results can be extended for other reference methods used for extrapolating the V-I curves.

REFERENCES

Aguiar R., Collares-Pereira M., 1991. Statistical properties of hourly global radiation. *Solar Energy,* Vol. 48, No. 3, pp. 157-167.

Aguiar R., Collares-Pereira M., 1992. TAG: A time-dependent, autorregressive, Gaussian model for generating synthetic hourly radiation. *Solar Energy,* Vol. 49, No. 3, pp. 167-174.

Alawi S.M., Hinai H.A., 1998. An ANN-based approach for predicting global radiation in locations with no direct measurement instrumentation. *Renewable Energy*, Vol. 14, No. 1-4, pp. 199–204.

Agarwal M., 1997. A systematic classification of neural-network-based control. *IEEE Control Systems Magazine,* Vol. 17, No. 2, pp. 75-93.

Bahgat A.B.G., Helwa N.H., Ahamd G.E., El Shenawy E.T., 2004. Estimation of the maximum power and normal operating power of a photovoltaic module by neural networks. *Renewable Energy,* Vol. 29, pp. 443-457.

Bahgat A.B.G., Helwa N.H., Ahamd G.E., El-Shenawy E.T., 2005. Maximum power point tracking controller for PV systems using neural networks. *Renewable Energy,* Vol. 30, pp. 1257-1268.

Balouktsis A., Tsalides Ph., 1986. Stochastic simulation model of hourly total solar radiation. *Solar Energy,* Vol. 37, No. 2, pp.119-126.

Barra L., Cataloni S., Fontana F., Lavorante F., 1984. An analytical method to determine the optimal size of a photovoltaic plant. *Solar Energy,* Vol. 33, No. 6, pp. 509-514.

Barron R., 1993. Universal approximation bounds for superpositions of a sigmoidal function. *IEEE Transactions on Information Theory*, Vol. 39, pp. 930-945.

Bartoli B., Cuomo V., Fontana F., Serio C., Silvestrini V., 1984. The design of photovoltaic plants: An optimization procedure, *Applied Energy,* Vol. 18, pp. 37-47.

Blaesser G., Munro D., 1995. Guidelines for the assessment of photovoltaic plants. *Document C. Issue 2. Report EUR 16340 EN.* Commission of the European Communities, Joint Research Centre of Ispra.

Bochud F., Nygard-Ferguson M., Scartezzini J.L., 1991. Stochastic models of hourly data of solar radiation and ambient temperature. *Proceedings of ISES World Renewable Energy Conference on CD-ROM,* Denver, Colorado, USA.

Caamaño E., Lorenzo E., Zilles R., 1999. Quality control of wide collections of PV modules: lessons learned from the IES experience. *Progress in Photovoltaics: Research and Applications,* Vol. 7, pp. 137-149.

Chapman R.N., 1987. *Sizing handbook for stand alone photovoltaic / storage systems*. Sandia National Laboratories, April 1987.

Curtiss PS, Brandemuehl MJ, Kreider JF., 1995. Energy management in central HVAC plants using neural networks. *In:* Haberl J.S., Nelson R.M., Culp C.C. (editors). The use of artificial intelligence in building systems. *ASHRAE*, pp. 199-216.

Cybenko G., 1989. Approximation by superpositions of a sigmoidal function. *Mathematics of Control, Signals and Systems,* Vol. 2, pp. 303-314.

Dagleman L. O., 1976. A weather simulation model for building analysis, ASHRAE Symposium on Weather Data, Seattle, WA Annual Meeting, June 1976. *ASHRAE Transactions*, pp. 435-447.

Egido M.A., Lorenzo E., 1992. The sizing of stand alone PV systems: a review and proposed method. *Solar Energy Materials and Solar Cells,* Vol. 26, pp. 51-69.

Excell R. H. B., 1981. A mathematical model for solar radiation in south-east Asia. *Solar Energy,* Vol. 26, pp. 161-168.

Fausset L., 1994. *Fundamentals of neural networks*. Prentice Hall.

Funahashi K., 1989. On the approximate realisation of continuous mappings by neural networks. *Neural Networks,* Vol. 2, pp. 183-192.

Gansler, R. Klein, S.A. Beckman, W.A., 1994. Assessment of the accuracy of generated meteorological data for use in solar energy simulation studies. *Solar Energy,* Vol. 53. No. 3, pp. 279-287.

Graham V.A., Hollands K.G.T, Unny T.E., 1988. A time series model for K_t with application to global synthetic weather generation. *Solar Energy,* Vol. 40, No. 3, pp. 269-279.

Graham V.A., Hollands K.G.T, 1990. A method to generate synthetic hourly solar radiation globally. *Solar Energy,* Vol. 44, No. 6, pp. 333-341.

Goh T., Tan K., 1977. Stochastic modelling and forecasting of solar radiation data. *Solar Energy,* Vol. 19, No. 6, pp.755-757.

Haykin S., 1994. *Neural networks. A comprehensive foundation*. Macmillan Publishing Company.

Hiyama T., 1995. Identification of optimal operating point for PV modules using neural networks for real time maximum power tracking control. *IEEE. Transaction on Energy Conversion*. Vol. 10, No. 2, pp. 360-367.

Hiyama T., Kouzuma S., Imakubo T., 1995. Evaluation of neural networks based real time maximum power tracking controller for PV system. *IEEE Transaction on Energy Conversion,* Vol. 10, No. 3, pp. 543-548.

Hittle D.C., Pedersen C.O., 1981. Periodic and stochastic behaviour of weather data. *ASHRAE Transactions,* Vol. 87, No. 2, pp. 545-557.

Hontoria L., Riesco J., Zufiria P.J., Aguilera J., 1999. Improved generation of hourly solar irradiation artificial series using neural networks. *5th International Conference on Engineering Applications of Neural Networks, EANN99,* pp. 87-92.

Hontoria, L. Riesco, J. Zufiria, P. and Aguilera, J. 2000. Application of neural networks in the solar radiation field. Obtainment of solar radiation maps. *16th European Photovoltaic Solar Energy Conference and Exhibition.* Glasgow, Vol. 3, Paper VD1.35, pp. 2539-2542.

Hontoria L., Aguilera J. ,Riesco J., Zufíria P., 2001. Recurrent neural supervised models for generating solar radiation synthetic series. *Journal of Intelligent and Robotic Systems,* Vol. 31, pp. 201-221.

Hontoria L., Aguilera J., Zufiria P.J., 2002. Generation of hourly irradiation synthetic series using the neural network multilayer perceptron. *Solar Energy,* Vol. 72, No. 5, pp. 441-446.

Hontoria L., Aguilera J., Zufiria P.J., 2003. A tool for obtaining the LOLP curves for sizing off-grid photovoltaic systems based in neural networks. *Proceedings of the 3rd World Conference on Photovoltaic Solar Energy Conversion on CD-ROM,* Osaka, Japan.

Hontoria L., Aguilera J., Zufiria P.J., 2005a. A new approach for sizing stand alone photovoltaic systems based in neural networks. *Solar Energy,* Vol. 78, No. 2, pp. 313-319, (Special Issue ISES Solar World Congress 2003).

Hontoria L., Aguilera J., Nofuentes G., Almonacid F., de la Casa J., 2005b.Contribution to quality control of PV modules: a new standard test conditions (stc) V-I curve conversion method using neural networks. *Proceedings of World Renewable Energy Congress on CD-ROM,* Aberdeen, United Kingdom.

Hornik K., Stinchcombe M., White H., 1989. Multilayer feedforward networks are universal approximators. *Neural Networks*, Vol. 2, No. 5, pp. 359-366.

Hush D. R., Horne B. G., 1993. Progress in supervised neural networks. What's new since Lippmann. *IEEE Signal Process Magazine,* Vol. 10, pp. 8-31.

Iqbal, M., 1983. *An introduction to solar radiation.* Academic Press.

Kajl S., Roberge M.A., Lamarche L., Malinowski P., 1997. Evaluation of building energy consumption based on fuzzy logic and neural networks applications. *Proceedings of CLIMA 2000 Conference on CD-ROM, Brussels,* Belgium.

Kalogirou S.A., Panteliou S., Dentsoras S., 1996. Modelling of solar domestic water heating systems using artificial neural networks. *Solar Energy,* Vol. 65, No. 6, pp. 335-342.

Kalogirou S.A., Neocleous C.C., Schizas C.N., 1997. Heating load estimation using artificial neural networks. *Proceedings of CLIMA 2000 Conference on CD-ROM, Brussels,* Belgium.

Kalogirou S., Eftekhari M., Pinnock D., 1999. Prediction of air flow in a single-sided naturally ventilated test room using artificial neural networks. *Proceedings of Indoor Air'99,* The 8th International Conference on Indoor Air Quality and Climate, *Edinburgh*, Scotland, Vol. 2, pp. 975-980.

Kalogirou, S.A., 2000. Applications of artificial neural-networks for energy systems. *Applied Energy,* Vol. 67, No. 1-2, pp. 17-35.

Kalogirou, S.A., 2001. Artificial neural networks in renewable energy systems applications: a review. *Renewable and Sustainable Energy Reviews,* Vol. 5, No. 4, pp. 373-401.

Kemmoku Y., Orita S., Nakagawa S., Skakibara T., 1999. Daily insolation forecasting using a multistage neural network. *Solar Energy,* Vol. 66, No. 3, pp. 193-199.

Kohonen, T., 1995. *Self-organising maps*. Springer Verlag, Berlin Heidelberg.

Kreider J.F., Wang X.A., 1995. Artificial neural network demonstration for automated generation of energy use predictors for commercial buildings. *In:* Haberl, J.S., Nelson,

R.M., Culp C.C. (editors). The use of artificial intelligence in building systems. *ASHRAE*, pp. 193-198.

Lalot S., 2000. Identification of the time parameters of solar collectors using artificial neural networks. *Proceedings of Eurosun 2000 on CD-ROM. Copenhagen*, Denmark.

Lippmann R.P., 1987.An introduction to computing with neural nets. *IEEE ASSP Magazine*, pp. 4-22, April 1987.

Liu B. Y. H. and Jordan R. C. 1960. The interrelationship and characteristics distribution of direct, diffuse and total solar radiation. *Solar Energy*, Vol. 4, No. 3, pp. 1-19.

Lorenzo E., Araujo G. L., Cuevas A., Egido M. A., Miñano J. C., Zilles R. 1994. Electricidad solar. Ingeniería de los sistemas fotovltaicos. Ed. Progensa.

Mellit A., Benghanem M., Hadj Arab A., Guessoum A., 2005. An adaptive artificial neural network model for sizing stand-alone photovoltaic systems: application for isolated sites in Algeria. *Renewable Energy,* Vol. 30, No. 10, pp. 1501-1524.

Mohandes M., Balchonaim A., Kassas R.S., Halawani T.O., 1998. Estimation of global solar radiation using artificial neural networks. *Renewable Energy,* Vol. 14, No. 1-4, pp.179-184.

Mustacchi C., Cena V. and Rocchi M., 1979. Stochastic simulation of hourly global radiation sequences. *Solar Energy,* Vol. 23, No. 1, pp. 47-51.

Narendra, K.S., Parthasarathy, K., 1991. Gradient methods for the optimisation of dynamical systems containing neural networks. *IEEE Transactions on Neural Networks,* Vol. 2, No. 2, pp. 252-262.

Negnevitsky M., Le T.L., 1995. Artificial neural networks application for current rating of overhead lines. *IEEE Transactions on Neural Networks,* pp. 418–222.

Rumelhart D. and MacClelland J.L., 1986. Learning internal representations by error backpropagation. *Chapter 8 from Parallel Distributed Processing,* Vol. 1: Foundations. The MIT Press.

Sfetsos A. and Coonick H., 2001. Univariate and multivariate forecasting of hourly solar radiation with artificial intelligence techniques. *Solar Energy,* Vol. 68. No. 2, pp. 169-178.

Sidrach-de-Cardona M., Mora López Ll., 1998. A simple model for sizing stand alone photovoltaic systems. *Solar Energy Materials and Solar Cells,* Vol. 55, No. 3, pp. 199-214.

Sidrach-de-Cardona M., Mora López L., 1999. A general multivariate qualitative model for sizing stand alone photovoltaic systems. *Solar Energy Materials and Solar Cells,* Vol. 59, No. 3, pp. 185-197.

Sözen A., Arcaklioglu E., Ozalp M., Caglar N., 2005. Forecasting based on neural network approach of solar potential in Turkey. *Renewable Energy,* Vol. 30, pp. 1075-1090.

Takashi H., Imakubo T., 1995. Identification of optimal operating point of PV modules using neural networks for real time maximum power tracking control. *IEEE Transactions on Energy Conversion,* Vol. 10, No. 2, pp. 360-367.

Veerachary M., Yadaiah N., 2000. ANN based peak power tracking for PV supplied motors. *Solar Energy,* Vol. 69, No. 4, pp. 343-350.

Wagner A., 2003. Photovoltaic measurement relevant to the energy yield. *Proceedings of the 3rd WCPEC on CD-ROM,* Osaka, Japan.

Weigend A.S. Rumelhart D.E., Huberman B.A., 1990. Back-propagation, weight-elimination and time series prediction. *Proceedings of the 1990 Connectionist models Summer School on CD-ROM.* Morgan Kaufman.

Zufiria P.J., 1995. An overview of dynamic system control using neural networks. *Neural Networks for Chemical Engineers* (Editor: Bulsari A.B.). Elsevier, pp. 385-408, ISBN 0-444-82097-3.

Zufiria P.J., Vázquez A., Riesco J., Aguilera J., Hontoria L., 1999. *A neural network approach for generating solar irradiation artificial series.* Notes in Computer Science 1607, Mira Juan V. Sánchez-Andrés (Editors), Applications of Bio-Inspired Artificial Neural Networks. Work-Conference on Artificial and Natural Networks, II-874-883, Springer Verlag, Berlin Heidelberg.

In: Artificial Intelligence in Energy and Renewable Energy Systems ISBN 1-60021-261-1
Editor: Soteris Kalogirou, pp. 201-233

Chapter 6

APPLICATION OF CONTROL ALGORITHMS FOR WIND SPEED PREDICTION AND ACTIVE POWER GENERATION

Arantxa Tapia, Gerardo Tapia and Patxi Flores
The University of the Basque Country (Polytechnic Faculty)
Donostia-San Sebastián, Spain

ABSTRACT

The main objective of this chapter is to propose a new method for predicting wind speed using artificial neural networks. It should be noted that the method provides predictions in time-scales that can vary from a few minutes to an hour. This is needed because the proportion of wind energy in energy distribution is increasing and has to compete with other energy sources that are not so variable in terms of generated active power. Furthermore, the deregulation of the electric power industry has created new requirements for producers in several electricity markets. In addition to the traditional monitoring, control, dispatch, analysis, optimisation and planning tools which are available for power systems, energy producers require tools to manage their physical and financial assets so as to participate in different markets as well as decision support tools to participate in a very competitive environment.

It is important to consider that nowadays, active power demand can vary quite rapidly because different sources of electricity generation are available. In the case of wind energy, wind speed prediction is an important tool that helps producers to make the best decisions when selling the energy they produce. These decisions are crucial in the electricity market, because of the economic benefits for producers and as a consequence their profitability depends on them. The wind speed prediction algorithm presented in this work is based on artificial neural networks. Two types of wind data have been used to test the algorithm. In the first type, data were collected from an area that is not very windy whereas in the second type, data were collected from a real wind farm located in Navarre (North of Spain) and the values vary from very low to high speeds. Even though the algorithm was not tested with typical wind speed values measured on offshore wind farm applications, from the first set of results presented in this chapter it can be concluded that the algorithm is also suitable for estimating average wind speed values in these cases.

Finally, a very generic algorithm for the control of active power generation in a wind farm is presented. By taking into account the designed artificial neural network (ANN) model prediction results for the wind speed at the wind farm, the active power generation policy can be established by considering two different situations:

- Optimisation of the generation; tracking the path of the maximum active power generation points for the best economic exploitation conditions. This means that all the energy produced can be sold.
- A constant generation reference fixed by the buyer and agreed beforehand (one hour) with the power utility.

Keywords: Short-time wind power forecasting, prediction algorithm, artificial neural network, wind power generation, double fed induction machine.

NOMENCLATURE

C_p	Power coefficient
d_j	Desired output
N	Number of generators in a wind farm
P_{Nom}	Nominal power (W)
P_m	Mechanical power (W)
P_{opt}	Maximum (optimum) power (W)
P_u	Utility power needs (W)
$P_{s\,ref}$	Available active power at the wind farm (W)
$P_{s\,max\,i}$	Maximum active power generated by the i generator (W)
$Q_{s\,ref}$	(= $Q_{n\,total\,ref}$) Reactive power that the wind farm has to generate/absorb (set-point) (VAR)
$Q_{n\,total}$	Really generated/absorbed reactive power at a wind farm (VAR)
$Q_{s\,max\,i}$	(= $Q_{i\,ind\,max}$ = $Q_{i\,cap\,max}$) Maximum reactive power that the i generator can generate/absorb (VAR)
r	Blade radius (m)
v	Wind speed (m/s)
W_j	Weight vector
X	Input vector

Greek

β	Pitch angle
λ	Tip speed ratio
ρ	Air density (g/m^3)
ω_r	Rotational speed (rad/s)

ABBREVIATIONS

ANN	Artificial Neural Network
DFIG	Double Fed Induction Generator
DFIM	Double Fed Induction Machines
LBF	Linear Basic Function
OPF	Optimal Power Flow
PV	Photovoltaic
PWM	Pulse Wide Modulation
SVM	Support Vector Machines

1. INTRODUCTION

Distributed energy generation units became one of the most important energy sources during the last few years. Even though these generation units bring important advantages for utilities and customers, the problems associated to the total amount of power connected to a distribution network are continuously increasing. The most important ones are problems related to the lack of regulation of frequency and voltage of the generation units, problems related to the non-dispatchable characteristic of most of the dispersed generation and problems related to assets management.

Several types of electric power generation are emerging with various degrees of development and future prospects (co-generating units, PV cells, fuel cells, etc), but today, one type of such systems, namely wind farms, is increasing its presence in a continuous and amazing way. This is due to several reasons, i.e., wind generators are available in a useful capacity range, they have zero fuel cost, their production is environmentally clean and many more.

In order to avoid the fact that dispersed and not controlled generation adds risks to the exploitation of electrical systems and decreases the quality of electrical service (electrical variables out of the exploitation range, risk of outages, etc.), utilities are trying to properly locate new generation units and more recently, to control the voltage in new types of wind driven generators.

Wind energy is one of the most important and promising sources of renewable energy all over the world, mainly because it is considered to be non-polluting and is economically viable. At the same time, there has been a rapid development of related wind turbine technology. Moreover, the deployment of wind farms is linked to the particular geographical characteristics of the sites selected according to the mean wind speed registered in a year. Wind farm power generators have to work under sudden changes of wind speed trying always to obtain maximum electrical power.

Different wind speeds can cause different rotor angular speeds and consequently, produce different amounts of active/reactive power on each generator of the wind farm. This behaviour introduces various problems in the control of the electricity grid.

Nevertheless, this kind of electric power generation usually causes problems in the electrical system it is connected to, because of the lack or scarcity of control on the produced active and reactive power. Several designs and arrangements have been implemented so as to cope with this difficulty.

As far as variable-speed generation is concerned, it is necessary to produce constant-frequency electric power from a variable-speed source. This can be achieved by means of synchronous generators, provided that a static frequency converter is used to interface the machine to the grid. An alternative approach consists in using a wound-rotor induction generator fed with variable frequency rotor voltage. This allows fixed-frequency electric power to be extracted from the generator stator. Consequently, the use of double fed induction machines (DFIM) is receiving increasing attention for wind generation purposes. One of the main advantages of these generators is that, the rotor current is governed by applying stator-flux-oriented vector control, carried out using commercial double-sided Pulse Wide Modulation (PWM) inverters, which results in decoupled control of stator side active and reactive power.

The ability to generate electricity with power factors different than one ($\cos\varphi \neq 1$) would reduce the costs of introducing additional capacitors for reactive power regulation and is especially advantageous for both producers and distributors in charge of transmission systems.

In order to obtain the maximum active power from the wind speed driving the blades of the generators, double fed induction machines have been used in wind farms built lately. But, the models of these generators are complicated since they have to consider many different variables and consequently, the modelling of a complete wind farm could become a very complicated task as the number of generators in the wind farm increases.

Even though the production of wind energy has been increasing during the last decade, power utilities have not perceived the connection of dispersed independent power producers to electrical networks as a threat until recently, mainly due to the fact that power from such producers represent only a small fraction of the total installed power. However, this scenario has varied drastically over the last few years as wind energy has grown to be one of the most important and promising sources of renewable energy (Ackermann and Saoder, 2002; Bourillon, 1999). In particular, the increased expectations for wind power in Spain are certainly remarkable, in view of the fact that 20GW associated to this energy source are planned to be installed in the Spanish power system by 2010. An important amount of this energy will be supplied to the transport network, while the rest will be transferred to the distribution network, complying with administrative and technical directives devised to ensure the quality of generated power. A similar expansion of wind energy systems is planned in many other countries of the world.

One aspect related to power quality is the management of the reactive power exchanged between wind farms and the grid. From this point of view, power factors lower than 0.86 will not be permitted for asynchronous generators at their rated power, so that capacitor banks or equivalent reactive power sources, together with protection devices to ensure their appropriate disconnection, will need to be provided. As far as double fed induction generators (DFIGs) and synchronous generators are concerned, their power factors will have to be kept between 0.8 and 1 — leading or lagging — at their connecting point and devices to ensure synchronism will also be compulsory. In addition, it is essential to take into consideration the fact that wind farms made up of DFIGs provide excellent results in sites where, owing to their specific orography, wind conditions are extremely variable. Finally, it is important to consider that these generators are able to optimise the active power production at each wind speed, while independently controlling the generated or consumed reactive power (Bhowmik *et al.*, 1999; Spee *et al.*, 1995; Peña *et al.*, 1996; Tapia *et al.*, 2003; 2004).

Conveniently exploiting this feature by means of adequate local control algorithms, the power factor of each DFIG can be modified independently so that it complies with the technical directives imposed by power utility companies. In fact, the most widespread mode of operation up to now consisted of tuning the local control system of each DFIG individually, so that its corresponding reactive power is regulated around zero.

Another interesting aspect of wind energy production is related to the non-predictable characteristics of the wind speed. Taking this into account and the fact that wind is a result of solar radiation, it is important to consider the heating of the air that occurs owing to this radiation. As the Earth surface is not heated uniformly (the heating effect depends mainly on the latitude), pressure differences occur in the atmosphere and the air flux tries to balance them out, causing the movement of an important mass of air – the wind.

Based on this principle, several wind farms have been built with hundreds of MW installed. Wind is considered however to be one of the most difficult meteorological phenomena to predict and this is required in order to optimise the operation of wind farms. Therefore, the use of statistical methods for reliable results (with errors maintained between pre-defined limits) becomes necessary in order to predict more accurately the wind speed hitting the blades of the generators located in a wind farm in very short time periods (hours). Combining these predictions with advanced control algorithms, the efficiency of the wind farm could be improved. The total amount of installed wind power worldwide is approximately 40.000 MW, therefore, the economic benefits obtained using modern technology, combining prediction and control, could reach important values (Fabbri *et al.*, 2005).

Some of the recently published papers refer to the estimation of wind speed using several algorithms, such as the neural network algorithm based on support vector machines (SVM) and comparing its performance with the multilayer perceptron algorithm (Mohandes *et al.*, 2004). The most typically used artificial neural network (ANN) is the one based on multilayer structure which has been tested by comparing its results with an autoregressive model (Mohandes, 1998) and the traditional statistical time-series analysis (More and Deo, 2003). A number of wind speed predictors based on ANNs are compared and classified by Kretsschmar *et al.* (2004) taking into account several data collected in Switzerland. ANNs have also been used in several types of renewable energy applications (Kalogirou, 2001).

This chapter proposes to move one step further in active/reactive power regulation so that, in addition to operating its DFIGs independently, the wind farm is also controlled as a whole. In this way, wind farms would not only be able to reduce the disturbances they cause on the grid and so comply with technical directives, but could also contribute to the control of the voltage profile of the electrical network. This possibility arises from a consideration of the theoretical idea of using electricity power plants as continuous sources of reactive power for secondary voltage control strategies (Nakamura and Okada, 1969; Narita and Hamman., 1973; Mamandur, 1982; Denzel *et al.*, 1988; Palanichamy and Srikrishna, 1989; Paul *et al.*, 1990; Arcidiacono *et al.*, 1990; Gubina and Gubina, 1997).

In this chapter, the ideas described are applied to a specific case of a wind farm made up of DFIGs. Reactive power control is achieved by simply adding to the existing control algorithm an extra control-loop to provide a power factor set-point for every DFIG local control system. The complete production algorithm is implemented in the central computer of the wind farm. Moreover, taking into consideration the fact that the capability of wind farms to produce or consume reactive power increases as the amount of generated active power decreases, it follows that they can contribute more significantly to the maintenance of the grid voltage profile at low wind speeds (Tapia *et al.*, 2002).

Thus, although a reduction in wind speed causes wind farm owners to make smaller profits, such a loss could be balanced in the near future if secondary voltage control is paid by power utility companies as an added service. Furthermore, in this hypothetical scenario these utilities might decide, depending on the grid requirements, to favour secondary voltage control rather than trying to optimise the active power generation even at high wind speeds.

Finally, the geographical location of wind farms is another fundamental question. Because wind energy is disperse, intermittent and not very regular in terms of the amount of generated power it is important to consider that when a "good location" is identified and used, a generator can work 6000 hours per year and the amount of energy produced can be similar to the energy produced by a conventional generator working at full power for 2200 hours per year. Therefore, before a wind farm is built, the local wind conditions must be accurately evaluated and the wind speed predictions must be thoroughly analysed. The method presented in this chapter is useful for these predictions and additionally it can be used for wind energy optimisation.

2. Problem Description

The study described in this chapter deals with the development of a simple and robust algorithm that analyses short-term wind power forecasting. The algorithm is integrated into the configuration of a power management system which considers all elements of the wind generating system in order to optimise the power flow and to prevent possible outages. All these elements are connected by means of a local information network whose role is to coordinate in a better way the functional parameters of each generator in a farm and to evaluate the consequences of disturbances in the circuit or of different consumer regimes, i.e., to supervise all elements that make up the wind farm power plant. By establishing the local network, it is possible to:

a) Improve the modeling of generators, consumers and network elements included in the wind farm,
b) Integrate new information related to the global functional parameters of power transmission and distribution systems,
c) Manage energy resources intelligently by including some forecast information in the model developed,
d) Develop new control strategies by taking into consideration the magnitudes of production and consumption.

Currently, in several countries (such as Spain) a large amount of energy is produced by hydro-electric plants but to cover the overall electricity requirements the electric companies have to compensate the fluctuations in demand by including other types of energy production. As wind energy varies according to the weather, depending on the wind speed hitting the generator blades, the possibility of predicting the wind energy production in the following hour becomes crucial for wind farm owners in order to allow them to work effectively on the electricity market. These predictions help producers to take decisions about the sale of energy and thus to increase production and profits. If an accurate prediction of the wind speed for the following hour can be calculated, the total amount of active power that can be produced by each generator on a wind farm can be determined and therefore, the amount of energy that could be sold during the following hour would be known.

Another important consideration is that the wind energy generated by wind farms and their operation and connection to the grid are governed by a system of switches and protections. The protections provide the correct working conditions at individual unit level (wind generators), for consumers and for connection to the grid. The above mentioned conditions can be classified in three different situations:

i) Excess power is generated; in this case, if the excess power could not be delivered to the grid, the power generation has to be reduced and the available wind power is not fully captured.
ii) Generated power meets the consumer demands; in this case all wind energy generated is consumed.
iii) The generated power is less than that required by consumers; in this case the deficit has to be covered from other sources.

In order to analyse the amount of energy that can be produced by a generator (produced active power), the wind speed prediction for the wind farm location and some aerodynamic test results of the generator are needed. So, it is important to consider the mechanical power, P_m, developed by a wind turbine (Bhowmik *et al.*, 1999; Spee *et al.*, 1995; Peña *et al.*, 1996, Tapia *et al.*, 2003; 2004), which depends directly on the blade radius, the power coefficient and the wind speed hitting the blades of the generator. Thus:

$$P_m = \frac{\pi}{2} C_p(\lambda, \beta) \rho r^2 v^3 \tag{1}$$

and

$$\lambda = \frac{\omega_r r}{v} \tag{2}$$

As an example, the power coefficient of a 660kW generator (aero-dynamical test results) typically installed on wind farms is shown in Fig. 1. This illustrates the fact that for any given wind speed there is a rotation speed, ω_r, which generates the maximum power $P_m = P_{opt}$. Therefore, following all the maximum power generation points, the optimum power curve can be obtained.

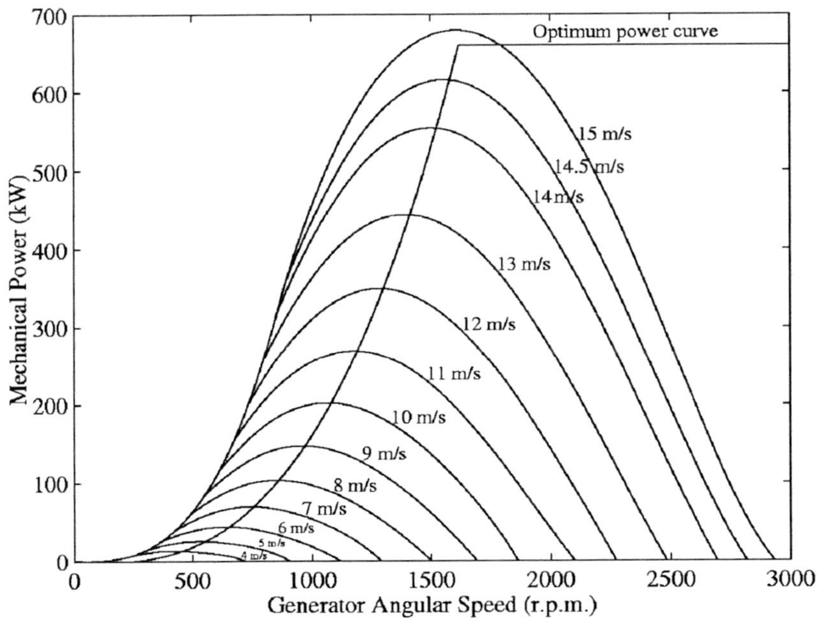

Fig. 1. Curve of optimum Cp for a 660kW turbine

Moreover, it is important to take into account the fact that when wind speed changes, the rotation speed varies in response to the change. But, since a reliable measurement of the wind speed hitting the blades of the generator cannot be guaranteed, the equation for obtaining the target power, P_{opt}, can be modified by eliminating the dependence on wind speed. Thus:

$$P_m = P_{opt} = \left(\frac{\pi}{2}\right) C_{p\,opt} \rho r^2 \left(\frac{r}{\lambda_{opt}}\right)^3 \omega_r^3 \quad (3)$$

As can be seen from Eq. (3) the generated power is a proportional function of the rotor speed and follows the cubic law:

$$P_{opt} = k\,\omega_r^3 \quad (4)$$

But, even though the generated power can in theory be unlimited, it is in fact limited because wind turbine components can only be operated within their mechanical and electrical constrains. Consequently, when the reference value of P_{opt} is greater than the nominal value of the electric turbine, P_{Nom}, the pitch control is activated (to generate the nominal value) and the optimum active power reference is not followed until $P_{opt} \leq P_{Nom}$.

The economic benefits of the active power generation would be increased (based on the optimum power curve) using the algorithms analysed in this work and by taking into account the expected evolution of the market regulatory framework. The economic benefits could also be increased if the reactive power is remunerated.

Additionally, the impact of implementing wind speed forecasts in the control algorithms will be analysed so that wind energy production can be used in the new electricity market. This means that the possibility of predicting wind speed has to be considered. There are a few studies in literature related to the very short-range local wind speed predictions which are based on non-statistical paradigms of the meteorological variables that are the most influential (Beyer *et al.*, 1994; Bossanti, 1985; Masón, 1982; Sfetsos, 2002). The forecasting methods mentioned in the literature however have a number of disadvantages:

(1) They rely on the knowledge and experience of a meteorologist.
(2) They are carried out by interpolating measured data from a few sources located over extensive areas and over long periods of time, instead of linking the problem to a source, a geographical point and an immediate period.

Different prediction techniques have been developed over the last couple of decades in order to predict the wind speed with several levels of resolution. In some of these techniques atmospheric pressure, temperature gradients, air humidity and other factors that contribute to the movement of air were not taken into account in order to simplify the prediction method, but the obtained prediction is not accurate enough.

An accurate forecast of wind speed is crucial for energy production purposes. In this work, a neural network methodology is applied to construct a meteorological forecasting tool which predicts the local short-, medium- and long-term wind speed. The advantages of the neural networks method over other prediction methods, such as the autoregressive integrated moving average (ARIMA), Kalman filters, exponential smoothing or systems of learning based on rules, are basically the following:

i) It is an entirely numeric method, as opposed to the implicit symbolism in the systems of rules,
ii) It does not require any previous knowledge of the system that is to be predicted,
iii) It is characterised by robustness and tolerance to noise.

On the other hand, the disadvantages of the neural networks method are well-known and can be summarised as:

i) They do not contribute any information about the background to the problem, while rule-based systems are more easily comprehensible and,
ii) In the majority of cases they do not permit any incremental training.

Considering the advantages and disadvantages mentioned above, it is important not to forget that the method used in this study must evaluate and predict the wind speed in one hour periods (good enough for the purposes of wind power generation) and it is based on one of the best known techniques in wind energy analysis which can also be applied to other types of renewable energy evaluations, the Artificial Neural Networks method – ANN (Beyer *et al.*, 1994; Bishop, 1996; Bossanyi, 1985; Masón, 1982; Sfetsos, 2002; Watson *et al.*, 1993; Murphy and Daan, 1985; Kalogirou, 2001).

As mentioned in the introduction, wind speed is a difficult parameter to predict but if this is done correctly it can be used as a method for achieving better levels of production at a wind farm and consequently to improve the owner's profits. Additionally, if the method employed is not expensive and is easy to use and maintain, a continuous improvement in the algorithm can be achieved and the benefits obtained are increased while, at the same time the owner's satisfaction is increased.

So, based on the considerations outlined above, this chapter describes the control algorithm developed for wind speed prediction purposes as a starting point for the necessary studies in the field of wind energy production optimisation and its connection to the electricity network.

3. PREDICTION ALGORITHM

The algorithm selected for wind speed prediction presented in this chapter is based on the feed-forward method, which means that the information is transferred in only one direction, from the input layer to the hidden and then to the output layers. Therefore, the algorithm developed is quite simple and the computational needs for its implementation are quite low and consequently it is capable of being inserted into the control hardware of the wind generators that are currently being used.

Additionally, both the network function and the propagation/activation function must be defined before the complete algorithm is developed. In the former the most typical function is the linear basic function (LBF). It measures all the inputs from the neuron and adds them in a weighted way as shown in Eq. (5).

$$x_j(Y,W_j)=\sum_{i=1}^{n} y_i w_{ij} \tag{5}$$

This is a first order function whose implementation is very simple and consequently, it is adopted in the algorithm described in this chapter. Concerning the activation function and even though the linear functions are easier to implement than the non-linear ones (not proportional to the input signals); the latter are the most widely used functions. Taking this factor into account, the sigmoidal function, shown in Fig. 2, was selected as the most appropriate for the wind prediction algorithm because:

i) It can work with weak signals,
ii) It is derivable $f'(x)=f(x)\cdot(1-f(x))dx$ and
iii) Its curvature is similar to the optimum active power generation curve of wind generators (see Fig. 1).

The sigmoidal function can be described as:

$$y_j=\frac{1}{(1+e^{-x_j/\sigma})} \tag{6}$$

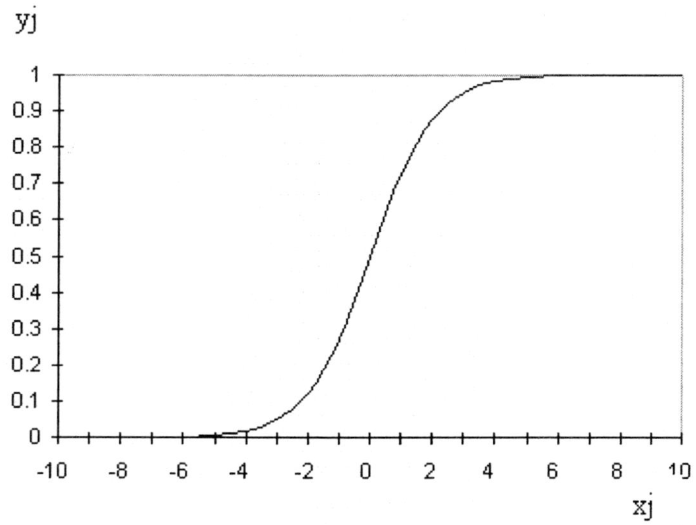

Fig. 2. Sigmoidal function with σ = 1

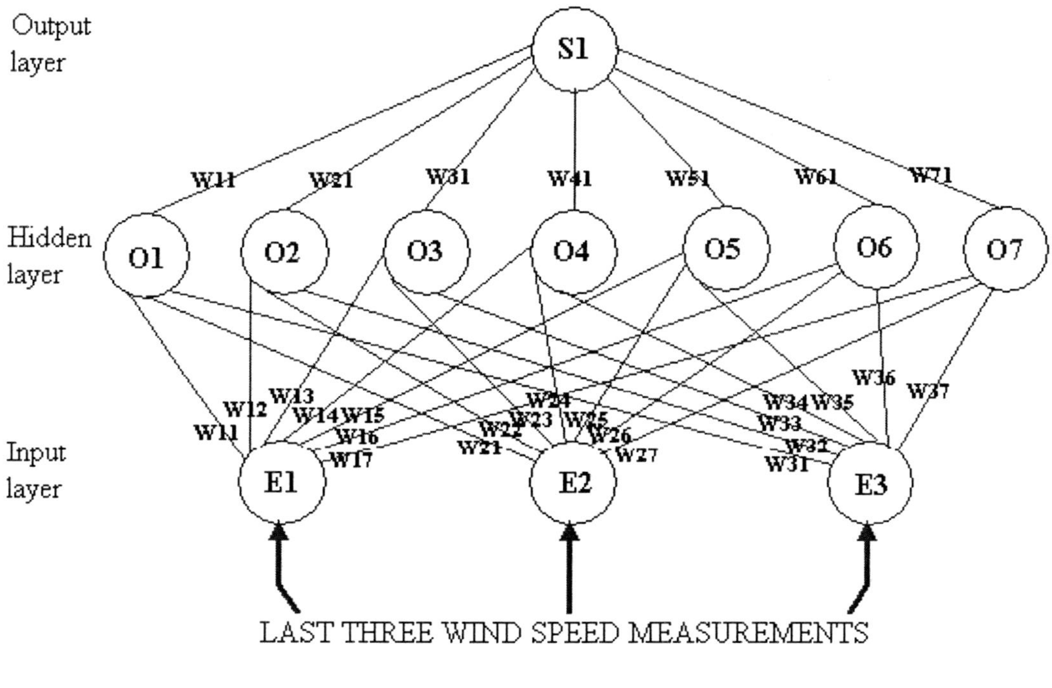

Fig. 3. Neural network architecture employed

The final factor to be taken into consideration before the prediction algorithm can be developed is the learning method to be used in the system. In this case, the back-propagation method has been selected. This is used to adjust the connection weight values in the network through a backward propagation of the error (from the final layer to the initial one) in order to minimise it and to comply with the minimum quadratic error criterion established beforehand by the user. The network architecture is shown in Fig. 3. The input layer comprises three neurons, corresponding to the last three wind speed measurements, there are seven neurons in the hidden layer and there is only one neuron in the output layer, corresponding to the present wind speed.

Therefore, the learning method applied in the algorithm presented in this chapter can be described with the following features:

- It is a non recurrent method
- It uses a linear propagation function
- The activation function is a sigmoidal function
- The different layers are totally interconnected
- Its operation is totally synchronous

Finally, the above mentioned learning algorithm can be implemented, after defining the quadratic error value that must be achieved, via the following steps:

i) Make the adjustment of the connection weight values a random process.
ii) Present a normalised input pattern and calculate the output.
iii) Activate the weighted input and its output with the j unit of the output layer and the i units of the hidden layer, Fig. 4, by means of Eq. (7) and (8).

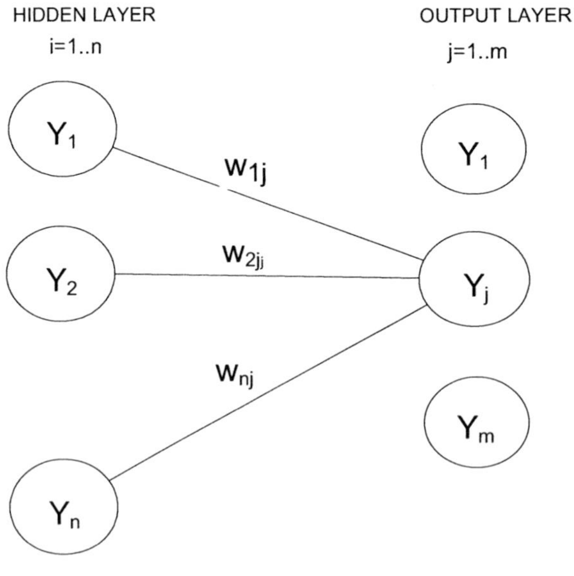

Fig. 4. Activation method

$$x_j = \sum_{i=1}^{n} (y_i \cdot w_{ij}) \tag{7}$$

$$y_j = \frac{1}{1+e^{-x_j}} \quad (8)$$

iv) Estimate the error when all the output unit activities have been computed. In this case, a quadratic function for the individual error of each unit has been used.

$$E = \frac{1}{2}\sum_j (d_j - y_j)^2 \quad (9)$$

v) Modifications in the error calculation are evaluated and those changes occur at a specific speed when the activity of the output unit changes.

$$EA_j = \frac{\partial E}{\partial y_j} = d_j - y_j \quad (10)$$

vi) Modifications in the error calculation are evaluated and those changes occur at a specific speed when the total input of the output unit changes.

$$\delta_j = \frac{\partial E}{\partial x_j} = \frac{\partial E}{\partial y_j}\frac{\partial y_j}{\partial x_j} = EA_j y_j (1-y_j) = y_j(1-y_j)(d_j - y_j) \quad (11)$$

vii) Modifications in the error calculation are evaluated and those changes occur at a specific speed when the connection weight of the output unit varies.

$$EP_{ij} = \frac{\partial E}{\partial w_{ij}} = \frac{\partial E}{\partial x_j}\frac{\partial x_j}{\partial w_{ij}} = \delta_j y_i \quad (12)$$

$$EA_i = \frac{\partial E}{\partial y_i} = \sum_{j=1}^{m}\left(\frac{\partial E}{\partial x_j}\frac{\partial x_j}{\partial y_i}\right) = \sum_{j=1}^{m}(\delta_j \cdot w_{ij}) \quad (13)$$

viii)Modify the above-mentioned weight with the error variation ratio against a connection weight to minimise the output error. In this algorithm, the generalised delta rule has been used to modify the weights.

$$\delta_i = EA_i y_i (1-y_i) = y_i(1-y_i)\sum_{j=1}^{m}(\delta_j w_{ij}) \quad (14)$$

$$EP_{hi} = \delta_i \cdot y_h \quad (15)$$

$$\Delta w_{ij} = \eta \delta_j y_i ;\ 0.01 \le \eta \le 1.0 \quad (16)$$

$$w_{ij}(t+1) = w_{ij}(t) + \Delta w_{ij} \quad (17)$$

4. Validation of the Prediction Algorithm

In order to test the validity of the ANN-based algorithm for wind speed prediction the following procedure is used. The first step is related to the ANN training. This is carried out by establishing different quadratic error values so as to obtain the minimum global error. Several tests have to be performed to find the minimum global error by considering both the global error and the epochs needed (and thus the time required) to finish the training of the network. Therefore, it is required to find a balance between the established quadratic error, the global error obtained and the time spent to finish the network training.

Finally, the verification process is carried out using real wind speed data and several prediction situations have to be evaluated in order to validate the generic algorithm used.

4.1. Experimental Results

Once the validation process is defined, the performance of the prediction algorithm was tested using two different sets of wind speed data. The first data set correspond to wind speed measurements taken every second for a number of hours at a wind farm located in Navarre. Data for one complete hour, i.e., 3,600 measurements, correspond to one sequence. In Fig. 5 a typical sequence is presented and as can be seen the wind speed change from very low to high values in a very short time. The most interesting aspect of these data sequences is that they were measured with a high frequency, one sample per second and therefore, the ANN can be trained to provide predictions at very short time intervals. It should be mentioned that the wind speed has been measured away from the wind turbine clusters and consequently the turbulence created by the wind generator blades is not affecting the measurements. Additionally, even though a large number of sequences were available (more than fifty) only ten of them have been used, five for the network training and five for validation, i.e., 18,000 patterns for each.

The second set corresponds to continuous measurements collected over four months (June-September 2002), in an area that is not so windy (Zizurkil) in the North of Spain, see Fig. 6 (EUSKALMET, Servicio Vasco de Meterología). In this case, as can be observed the wind speed varies from 1m/s to 5m/s, Fig. 6, and these values are classified as low values for wind energy generation purposes. Even though this site could not be considered as a proper location for a wind farm and for wind generation purposes, the frequency of measurements is very low – 1 measurement per hour (average value of wind speed recorded every two minutes) – and consequently, the tests developed can help to evaluate the performance of the algorithm. The wind speed predictions developed with these data were evaluated over longer periods than those in the first case and so the performance of the ANN-based algorithm in the two different types of data was analysed. In this case, a total of 3000 data patterns were available as shown in Fig. 6. Patterns corresponding to 1460 hours are used to train and validate the network; patterns for 1000 hours are used for the training and 460 for the validation. So, the collected data are divided in two groups (training and validation) corresponding to about two months of measurements for each group.

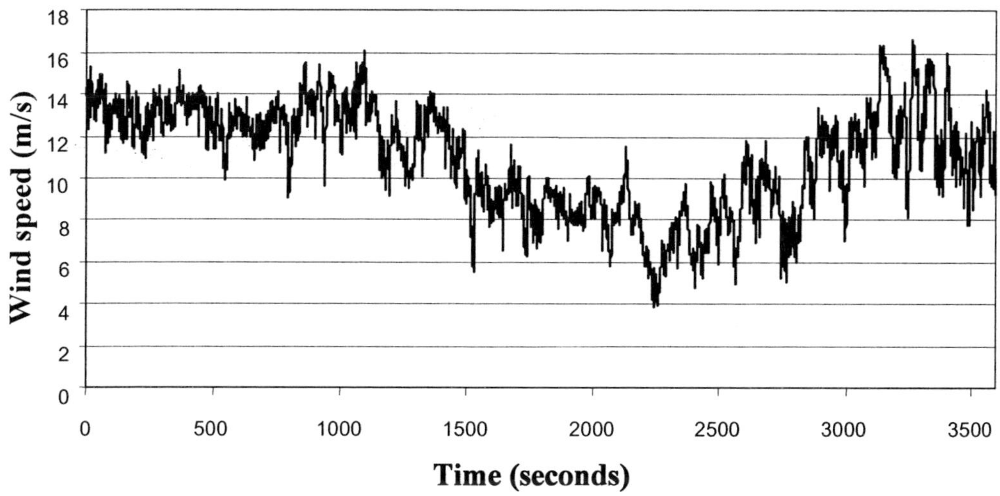

Fig. 5. Wind speed measured in Navarre

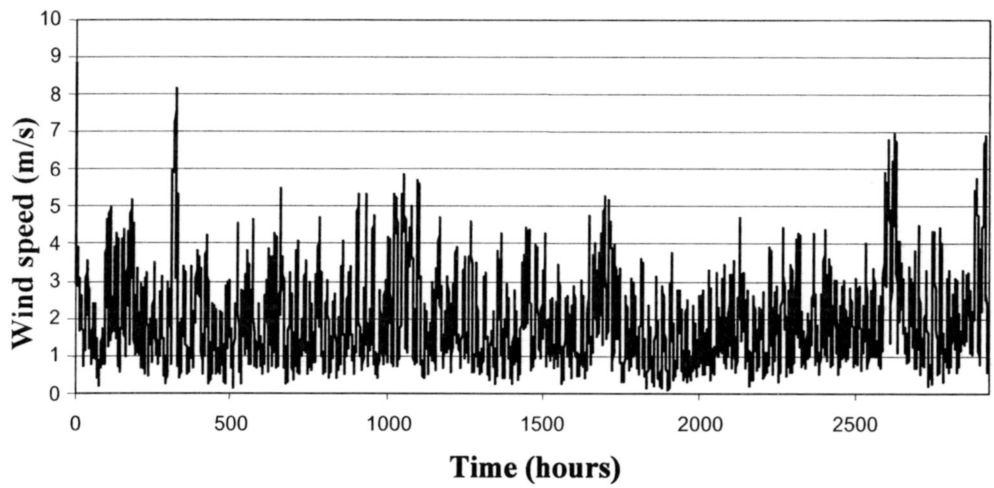

Fig. 6. Wind speed measured in Zizurkil

Finally, it is important to point out that nowadays one hour predictions are considered for wind energy production purposes but this scenario is continuously changing and thus, shorter time periods have been considered in the algorithm development. Consequently, if the developed algorithm proves to be valid under both conditions (on a second- or hour-basis) it would not be necessary to change the algorithm even if new shorter time periods are used in the near future.

4.2. Case 1: Measurements in a Real Wind Farm-Navarre

The ANN training was carried out by using the training data set (5 sequences) described above (one sequence is shown in Fig. 5) and establishing different quadratic error values to obtain the minimum global error. First, a permitted quadratic error value, between 0.01 and 0.001 was introduced and the final global error obtained was evaluated. If the permitted error value is quite near 0.001, the training process will be longer and more exact but if it is near 0.01, the training is shorter but it is not so accurate.

In Table 1, the relation between the established quadratic error and the real error obtained is compared, for a number of training epochs.

Table 1. Error in the ANN with different quadratic error values

Test number	Training epochs	Permitted minimum quadratic error	Obtained error
1	1008	0.005	0.102106
2	1368	0.004	0.086384
3	1040	0.003	0.071022
4	3574	0.002	0.046677
5	5927	0.00108	0.025341
6	10130	0.00105	0.025268
7	Training not finished	0.001	-

From the results presented in Table 1 it can be concluded that an error of 0.025341 is considered as acceptable because by reducing the permitted quadratic error from 0.00108 to 0.00105 the global error is not significantly reduced (from 0.025341 to 0.025268) but the training process becomes quite long (from 5,927 to 10,130 epochs). If the quadratic error is smaller than 0.00105 the improvement obtained in the overall error will not be substantial when compared to the duration of the training process, which would be too long (more than two minutes) for the whole control algorithm purposes. It should be noted that the time required for the training process has been analysed under the best conditions, with only the present algorithm running on the PC and the PC used is quite fast with high memory capability. In an actual situation, the control algorithm would be working on a small computer in the wind farm at the same time with other algorithms that are needed for the wind farm operation in order to produce energy in a proper way. The other algorithms are used for communication purposes, protection of wind turbines, production policy implementation and many more. So the time required for training is very important.

The error evolution in the ANN used with the above mentioned 0.00108 quadratic error value can be observed in Fig. 7. As can be seen from epoch 2500 there is no substantial improvement in the error.

Even though the statistical criterion R^2 could be considered as a good measurement of the accuracy of the obtained performance, it does not give additional information about the predictive capability of the developed algorithm and it has not been used during the validation process; only quadratic values have been evaluated during the whole process. This means that in some cases, the global minimum error could not be obtained but if a pre-established minimum error is achieved, the developed prediction could be assumed to be acceptable.

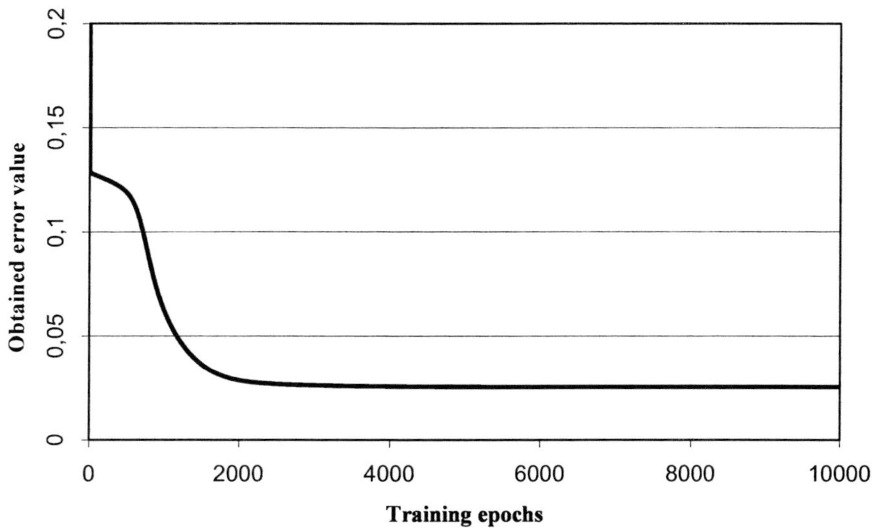

Fig. 7. Evolution of the error during the ANN training

Once the value of the permitted quadratic error is established, the training process is considered as finished. A comparison of the actual and ANN predicted wind speed for the training data set is shown in Fig. 8 (sample). A sample of the results obtained in the verification process appear in Fig. 9. All wind speed data values included in the training and validation data sets are measured at the real wind farm.

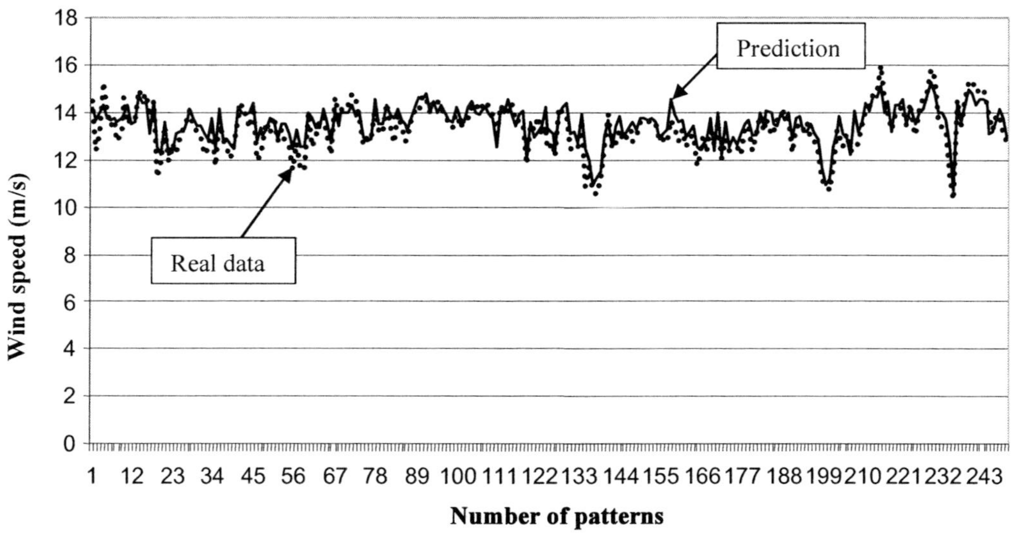

Fig. 8. Comparison of the actual and ANN predicted wind speed-Training set (sample)

As can be observed from the verification results, Fig. 9, the neural network prediction is quite acceptable and the algorithm developed shows quite good accuracy even though the wind speed varies considerably from second to second. Finally, as the results obtained by training the network with data collected every second are quite acceptable, it can be concluded that the algorithm could be used for predicting on an hourly basis the wind speed for the following hour in the same area by taking into account the set of wind speed data from the preceding three hours. Therefore, by means of the developed ANN architecture and by using the last three wind speed values (of the last three hours) as inputs, the wind speed of the next hour can be predicted. Then, based on that predicted wind speed value, the active and reactive power that can be produced by a wind farm could be foreseen and the average production capability of a wind farm would be used in the electriciy market. This hour-basis situation is also tried with data collected every hour in case two presented below.

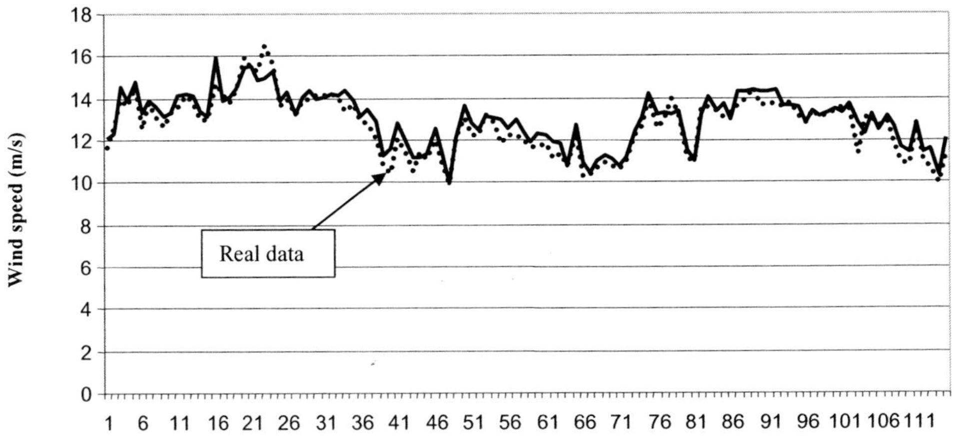

Fig. 9. Comparison of the actual and ANN predicted wind speed-Verification set (sample)

4.3 Case 2: Measurements in an Area that is Not so Windy - Zizurkil

The second case analysed is quite different from the first one from the wind speed variation point of view. Nevertheless, using the same method that has been explained for the wind speed data measured in the first case, the obtained results are summarised in Table 2 and Figs. 10 and 11.

Table 2. ANN Error with different quadratic error values

Test number	Training finished (epoch)	Permitted minimum quadratic error	Obtained error
1	1053	0.005	0.108065
2	1276	0.004	0.097511
3	1613	0.003	0.084982
4	2748	0.0025	0.06955
5	3841	0.0023	0.066133
4	13512	0.002	0.059594
5	45917	0.0019	0.057154
7	Training not finished	0.0015	-

In this case, a 0.002 quadratic error value is considered as acceptable because, even though the global error obtained is smaller when the quadratic error is reduced from 0.002 to 0.0019, the training process becomes too long to be considered feasible. A comparison between the actual and ANN predicted wind speed for the training data set which comprise wind speed data from April and May 2002 appear in Fig. 10 (sample).

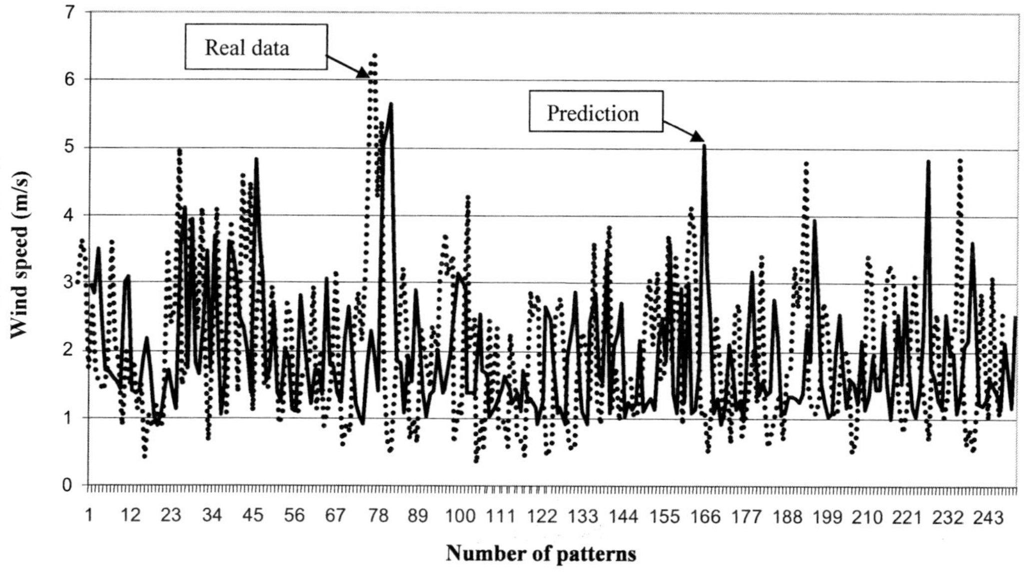

Fig. 10. Comparison between the actual and ANN predicted wind speed-Training set (data from April to May 2002-sample)

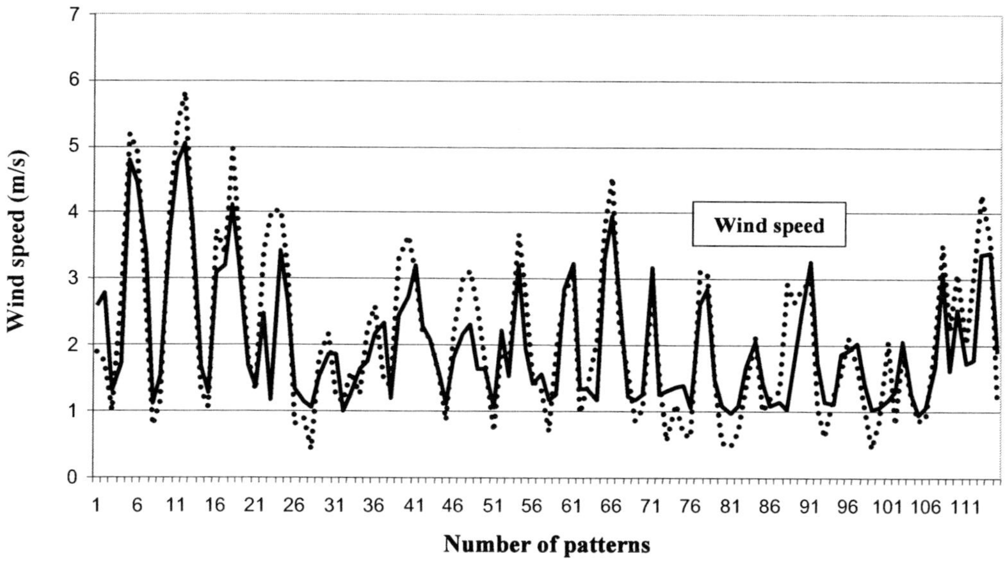

Fig. 11. Comparison between the actual and ANN predicted wind speed-Verification set (data from June to July 2002-sample)

Then the verification of the ANN was evaluated and a sample of the results obtained is shown in Fig. 11. As can be seen there is a good match between the predicted and actual wind speed. Consequently, the developed algorithm can be considered as acceptable for wind speed prediction purposes. It is important to remember that a periodic training of the network would also be convenient in order to ensure better predictions.

As can be observed from Fig. 11, the results obtained in the verification process tracked the real data reference quite correctly and therefore, the prediction process can be considered as acceptable. It should be noted that the real data used in the verification process are totally different from the data used in the training process.

The way that the developed architecture is used in the wind farm control algorithm is to input the last three samples of wind speed data (corresponding to the measurements obtained during the final three hours) to the ANN and the ANN will return the expected wind speed for the following hour. The predicted wind speed will then be used to establish the active power production policy in the wind farm. This is shown in the next section.

5. Active Power Generation

The concept of optimising active power generation is only useful for variable speed generators, which are the most widely used type of generator in wind farms built recently. With double fed induction machines (DFIG), two types of control algorithms must be developed; the independent active and reactive power control, one for each loop (active and reactive power generation), that can be changed and adapted to different exploitation situations within the distribution network.

By considering the designed ANN prediction results, the impact of implementing wind speed forecasts to the control algorithms is examined, so that wind energy production can be used in the new electricity market. The active power generation policy can be established by taking into consideration two different situations:

- *Optimisation of the wind power generation.* Tracking the path of the maximum active power generation points for the best economic exploitation conditions and maximising profits for the wind farm owner. This means that all the energy produced can be sold, otherwise the generation has to be reduced and the available wind power is not fully exploited.
- *Generated power meets consumer demand.* Constant reference fixed by the buyer and agreed beforehand (one hour) with the power utility. Although reduction of wind speed means that wind farm owners have lower incomes such a loss could be balanced in the near future if secondary voltage control is paid for, as an added service by utility companies. Furthermore, in this hypothetical scenario utilities might decide, depending on the grid requirements, to favour secondary voltage control rather than optimisation of active power generation even at high wind speeds.

These two situations are considered in the following subsections which are related to the analysis and design of different controllers in order to comply with the objectives defined in each case. Additionally, it has to be pointed out that taking into account the actual configuration of wind farms such as electronic components, generators installed, existing protection systems, etc., the solution developed must be simple, effective and robust, so as to be able to be installed in real systems. Thus, the methodology used to complete the work must be based on the following objectives:

- Analysis of the dynamic performance of a double fed induction generator (660kW) in order to calculate its active and reactive power generation capability (Tapia *et al.*, 2003; Tapia and Tapia, 2005; Vas 1990; 1998).
- Development of a precise mathematical model for a DFIG and validation of the model with real data (Tapia *et al.*, 2003; Tapia and Tapia, 2005; Vas, 1990; 1998).
- Design of a traditional Proportional Integral (PI) and/or another type of control law to ensure the optimisation of the active power generation in a DFIG. In this way, the trajectory described by the maximum generation for each wind speed will be followed (Bhowmik *et al.*, 1999; Tapia *et al.*, 2006).
- Design of a proportional control law to regulate the reactive power generation (Tapia *et al.*, 2001a; 2001b; 2002; 2004).
- Implementation of the designed control laws (active and reactive) together with the wind speed prediction algorithm on the developed and validated model and analysis of their dynamic performance.
- Analysis of the possibility of installing the designed laws on a real generator, by considering the advantages/disadvantages and the modifications needed.

5.1. Optimisation of the Energy Production

This is the most typical case from the two situations described above and the way current systems are normally working so as to obtain maximum profits. The optimum generation curve in Fig. 1 must always be tracked when working under this regime. In the past few years several algorithms have been developed to ensure this tracking and interesting results have been recorded (Bhowmik *et al.*, 1999; Spee *et al.*, 1995; Leithead and Connor, 2000; Tapia *et al.*, 2006). Therefore, with the development of special algorithms that follow a variable reference in order to get the P_{opt} value in every case, important economic benefits could be obtained. The control algorithms applied for this purpose must always ensure the correct tracking of this reference curve at any time.

As it has been mentioned above, by following the maximum generation curve, the production policies implemented on each wind generator take the wind farm exploitation to an important level of profitability, but they can only be used when the power utility (owner of the network that the farm is connected to) guarantees that all the energy produced can be supplied to the distribution network, i.e., all the production can be sold.

5.2. Constant Reference Fixed by the Utility

This second situation is related to a new production/generation strategy. In the deregulated electric market the utilities do not have to buy all the "green energy" produced. Therefore, depending on the levels of electricity consumption and the specific needs (peak hours with high consumption, high consumption in local factories, winter-time demands, etc.), utilities can buy energy from different sources.

In order to avoid the fact that dispersed generation (or generation that is not controlled) adds risks to the exploitation of electrical systems and decreases the quality of the electrical service (electrical variables out of the exploitation range, risk of outages, etc.), Spanish utilities have been trying to control the reactive power of wind farms consisting of double fed induction machines, so that the voltage level control of distribution networks can be achieved (Tapia *et al.*, 2001a; 2001b; 2002; 2004). It is important to mention that in this kind of generators, both active and reactive power control can be attained and therefore, they can become an important continuous reactive power source for utilities (Nakamura and Okada, 1969; Narita and Hamman, 1973; Mamandur, 1982; Denzel *et al.*, 1988; Palanichamy and Srikrishna, 1989; Paul *et al.*, 1990; Arcidiacono *et al.*, 1990; Gubina and Gubina, 1997).

Under these conditions wind energy can contribute to the grid in several ways, i.e., by voltage level maintenance for the grid - through reactive power control (Tapia *et al.*, 2001a; 2001b; 2002), active power production, etc. But the main problem is related to the non-predictable characteristic of wind energy, because in the electricity market the energy selling process must be completed one hour before the energy is produced.

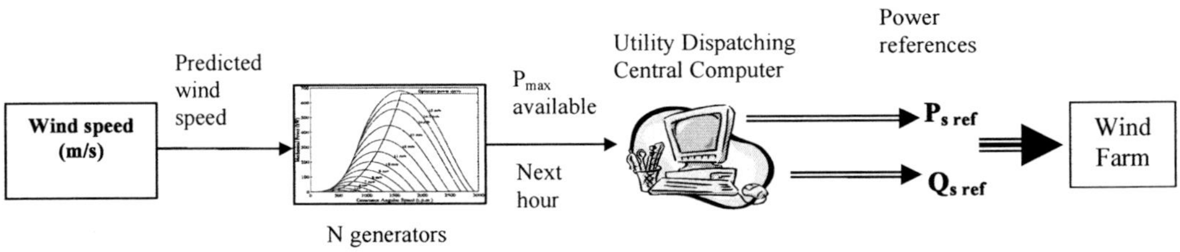

Fig. 12. Active and reactive power reference establishment

So, if an effective prediction process is developed, wind energy would be sold with "enough" production guarantee. Such a prediction process is shown schematically in Fig. 12. The first stage of the process includes the application of the prediction algorithm based on the ANN described in section 4. Once the wind speed prediction is finished, taking into account the optimum generation curve (shown in Fig. 1) for each generator and the predicted wind speed, the amount of active power that each generator would be able to produce could be determined. Consequently, the maximum amount of active energy that the wind farm would be capable to produce (one hour after the prediction has been developed) could be calculated. Finally, by considering the definition of the active power generation reference and the reactive power production policy, the wind farm control algorithm can be designed.

5.3. Control Algorithm Design

The algorithm that is needed to develop the complete control of a wind farm is related to the control of both the generated active and reactive power. Once the wind farm made up of double fed induction machines has been completely modeled, a control strategy for the reactive power control can be implemented. At this point, the operation limitations of the generator have to be considered in order to ensure a correct dynamic performance of the generator and a sufficient security-level from the maintenance point of view.

The developed control algorithm tries to distribute the reactive power that the farm has to generate/absorb in a proportional way, taking always into account the instantaneous active power generation level of each generator, thus:

$$Q_{s\ ref\ i} = \frac{Q_{s\ ref}}{\sum_{i=1}^{n} Q_{s\ max\ i}} * Q_{s\ max\ i} \qquad (18)$$

Operating Conditions

In order to analyse the operating conditions of a generator some operative restrictions must be considered. Initially, the reactive power generation/absorption capability for each generated active power of the modelled generator (a 660 kW generator) must be considered, i.e, the typical P/Q curves depending on the ambient temperature.

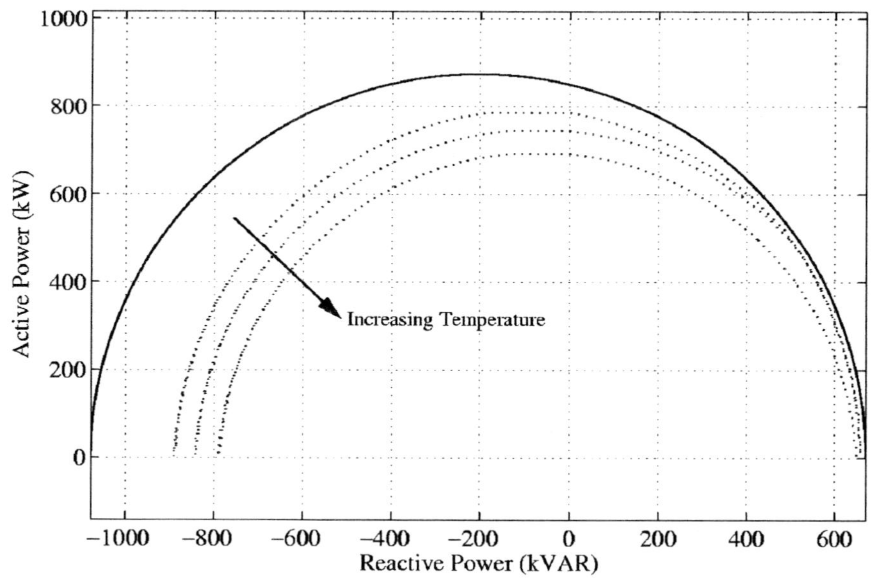

Fig. 13. P/Q curves for the generator + inverter for various ambient temperatures (30 °C, 40 °C and 50 °C)

Subsequently, the currents that can be driven by the double-sided pulse wide modulation (PWM) must be analysed to obtain the operative conditions. By performing a series of tests on the actual wind turbine the curves appearing in Fig. 13 have been obtained.

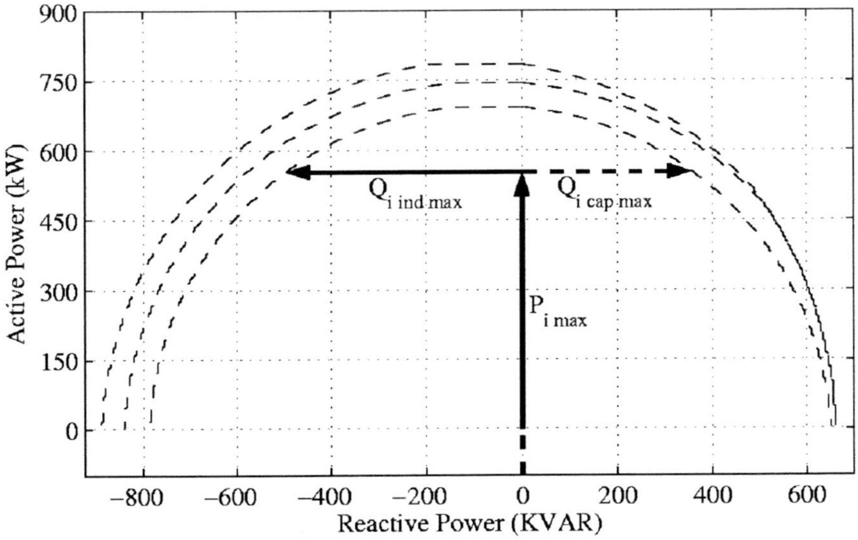

Fig. 14. Limitation for the P/Q operation in the DFIM generator

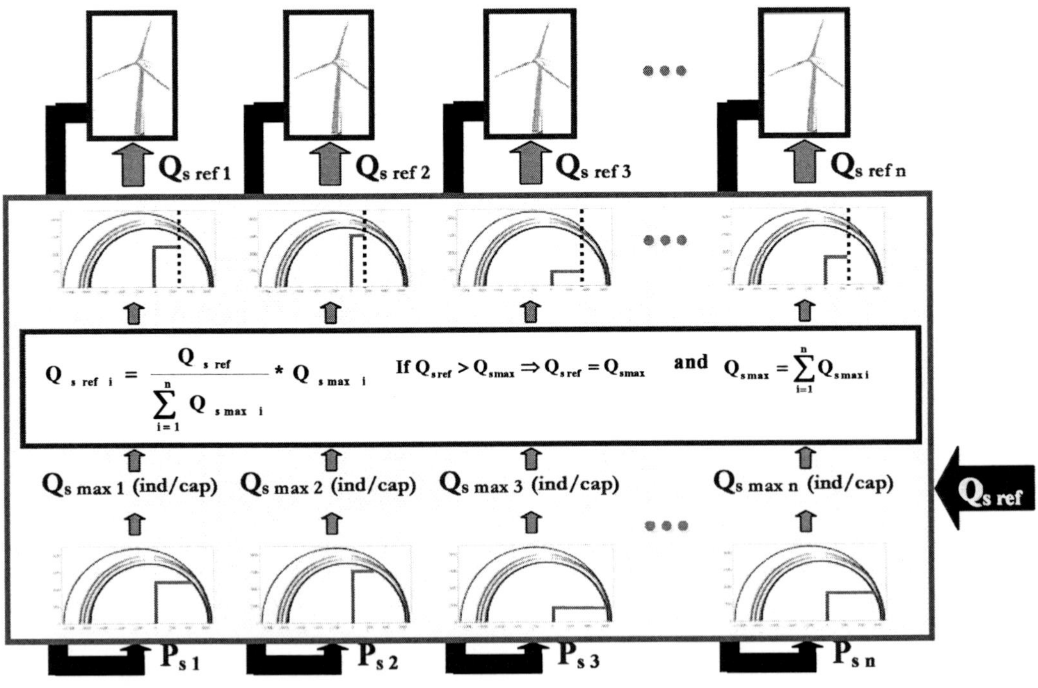

Fig. 15. Active and reactive power generation control system of a wind farm

It can be seen that the outer curve, corresponding to the limitations imposed by the inverter, are less restrictive than those fixed by the generator itself. Consequently, it must be ensured that the operation range of each generator, developed active and reactive power (P_i, Q_i) by means of the wind speed hitting the blades of the generator, is always maintained inside the most restrictive curve ($P_{i\ max}$, $Q_{i\ max}$), the one obtained for the hotter ambient temperature, (see Fig. 14).

The proportional distribution algorithm shown in Fig. 15, has to measure the working conditions of each generator, the instantaneously generated active power, and then, it has to obtain the reactive power generation limit of each machine $Q_{i\ ind\ max}/Q_{i\ cap\ max}$ (according to Fig. 14). The reactive power control has been widely analysed (Bhowmik *et al.*, 1999; Spee *et al.*, 1995; Peña *et al.*, 1996; Tapia *et al.*, 2001a; 2001b; 2002; 2004; Murphy and Daan, 1985) and its validity has been proved. In this work the algorithm described in (Tapia *et al.*, 2003) will be used to regulate the reactive power generated in a farm.

Moreover, the active power has always been controlled in order to obtain maximum economic profit but in this new situation, the active power reference will depend on the generation capability (related to the wind speed) and the utility power needs, P_u. Consequently, the algorithm that is implemented at a wind farm, made up of N wind turbines, can be described as follows:

(1) Predict wind speed for the wind farm during the following hour using the developed ANN algorithm.
(2) Calculate the maximum active power that can be obtained during the next hour, P_{max}, by using Fig. 1 and considering the number of generators.
(3) Compare the obtainable active power amount, P_{max} and the utility needs P_u.
(4) If $P_{max} < P_u$, the active power reference for the farm is the maximum available and $P_{max} = P_{s\ ref}$. Therefore the following procedure is followed:

a) Measure the rotor angular speed, ω_r in the generator/turbine.
b) Calculate $P_{s\ max\ i}$, using Fig. 1.
c) Apply the Proportional Integral (PI) or any other algorithm to follow the optimum generation curve, shown in Fig. 1.
d) Calculate the reactive power reference for the farm, $Q_{s\ ref}$, by considering the active/reactive power (P/Q) limits of the generators used (Fig. 14).
e) Calculate the reactive power reference for each generator by applying the proportional distribution algorithm (Tapia *et al.*, 2003), Eq. (18).
f) Send the reactive power set-point obtained from the Optimal Power Flow (OPF) algorithm in the central computer of the utility to each generator simultaneously.
g) Compare the reactive power generated from the farm with the reactive power reference, $Q_{s\ ref}$.

(5) If $P_{max} > P_u$, the active power reference for the farm will be fixed by the utility and $P_u = P_{s\ ref}$. Therefore the following procedure is followed:

a) Calculate the available reactive power $Q_{s\ ref}$ by taking into account the P_u value and the P/Q curves of the generators (Bourillon, 1999; Bhowmik *et al.*, 1999; Palanichamy and Srikrishna, 1989; Paul *et al.*, 1990).
b) Calculate and send the active and reactive power references for each generator,

$$Q_{s\ ref\ i} = Q_{s\ ref} / N \ ; \ P_{s\ ref\ i} = P_{s\ ref} / N \tag{19}$$

c) Apply the PI control algorithms both for the active and reactive power control loops in order to obtain the references $Q_{s\,ref\,i}$ and $P_{s\,ref\,i}$, which will be lower than the maximum achievable for the predicted wind speed.

It is important to take into account the fact that in both cases if the references for the active and reactive power cannot be achieved, the active power reference tracking is considered more important and consequently, this is the reference that will always be maintained. If under special situations in a deregulated market, the reactive power generation/absorption is better paid than the active power generation, then the reference tracking will be the opposite, i.e., the reactive power tracking will be considered more important than the active power reference tracking. On Fig. 16, a real implementation of the algorithm is presented. There it can be seen that different amounts of reactive power could be obtained by always looking for an optimum active power production. In this case, considering the ANN active power production for the next hour, the amount of active and reactive energy that could be sold would be known beforehand and the utility would work with the Optimal Power Flow (OPF) algorithm in order to maintain correct voltage levels in the grid where the wind farm is connected.

6. Real Implementation Considerations

According to the actual procedures for the installation and operation of renewable energy sources, wind farms must maintain a unity power factor and receive in return a financial incentive. In this way, in operational procedures related to the Voltage Secondary Service, the operation rules for the different agents involved in electricity networks (generators, distributors, etc.) are defined, so that the voltage profile in the transmission-distribution border points will be maintained within the limits.

At the moment wind farms are not required to implement this voltage control and consequently, once in service, they modify the reactive power fluxes in the network and the power factor at the border points where the distribution companies make their measurements to analyse the dynamic performance of the service, but this may change. If the voltage values of the electric grid fall outside the required limits, the wind farm owner should be penalized for the above mentioned incorrect performance.

In this context, the initial performance results obtained from the designed control-law implementation (active and reactive power regulation) have proved the accuracy of the DFIM models developed and the efficiency of the control law. The simulated performance results have been compared to the results obtained from real tests developed in a farm located in Navarre consisting of DFIM generators. The control centre of the distribution company generates a reactive power set point based on an OPF tool and sends it to the wind farm control centre. Consequently, since the control-algorithm has already proved its validity and the ANN performance has already been tested (good performance for intervals of 1 second and for intervals of 1 hour), it can be concluded that the implementation of the ANN together with the control algorithm is quite simple to develop and optimal results could be expected.

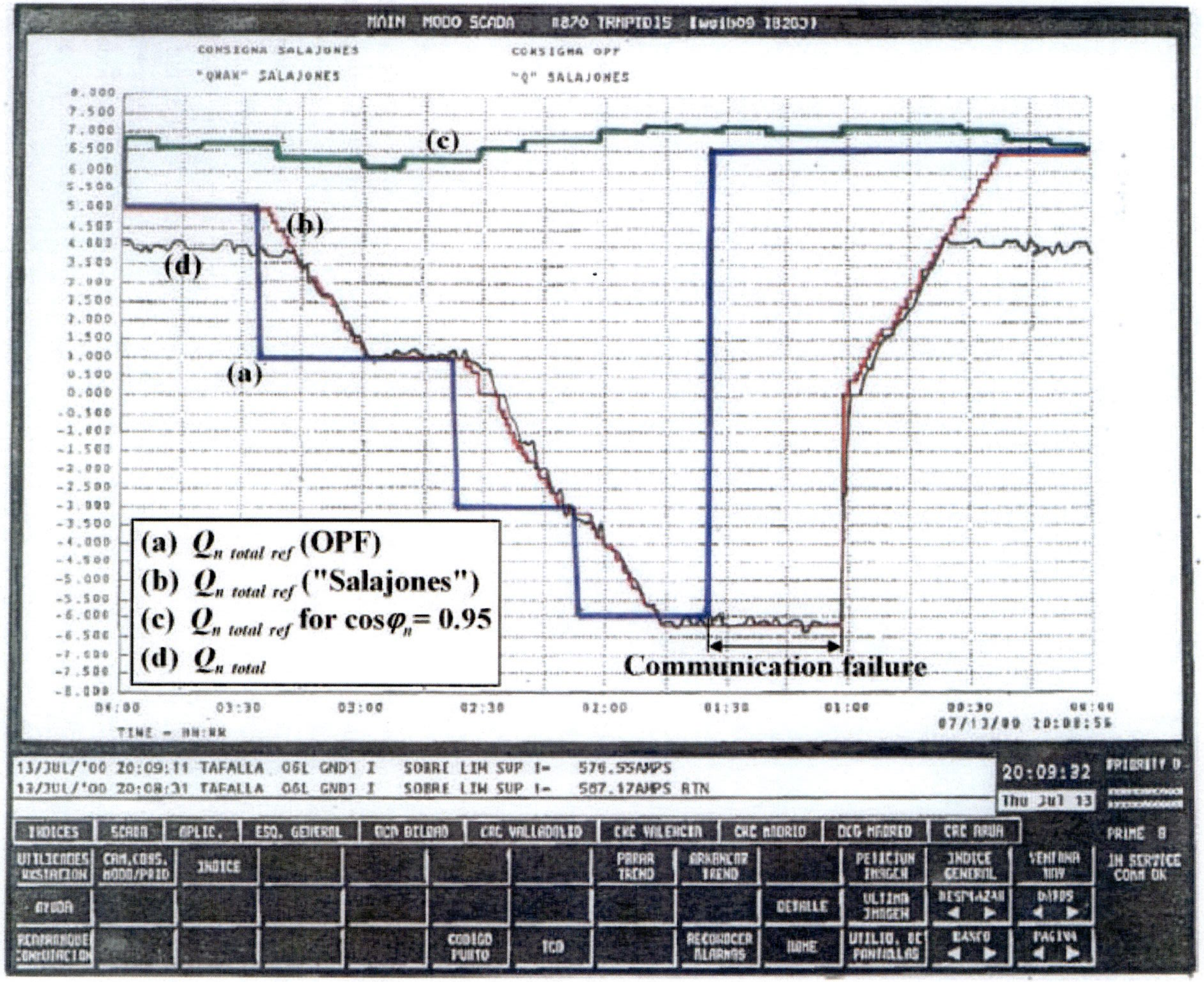

Fig.16. Remote active/reactive power control in a farm

The evaluation of different tests has proved that the set point tracking depends mainly on the communication links and the time cycles of the applications used in the distribution company control centre. It has to be pointed out that neither a high accuracy nor high speed are required since the daily voltage variation due to the changes in demand is slow (see Fig. 16).

In this case, the wind farm net reactive power set-point is updated by an OPF algorithm running at the operation central office —DCD— that Spanish utility IBERDROLA, S.A. owns in Bilbao. Such a set-point is then transmitted to the central computer in "Salajones", a wind farm located about 200 km away from Bilbao DCD. The results corresponding to one of the earliest remote control experimental tests, carried out for 4 hours over "Salajones", are shown in Fig. 16. The time progress of four different curves, related to wind farm net reactive power, can be observed. The ordinate axis range varies from -8 (at the bottom) to +8 MVAR (at the top), where positive and negative values indicate that reactive power is being generated or consumed, respectively. Curve (a) represents the net reactive power set-point that the OPF algorithm, running at Bilbao DCD, fixed for "Salajones".

The stepped curve (b) corresponds to the actual $Q_{n\ total\ ref}$ set-point received by the wind farm central computer. In fact, during this test, set-points derived by the OPF are not sent to the wind farm at once; these set-points are calculated by asking it to produce or to absorb 200 additional KVARs each minute, hence arranging the OPF original set-points at regular intervals. Consequently, if, due to a set-point update, the net reactive power variation required by the OPF to the wind farm is substantial, long time intervals may elapse until the stepped set-point received by the central computer reaches that provided by the OPF algorithm. Therefore, during this test, the latter is updated at a frequency lower than 15 minutes. Curve (c) represents the net reactive power that "Salajones" should generate if a 0.95 capacitive net power factor was to be kept all the way through. Similarly, if the wind farm consumed identical levels of net reactive power instead of producing them, an inductive net power factor of 0.95 would be achieved.

Even though it is not reflected, the curve corresponding to the above mentioned reactive power consumption considering a 0.95 power factor and (c) would be symmetrical with respect to the horizontal line representing the absence of reactive power. Additionally, curve (c), which is updated every 10 minutes, is a measure of the net active power that is being generated by the wind farm. Note that it would be sufficient to divide the newly updated value of curve (c) by tan (arccos 0.95) = 0.3287, in order to derive the level of $P_{n\ total}$ at that very instant. Finally, curve (d) corresponds to the time progress of the actual net reactive power exchanged between "Salajones" and the grid. Excluding both the initial and final parts of the test, curve (d) responds to the stepped set-point (b) satisfactorily. In contrast, during the first 43 and the last 40 minutes of the experiment, the level of generated net reactive power remains around 4 MVAR, revealing that the wind farm fails to reach higher capacitive set-points. In the course of those two time intervals, curve (c) fluctuates between 6.5 and 7 MVAR, which indicates that "Salajones" must have generated around $6.5/0.3287 \approx 19.8$ and $7/0.3287 \approx 21.3$ MW. Considering that the rated value of the wind farm global net active power is 21.78 MW, it follows that the net active power generated by the wind farm DFIGs must be considerably close to their corresponding rated value.

According to their P-Q curve, DFIGs generating active powers close to their rated values are not capable of reaching the reactive power levels required at the beginning and at the end of this experiment. As a result, they operate on their saturation limit, producing around 4

MVAR reactive power in total. Nevertheless, although the active power level generated by "Salajones" between the second hour and a quarter and the third hour is similar, the 6.2 MVAR inductive set-point demanded is reached with no trouble. This is a sign of the inductive nature of DFIGs, which becomes evident by noting the eccentricity of their P-Q load curves (see Fig. 15). Around the second and a half hour, a communication failure occurs while transmitting the set-point to the wind farm central computer. Accordingly, even though both the nature and value of the set-point fixed by the OPF algorithm —curve (a)— change, the stepped one received by the wind farm —curve (b)— keeps its previous value of 6.2 MVAR. About half an hour later, as a consequence of resetting the communication, the stepped set-point varies instantaneously from 6.2 to 0 MVAR. Thanks to this communication failure between Bilbao DCD and "Salajones", it is proved that the wind farm is able to track adequately even larger reactive power set-point step changes.

From the economic point of view, it has to be mentioned that the implementation cost of the scheme described is very low. This is due to the fact that the electronic technology associated to the DFIM machines allows the independent control of the generated active and reactive power by just modifying the control algorithm that is installed in the electronic card (power factor $\neq 0$), with no additional costs for the installation of capacitor or reactance batteries. The developed algorithm runs locally in the wind farm control centre and needs the measurements of the active power generated by each windmill to validate the developed active power prediction, to define the wind production policy and to calculate the new power factor set points.

The control signals exchanged between the distribution company and the wind farm control centre include the number of generators in service and the maximum reactive power available at each moment. These data must be sent to feed the Optimal Power Flow (OPF) algorithm in order to obtain the new reactive power set point. Therefore, wind farms consisting of DFIMs are an important tool to develop voltage regulation. The above mentioned generators are strongly recommended for new wind farms from the power system point of view, especially in areas with large wind generation concentration. The Spanish voltage regulation structure in the near future must include wind farms as conventional generators and new regulation methods must be developed to use the available reactive power generation and absorption capability of DFIM based wind farms.

At the moment, as the amount of dispersed generation is continuously growing (wind farms), different restrictions or penalizations may come into force to avoid over costs in the power system due to the generation share of non-regulating units. As participation in voltage control and other network services will be required in the near future, wind farms with voltage control capability will be in a better position to be accepted in the system, in terms of planning, installation and operation.

Finally, as new power regulation systems related to the renewable energy sources are being applied in different countries, more research in this field would be required, related to the generated active power and its "quality" together with the economic aspects of wind farm exploitation.

7. CONCLUSIONS

A complete control strategy to manage the net active and reactive power exchanged between the grid and wind farms made up exclusively of DFIGs is suggested in this chapter. During the experimentation two stages were considered, in the first the wind farm net active and reactive power set-point were entered manually in its central computer — local control. At the second stage, the power set-points updated by an OPF algorithm running at the central operating office of the utility company, is transmitted to the wind farm's central computer — remote control. The latter is expected to be the most common operation mode in the near future.

Currently, the most usual practice consists in tuning the local control system of each DFIG independently so that it generates active power with unit net power factor. This obviously implies that the amount of reactive power consumed by the different lines and transformers in the wind farm is not compensated, unless reactive power sources, such as capacitor banks, are installed. However, by taking advantage of the reactive power generation ability of DFIGs, the use of such costly discrete compensation components can be avoided. Moreover, wind farms themselves can even be employed as continuous sources of reactive power that may track set-points provided by secondary voltage control strategies, as evidenced in this chapter. The initial parameter required to implement the control algorithm suggested in this chapter is based on ANNs.

The ANN presented in this chapter based on the feed-forward method, was evaluated with real data measured at two different locations in the Basque country (Spain) and demonstrated a correct dynamic performance in all evaluation tests. This means that the performance of the model for wind speed forecasting has been demonstrated as appropriate for the short-term predictions needed in the typical situations found in a deregulated electric market.

Additionally, if the ANN is used to predict the wind speed at a wind farm, the active and reactive power generation policies can be established in order to obtain the maximum economic benefits for both producers and power utilities.

It is worth mentioning that for the wind farm operator the possibility of calculating the wind power availability in advance is especially interesting from the energy supply point of view. Thus, wind power generation could be considered as a reserve power supply for different consumption situations in the electric grid which can be combined with other traditional energy sources.

Finally, working under several active and reactive power reference situations in the deregulated market, the wind farm owner will be able to recover the investment in new farms even though the "green energy" price is not as high in the future as it is today.

ACKNOWLEDGEMENTS

The authors are especially grateful to the induction generator manufacturing company, INDAR, the control systems company, INGETEAM and the power utility IBERDROLA for their contribution in providing the characteristics of the electric machines and all the test results needed for this research work. Diputaci6n Foral de Gipuzkoa, a local institution has funded part of this work.

REFERENCES

Ackermann T., Säoder L., 2002. An overview of wind energy-status, *Renewable and Sustainable Energy Reviews*, Vol. 6, pp. 67–128.

Arcidiacono V., Corsi S., Natale A., Raffaelli C., Menditto V., 1990 New developments in the application of ENEL transmission system voltage and reactive power automatic control, *Proceedings of CIGRE'90*, Paris, France, Report 38/39-06.

Beyer H.G., Degner T., Hausmann J., Hoffmann M., Rujan P., 1994. Short-term predictions of wind speed and power output of a wind turbine with neural networks, *Proceedings of the 2nd European Congress on Inteligent Techniques and Soft Computing*, Aachen, Germany, pp. 20-23.

Bhowmik S., Spee R., Enslin J.H.R., 1999. Performance optimization of doubly-fed wind power generation systems, *IEEE Transaction on Industry Applications*, Vol. 3, No. 4, pp. 949-958.

Bishop C.M., 1996. *Neural Networks for Pattern Recognition*, Clarendon Press.

Bossanyi EA., 1985. Short-term wind prediction using Kalman filters, *Wind Engineering*, Vol. 9, pp.1-8.

Bourillon C., 1999. Wind energy —clean power for generations, *Renewable Energy*, Vol. 16, pp. 948–953.

Denzel D., Edwin K.W., Glavitsch H., Graf F.R., 1988. Optimal Power Flow and its real time application at the RWE energy control centre, *Proceedings of. CIGRE'88*, Paris, France, Report 39-19.

EUSKALMET, Servicio Vasco de Meteorología, Av/ Duque de Wellington, No.2. 01010 Vitoria – Gazteiz.

Fabbri A., Gomez T., Roman S., Abbad J.R., Quezada V.H.M., 2005. Assessment of the cost associated with wind generation prediction errors in a liberalizad electricity market, *IEEE Transactions On Power System,* Vol. 20, No. 3, pp. 1440-1446.

Gubina A., Gubina F., 1997. Alternative approach to secondary voltage control: a solution with local ANN controllers, *Proceedings of the 10th International Conference on Power System Automation and Control (PSAC'97)*, Ljubljana, Slovenia, pp. 67–72.

Kretzschmar R., Eckert P., Cattani D., Eggimann F., 2004. Neural network classifiers for local wind prediction, *Journal of Applied Meteorology*, Vol. 43, No. 5, pp. 727-738.

Kalogirou S.A., 2001. Artificial neural networks in renewable energy systems applications: a review, *Renewable and Sustainable Energy Reviews*, Vol. 5, pp. 373–401.

Leithead W.E., Connor B., 2000. Control of variable speed wind turbines: design task, *International Journal of Control*, Vol. 73, No. 13, pp. 1189–1212.

Mamandur K.R.C., 1982. Emergency adjustments to VAR control variables to alleviate over-voltages, under-voltages and generator VAR limits, *IEEE Transactions on Power Apparatus and Systems*, Vol. PAS-101, No. 5.

Masón I.B., 1982. A model for assessment of weather forecasts. *Australian Meteorologic Magazine*, Vol. 30, pp. 291-303.

Mohandes M.A., Halawani T.O., Rehnam S., Hussain A.A., 2004. Support vector machines for wind speed prediction, *Renewable Energy,* Vol. 29, No. 6, pp. 929-947.

Mohandes M.A., Rehnam S., Halawani T.O., 1998. A neural networks approach for wind speed prediction, *Renewable Energy,* Vol. 13, No. 3, pp. 345-354.

More A., Deo M.C., 2003. Forecasting wind with neural networks, *Marine Structures*, Vol. 16, No. 1, pp. 35-49.

Murphy A.H., Daan H., 1985. *Forecast evaluation. Probability, statistics and decision making in the atmospheric sciences*, Boulder, CO: Westview Press.

Nakamura Y., Okada T., 1969. Voltage and reactive power control by dividing a system into several blocks, *IEE in Japan*, Vol. 89, No. 12, pp. 75–82.

Narita S., Hamman M.S.A.A., 1973. Multicomputer control of system voltage and reactive power on real-time basis, *IEEE Transactions on Power Systems*, Vol. PAS-92, No. 1.

Palanichamy C., Srikrishna K., 1989. An elegant approach to optimal real and reactive power dispatch, *Electric Power System Research*, Vol. 16, pp. 173–181.

Paul J.P., Corroyer C., Jeannel P., Tesseron J.M., Maury F., Torra A., 1990. Improvements in the organization of secondary voltage control in France, *Proceedings of CIGRE'90*, Paris, France, Report 38/39-03.

Peña R.S., Clare J.C., Asher G.M., 1996. Vector control of a variable speed doubly-fed induction machine for wind generation systems, *EPE Journal*, Vol. 6, No. 3–4, pp. 60–67.

Sfetsos A., 2002. A novel approach for the forecasting of mean hourly wind speed time series, *Renewable Energy*, Vol. 27, pp. 163-174.

Spee R., Bhowmik S., Enslin J.H.R., 1995. Novel control strategies for variable-speed doubly fed wind power generation systems, *Renewable Energy*, Vol. 6, No. 8, pp. 907–915.

Tapia A., Tapia G., Ostolaza X., Sáenz J.R., Criado R., Berasategui J.L., 2001a. Reactive power control of a wind farm made up with doubly fed induction generators (I), *Proceedings of the IEEE Porto Power Tech'2001*, September 2001. (Vol. IV), DSR1-054.

Tapia A., Tapia G., Ostolaza X., Zubia I., Sáenz J.R., 2001b. Reactive power control of a wind farm made up with doubly fed induction generators (II), *Proceedings of the IEEE Porto Power Tech'2001*, September 2001. (Vol. IV), DSR1-055, Oporto.

Tapia G., Tapia A., Berasategui J.L., Saenz J.R., 2002. Voltage regulation of distribution networks through reactive power control, *IFAC Publications* – Elsevier Science Ltd., ISBN 008 044184, *Proceedings of the 15th IFAC World Congress 2002*, pp. 377-382.

Tapia A., Tapia G., Ostolaza X., Sáenz J.R., 2003. Modelling and control of a wind turbine driven doubly fed induction generator, *IEEE Transaction on Energy Conversion*, Vol. 1, No. 2, pp.194-204.

Tapia A., Tapia G., Ostolaza X., 2004. Reactive power control of wind farms for voltage control applications, *Renewable Energy*, Vol. 29, No. 3, pp. 377-392.

Tapia G., Tapia A., 2005. Wind generation optimisation algorithm for a doubly fed induction generator, *IEE Proceedings on-Generation, Transmission and Distribution,* Vol. 152, No. 2, pp. 253–263.

Tapia G., Tapia A., Ostolaza J.X., 2006. Two alternative modelling approaches for the evaluation of wind farm active and reactive power performances, *IEEE Transactions on Energy Conversion*, in press.

Vas P., 1990. *Vector Control of AC Machines*, New York: Oxford University Press.

Vas P., 1998. *Sensorless Vector and Direct Torque Control*, New York: Oxford University Press.

Watson S.J, Landberg L., Halliday J.A., 1993. Wind speed forecasting and its application to wind power integration, *Proceedings of 15th British Wind Energy Association Conference*, pp. 291-298.

In: Artificial Intelligence in Energy and Renewable Energy Systems ISBN 1-60021-261-1
Editor: Soteris Kalogirou, pp. 235-268

Chapter 7

NEURO-FUZZY HAMMERSTEIN MODEL BASED PREDICTIVE CONTROL OF A SOLID OXIDE FUEL CELL

Francisco Jurado
University of Jaén, Department of Electrical Engineering
EPS Linares (Jaén), Spain

ABSTRACT

The solid oxide fuel cell (SOFC) power plant is known to be a potential alternative in the electric utility. However, the output voltage of the SOFC changes with the load variations. Model predictive control is part of a family of optimization based control methods, which are based on on-line optimization of future control moves.
The Hammerstein model configuration, which includes a nonlinear static block followed by a linear dynamic block, is applied to model the static and dynamic characteristics of the solid oxide fuel cell. This chapter investigates the use of a neuro-fuzzy method as a tool for identification of the nonlinear static block.
A neural network is designed to tune the gains of the fuzzy model based on the operating conditions of SOFC. In this chapter a model-based controller for the regulation of a solid oxide fuel cell is proposed. The performances using both the linear and neuro-fuzzy Hammerstein models are evaluated with constraints.

Keywords: Fuel cells, fuzzy control, predictive control, nonlinear systems.

NOMENCLATURE

C_p^s	Heat capacity of the cell unit (J/kg K)
E_o	Standard reversible cell potential (V)
$e(t)$	Error between the measured output and the set-point

F	Faraday's constant (C/kmol)
$\overline{h}_{ai}^{in}$ $\overline{h}_{ci}^{in}$	Anode (cathode) inlet partial molar enthalpies (J/mol)
$\overline{h}_i^s$	Partial molar enthalpies at stack temperature (J/mol)
I	Stack current (A)
i	Stack current density (A/m^2)
i_L	Limiting current density (A/m^2)
i_0	Exchange current density (A/m^2)
K_{H_2}, K_{H_2O}	Valve molar constants for hydrogen and water, (kmol/ s atm)
K_i	Valve molar constants
K_r	Constant dependent on Faraday's constant and number of electrons in the Reaction (kmol/s A)
M^s	Mass of the cell unit (kg)
N_i^{in}, N_i^o	Molar flow rates (mol/s) of the ith reactant at the cell input and output, respectively
N_{ta}^{in} N_{tc}^{in}	Anode (cathode) total inlet molar flow (mol/s)
N_i^r	Reaction rate (mol/s) of the ith reactant
N_{H_2}, N_{H_2O}	Molar flows of hydrogen and water, respectively, through the anode valve (kmol/s)
N_o	Number of cells in stack
n	Number of electrons participating in the reaction
P_{dc}	Stack dc power (W)
p_{H_2}, p_{H_2O}	Partial pressures of hydrogen and water (atm)
p_i	Partial pressure of the ith reactant (atm)
R	Gas constant (8.31 J/mol K)
R_{ai} R_{ci}	Anode (cathode) total rate of production of species (mol/s)
r	Ohmic resistance (Ω m^2)
T_f	Finite time
T^0	Temperature constant (K)
T^s	Stack solid average temperature (K)
V	Compartment volume (m^3)
V_{dc}	Cell voltage (V)
V_e	Volume of the cell unit (m^3)
V_o	Open-circuit reversible potential (V)
x_i	Mole fractions of species
x_{ai}^{in} x_{ci}^{in}	Anode (cathode) inlet mole fractions

Greek

α	Electron transfer coefficient of the reaction at the electrode

α_r, β_r	Ohmic resistance constants
β	Scaling factor
η_{act}	Activation losses (V)
η_{con}	Concentration losses (V)
ξ	Total gas components in anode or cathode
τ_{H2}	Time constant associated with the hydrogen flow and is a function of temperature (s).

ABBREVIATIONS

ARX	AutoRegression with eXogenous signal
DG	Distributed Generation
FH	Fuzzy Hammerstein
GPC	Generalized Predictive Control
ITAE	Integral of Time Absolute Error
LPV	Linear Parameter Varying system model
LTI	Linear Time Invariant model
MIMO	Multiple-Input, Multiple-Output system
MISO	Multi-Input, Single-Output system
MPC	Model Predictive Control
NAARX	Nonlinear Autoregressive moving Average with eXogenous signal
NARX	Nonlinear AutoRegressive with eXogenous signal
NF	Neuro Fuzzy
NN	Neural Network
RBF	Radial Base Functions
SISO	Single-Input, Single-Output system
SOFC	Solid Oxide Fuel Cell
TS	Takagi-Sugeno model

1. INTRODUCTION

Various technologies are being used in distributed generation (DG) applications with variable degree of success. Among those are: wind turbines, small-scale hydropower plants, biomass, microturbines, photovoltaic arrays, and fuel cells. A fuel cell is a device that directly converts the chemical energy of fuel to electric energy. Recent advances in the fuel cell technology significantly improved the technical and economical characteristics of this technology. Environmental friendliness, practically noise free operation, and very high efficiency combined with the forecasted shift to gaseous fuels make fuel cells a very sound competitor on the future electricity markets. In addition, it should be noted that fuel cell-based generators possess other important properties such as compact size, modularity and controllability.

Solid oxide fuel cells (SOFCs) have attracted considerable interest during the past decade as highly effective and environmentally acceptable sources of electrical energy. The main features of the SOFC are all solid-state construction and high-temperature operation. The combination of these features leads to a number of unique characteristics and advantages for this type of fuel cell, including flexibility in cell and stack designs, manufacturing processes, and power plant sizes.

Padullés *et al.* (2000a) develop a SOFC model which includes species dynamics, but it does not consider temperature dynamics. Hall and Colclaser (1999) modelled a 3-kW SOFC but they did not take into account dynamics of the chemical species. Achenbach (1994) develops a mathematical model of a planar SOFC, which concentrates on effects of temperature changes on output voltage response. Temperature dynamics is modelled in a three-dimensional (3-D) vector space. The same author investigated the transient behaviour of a stand-alone SOFC caused by a load change. It shows that the relaxation time of the output voltage is highly related to the effect of temperature dynamics. The application of the autoregression with exogenous signal (ARX) identification algorithm to compute linear system models is presented by Jurado (2004a). The nonlinear autoregressive exogenous (NARX) approach is used to analyze the dynamics of this fuel cell (Jurado, 2005).

For nonlinear dynamic systems, the conventional techniques of modelling and identifications are difficult to implement and sometimes impracticable. However, other techniques based on fuzzy logic are more and more used for modelling this kind of process (Chen *et al.*, 1994). Among the different fuzzy methods, the Takagi-Sugeno model (TS) has attracted most attention (Takagi and Sugeno, 1985). In fact, this model consists of if-then rules with fuzzy antecedents and mathematical functions in the consequent part. The task of system identification is to determine both the nonlinear parameters of the antecedents and the linear parameters of the rules consequent. A fuzzy logic control of three-phase inverters for fuel cell systems is presented by Jurado (2004b).

Takagi-Sugeno fuzzy models are suitable to model a large class of nonlinear systems (Babuška, 1998; Setnes *et al.*, 1998). Fuzzy modelling and identification from measured data are effective tools for the approximation of uncertain nonlinear systems. So far, most attention has been devoted to single-input, single-output (SISO) or multi-input, single-output (MISO) systems. Recently, methods have also been proposed to deal with multi-input, multi-output (MIMO) systems (Kouatli, 1994; Lu *et al.*, 1997; Gegov, 1997).

The Hammerstein models are special kind of nonlinear systems where the nonlinear block is static and is followed by a linear system. These models have applications in many engineering problems and therefore, identification of Hammerstein models has been an active research area for many years. There exist a large number of research papers in the literature on the topics of Hammerstein model identifications.

Existing methods for Hammerstein model identification can be roughly divided into four categories: the iterative method (Narendra and Gallman, 1966; Stoica, 1981), the over-parameterization method (Bai, 1998; Boutayeb *et al.*, 1996; Chang *et al.*, 1971; Hsia, 1976), the stochastic method (Billings and Fakhouri, 1978; Greblicki, 1992; Pawlak, 1991) and the separable least-squares method (Bai, 2001). In most cases, the structure of the nonlinearity is assumed to be known. Otherwise, identification becomes a structural estimation problem (Haber and Unbehauen, 1990).

The recent growth in attention to neural networks (NNs) has led to many suggestions for combined use of fuzzy logic and neural networks in intelligent control. It is shown that by tuning the gains of the fuzzy Hammerstein model with the aid of a NN, optimum identification of the system can be achieved in a wide range of operating conditions compared to fix gain fuzzy Hammerstein model.

However, the output voltage of the solid oxide fuel cell changes with the load variations. The amount of fuel flow can be controlled according to the current. The current is proportional to the terminal load. It is evident that an increase in the load increases the current, which in turn decreases the output voltage of the fuel cell. The increase in current makes fuel flow rate bigger, and augments power transfer from the SOFC to the load.

Model predictive control (MPC) has been an active field of research during the last three decades, driven both by numerous successful applications of the technology (Richalet, 1993; Qin and Badgwell, 1997; 2000) and by the research interests of the academia. The main reason of this success is the ability of MPC to control multivariable systems under constraints in an optimal way. In model predictive control, the control action is computed by solving an optimization problem on line in each sampling period. This is the main difference from conventional control, where a precomputed control law is employed. Many applications of MPC based on fuzzy prediction models have been reported (Fischer *et al.*, 1998; Hu and Rose, 1999; Abonyi *et al.*, 2001; Mollov *et al.*, 2004).

MPC is part of a family of optimization-based control methods, which is based on on-line optimization of future control moves. This work proposes a model-based controller for the regulation of a SOFC.

The first step in designing an MPC system is the derivation of a model that the controller will use for the optimization. This model should be as accurate as possible, while being simple enough to allow for repeated calculations during the optimization. The inputs to the SOFC model are the average temperature, the fuel flow, the air flow, and the current. The output of the SOFC model is the voltage.

The chapter is organized as follows. In Section 2, general principles of SOFC are explained. In Section 3, fuzzy models are introduced. Section 4 presents the Hammerstein model. Identification of the linear part and identification of the nonlinear part are discussed in Section 5. The MPC is formulated in Section 6. Simulation examples illustrating the performance of SOFC are presented in Section 7, and finally, conclusions are provided in Section 8.

2. Solid Oxide Fuel Cell Dynamic Model

The proposed stack model is based on the following assumptions:

(1) Stack is fed with hydrogen and air, therefore the fuel processor dynamics is not included.
(2) A uniform gas distribution among cells is assumed, since there is a small deviation of the gas distribution among the cells.
(3) There is no heat transfer among cells. Each cell has the same temperature and current density (Padullés *et al.*, 2000a).
(4) The channels that transport gases along the electrodes have a fixed volume, but their lengths are small, so that it is only necessary to define one single pressure value in

their interior. The ratio of pressures between the interior and exterior of the channel is large enough to consider that orifice is choked (Padullés *et al.*, 2000a; 2000b).

Fig. 1 shows the dynamic model of SOFC, along with its major chemical reactions.

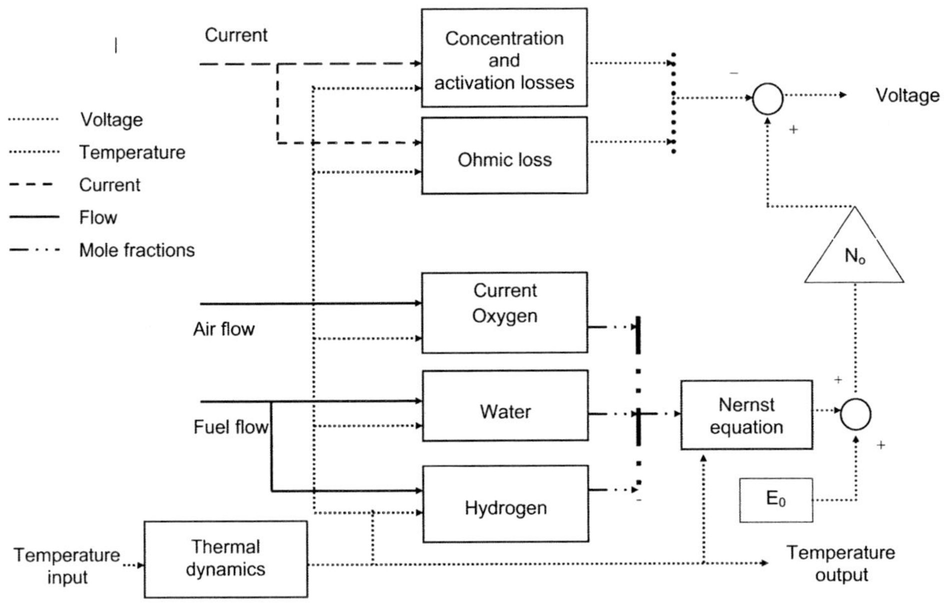

Fig. 1. Dynamic model of SOFC

2.1. Electrochemical Model

The change in concentration of each species that appears in the chemical reactions can be written in terms of input and output flow rates and exit molarities due to the following chemical reaction (Lukas *et al.*, 1999; U.S. Department of Energy, 2000):

$$\frac{V}{RT}\frac{d}{dt}p_i = N_i^{in} - N_i^{o} - N_i^{r} \tag{1}$$

where

V	compartment volume (m^3)
N_i^{in}, N_i^{o}	Molar flow rates (mole/s) of the ith reactant at the cell input and output, respectively
N_i^{r}	reaction rate (mole/s) of the ith reactant
T	cell temperature (K)
p_i	cell pressure (atm)
R	gas constant (8.31 J/mole K)

In agreement with the basic electrochemical relationships, the molar flow that reacts can be calculated as:

$$N_i^r = \frac{N_0 I}{2F} = 2K_r I \tag{2}$$

where

F	Faraday's constant (C/kmol)
I	stack current (A)
K_r	constant dependent on Faraday's constant and number of electrons in the reaction (kmol/s A)
N_o	number of cells in stack

The cell utilization (u) is defined through the input and output hydrogen flow rates $\left(N_{H_2}\right)$ as follows:

$$u = \frac{N_{H2}^{in} - N_{H2}^{o}}{N_{H2}^{in}} \tag{3}$$

For orifice that is choked (Lukas *et al.*, 1999), molar flow of any gas through the valve is proportional to its partial pressure inside the channel according to the following expressions (Padullés *et al.*, 2000a):

$$\frac{N_{H_2}}{p_{H_2}} = K_{H_2} \quad \frac{N_{H_2O}}{p_{H_2O}} = K_{H_2O} \tag{4}$$

where

N_{H_2}, N_{H_2O}	molar flows of hydrogen and water, respectively, through the anode valve (kmol/s)
p_{H_2}, p_{H_2O}	partial pressures of hydrogen and water (atm)
K_{H_2}, K_{H_2O}	valve molar constants for hydrogen and water, (kmol/ s atm)

Considering the hydrogen partial pressure,

$$\frac{V}{RT}\frac{d}{dt}p_{H_2} = N_{H_2}^{in} - N_{H_2}^{o} - 2K_r I \tag{5}$$

Applying the Laplace transformation to the above equations and isolating the hydrogen partial pressure, yields the following expressions:

$$p_{H_2} = \frac{\frac{1}{K_{H_2}}}{1+\tau_{H_2}\cdot s}\left(N_{H_2}^{in} - 2K_r I\right) \tag{6}$$

$$\tau_{H2} = \frac{V}{K_{H2}RT} \tag{7}$$

where τ_{H_2}, expressed in seconds, is the time constant related to the hydrogen flow and is a function of temperature.

2.2. Thermal Model

The fuel cell power output is closely related to the temperature of the cell unit. The heat storage in the thin fuel unit gas or oxidant gas layer is neglected. The thin fuel unit or oxidant gas layers are lumped to the cell unit and gas layers are assumed to have the same temperature as the cell unit (Hall and Colclaser, 1999; Achenbach, 1994; 1995). The energy balance equation for each cell unit is as follows:

$$M^s C_p^s \frac{dT^s}{dt} = N_{ta}^{in}\left[\sum_{i=1}^{\xi} x_{ai}^{in}\left(\overline{h}_{ai}^{in} - \overline{h}_i^s\right)\right] - \sum_{i=1}^{\xi} \overline{h}_i^s R_{ai} + N_{tc}^{in}\left[\sum_{i=1}^{\xi} x_{ci}^{in}\left(\overline{h}_{ci}^{in} - \overline{h}_i^s\right)\right] - \sum_{i=1}^{\xi} \overline{h}_i^s R_{ci} - P_{dc} \tag{8}$$

where

$M^s C_p^s$	stack solid mass-specific heat product (J/K)
T^s	stack solid average temperature (K)
$N_{ta}^{in}\ N_{tc}^{in}$	anode (cathode) total inlet molar flow (mol/s)
$x_{ai}^{in}\ x_{ci}^{in}$	anode (cathode) inlet mole fractions
$\overline{h}_{ai}^{in}\ \overline{h}_{ci}^{in}$	anode (cathode) inlet partial molar enthalpies (J/mol)
$\overline{h}_i^s$	partial molar enthalpies at stack temperature (J/mol)
ξ	total gas components in anode or cathode
P_{dc}	stack dc power (W)
$R_{ai}\ R_{ci}$	anode (cathode) total rate of production of species (mol/s)

Under the ideal gas supposition the partial molar enthalpies are calculated using:

$$\overline{h}_i = \overline{h}_i^{ref} + \int_{T_{ref}}^{T} c_{p,i}(u)\,du \tag{9}$$

and coefficients of the specific heats $c_{p,i}$ are given by:

$$c_{p,i} = a_i + b_i T + c_i T^2 + d_i T^3 \tag{10}$$

Values of the coefficients of Eq.(10) can be found in standard reference tables. The reference enthalpy stands for energy at standard reference temperature (298 K) and considers the heat of formation for each gas species to account for energy change on chemical reaction.

2.3. Nernst's Equation

Applying Nernst's equation and Ohm's law (taking into account ohmic, concentration, and activation losses), the stack is connected in series and the stack output voltage is represented as follows (Lukas *et al.*, 1999; 2001; U.S. Department of Energy, 2000):

$$V_{dc} = V_o - rI - \eta_{act} - \eta_{con} \tag{11}$$

$$V_o = N_o\left(E_o + \frac{RT}{2F}\left[\ln\frac{x_{H_2}x_{O_2}^{0.5}}{x_{H_2O}}\right]\right) \tag{12}$$

where

V_o	open-circuit reversible potential (V)
E_o	standard reversible cell potential (V)
x_i	mole fractions of species
r	ohmic resistance (Ω m^2)
N_o	number of cells in stack
η_{act}	activation losses (V)
η_{con}	concentration losses (V)

For the reason that the reactant is consumed at the electrode by electrochemical reaction, there is a loss of potential due to the inability of the surrounding material to maintain the initial concentration of the bulk fluid. Concentration loss equation is given by (Hall and Colclaser, 1999; Lukas *et al.*, 1999; 2001; U.S. Department of Energy, 2000):

$$\eta_{con} = \frac{RT}{nF}\ln\left(1 - \frac{i}{i_L}\right) \tag{13}$$

where

i	stack current density (A/m^2)
i_L	limiting current density (A/m^2)
n	number of electrons participating in the reaction

Activation polarization is existent when the rate of an electrochemical reaction at an electrode surface is controlled by sluggish electrode kinetics (U.S. Department of Energy, 2000). Activation loss equation is given by:

$$\eta_{act} = \frac{RT}{\alpha nF} \ln\left(\frac{i}{i_0}\right) \tag{14}$$

where

i_0	exchange current density (A/m^2)
α	electron transfer coefficient of the reaction at the electrode

α is considered to be the fraction of the change in polarization that leads to a change in the reaction rate constant and its value is usually 0.5 for the fuel cell applications.

Tafel plots provide a visual understanding of the activation polarization of a fuel cell. They are used to measure the exchange current density, given by the extrapolated intercept at $\eta_{act} = 0$ which is a measure of the maximum current that can be extracted at negligible polarization, and the slope representing the transfer coefficient.

The usual form of the Tafel equation that can be easily expressed by a Tafel Plot is:

$$\eta_{act} = a + b \ln i \tag{15}$$

where $a = (-RT/\alpha nF)\ln i_0$ and $b = RT/\alpha nF$.

The term b is called the Tafel slope, and is obtained from the slope of a plot of η_{act} as a function of ln i. The Tafel slope for an electrochemical reaction is about 110 mV/decade at room temperature. Decade is Log current density (mA/cm^2).

Ohmic losses occur because of resistance to the flow of ions in the electrolyte and resistance to the flow of electrons through the electrode materials. This resistance is dependent on the cell temperature and is obtained by (Lukas *et al.*, 2001):

$$r = \alpha_r \exp\left[\beta_r\left(\frac{1}{T^o} - \frac{1}{T}\right)\right] \tag{16}$$

3. Fuzzy Models

Fuzzy systems can be used to model human knowledge in engineering problems. This knowledge may be classified into two categories: conscious knowledge explicitly expressible in words, and subconscious knowledge that a human expert transforms into actions but cannot be explained in words.

Conscious knowledge can be expressed in terms of fuzzy IF-THEN rules and implement the rules in fuzzy systems. For subconscious knowledge, the human expert can be considered as a black box. Thus, subconscious knowledge is represented by a set of input-output pairs. Hence, a problem of fundamental importance is to construct fuzzy systems from input-output pairs. To solve this problem, it is necessary to determine the membership functions of the fuzzy input and output sets, and define a fuzzy rule base.

Some of the most important applications of fuzzy theory have been focused on control problems where human operators, who intuitively know the behaviour of the system, provide the best control. The human operator however may not always be able to satisfactorily control a process. Then, the control strategy may take the form of a set of situation-action pairs, known to the human operator, among which the control system can be designed. This provides the core knowledge for the system, but usually does not completely describe it. To perfect this description, one must define fuzzy sets representing the range of control parameters or sensor fields. The system description is then finely tuned by adjusting the fuzzy sets and the rule base.

Many investigators have developed approaches to determine fuzzy sets and rule bases. For example, Manoranjan *et al.* (1995) proposed a systematic approach for determining the grades of membership of fuzzy subsets that represent the behaviour of a control practitioner. Abe and Lan (1995) formulated a method for extracting fuzzy rules directly from numerical input-output data for pattern classification and function approximation. Homaifar and McCormick (1995) analyzed the applicability of genetic algorithms in the simultaneous design of membership functions and rule sets for fuzzy logic controllers. Ishibuchi *et al.* (1995) used genetic algorithms for selecting a small number of significant fuzzy IF-THEN rules to construct a compact fuzzy classification system with high classification power. Kosko (1992) proposed an approach by combining neural networks and fuzzy systems. Perneel *et al.* (1995) applied genetic algorithms and neural networks as optimization methods to improve the performance of the fuzzy decision-making system for heuristic search algorithms. Rovatti *et al.* (1995) proposed a new methodology for the minimization of a given set of fuzzy rules. Rovatti and Guerrieri (1996) also developed an on-line adaptive algorithm, which learned the extent to which inclusion of a rule in the rule set significantly contributed to the reproduction of the target behaviour for system identification.

Fuzzy logic systems comprise four main components; the fuzzifier, the fuzzy rule base, the fuzzy inference engine and the defuzzifier. The main difference between the Mamdani and Sugeno fuzzy systems lies in the consequents of fuzzy rules (Takagi and Sugeno, 1985; Mamdani and Assilian, 1999). As was mentioned by Mendel (1995), many defuzzifiers were proposed in the literature. However, there is no scientific base for any of them. Consequently, defuzzification is an art rather than a science. One criterion for the choice of a defuzzifier is computational simplicity. For a Sugeno fuzzy model, the consequence of each fuzzy rule is a constant, and defuzzification in such a model is made using the *centre of gravity* method.

3.1. Fuzzy Model Structures

This chapter mainly deals with a Takagi-Sugeno (TS) fuzzy model proposed by Takagi, Sugeno, and Kang (Takagi and Sugeno, 1985; Sugeno and Kang, 1986) to develop a systematic approach for generating fuzzy rules from a given input-output data set. In this section the structure of this model and the related modelling paradigms are presented.

The TS model is a combination of a logical and a mathematical model. This model is also formed by logical rules consisting of a fuzzy antecedent and a mathematical function as consequent part. The antecedents of fuzzy rules partition the input space into a number of fuzzy regions, while the consequent functions describe the system behaviour within a given region:

$$R_j: \text{ If } z_1 \text{ is } A_{1,j} \text{ and}\ldots\text{and } z_n \text{ is } A_{n,j} \text{ then } y = f_j\left(q_1,\ldots,q_m\right), \tag{17}$$

where $\boldsymbol{z} = \left[z_1,\ldots,z_n\right]^T$ is the n-dimensional vector of the antecedent variables, and $\boldsymbol{z} \in \boldsymbol{x}$, $\boldsymbol{q} = \left[q_1,\ldots,q_m\right]^T$ is the m-dimensional vector of the consequent variables $\boldsymbol{q} \in \boldsymbol{x}$, where $\boldsymbol{x}$ denotes the set of all inputs of the $y = f\left(\boldsymbol{x}\right)$ model. $A_{i,j}\left(z_i\right)$ denotes the antecedent fuzzy set for the ith input. The antecedents of fuzzy rules partition the input space into a number of fuzzy regions, while the $f_j\left(\boldsymbol{q}\right)$ consequent functions describe the system behaviour within a given region.

With respect to the antecedent membership functions and the structure of the rules, three typical ways of partitioning the input-space can be obtained. These divisions are the grid partition, the tree partition, and scatter partition.

Usually, the f_j consequent function is a polynomial the input variables, but it can be any arbitrarily chosen function that can appropriately describe the output of the system within the region specified by the antecedent of the rule. When $f_j\left(\boldsymbol{q}\right)$ is a first-order polynomial:

$$f_j\left(\boldsymbol{q}\right) = p_j^0 + p_j^1 q_1 + \ldots + p_j^m q_m = \sum_{l=0}^{m} p_j^l q_l, \text{ where } q_0 = 1 \tag{18}$$

the resulting fuzzy inference system is called first-order Takagi-Sugeno or simply a Takagi-Sugeno fuzzy model. If $f_j\left(\boldsymbol{q}\right)$ is a constant (also called singleton models), $f_j = p_j^0$, we have a zero-order Takagi-Sugeno or singleton fuzzy model, which is a special case of the linguistic fuzzy inference systems (Grauel and Mackenberg, 1997) and the TS fuzzy model ($m = 0$).

In singleton models, the nonlinear mapping is represented by a set of fuzzy rules in the following form:

$$R_j: \text{If } \boldsymbol{z} \text{ is } A_i\left(\boldsymbol{z}\right) \text{ then } y = p_j^0 \tag{19}$$

where $i = 1,\ldots,N_r$ is the rule index and $A_i\left(\boldsymbol{z}\right)$ is a fuzzy set defined on $\boldsymbol{z}$. The output of the fuzzy model is computed as the weighted average of the rule consequents p_j^0, as depicted in Fig. 2 and estimated by:

$$y = \frac{\sum_{i=1}^{N_r} A_i(\mathbf{z}) p_j^0}{\sum_{i=1}^{N_r} A_i(\mathbf{z})} \tag{20}$$

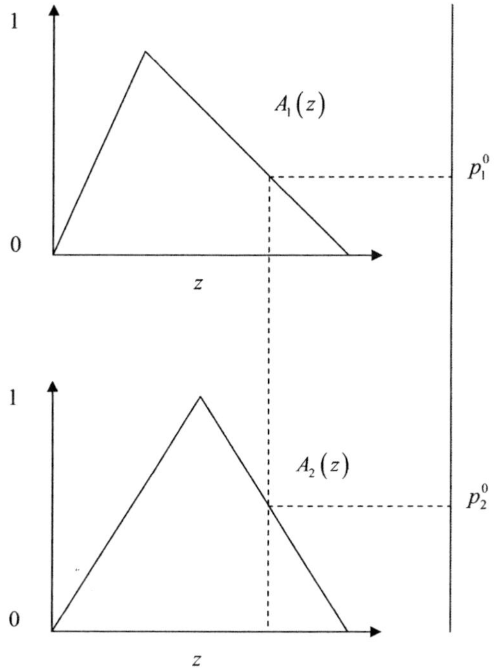

Fig. 2. Inference method of the Takagi-Sugeno fuzzy model

3.2. Fuzzy Models of MIMO Systems

Recently, methods for multiple-input, multiple-output systems (MIMO) have been investigated (Babuška *et al.*, 1998). In this article the multivariable system is modelled by a set of coupled MISO fuzzy models. Another approach is followed in this chapter. The nonlinear multivariable process is represented by a MIMO fuzzy model, which consists of a set of local linear MIMO ARX models whose operating regions are defined by fuzzy sets.

It is assumed that the multiple-input, multiple-output (MIMO) dynamic process can be represented by the following nonlinear vector function:

$$\hat{\mathbf{y}}(k) = f\left(\mathbf{y}(k-1), \ldots \mathbf{y}(k-n_a), \mathbf{u}(k-n_d), \ldots, \mathbf{u}(k-n_b-n_d)\right) \tag{21}$$

where, $\boldsymbol{y}=\left[y_1,\ldots y_{n_y}\right]^T$ is an n_y-dimensional output vector and $\boldsymbol{u}=\left[u_1,\ldots u_{n_u}\right]^T$ is an n_u-dimensional input vector, n_a and n_b are maximum lags considered for the outputs and inputs, respectively, n_d is the minimum discrete dead time, and f represents the multivariable nonlinear model.

This MIMO NARX model is represented by a Takagi-Sugeno fuzzy model:

$$R_j : \text{If} \quad z_1 \quad \text{is} \quad A_{1,j} \quad \text{and}\ldots\text{and} \quad z_n \quad \text{is} \quad A_{n,j} \quad \text{then} \tag{22}$$

$$\hat{\boldsymbol{y}}(k)=\sum\nolimits_{i=1}^{n_a}\mathbf{A}_i^j\boldsymbol{y}(k-1)+\sum\nolimits_{j=1}^{n_b}\mathbf{B}_i^j\boldsymbol{u}(k-i-n_d)+\boldsymbol{c}^j$$

where $\boldsymbol{z}=\left[z_1,\ldots z_n\right]$ is a "scheduling" vector, which is usually a subset of the previous process inputs and outputs, given by:

$$\boldsymbol{z}\in\left\{y_1(k-1),\ldots y_1(k-n_a),\ldots,y_{n_y}(k-n_a),u_1(k-n_d),\ldots,u_{n_u}(k-n_b-n_d)\right\} \tag{23}$$

The matrices $\mathbf{A}_i^j$, $\mathbf{B}_i^j$ and the vector $\boldsymbol{c}^j$ represent the jth local linear multivariable model, where the operating regime of this model is defined by the antecedent part of the jth fuzzy rule. If there is no *a priori* knowledge regarding the order of the nonlinear system, it can be directly estimated from input-output data (Bomberger and Seborg, 1998).

The proposed fuzzy model can be seen as a multivariable linear parameter varying system model (LPV), where at the $\boldsymbol{z}$ operating point, the fuzzy model represents the following linear time invariant (LTI) model:

$$\hat{\boldsymbol{y}}(k)=\sum\nolimits_{i=1}^{n_a}\mathbf{A}_i(\boldsymbol{z})\boldsymbol{y}(k-1)+\sum\nolimits_{j=1}^{n_b}\mathbf{B}_i(\boldsymbol{z})\boldsymbol{u}(k-i-n_d)+\mathbf{c}(\boldsymbol{z}) \tag{24}$$

$$\mathbf{A}_i(\boldsymbol{z})=\sum\nolimits_{j=1}^{N_r}\beta_j(\boldsymbol{z})\mathbf{A}_i^j, \quad i=1,\ldots,n_a \tag{25}$$

$$\mathbf{B}_i(\boldsymbol{z})=\sum\nolimits_{j=1}^{N_r}\beta_j(\boldsymbol{z})\mathbf{B}_i^j, \quad i=1,\ldots,n_b \tag{26}$$

$$\mathbf{c}(\boldsymbol{z})=\sum\nolimits_{j=1}^{N_r}\beta_j(\boldsymbol{z})\mathbf{c}^j \tag{27}$$

where $0\le\beta_j(\boldsymbol{z})\le 1$ is the normalized truth value of the jth rule,

$$\beta_j(\boldsymbol{z})=\frac{\prod_{i=1}^{n}A_{i,j}(z_i)}{\sum_{j=1}^{N_r}\prod_{i=1}^{n}A_{i,j}(z_i)} \tag{28}$$

4. Hammerstein Models

In the previous Section a fuzzy model was used to identify the behaviour of a nonlinear MIMO system. In this Section the fuzzy Hammerstein model structure is introduced. A special case of the nonlinear autoregressive moving average with exogenous input model (NAARX) is the Hammerstein model, which is a series combination of a memoryless nonlinearity and linear dynamics.

4.1. Problem Formulation

A multivariable Hammerstein model is symbolized in Fig. 3. The model consists of a zero-memory nonlinear element $f(.)$ in cascade with a LTI system with transfer function matrix $G(q) \in H_2^{n_y \times n_u}(\mathbb{T})$.

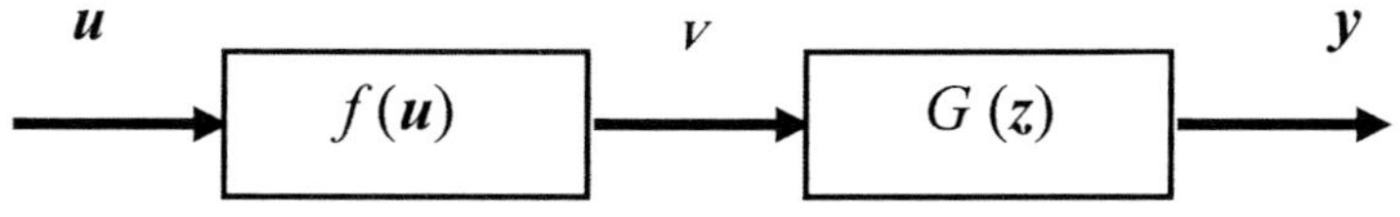

Fig. 3. Multivariable Hammerstein model

The input–output relationship is then given by:

$$y_k = G(q) f(u_k) \tag{29}$$

where $y_k \in \mathbb{R}^{n_y}$, $u_k \in \mathbb{R}^{n_u}$, and $v_k \in \mathbb{R}^{n_u}$ are the system output, input, and the transformed input vectors at time k, respectively. It can be assumed that the nonlinear block can be specified as:

$$f(u_k) = \sum_{h=1}^{N_r} p_h f_h(u_k) \tag{30}$$

where $f_h(.) = \mathbb{R}^{n_u} \to \mathbb{R}^{n_u} \left(i = 1,\ldots,N_r\right)$ are known nonlinear base functions, and $p_h \in \mathbb{R}^{n_u \times n_u} \left(i = 1,\ldots,N_r\right)$ are unknown matrix parameters. Typically, the nonlinear functions $f_h(.)$ are polynomials that allow the representation of smooth nonlinearities, but they can also be radial base functions (RBF) or base functions generated by translations and dilations of a mother function (e.g., wavelets). Any smooth function in an interval can be represented with arbitrary accuracy by a polynomial of sufficiently high order.

From another point of view, the LTI system will be represented using rational orthonormal bases, given by:

$$G(q) = \sum_{l=0}^{p-1} b_l \mathcal{B}_l(q) \tag{31}$$

where $b_l \in \mathbb{R}^{n_y \times n_u}$ are unknown matrix parameters, and $\{\mathcal{B}_l(q)\}_{l=0}^{\infty}$ are rational orthonormal bases.

The identification problem is to estimate the unknown parameter matrices $p_h \in \mathbb{R}^{n_u \times n_u}$ $(i = 1,\ldots,N_r)$ and $b_l \in \mathbb{R}^{n_y \times n_u}$ $(l = 0,\ldots p-1)$ characterizing the nonlinear and the linear parts, respectively, from an *N*-point data set $\{u_k, y_k\}_{k=1}^{N}$ of observed input–output measurements.

The structure represented by Eqs. (29)–(31) of the identified Hammerstein model can be interpreted as an equivalent LTI model whose inputs are the actual inputs u_k filtered by the nonlinear base functions $f_h(.)$ used to represent the static nonlinearity in the Hammerstein model. This interpretation of the Hammerstein model is represented in the block-diagram of Fig. 4.

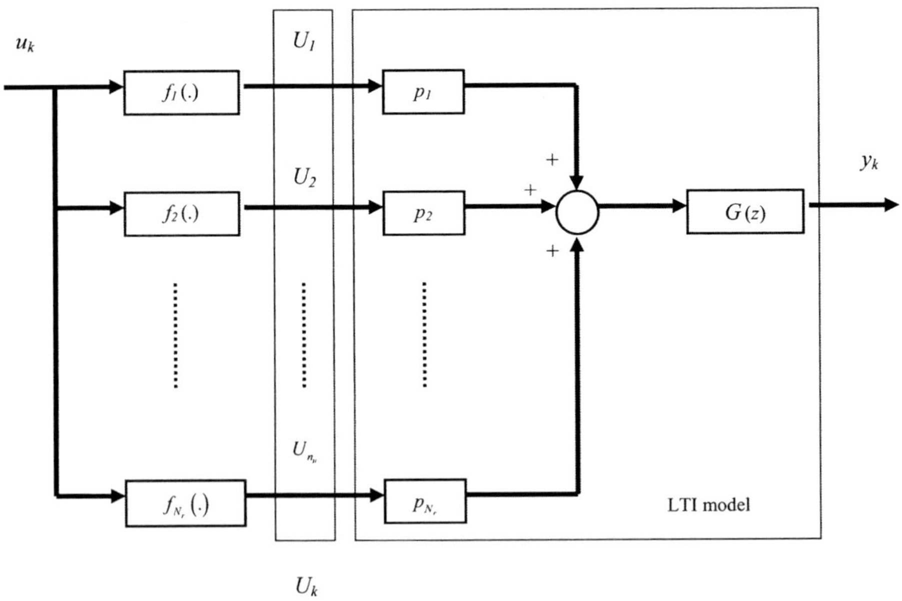

Fig. 4. Equivalent LTI model

4.2. Fuzzy Hammerstein Model

The fuzzy Hammerstein (FH) model consists of a series connection of a memoryless nonlinearity, *f*, and linear dynamics, *G*, (Fig. 3.), where $\boldsymbol{y} = \left[y_1,\ldots y_{n_y} \right]^T$ is the output vector, $\boldsymbol{u} = \left[u_1, \ldots u_{n_u}\right]^T$ the input vector, and $\boldsymbol{v} = \left[v_1, \ldots v_{n_u}\right]^T$ represents the transformed input variables (Abonyi and Babuška, 2000).

If the static nonlinearity is separately parameterized, $f(.)$ can be formulated as a set of functions $v_h = f_h(\boldsymbol{u})$ for $h = 1, \ldots, n_u$. In this chapter, the functions $f_h(\boldsymbol{u})$ are represented by zero-order Takagi-Sugeno (TS) fuzzy models formulated as a set of rules (Takagi and Sugeno, 1985):

$$R_j^h : \text{If } u_1 \text{ is } A_{1,j} \text{ and} \ldots \text{and } u_{n_u} \text{ is } A_{n_u,j} \text{then } v_h = p_j^h \tag{32}$$

From a given input vector, $\boldsymbol{u}$, the output of the fuzzy model, v_h, is inferred by computing the weighted average of the rule consequents:

$$v_h = \frac{\sum_{j=1}^{N_r} \beta_j(\boldsymbol{u}) p_j^h}{\sum_{j=1}^{N_r} \beta_j(\boldsymbol{u})} \tag{33}$$

The static nonlinearity is followed by a multivariable linear dynamic ARX model. Hence, the NAARX representation of the MIMO Hammerstein model is given by:

$$\hat{\boldsymbol{y}}(k) = \sum_{i=1}^{n_a} \mathbf{A}_i \boldsymbol{y}(k-i) + \sum_{i=1}^{n_b} \mathbf{B}_i f(\boldsymbol{u}(k-i-n_d)) \tag{34}$$

where $\boldsymbol{y}(k), \ldots, \boldsymbol{y}(k-n_a+1)$ and $\boldsymbol{u}(k-n_d), \ldots, \boldsymbol{u}(k-n_b-n_d+1)$ are the lagged outputs and inputs of the linear dynamic system, where n_a and n_b denote the maximum lags for the past outputs and inputs, and n_d is the discrete time delay. $\mathbf{A}_1, \ldots, \mathbf{A}_{n_a}$ and $\mathbf{B}_1, \ldots, \mathbf{B}_{n_a}$ are $n_y \times n_y$ and $n_y \times n_u$ matrices, respectively.

Hence, a compact form of the FH model that represents a SISO process is formulated as:

$$\begin{aligned}\hat{y}(k) &= \sum_{i=1}^{n_a} a_i y(k-i) + \sum_{i=1}^{n_b} b_i \sum_{j=1}^{N_r} \beta_j(u(k-i-n_d)) p_j \\ &= \sum_{i=1}^{n_a} a_i y(k-i) + \sum_{j=1}^{N_r} \sum_{i=1}^{n_b} b_i p_j \beta_j(u(k-i-n_d))\end{aligned} \tag{35}$$

The parameters a_i and b_i belonging to the linear dynamic model are called the "linear parameters", while the parameters p_j, belonging to the fuzzy model, are called the "nonlinear parameters". The structure of the resulting model is shown in Fig. 5, where q denotes the shift operator, i.e., $u(k)q^{-1} = u(k-1)$.

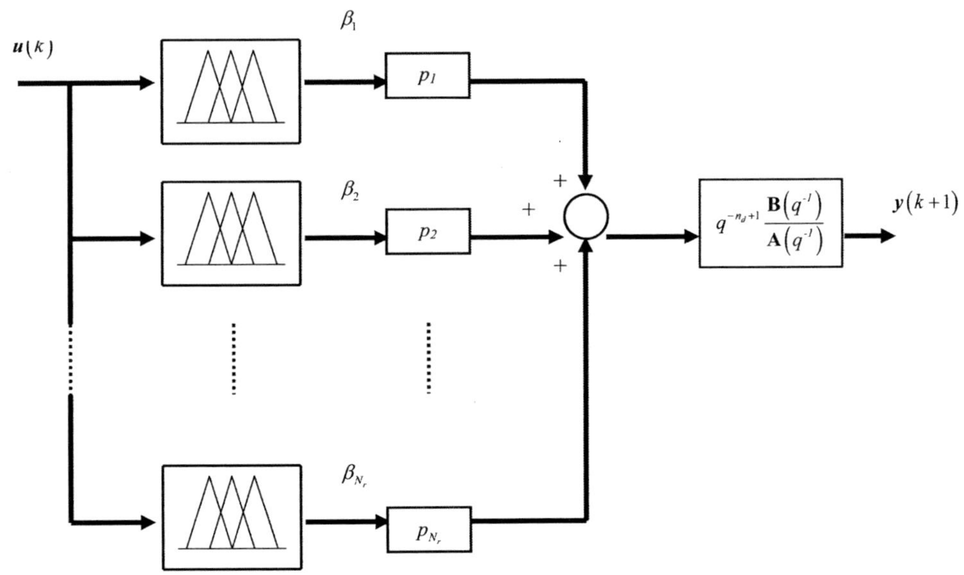

Fig. 5. Structure of the fuzzy Hammerstein model

5. Model Identification

5.1. Iterative Off-Line Identification

The iterative off-line identification is an extension of the algorithm proposed by Narendra and Gallman (1966). It is based on a standard alternating optimization procedure which works as follows. First, suppose the parameters, a_i, b_i, of the linear dynamic model are known. Then, the parameters, $\theta = \left[p_1, \ldots, p_{N_r} \right]$, of the nonlinear part can be estimated by solving the regression problem:

$$y_d = \mathbf{Q}_d \theta + \varepsilon \tag{36}$$

where ε denotes the zero-mean, normally-distributed modelling error. For N data pairs, the vector y_d and the matrix $\mathbf{Q}_d$ are given by:

$$y_d = \begin{bmatrix} y_d(1) \\ y_d(2) \\ \vdots \\ y_d(N) \end{bmatrix} \qquad \mathbf{Q}_d = \begin{bmatrix} \mathbf{q}_d(1) \\ \mathbf{q}_d(2) \\ \vdots \\ \mathbf{q}_d(N) \end{bmatrix} \tag{37}$$

with

$$y_d(k) = y(k) - \sum_{i=1}^{n_a} a_i y(k-1) \tag{38}$$

$$\mathbf{q}_d(k) = \left[\sum_{i=1}^{n_b} b_i \beta_1 \left(u(k-i-n_d)\right), \ldots, \sum_{i=1}^{n_b} b_i \beta_{N_r} \left(u(k-i-n_d)\right)\right] \tag{39}$$

The least-squares solution of Eq. (36) is:

$$\theta = \left[\mathbf{Q}_d^T \mathbf{Q}_d\right]^{-1} (\mathbf{Q}_d)^T \boldsymbol{y}_d \tag{40}$$

Given θ, the input to the linear part, $v(k)$, now can be computed as:

$$v(k) = \sum_{j=1}^{N_r} \beta_j \left(u(k)\right) p_j \tag{41}$$

The parameters of the linear dynamic model, $\theta_j = \left[a_1, \ldots, a_{n_a}, b_1, \ldots, b_{n_b}\right]^T$ can be estimated by solving the regression problem given by:

$$\boldsymbol{y}_l = \mathbf{Q}_l \theta_l + \varepsilon \tag{42}$$

where

$$\boldsymbol{y}_l = \begin{bmatrix} y(1) \\ y(2) \\ \vdots \\ y(N) \end{bmatrix} \qquad \mathbf{Q}_l = \begin{bmatrix} \mathbf{q}_l(1) \\ \mathbf{q}_l(2) \\ \vdots \\ \mathbf{q}_l(N) \end{bmatrix} \tag{43}$$

and

$$\mathbf{q}_l(k) = \left[y(k-1), \ldots, y(k-n_a), \sum_{j=1}^{N_r} \beta_j p_j \left(u(k-i-n_d)\right), \ldots, \sum_{j=1}^{N_r} \beta_j p_j \left(u(k-i-n_d)\right)\right] \tag{44}$$

The least-squares estimate of the linear parameters is found through the following equation:

$$\theta_l = \left[\mathbf{Q}_l^T \mathbf{Q}_l\right]^{-1} (\mathbf{Q}_l)^T \boldsymbol{y}_l \tag{45}$$

5.2. Neuro-Fuzzy Logic

Fuzzy logic provides an alternative approach, which allows one to visualize a model using a higher level of abstraction. This makes the fuzzy logic very attractive for ill-defined systems or systems with uncertain parameters (Zadeh, 1973).

The recent growth in attention to Neural Networks (NNs) has led to many suggestions for combined use of fuzzy logic and neural networks (Jang and Sun, 1995; Welstead, 1994; Lown *et al.*, 1997; Talaq and Al-Basri, 1999). In a multi-layer neural network of feed-forward type, input nodes record the features and pass activation values to the output layer through a hidden layer. The addition of the hidden layers to the two layer perceptron networks allows these networks to represent any continuous mapping from input to output (Bharath and Drosen, 1994; Fu, 1994). An appropriate training technique adjusts the connection weights of the network to improve the match between the output of the network and the correct results.

To design the neuro fuzzy (NF) model some variables, which can represent the dynamic performance of the system, should be chosen to be fed as the NN inputs. After the model inputs pass through appropriate gains or scaling factors are fed to the fuzzy model. The output of the fuzzy model is also scaled by passing through the output gain.

The gains of the fuzzy model are tuned with a NN, making the fuzzy model adaptable to changes in operating conditions. The fuzzy model is tuned by computing optimal gains, exploiting the NN. The NN is composed of three layers, as shown in Fig. 6.

The inputs to the NN are the average temperature, the fuel flow, the air flow, and the current. The average temperature is taken as the arithmetic average of that of the cathode inlet and cathode exhaust, under constant air flow. These inputs represent the operating condition of the system. For various sets of input data to the NN, the values of gains are searched sequentially using simulations.

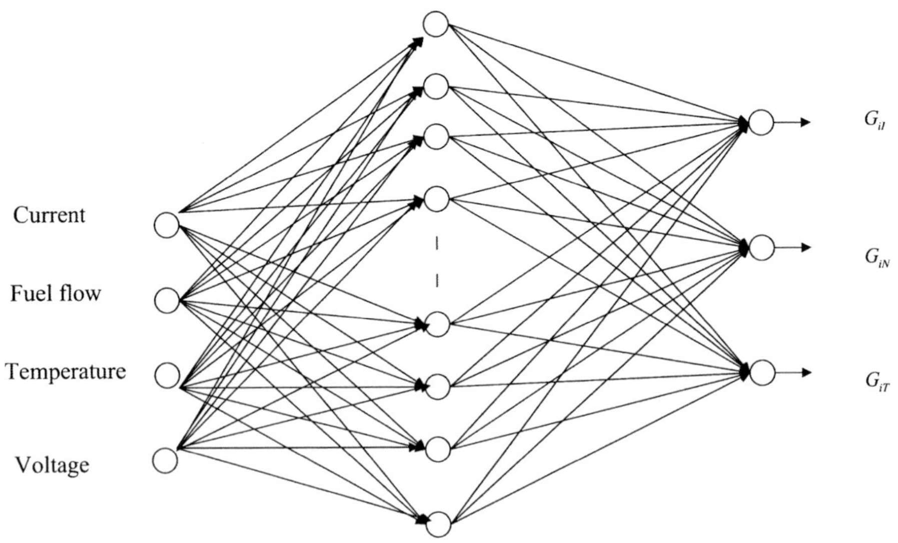

Fig. 6. Neural network architecture employed

An optimization technique called Levenberg-Marquardt method is used to train the neural network. The application of Levenberg-Marquardt is described in (Hagan and Menhaj, 1994; Hagan *et al.*, 1996). This algorithm appears to be the fastest method for training moderate-sized feedforward neural networks (up to several hundred weights). This method is more powerful than gradient descent (backpropagation training method), but requires more memory of the computer. Since the training process is done off-line this requirement does not degrade the performance of the system. The Levenberg-Marquardt update rule is given by:

$$\Delta W = \left(J^T J + \eta I\right)^{-1} J^T e \tag{46}$$

where J is the Jacobian matrix of derivatives of each error to each weight, η is a scalar quantity, and e is the error vector. If η is very large, the above expression approximates gradient descent, while if it is small then it becomes the Gauss-Newton method. Since the Gauss-Newton method is faster, but tends to be less accurate when near an error minimum, η is adjusted during training.

6. Model Predictive Control

6.1. Theoretical Background

Model predictive control (MPC) refers to a class of control algorithms in which a dynamic process model is used to predict and optimize system performance. MPC is rather a methodology than a single technique. The methodology of controllers belonging to the MPC family is characterized by the following strategy illustrated in Fig. 7.

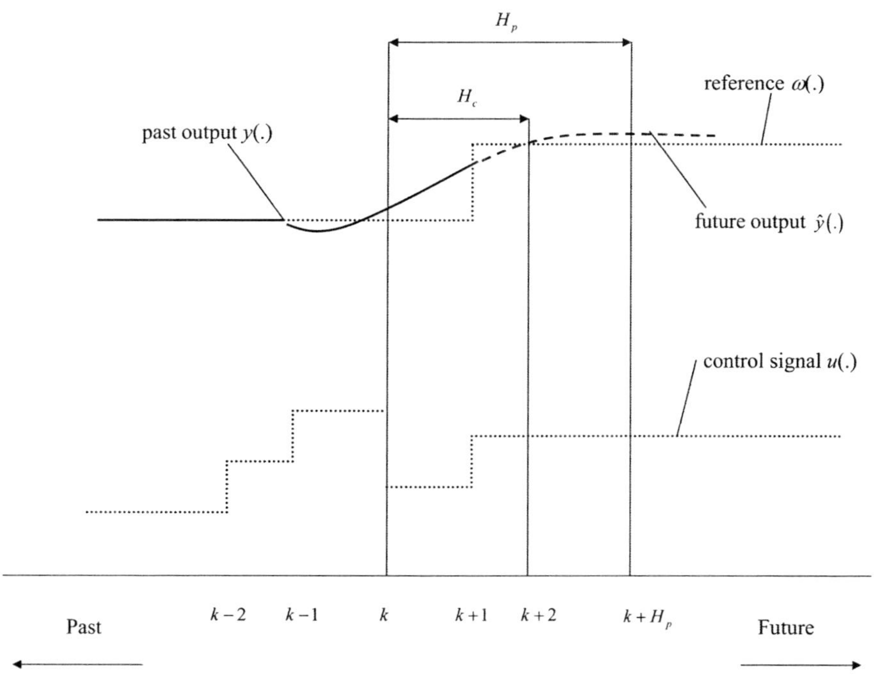

Fig. 7. Strategy of model predictive controller

As shown in Fig. 7, in MPC, the future outputs (fuel cell voltage) for a determined *prediction horizon* H_p are predicted at each instant k using a prediction model. These predicted outputs $\hat{y}(k+j), j=1,...,H_p$ depend on the state of the model at the current time k (given, for instance, by the past inputs and outputs) and on the future control signals $u(k+j)$.

The control signal (fuel flow) change only inside the *control horizon, H_c,* remaining constant afterwards, given by:

$$u(k+j)=u(k+H_c-1), \quad j=H_c,...,H_p \tag{47}$$

The set of control signals is calculated by optimizing a *cost function* in order to keep the process as close as possible to the *reference trajectory* (fuel cell voltage reference), $\omega(k+j)$, j = 1,..., H_p. This criterion usually requires a quadratic function of the errors between the predicted output signal and the reference trajectory. The control effort is included in the objective function in most cases. An explicit solution can be obtained if the criterion is quadratic, the model is linear and there are no constraints. Otherwise an iterative optimization method has to be used.

In practice all systems are subject to restrictions. The actuators have a limited field of action, as in the case of valves. Constructive constrains and safety or environmental concerns can cause limits in the system variables such as fuel flow or maximum temperatures and pressures. All of them lead to the introduction of constraints in the MPC problem. Usually, *input constraints* like:

$$u_{min} \leq u(k+j) \leq u_{max}, \quad j=1,...,H_c \tag{48}$$

$$\Delta u_{min} \leq \Delta u(k+j) \leq \Delta u_{max}, \quad j=1,...,H_c-1 \tag{49}$$

are hard constraints in the sense that they must be satisfied. Conversely, *output constraints* can be viewed as soft constraints because their violation may be necessary to obtain a feasible optimization problem:

$$y_{min} \leq y(k+j) \leq y_{max}, \quad j=j_1,...,H_p \tag{50}$$

where j_1 represents the lower limit for output constraint enforcement.

6.2. Linear Model Based Predictive Control

The basic idea is to use the linear model to predict the future system behaviour. This model is used throughout the entire prediction horizon. Even if this model is very accurate at the linearization point, its accuracy decreases over the prediction horizon. As a consequence, there may be a significant prediction error at $k+H_p$.

6.3. Fuzzy Hammerstein Model Based Predictive Control

Due to the relatively simple block-oriented structure, the application of Hammerstein models in MPC is more straightforward than the application of the general NARX or NAARX models. In this section, the FH model is implemented in MPC by inverting the fuzzy model that represents static nonlinearity (Fruzzetti *et al.*, 1997). As the remaining part of the prediction model is the linear dynamic part of the FH model, the MPC optimization can be solved by quadratic programming.

The combination of the inverse fuzzy model and the nonlinear system results in a transformed dynamical system. This system is linear if the system is of the Hammerstein type and the static nonlinearity is identical to the fuzzy model.

As the inversion of the single-input single-output and multiple-input single-output fuzzy model is a straightforward analytical procedure, the computational demand of the controller is quite comparable to the linear generalized predictive control (GPC). This is a significant advantage compared to other nonlinear models which require the use of nonlinear programming or linearization techniques.

In order to cope with the model-plant mismatch and also with disturbances (load changes), the internal model control (IMC) scheme is used (Garcia and Morari, 1982). The resulting scheme is depicted in Fig. 8.

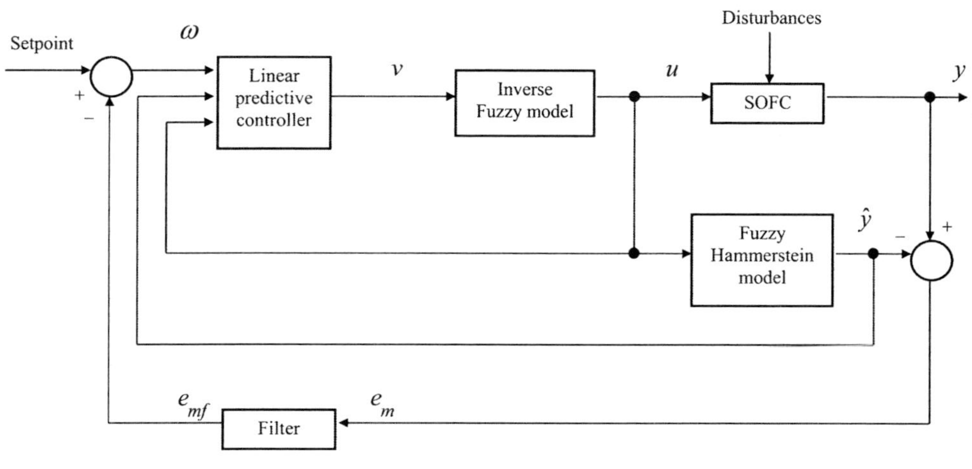

Fig. 8. Fuzzy Hammerstein model based predictive control scheme

6.4. Optimization

In general, the GPC algorithm computes the control sequence $\{\Delta u(k+j)\}$, $j=1,\ldots,H_c$, such that the following quadratic cost function is minimized:

$$J\left(H_{p1},H_{p2},H_c,\lambda\right)=\sum_{j=H_{p1}}^{H_{p2}}\left(\omega(k+j)-\hat{y}(k+j)\right)^2+\lambda\sum_{j=1}^{H_c}\Delta u^2(k+j-1) \tag{51}$$

Here, $\hat{y}(k+j)$ denotes the predicted system output, $\omega(k+j)$ the modified setpoint that is assumed to be known in advance, H_{p1} is the minimum costing horizon, H_{p2} is the maximum costing or prediction horizon, H_c is the control horizon, and λ is the move suppression coefficient.

7. RESULTS

7.1. Fuzzy Hammerstein Model

The inputs to the FH model are the average temperature, the fuel flow, the air flow, and the current. The output of the FH model is the voltage. The average temperature is taken as the arithmetic average of that of the cathode inlet and cathode exhaust, under constant air flow.

The linear subsystem in the Hammerstein model is represented using the rational orthonormal bases with fixed poles presented in (Ninness and Gustafsson, 1997; Gómez and Baeyens, 2004):

$$\mathcal{B}_l(q)=\left(\frac{\sqrt{1-\left|\xi_l\right|^2}}{q-\xi_l}\right)\prod_{i=0}^{l-1}\left(\frac{1-\bar{\xi}_i q}{q-\xi_i}\right), \qquad l\geq 1 \tag{52}$$

$$\mathcal{B}_0(q)=\left(\frac{\sqrt{1-\left|\xi_0\right|^2}}{q-\xi_0}\right) \tag{53}$$

where $\left(\xi_0,\xi_1,\ldots,\xi_{p-1}\right)$ are the poles of the bases.

In order to determine the model order of the linear subsystem, as well as initial guesses for the location of the poles of the bases, the same input–output data are used to identify a linear model of the process using a subspace method. System Identification Toolbox for use with MATLAB (Ljung, 2004) is used for the identification of the linear model. As a result of the identification process a fourth order model is estimated as the linear part of the Hammerstein model.

In this chapter the rules of the fuzzy system are designed based on the available *a priori* knowledge. The parameters of the membership and the consequent are based on the available input-output data. The membership functions presented in Fig. 9 are obtained by performing a number of tests and obtaining the values by trial and error.

Therefore, for a good model performance, the antecedent fuzzy sets on the input variables are designed. Figure 9 shows the membership functions corresponding to temperature, current and fuel flow. The nominal operating conditions of the column considered in this example are given in Table 1 (Kuipers, 1998; Singhal, 1999).

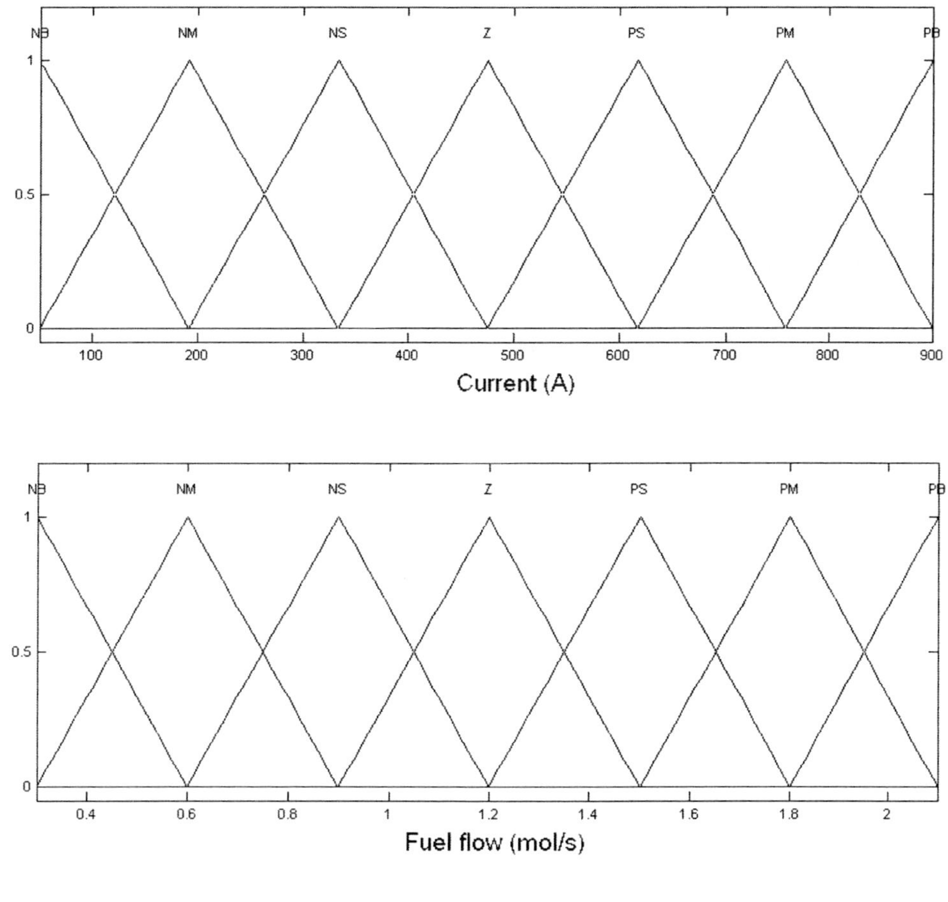

Fig. 9. Antecedent membership functions

For the purposes of identification, the white-box model of the SOFC was excited with band-limited white noise around the nominal value of the fuel flow (u_2), while all the other inputs were kept constant in their nominal values. Changes in the fuel flow were produced every 40 s, with a maximum amplitude of $\pm 100\%$ of the nominal value. A set of 12000 data was collected from the simulation with a sampling time of 0.1 s. The first 6000 data were used for the estimation of a Hammerstein model of the fuel cell, while the remaining 6000 data were used for validation purposes.

In order to tune the NF model, the current is scaled according to the following relation: $I^* = G_{iI} I$; the fuel flow is scaled in agreement with to the relation: $N^* = G_{iN} N$; and the temperature is scaled according to the relation: $T^* = G_{iT} T$. For the sake of simplicity, the outputs of the NF model are not scaled. In the aforementioned relations G_{iI} , G_{iN} and G_{iT} are the scaling factors or gains. The gains of the NF model are tuned with a NN, making the model adaptable to changes in operating conditions.

Table 1. Operating point data

Parameter	Value
Power	100 kW
Stack voltage	286.3 V
Stack current	300 A
Number of cells	384
Number of stacks	1
Open circuit voltage for each cell	0.935 V
Input fuel flow	1.2 10^{-3} kmol/s
Input air flow	2.4 10^{-3} kmol/s
Cell area	1000 cm^2
Cell temperature	1000 ºC
Transfer coefficient α	0.5
Ohmic resistance constant β_r	-2870
Ohmic resistance constant α_r	0.2
Temperature constant T^0	923 K
Limiting current	0.8 A/m^2
K_{H2}	8.43 10^{-4} kmol/(atm s)

A number of patterns related to the system conditions are assembled as training sets to ensure adequate training is presented to the NN during the training phase. The selection of the training features used to form the training pattern is of critical importance to the success of operating a NN. Techniques exist to determine the proper number of training examples from a data set. Haykin and Kosko (2001) suggest the following:

$$N \geq \frac{32W}{\varepsilon}\ln\left(\frac{32M}{\varepsilon}\right) \tag{54}$$

$$N > \frac{W}{\varepsilon} \tag{55}$$

A data set has been generated by computer simulation by varying the input signals. This data set comprised 1500 patterns. From these, 750 were used for learning purposes, 375 for test and 375 for validation, using a random selection with uniform distribution. The training performance of the NN was assessed by examining the root mean square error, assuming that when the error was less than 0.006, then the NN was considered to be satisfactorily trained.

The full model is a physics based component level model of this SOFC configuration, as described in Section 2 of the chapter. This is a very detailed model with high-fidelity which can model each component of the system.

The true (full model) and estimated (Hammerstein model) outputs are represented in Figs. 10-13. A step change in the stack current (from 300 to 400 A) is applied. The true (solid line) and estimated (dashed line) output voltage corresponding to the identified model are represented in Fig. 10.

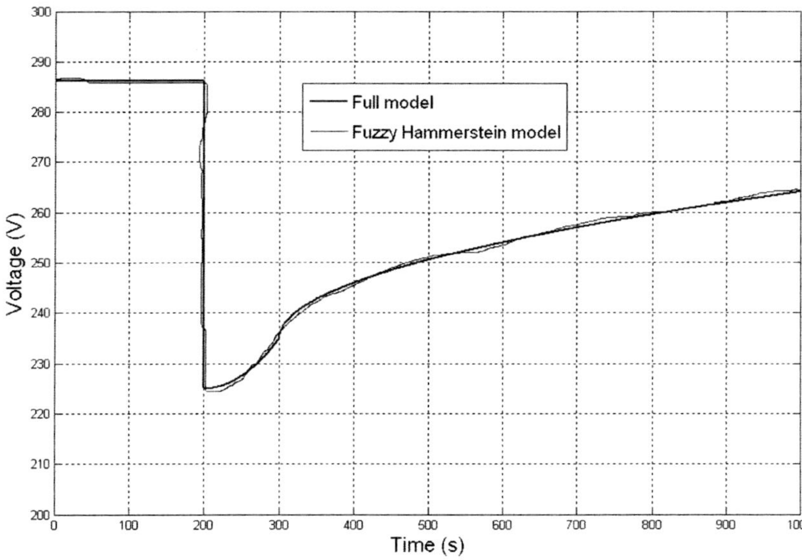

Fig. 10. Output stack voltage response due to stack current step increase

Figure 11 represents the cell temperature due to this step change in the stack current.

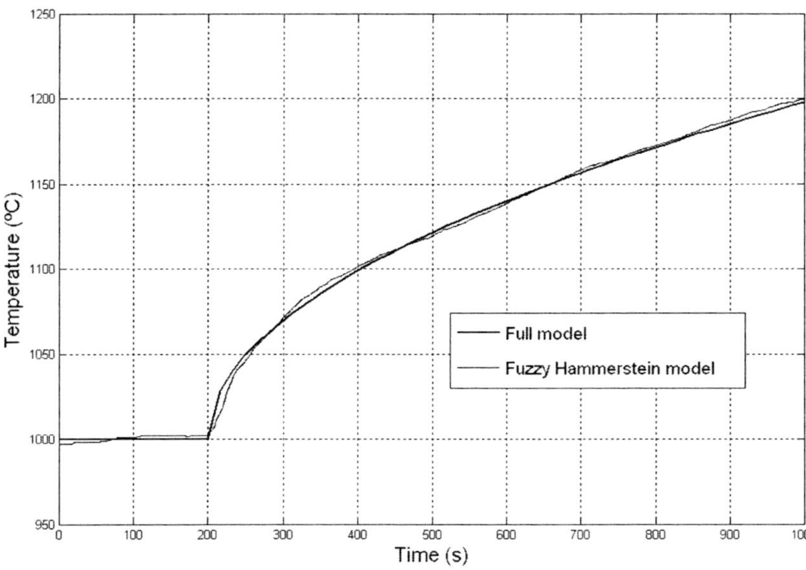

Fig. 11. Output stack temperature response due to stack current step increase

Finally, the voltage step response of the SOFC due to the change in fuel flow input from 1.2 to 1.5 kmol/s is depicted in Fig. 12.

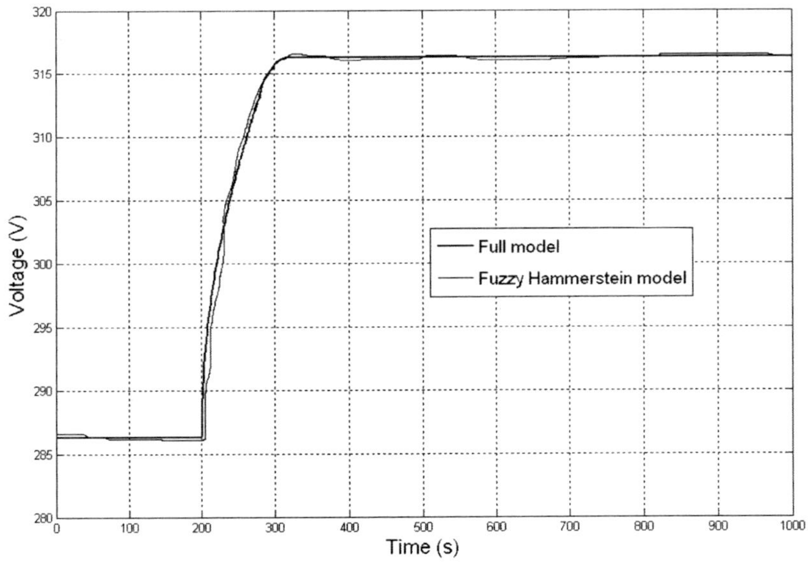

Fig. 12. Output voltage response due to fuel flow step change

7.2. Model Based Predictive Control

The amount of fuel flow can be controlled according to the current. The current is proportional to the terminal load. It is evident that an increase in the load increases the current, which in turn decreases the output voltage of the fuel cell. The increase in current makes fuel flow rate bigger. This increase augments power transfer from the SOFC to the load. Figure 13 shows the load step changes and Fig. 14 represents the output current due to these changes in the load.

Two different controller configurations are used to illustrate the performance of the SOFC:

- Linear MPC using a model obtained by ARX identification of the complete system model at nominal operating conditions (Jurado and Cano, 2004).
- Fuzzy Hammerstein MPC. The MATLAB implementation of quadratic programming is used (Coleman *et al.*, 2004).

The MPC parameters are selected according to the tuning rules given by Soeterboek (1992). The two methods use identical control parameters.

The prediction horizon is set at a sufficiently large value (50 sampling periods, i.e., 500 s) in such a way that the system steady-state response is partly present in the predicted system response. The control horizon is set to one sampling period. The use of shorter prediction horizons would produce manipulated variations mainly on the basis of the high frequency process dynamics, while with longer prediction horizons the controller would tend to behave like a steady-state controller. Furthermore, a longer control horizon would have dramatically increased the computational burden and made the controller still more difficult to implement in a real-time application because it would have been accompanied by an increased number of optimization variables.

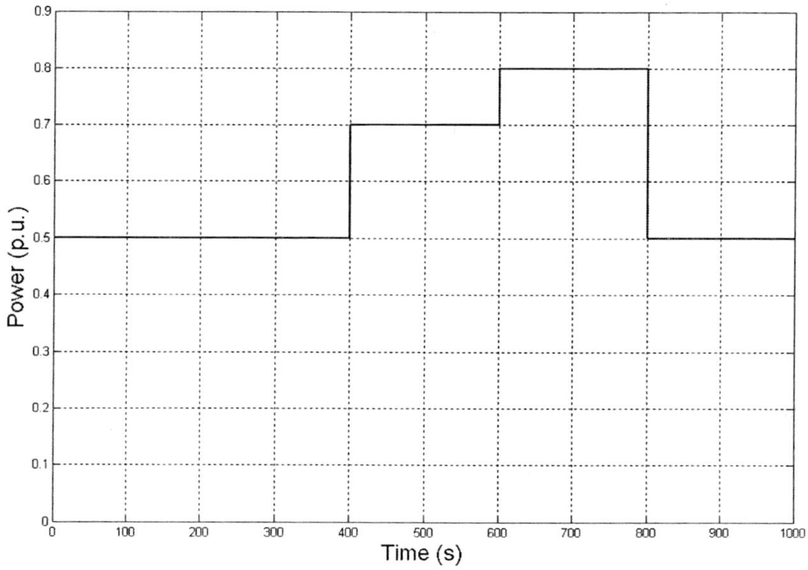

Fig. 13. Load step changes

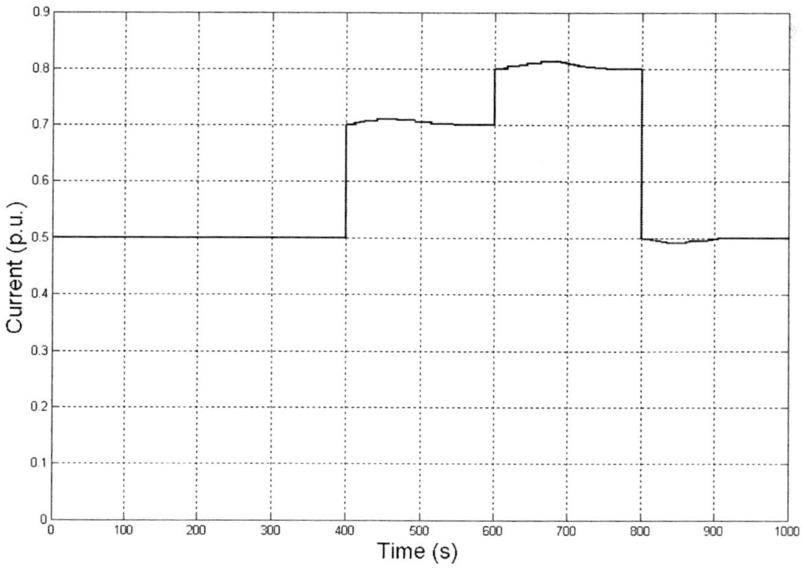

Fig. 14. Output current

The focus of the following simulations is to compare the performance of the MPC framework using the Hammerstein model and the linear model, as described in section 7.2.

The performance using both the linear and Hammerstein models is evaluated with constraints. Fig.15 shows the output voltage response of the SOFC due to a change in fuel flow input. To evaluate the performance of the control system, the integral of time absolute error (*ITAE*) will be used:

$$ITAE = \int_0^{T_f} \left[\beta \left| e(t) \right| t \right] dt \tag{56}$$

where T_f is a finite time chosen somewhat bigger than the rising time and around the settling time, $e(t)$ is the error which is the difference between the measured output and the set-point which is chosen as a unit step, and β is a scaling factor. The resulting ITAE using linear MPC is 1501.3, which is rather higher than that of using Hammerstein MPC (352.6).

Limiting fuel flow to a specific value produces a fixed output power. Output power will have a time delay in following the load power. This is due to the SOFC time constants.

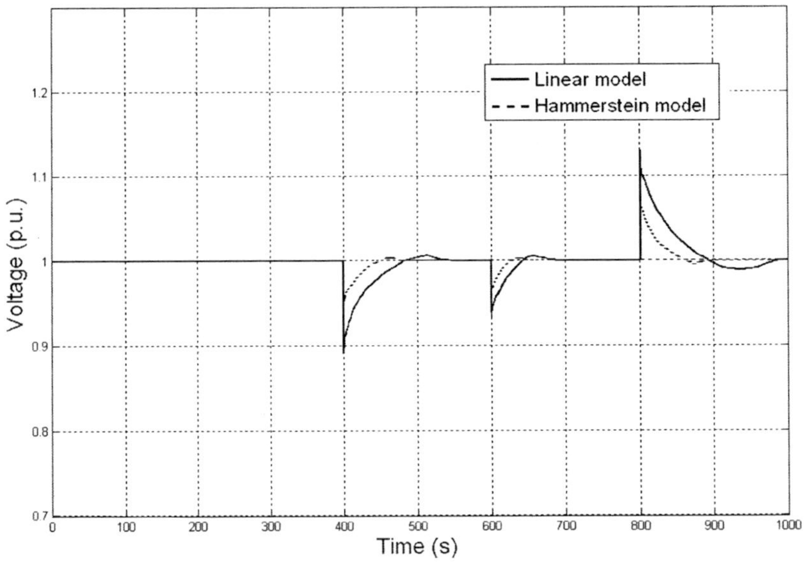

Fig. 15. Output voltage response

8. CONCLUSIONS

Modelling and identification of SOFC constitute an essential stage in practical control design and applications. Controllers commanding systems that operate at varying conditions or require high precision operation raise the need for accurate approach in modelling and identification.

Fuzzy model identification is an effective tool for the approximation of uncertain nonlinear systems on the basis of measured data. The identification of a fuzzy model using input-output data can be divided into two tasks; structure identification, which determines the type and number of the rules and membership functions, and parameter identification.

The Hammerstein model is a special kind of nonlinear systems which has applications in many engineering problems. This chapter describes a multivariable fuzzy Hammerstein model of a SOFC. It has been shown that by tuning the FH model with the aid of a neural network, optimum identification of the system can be achieved in a wide range of operating conditions.

This chapter has described a model based controller (MPC) for the regulation of a SOFC. The first step in designing an MPC system is the derivation of a model that the controller will use for the optimization. This model should be as accurate as possible, while being simple enough to allow for repeated calculations during the optimization.

An MPC is used to control the output voltage of the SOFC by manipulating the fuel flow to keep the voltage at a desired value. The performance of the closed-loop system improves when the Hammerstein model is used instead of the linear model.

REFERENCES

Abe, S., Lan M.S., 1995. Fuzzy rules extraction directly from numerical data for function approximation. *IEEE Transactions on Systems, Man. and Cybernetics*, Vol. 25, No. 1, pp. 119-129.

Abonyi J., Babuška R., 2000. Local and global identification and interpretation of parameters in Takagi-Sugeno fuzzy models. *IEEE International Conference on Fuzzy Systems*, Vol. 2, pp. 835 – 840.

Abonyi J., Nagy L., Szeifert F., 2001. Fuzzy model based predictive control by instantaneous linearization. *Fuzzy Sets and Systems*, Vol. 120, No. 1, pp. 109–122.

Achenbach E., 1994. Three-dimensional and time-dependent simulation of a planar SOFC stack. *Journal of Power Sources*, Vol. 49, No. 1-3, pp. 333-348.

Achenbach E., 1995. Response of a solid oxide fuel cell to load change. *Journal of Power Sources*, Vol. 57, No.1-2, pp. 105-109.

Babuška R., 1998. *Fuzzy Modelling for Control*. Kluwer Academic Publishers, Boston.

Babuška R., Roubos J.A., Verbruggen H.B., 1998. Identification of MIMO systems by input-output TS fuzzy models. *Proceedings of the IEEE International-Conference on Fuzzy Systems*, Vol. 1, pp. 657 – 662.

Bai E.W., 1998. An optimal two-stage identification algorithm for Hammerstein-Wiener nonlinear systems. *Automatica*, Vol. 34, No. 3, pp. 333-338.

Bai E.W., 2001. *Identification of Systems with Hard Input Nonlinearities*. In R. Moheimani (Ed.), Perspectives in Control, Springer, Berlin.

Bharath R., Drosen J., *Neural Network Computing*. New York: McGraw-Hill, 1994.

Billings S., Fakhouri S., 1978. Identification of a class of nonlinear systems using correlation analysis. *Proceedings of IEE*, Vol. 125, No. 7, pp. 691-697.

Bomberger J.D., Seborg D.E., 1998. Determination of model order for NARX models directly from input-output data. *Journal of Process Control*, Vol. 8, No. 5-6, pp. 459-468.

Boutayeb M., Rafaralahy H., Darouach M., 1996. A robust and recursive identification method for Hammerstein model. *Proceedings of the IFAC World Congress*, pp. 447-452.

Chang F.H.I., Luus R., 1971. A non-iterative method for identification using Hammerstein model. *IEEE Transactions on Automatic Control*, Vol. 16, No. 5, pp. 464-468.

Chen J.Q., Lu J.H., Chen L.J., 1994. An on line identification algorithm for fuzzy systems. *Fuzzy Sets and Systems*, Vol. 64, No. 1, pp. 63-72.

Coleman T., Branch M.A., Grace A., 2004. Optimization Toolbox, ver. 2.2. The MathWorks, Inc., Natick, MA.

Fischer M., Nelles O., Isermann R., 1998, Predictive control based on local linear fuzzy models. *International Journal of Systems Science*, Vol. 29, No. 7, pp. 679–697.

Fruzzetti K.P., Palazoglu A., McDonald K.A., 1997. Nonlinear model predictive control using Hammerstein models. *Journal of Process Control*, Vol. 7, No. 1, pp. 31-41.

Fu L., 1994. *Neural Networks in Computer Intelligence*. New York: McGraw-Hill.

Fuel Cell Handbook, 5th ed. 2000. EG and G Services Parsons Inc., U.S. Department of Energy.

Garcia C.E., Morari M., 1982. Internal model control: 1. A unifying review and some new results. *Industrial and Engineering Chemistry Process Design and Development*, Vol. 21, No. 2, pp. 308-323.

Gegov A.E., 1997. Multilayer fuzzy control of multivariable systems by active decomposition. *International Journal of Intelligent Systems,* Vol. 12, No.1, pp. 83-103.

Gómez J.C., Baeyens E., 2004. Identification of block-oriented nonlinear systems using orthonormal bases. *Journal of Process Control*, Vol. 14, No. 6, pp. 685–697.

Grauel A., Mackenberg H., 1997. Mathematical analysis of the Sugeno controller leading to general design rules. *Fuzzy Sets and Systems*, Vol. 85, No. 2, pp. 165-175.

Greblicki W., 1992. Nonparametric identification of Wiener system. *IEEE Transactions on Information Theory*, Vol. 38, No. 5, pp. 1487-1493.

Haber R., Unbehauen H., 1990. Structure identification of nonlinear dynamic systems—A survey on input/output approaches. *Automatica*, Vol. 26, No. 4, pp. 651-677.

Hagan M.T., Menhaj M., 1994.Training feedforward networks with the Marquardt algorithm. *IEEE Transactions on Neural Networks*, Vol. 5, No. 6, pp. 989–993.

Hagan M.T., Demuth H.B., Beale M.H., 1996. *Neural Network Design*. Boston, MA: PWS Publishing.

Hall D.J., Colclaser R.G., 1999. Transient modelling and simulation of tubular solid oxide fuel cells. *IEEE Transactions on Energy Conversion*, Vol. 14, No. 3, pp. 749-753.

Haykin S., Kosko B., 2001. *Intelligent Signal Processing*. IEEE Press, USA.

Homaifar, A., McCormick E., 1995. Simultaneous design of membership functions and rule sets for fuzzy controllers using genetic algorithms. *IEEE Transactions on Fuzzy Systems*, Vol. 3, No. 2, pp. 129-138.

Hsia T., 1976. A multi-stage least squares method for identifying Hammerstein model nonlinear systems. *Proceedings of the IEEE Conference on Decision and Control*, pp. 934-938.

Hu J.Q., Rose T., 1999. Generalized predictive control using a neuro fuzzy model. *International Journal of Systems Science*, Vol. 30, No. 1, pp. 117–122.

Ishibuchi, H., Nozaki K., Yamamoto N., Tanaka H., 1995. Selecting fuzzy if – then rules for classification problems using genetic algorithms. *IEEE Transactions on Fuzzy Systems,* Vol. 3, No. 3, pp. 260-271.

Jang J. S. R., Sun C.T, 1995. Neuro-fuzzy modelling and control. *Proceedings of the IEEE*, Vol. 83, No. 3, pp. 378-406.

Jurado F., 2004a. Modelling SOFC plants on the distribution system using identification algorithms. *Journal of Power Sources*, Vol. 129, No. 2, pp. 205-215.

Jurado F., 2004b. Fuzzy logic applied to the inverter of a SOFC power plant. *Fuel Cells – From Fundamentals to Systems*, Vol. 4, No. 4, pp. 378-387.

Jurado F, Cano A, 2004, Use of ARX algorithms for modelling micro-turbines on the distribution feeder, *IEE Proceedings-Generation Transmission and Distribution*, Vol. 151, No. 2, pp. 232-238.

Jurado F., 2005. Experience with nonlinear model identification of fuel cell plants. *Fuel Cells – From Fundamentals to Systems*, Vol. 5, No. 1, pp. 105-114.

Kosko, B., 1992. *Neural Networks and Fuzzy Systems: A Dynamic Systems Approach to Machine Intelligence*. Englewood Cliffs, NJ: Prentice-Hall.

Kouatli I.M., 1994. A simplified fuzzy multivariable structure in a manufacturing environment. *Journal of Intelligent Manufacturing*, Vol. 5, No. 6, pp. 365-387.

Kuipers J.A., 1998. *Proceedings of the Fuel Cell Seminar*, pp. 450.

Ljung L., 2004. *System Identification Toolbox*, ver. 6.0.1. The MathWorks, Inc., Natick, MA.

Lown M., Swidenbank E., Hogg B.W., 1997. Adaptive fuzzy logic control of a turbine generator system. *IEEE Transactions on Energy Conversion*, Vol. 12, No. 4, pp. 394-399.

Lu Y.Z., He M., Xu C.W., 1997. Fuzzy modelling and expert optimization control for industrial processes. *IEEE Transactions on Control Systems Technology*, Vol. 5, No. 1, pp. 2-12.

Lukas M.D., Lee K.Y., Ghezel-Ayagh H., 1999. Development of a stack simulation model for control study on direct reforming molten carbonate fuel cell power plant. *IEEE Transactions on Energy Conversion*, Vol. 14, No. 4, pp. 1651-1657.

Lukas M.D., Lee K.Y., Ghezel-Ayagh H., 2001. An explicit dynamic model for direct reforming carbonate fuel cell stack. *IEEE Transactions on Energy Conversion*, Vol. 16, No. 3, pp. 289-295.

Mamdani E.H., Assilian S., 1999. An experiment in linguistic synthesis with a fuzzy logic controller. *International Journal of Human-Computer Studies*, Vol. 51, No. 2, pp. 135-147.

Manoranjan, V.S., de Sam Lazaro A., Edwards D., Athalye A., 1995. A systematic approach to obtaining fuzzy sets for control systems. *IEEE Transactions on Systems, Man. and Cybernetics*, Vol. 25, No. 1, pp. 206-213.

Mendel J.M., 1995. Fuzzy logic systems for engineering: A tutorial. *Proceedings of the IEEE*, Vol. 83, No. 3, pp. 345–377.

Mollov S., Babuška R., Abonyi J., Verbruggen H., 2004. Effective optimization for fuzzy model predictive control. *IEEE Transactions on Fuzzy Systems*, Vol. 12, No. 5, pp. 661-675.

Narendra K., Gallman P., 1966. An iterative method for the identification of nonlinear systems using a Hammerstein model. *IEEE Transactions on Automatic Control*, Vol. 11, No. 3, pp. 546-550.

Ninness B., Gustafsson F., 1997. A unifying construction of orthonormal bases for system identification. *IEEE Transactions on Automatic Control*, Vol. 42, No. 4, pp. 515–521.

Qin S.J., Badgwell T.A., 1997. An overview of industrial model predictive control technology. *Proceedings of the International Conference of Chemical Process Control*, pp. 232–256.

Qin S.J., Badgwell T.A., 2000. An overview of nonlinear predictive control applications. In Allgöwer F. and Zheng A. (Eds.), Nonlinear Model Predictive Control. ser. *Progress in Systems and Control Theory*, MA: Birkhäuser, Vol. 26, pp. 369–392.

Padullés J., Ault G.W., McDonald J.R., 2000a. An integrated SOFC plant dynamic model for power systems simulation. *Journal of Power Sources*, Vol. 86, No. 1-2, pp. 495-500.

Padullés J., Ault G.W., McDonald J.R., 2000b. An approach to the dynamic modelling of fuel cell characteristics for distributed generation operation. *Proceedings of the IEEE/PES Winter Meeting*, Vol. 1, pp. 134 -138.

Pawlak M., 1991. On the series expansion approach to the identification of Hammerstein systems. *IEEE Transactions on Automatic Control*, Vol. 36, No. 6, pp. 763-767.

Perneel, C., Themlin J. M., Renders J. M., Acheroy M., 1995. Optimization of fuzzy expert systems using genetic algorithms and neural networks. *IEEE Transactions on Fuzzy Systems,* Vol. 3, No. 3, pp. 300-312.

Richalet J., 1993. Industrial applications of model based predictive control. *Automatica*, Vol. 29, No. 5, pp.1251–1274.

Rovatti, R., Guerrieri R., Baccarani G., 1995. An enhanced two-level boolean synthesis methodology for fuzzy rules minimization. *IEEE Transactions on Fuzzy Systems*, Vol. 3, No. 3, pp. 288-299.

Rovatti, R., Guerrieri R. 1996. Fuzzy sets of rules for system identification. *IEEE Transactions on Fuzzy Systems,* Vol. 4, No. 2, pp. 89-102.

Setnes M., Babuška R., Verbruggen H.B., 1998. Rule-based modelling: Precision and transparency. *IEEE Transactions on Systems, Man. and Cybernetics*, Vol. 28, No. 1, pp. 165-67.

Singhal S.C., 1999. Progress in tubular solid oxide fuel cell technology. *Proceedings of the Solid Oxide Fuel Cells Conference*, pp. 39-51.

Soeterboek R.M., 1992. *Predictive Control: A Unified Approach*. Upper Saddle River, NJ: Prentice-Hall.

Stoica P., 1981. On the convergence of an iterative algorithm used for Hammerstein system identification. *IEEE Transactions on Automatic Control*, Vol. 26, No. 4, pp. 967-969.

Sugeno M., Kang G.T., 1986. Fuzzy modelling and control of multilayer incinerator. *Fuzzy Sets and Systems*, Vol. 18, No. 3, pp. 329-345.

Takagi T., Sugeno M., 1985. Fuzzy identification of systems and its application to modelling and control. IEEE *Transactions on Systems, Man. and Cybernetics*, Vol. 15, No. 1, pp. 116-132.

Talaq J., Al-Basri F., 1999. Adaptive fuzzy gain scheduling for load frequency control. *IEEE Transactions on Power Systems*, Vol. 14, No.1, pp. 145-150.

Welstead S., 1994. *Neural Network and Fuzzy Logic Applications in C/C++*. Chichester: John Wiley and Sons, Inc.

Zadeh L.A., 1973. Outline of a new approach to the analysis of complex systems and decision processes. *IEEE Transactions on Systems, Man. and Cybernetics*, Vol. 3, No.1, pp. 28-44.

In: Artificial Intelligence in Energy and Renewable Energy Systems ISBN 1-60021-261-1
Editor: Soteris Kalogirou, pp. 269-307

Chapter 8

THE APPLICATION OF FUZZY SYSTEMS TO CONTROL AND MODELLING OF BUILDING SYSTEMS

Chris P. Underwood
School of the Built Environment,
Northumbria University, Newcastle upon Tyne, UK

ABSTRACT

Design and problem-solving concerning energy and renewable energy in buildings is frequently as much to do with art as science. As a consequence the pursuit of good comfort, minimum energy use and minimum maintenance is often easier to describe using semi-qualitative norms than through the application of precise metrics. Fuzzy inference methods open up an opportunity to develop solutions to problems based on imprecise data, mixed qualititative and quantitative data and heuristic rules-of-thumb. In this chapter the application of fuzzy inference to energy systems is explored with particular reference to buildings. It will be established that successful solutions to many building energy-related problems can be achieved using these methods with limited or incomplete data to problems as diverse as system selection (decision-making), energy control systems and energy modelling. A thorough review of the literature concerning these applications is presented followed by a set of case studies designed to highlight the potential of fuzzy inference when applied to a variety of building energy related problems.

Keywords: Fuzzy logic, control, modelling, building systems, energy

NOMENCLATURE

A, B	Fuzzy set
$\overline{A}$	Complement of the general fuzzy set, A
E	General sum-error-squared term (*)
H	Fuzzy cluster partition entropy

i	Cluster index
k	Constant (general) (*)
K	Learning rate (*)
n	Sum of all data points
P_i	i^{th} membership function parameter (*)
RH	Relative humidity (%)
T	Temperature (°C)
u	Control signal
U	Universe of discourse (*)
w	Membership function width parameter (*)
x	Crisp input (*)
x_c	Membership function crisp centre value (*)
y	Crisp output (*)

Greek

Δ	Change (*)
μ	Degree of membership
μ_A	Degree of membership in the fuzzy set, A
μ_B	Degree of membership in the fuzzy set, B
$\mu_{A\cup B}$	Union degree of membership for the fuzzy sets, A, B
$\mu_{A\cap B}$	Intersection degree of membership for the fuzzy sets, A, B
μ_I	Degree of membership in the Ith data cluster

* implies that units are application-dependent

ABREVIATIONS

ANFIS	Adaptive Neuro-Fuzzy Inference System
ANN	Artificial Neural Network
AR	AutoRegression
C-Bio	Centralised Biofuel
CFD	Computational Fluid Dynamics
CHP	Combined Heat and Power
D-Bio	Distributed Biofuel
DoM	Degree of Membership
FIS	Fuzzy Input Set
FLC	Fuzzy Logic Controller
FOS	Fuzzy Output Set
FSI	Fuzzy Selection Index
HVAC	Heating, Ventilating and Air Conditioning
PID	Proportional plus Integral plus Derivative
PMV	Predicted Mean Vote
PV	PhotoVoltaic

RES	Renewable Energy System
T-S	Takagi-Sugeno
WT	Wind Turbine

1. INTRODUCTION

A fuzzy inference system is merely a method for mapping input variables onto output variables. Whereas conventional methods for establishing relationships between variables use deterministic theory or observable data, the term "fuzzy" is used in fuzzy inference because the mapping is carried out using a set of heuristic rules. Thus the traditional difficulties of setting out complex theory or obtaining data of adequate quality for model-fitting are replaced by rules that are based on experience or pragmatic human reasoning. Of essence, fuzzy inference is interpolative in that it is driven by degrees-of-membership that the input has of a set of membership functions and, as such, it is completely free from the true/false limitations imposed by traditional Boolean systems. This gives fuzzy inference methods a high potential for applications that contain uncertainty or are ill-defined. The development of fuzzy inference methods can be traced back to the classical work of Lotfi Zadeh during the 1960s and 1970s (Zadeh, 1973).

As far as applications to building energy problems are concerned, fuzzy systems have been receiving attention throughout the past 15 years. The earliest work focused on expert system applications (see for example Dexter and Trewhella, 1990) leading on to a significant amount of work dedicated to heating, ventilating and air conditioning (HVAC) control systems (see for example So *et al.,* 1997) and, more recently, renewable energy systems (for example Gouda *et al.,* 2005). During the past five years, most of the energy-related interest in fuzzy systems has focused on fuzzy modelling (for example He *et al.,* 2005).

This chapter concerns itself with the application of fuzzy systems to energy and renewable energy in buildings. It starts with a review of some of the underlying principles of fuzzy sets, fuzzy inference and fuzzy inference methods and follows this with a detailed review of the literature relating to applications of fuzzy inference to expert (decision-making) systems, control systems and data modelling. Finally, a set of case studies designed to demonstrate the potential of fuzzy systems when applied to these three application themes is presented.

2. FUZZY INFERENCE SYSTEMS

A fuzzy inference system consists of three elements as shown in Fig. 1:

- An (input) fuzzifier
- A fuzzy inference mechanism
- An (output) defuzzifier

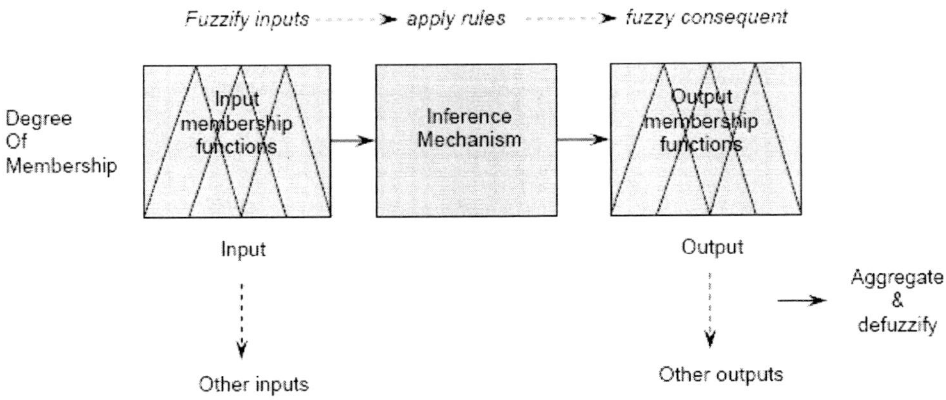

Fig. 1. The fuzzy inference system

The fuzzifier accepts one or more inputs as crisp points (i.e., precise numerical values) and maps these to degrees of membership of one or more fuzzy input sets (FIS). The fuzzy inference mechanism then maps the resulting FIS onto a single aggregated fuzzy output set (FOS) using rule-based reasoning and the defuzzifier translates the aggregated FOS into a crisp value. Thus the fuzzifier and fuzzy inference mechanism form a fuzzy antecedent from which a fuzzy consequent is deduced.

2.1. Membership Functions and Fuzzy Sets

A universe of discourse, U, represents all possible elements of a particular class of elements. If x is an element in U then $x \in U$. A fuzzy set in a universe of discourse can be expressed as a membership function $\mu_A(x)$ and can take values in the interval (0,1) whereas a classical set has membership functions whose values can be only either 0 or 1. Thus a fuzzy set can be thought of as an interpolative form of classical set.

A fuzzy set, A, in U is defined as a set of ordered pairs:

$$A = \{x, \mu_A(x)(x \in U)\} \tag{1}$$

For two fuzzy sets, A and B, in the same universe of discourse, U, the union of A and B ($A \cup B$) is defined by the membership function:

$$\mu_{A \cup B}(x) = \max\left[\mu_A(x), \mu_B(x)\right] \tag{2}$$

thus $\mu_{A \cup B}(x)$ is the smallest fuzzy set containing both A and B (e.g. Fig. 2):

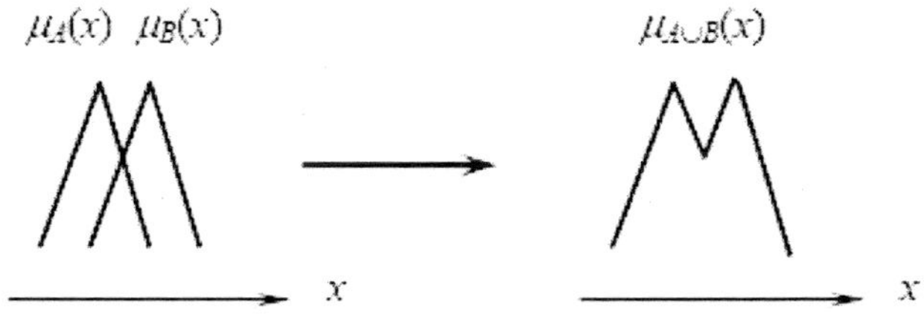

Fig. 2. Fuzzy union

The intersection of *A* and *B* ($A \cap B$) in *U* is given by the membership function:

$$\mu_{A \cap B}(x) = \min\left[\mu_A(x), \mu_B(x)\right] \tag{3}$$

thus $\mu_{A \cap B}(x)$ is the largest fuzzy set containing both *A* and *B* (e.g. Fig. 3):

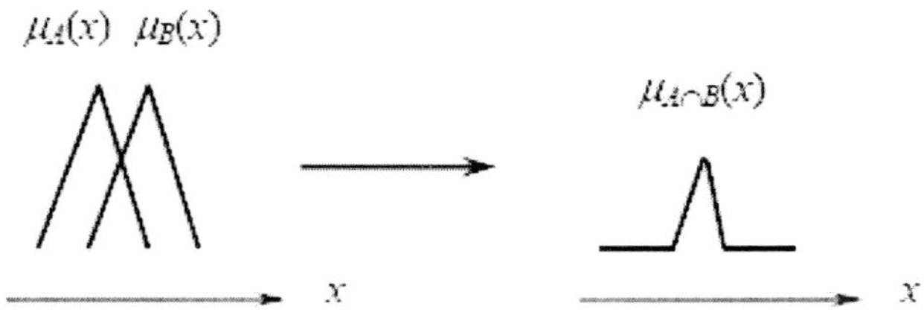

Fig. 3. Fuzzy intersection

A and *B* are equal if, and only if, $\mu_A(x) = \mu_B(x), \forall_{x \in U}$.

The complement of *A* (denoted $\bar{A}$) is a fuzzy set whose membership function is:

$$\mu_{\bar{A}}(x) = 1 - \mu_A(x) \tag{4}$$

In practice a union of fuzzy sets is interpreted as the fuzzy operator OR, intersection as the fuzzy operator AND, and fuzzy complement as the fuzzy operator NOT. Whilst most fuzzy inference applications interpret these operators in these precise ways there are other possibilities (e.g. products, probabilities).

The mapping of a crisp point onto a FIS is achieved through fuzzy membership functions. Figure 4 illustrates the three most common membership function types used. Again, many other membership function types are possible including irrational shapes for specific applications. However there are no procedural rules for the selection of membership functions other than making the simplest choice that leads to acceptable results. Many applications work perfectly well with triangular membership functions. The important thing is to structure the membership functions so as to yield smooth interpolation between them.

In some instances, a fuzzy singleton form of membership is useful in that it can make subsequent fuzzy inference more immediately transparent. The singleton fixes membership at unity at a specified crisp point and at zero for all other inputs. The Gaussian, trapezoid and triangular membership functions have the advantage that they can intrinsically achieve a certain amount of input noise suppression.

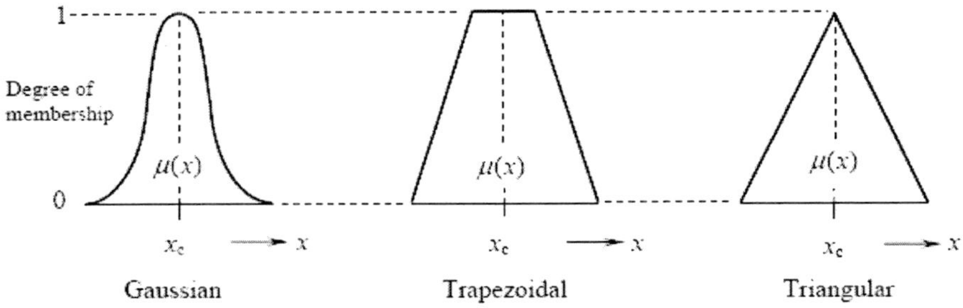

Fig. 4. Typical membership functions

2.2. Mamdani Inference

The inference mechanism is the heart of a fuzzy inference system. It consists of a rulebase of IF-THEN rules and a means for translating the (one or more) fuzzy input sets onto an aggregated fuzzy output set based on these rules.

Each rule in the rulebase takes the following general form (AND or OR will appear if there is more than one input):

$$\text{Rule: IF } (x_1 \text{ is } A_{x_1}) \text{ AND } (x_2 \text{ is } A_{x_2}) \text{ THEN } (y \text{ is } B_y) \qquad (5)$$

(a n t e c e d e n t) (c o n s e q u e n t)

where A_{x_1}, A_{x_2} are fuzzy subsets of the universes of discourse U_{x_1}, U_{x_2} (i.e., $A_{x_1} \in U_{x_1}$ and $A_{x_2} \in U_{x_2}$) and B_y is a subset of the output space, U_y (i.e., $B_y \in U_y$).

Conventionally, the most common and enduring inference method is Mamdani inference (Mamdani and Assilian, 1975) mainly because it is intuitive. First, AND and OR combinations in the antecedents of the rules are resolved using normal Boolean logic with fuzzy membership degrees preserved with the min operator used to resolve AND and the max operator used to resolve OR.

Next, the result from the (resolved) antecedent is "fired" at the consequent with equality of degree. That is, the consequent enjoys the same degree of fuzzy membership as the antecedent. Of essence, Mamdani inference means that the resulting output is one or more fuzzy sets. The following simple illustration shows how a single rule can thus be inferred. Consider the fuzzy system of Fig. 5(a) consisting of two fuzzy input sets, U_1 and U_2 and fuzzy output set, V.

Suppose the following rule applies, where x denotes a crisp input:

$$\text{IF } \left(x(U_1) \text{ is } A_{11}\right) \text{ AND } \left(x(U_2) \text{ is } A_{22}\right) \text{ THEN } \left(V \text{ is } B_1\right) \qquad (6)$$

Suppose also that, on some occasion, crisp input points lie at $x(U_1) = -0.5$ and $x(U_2) = 0.6$. From the fuzzy inference of Fig. 5(b) the membership degrees of $A_{11} = 0.5$ and $A_{12} = 0.8$ are obtained. The resolved antecedent will therefore be,

$$\min(A_{11}, A_{22}) = \min(0.5, 0.8) = 0.5 \tag{7}$$

This is fired at the fuzzy output set resulting in a trapezoidal patch of the B_1 output fuzzy subset with a height of 0.5 representing the required fuzzy output (Fig. 5(b)).

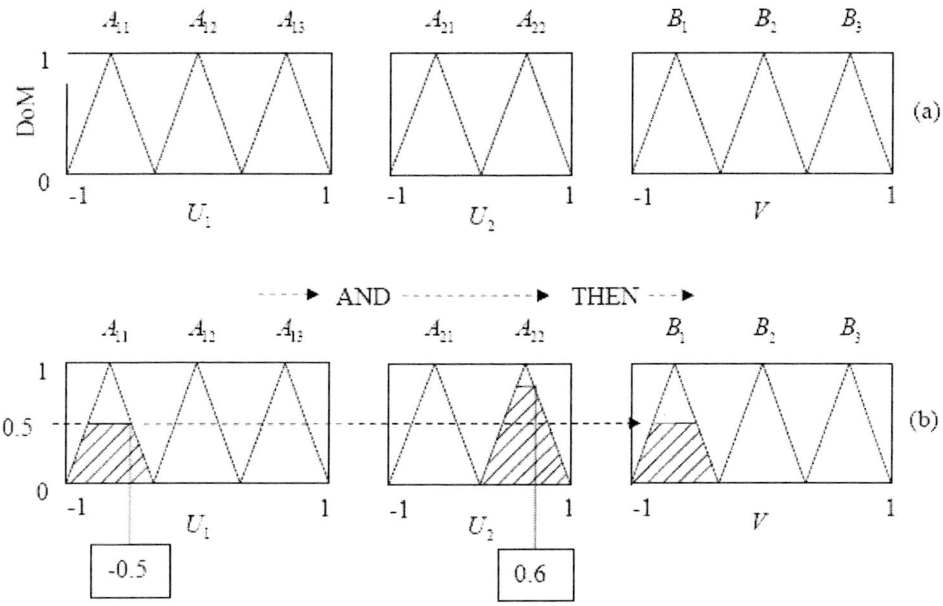

Fig. 5. Fuzzy inference illustration (a) Input/output fuzzy sets (b) Inference

2.3. Defuzzification

The fuzzy output arising from Mamdani's method is now translated to a crisp point by defuzzification. In practice, the output will be an aggregate fuzzy set consisting of all patches of fuzzy outputs fired from each rule added together to form a polygon. The crisp point can be deduced using a number of methods – the average of the polygon of the aggregated set; the largest maximum; the smallest maximum; the centroid, to name four such methods. The most common is the centroid in which the centre of the area of the aggregate polygon is located. This can be obtained by finding the weighted sum of the centroids that arise when the polygon is partitioned into a set of triangles. The centroid of a triangle is simply the average of its three vertices. Thus the sum of the centroids of the resulting triangles weighted by their areas normalized by the whole polygon area gives the overall centroid. Crisp point defuzzification for the previous illustration is shown in Fig. 6.

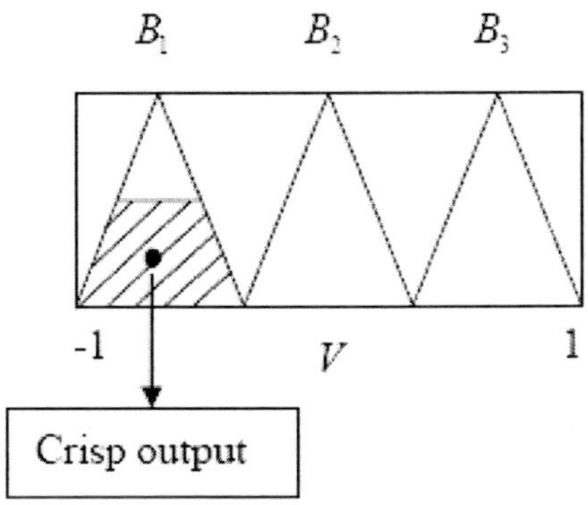

Fig. 6. Defuzzification to a crisp output

2.4. Sugeno Inference

An alternative method of inference is due to Sugeno (1985), referred to as Sugeno inference (or sometimes Takagi-Sugeno inference). In this method, input fuzzification and the application of rules is as before but instead of outputting to fuzzy membership functions the outputs take the form of constants or linear functions of the input variables. Thus a typical rule with Sugeno inference may take the form:

$$\text{Rule: IF } (x_1 \text{ is } A_{x_1}) \text{ AND } (x_2 \text{ is } A_{x_2}) \text{ THEN } y = k \tag{8}$$

(where k is a constant).

The result is a set of output singletons arrived at by multiplying the applied input by k, weighting the result according to the input membership functions, and then aggregating all the output singletons by averaging. This method has appeal for fuzzy control applications because it represents a convenient way to do gain-scheduling control. It also has appeal in multivariable control applications where interactions between input variables can be conditionally compensated for.

2.5. Adaptive Neuro-fuzzy Inference Systems

Membership functions are defined by parameters whose values determine their precise shape. For example, a Gaussian membership function is defined by a crisp value at its centre and a width parameter. Conventionally, the setting of these parameters is largely intuitive. Adaptive neuro-fuzzy inference systems (ANFIS) attempt to fit the best parameters to the membership functions by learning from the features of an input data set. An ANFIS adopts a network-based structure which is similar to that used in an artificial neural network (ANN). In common with ANN techniques, a learning method sweeps through the network and constantly updates the values of the parameters until some pre-defined error criterion has been met. Figure 7 illustrates a typical ANFIS network structure.

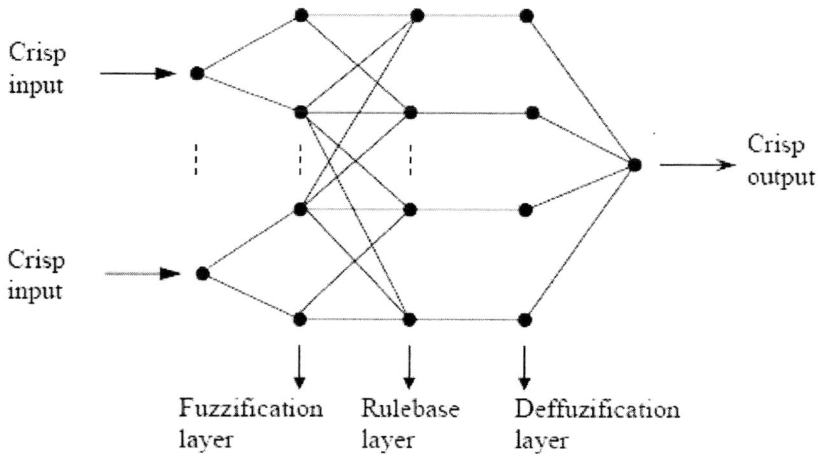

Fig. 7. Typical ANFIS network structure

Training is carried out using backpropagation or a combination of backpropagation and some other method such as least squares. An input data set and its corresponding outputs are usually split into two or sometimes three sets. The first of these is used as training data, the second for model validation and the third for model testing. Inputs from the first set enter the network and outputs are generated by using an initial approximation for the membership function parameters in the fuzzification layer. The resulting outputs and their targets (from the output data in the first data set) are used to generate errors which are then propagated back through the network from the output side to the input side resulting in error values in membership function parameters. The membership function parameters are then adjusted until a minimum error criterion (e.g. minimum sum-square-error) is met. A common approach is to change parameters using a constant learning rate which achieves a gradient descent in sum-square-error (membership function parameters change in the opposite direction to the error gradient):

$$\Delta(P_i) = -K\frac{\partial E}{\partial P_i} \qquad (9)$$

(where $\Delta(P_i)$ is the required change in the ith membership function parameter, E is the current sum-square-error value and K is the learning rate). To avoid over-training (a condition in which the trained network faithfully reproduces the training data outputs but begins to loose generality when applied to other input data), the second data set is used to validate the model by checking for any turning points in error arising from comparisons of generated outputs when applied to the second data set. Once a valid model has been obtained, a third data set can optionally be used for final independent testing.

ANFIS methods are ideally suited to data modelling applications in which they enjoy some of the advantages of artificial neural networks together with the advantages of dealing with uncertain data sets that lie at the heart of fuzzy applications.

2.6. Clustering

When large data sets are to be used it can be helpful to look for natural groups or patterns in the data. If groups or clusters of this kind can be found to exist, ANFIS modelling can be arranged to establish degrees of membership of all the data points to each specific group of data with the potential of reducing the number of modelling membership functions required and, in so doing, generate a more efficient network with lower computational demand. Clustering techniques are methods that can be applied to very large data sets in order to find these groups or patterns.

The dominant method of clustering used in fuzzy data modelling is fuzzy c-means clustering (Bezdek, 1981). In this method, each data point belongs to a cluster to a degree that is specified by its membership grade. A datum point, x, in a set of n data points divided into c clusters must simultaneously satisfy:

$$\mu_i(x) \in 0,1 \quad \& \quad \sum_{i=1}^{c} \mu_i(x) = 1 \quad \& \quad 0 < \sum_{x=1}^{n} \mu_i(x) < n \tag{10}$$

which implies that a membership of 1 in one cluster requires that it has zero membership in the others. A partitioning of all n points of data such that these conditions are satisfied represents fuzzy c-means clustering. The number of clusters possible in a large data set can be obtained by evaluating a partition entropy criterion (Zimmerman, 1994):

$$H = \frac{1}{n} \sum_{i=1}^{c} \sum_{x=1}^{n} \mu(x) \log \mu(x) \tag{11}$$

By decreasing the number of clusters from some initial estimate until the variation in H becomes non-negligible reduces the number of clusters to a minimum.

The membership grade of each datum point in a cluster is related to its distance from the cluster centre data point. The network to establish cluster centre datum points and membership grades for all other points in each cluster may be achieved by starting with approximate centre points and, hence, approximate membership grades for the enclosed data. Cluster centres and membership grades are then iteratively updated by minimising an objective function until the correct cluster centre and membership grades are obtained.

3. Review of Applications

As fuzzy systems have matured they have demonstrated utility in three main categories of application:

- Control
- Decision making or expert systems
- Data modelling

The following gives a review of progress in applications of relevance to energy and renewable energy with specific reference to buildings.

3.1. HVAC Control

Control represents what is arguably the most popular branch of application for fuzzy systems largely because intuitive solutions to a wide range of control problems are possible instead of the (frequently intractable) difficulties associated with conventional control system tuning and commissioning. For example the uncertainties posed by the selection of acceptable tuning parameters for the classical proportional-plus-integral-plus-derivative (PID) controller in applications containing mixed external disturbance patterns, nonlinearities and delay times, can be completely (or at least partly) circumvented through the application of fuzzy control in some way.

HVAC control has received widespread attention mainly because control objectives tend to be imprecisely defined coupled with the fact that a wealth of experiential knowledge exists about how control objectives should respond in given disturbance situations. A variety of fuzzy control system architectures is possible – Fig. 8 gives some typical approaches.

Huang and Nelson (1991) were among the first to apply fuzzy logic methods to HVAC plant control. They combine a PID controller with a fuzzy logic controller (FLC) and apply it to a general second order plant model. Fuzzy input sets are developed for control error, integral error and derivative error (essentially the terms of a PID controller) and an output set for the control signal. Mamdani max-min implications are used with centroid defuzzification of the resulting aggregate fuzzy output set. Results obtained using computer simulation show that the combined controller can virtually eliminate all of the damped oscillation common with tuned conventional PID control as well as to achieve faster response.

In later work, Huang and Nelson (1994a) give detailed consideration to rule development of FLCs for HVAC plant. They then go on to test the resulting FLC experimentally (Huang and Nelson, 1994b). In this work, error feedback and error rate are used to form triangular fuzzy input sets. They identify the major influence of the rulebase in the frequent problem of oscillation at (or near) zero error experienced by many FLCs (and demonstrated in the illustrative temperature control problem considered earlier). To find a remedy, the rule base is set out on a linguistic plane in which the fuzzy subsets of both universes of discourse (error and rate) are "plotted" against one another forming a matrix of input membership. Thus it is possible to identify desirable rule trajectories on a linguistic plane and this led to the elimination of near zero error oscillation in the earlier work. In their subsequent experimental work, a FLC is implemented for the control of a steam-air heating coil in an air handling plant. Results show that the FLC achieves superior performance when compared with tuned PID control, most particularly exhibiting far less oscillation than the latter. Recognising that conventional PID controllers can have excellent response quality characteristics providing that tuning variables can be matched to plant operational modes, Ardehali *et al.* (2004) use fuzzy inference to tune the settings of a conventional PID controller for heating and cooling energy inputs resulting in reduced energy and better comfort across the plant operational range.

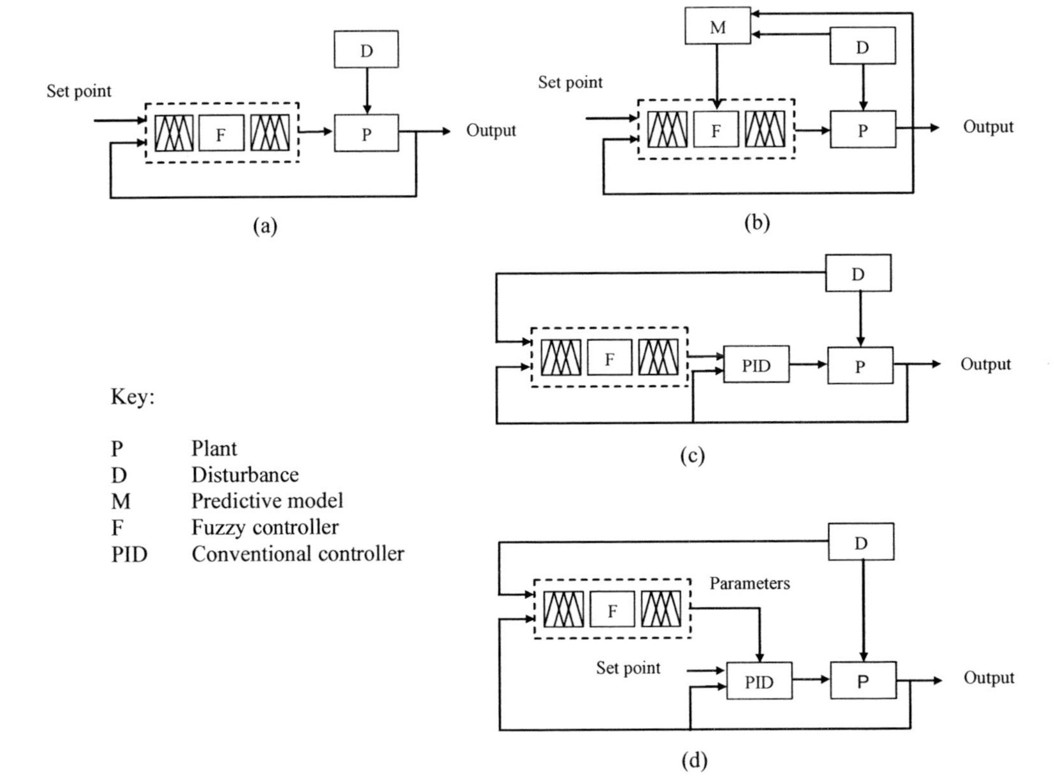

Fig. 8. Typical fuzzy/hybrid control architectures (a) Static fuzzy (b) Adaptive fuzzy (c) Scheduling fuzzy-PID (d) Adaptive fuzzy-PID

So *et al.* (1994) apply a FLC to the control of air handling plant in VAV systems. Using triangular fuzzy input sets for error and error rate, FLC signals are generated for the air handling plant fresh air damper, fan control, cooling coil, humidifier and reheat coil. Results are compared with tuned and detuned PID control (the detuned controller enabling robustness to be considered), using computer simulation. The results show that the FLC compares well with tuned PID control but was more robust and the FLC was superior (in terms of rate of response and offset) than detuned PID control. In later work, they develop a self-learning FLC using an artificial neural network (ANN) based on the same air handling control problem as their earlier work (So *et al.,* 1997). The ANN is used to monitor the plant and update the parameters of the FLC which permits robustness in spite of changes in operating conditions and nonlinearities. Other work related to the fuzzy control of specific items of HVAC plant are due to Aprea *et al.* (2004) who consider speed control of refrigeration plant and Li *et al.* (2005) who deal with cascade control of a gas engine heat pump. In the former, control error and rate of change of control temperature form inputs whose AND combinations in rule formation are treated using a product operator instead of the conventional min operator for improved interpolative reasoning. The output generated is the required speed of the compressor and a significant energy saving over conventional thermostatic control is reported. Li *et al.* (2005) consider the difficulties presented by a gas engine heat pump with significantly varying time constants ("stiff" system) under disturbance-rejection control. The problem relates to a rapid response from the engine system coupled with a relatively slow response from the heat pump. Two cascade control loops are applied – one responding to engine disturbance-rejection and the other applied to heat pump disturbance-rejection.

Numerous attempts to develop fuzzy controllers for the control of internal comfort criteria have been reported (Gouda *et al.*, 2001; Calvino *et al.*, 2004; Chu *et al.*, 2005; Dounis *et al.*, 1995; Lah *et al.*, 2005; Kolokotsa, 2003). Gouda *et al.* (2001) and Calvino *et al.* (2004) apply fuzzy control to target predicted mean vote (PMV) comfort criteria – the latter developing a hybrid fuzzy-PID controller in which control actions arising from Sugeno-inferences applied to control error and rate-of-error in a fuzzy controller are added to an integral-of-error term. Chu *et al.* (2005) develop a fuzzy controller which uses a least enthalpy estimator to generate set points in dry bulb temperature and relative humidity which acknowledges the inter-dependency between these two variables. Dounis *et al.* (1995) considers thermal and visual comfort resulting in a fuzzy controller which controls heating, cooling and window opening, shading and lighting and Lah *et al.* (2005) consider fuzzy control over a roller window blind in order to achieve simultaneous control over both temperature and illumination. Kolokotsa (2003) compares a variety of hybrid fuzzy-P/PI/PID controllers to regulate heating/cooling, window opening, shading and electric lights in which the extent of conventional control actions are determined from fuzzy membership functions. Again, PMV is used as a target comfort index. In related work a fuzzy control architecture is used for a variety of control objectives based on inputs from a smart card system in residential-type accommodation (Kolokotsa *et al.*, 2005). Alcalá *et al.* (2005) use weighted fuzzy rules for the integrated control of HVAC plant (again targeting PMV criteria) whilst minimising energy use. They adopt a genetic search method for rule weight selection.

3.2. Renewable Energy System Control

To date, most of the attention focused on fuzzy logic control applied to the specific case of renewable energy systems (RES) has been restricted to wind energy systems (Jurado and Saenz, 2002; Leclercq *et al.*, 2003; Miland *et al.*, 2006; Mohamed *et al.*, 2001) though there is growing interest in applications to other types of RES.

Jurado and Saenz (2002) consider the adaptive fuzzy control of a stall-regulated wind turbine with a back-up biogas generator. Inputs to the controller are the engine speed and output voltage and their time rates-of-change. Weights are placed on these input signals with the values of the weights tuned by an artificial neural network thus introducing adaptivity to the control. Leclercq *et al.* (2003) also consider fuzzy control applied to a wind turbine with a back-up generator in a system which incorporates a flywheel energy store. The control strategy seeks to enable the plant to compensate for rapid variations in power generated by the wind turbine. A fuzzy logic supervisor is applied which determines the required power level from the wind turbine and flywheel combination by using the flywheel speed and wind turbine active power as inputs. Miland *et al.* (2006) deal with the control of numerous wind turbines coupled to an electrolyser for hydrogen storage. The strategy they adopt is to use the variations in active power flow into the network from the wind turbines in order to effect control over power frequency. The fuzzy logic controller uses the power network frequency and its time rate-of-change as inputs and generates a required load change to maintain the required supply frequency. The fuzzy controller is self-tuning in that adjustments can be made to input membership functions during an initial learning period. Mohamed *et al.* (2001) deal with the fuzzy control of the grid-interaction of a wind energy conversion system which seeks to identify and extract maximum power from the wind for transfer to the utility grid.

Relatively limited attention has been given to the control of solar renewable systems using fuzzy methods. In what is a fuzzy modelling application, Sözen *et al.* (2004) use a conventionally structured fuzzy inference system to model the main output variables of a solar-powered absorption cycle refrigeration plant. Compared with an analytical model, the simpler fuzzy approach gives good results which suggests that the method has potential for applications in model-following control. Luk *et al.* (1999) investigate the control over parabolic mirror solar collectors used for power generation. They use a genetic search method to establish rule weights for a fuzzy logic controller with results based on simulations suggesting improved robustness performance due to fuzzy control. Patcharaprakiti *et al.* (2005) apply a fuzzy controller to maximum power point tracking of a grid-connected photovoltaic system. Control error with respect to maximum possible power and its time rate-of-change inform an inverse fuzzy model which consists of a set of relational operators which are used to adjust the fuzzy rule base and input membership functions. The resulting adaptive fuzzy controller is shown to generate higher solar power outputs than could be achieved with a static fuzzy controller. The motivation for this approach was that the system was expected to operate over a very wide range with unreliable expert knowledge considered to be intractable for a static (fixed-parameter) fuzzy controller. Gouda *et al.* (2005) use an artificial neural network to predict a future time row of internal air temperature in a passive solar heating application. The predictive model output is used as one of several inputs to a static fuzzy controller resulting in a "quasi-adaptive" solution which is shown to give improved control performance over conventional PID control.

Other work on fuzzy logic control applied to renewable energy systems has focused on fuel cells and electrolysers mainly from the viewpoint of energy storage control. Bilodeau and Agbossou (2005) apply fuzzy logic to energy storage control in a photovoltaic/wind turbine hybrid renewable energy system with short term storage provided by batteries and long term storage provided by electrolysis-to-hydrogen. The fuzzy control objective is to reduce energy transfers from short term to long term storage. Control inputs are net power flow and the battery state-of-charge whilst the output is a required power set point (a positive value used to excite the fuel cell on the discharge side of the hydrogen storage system; a negative value used to excite the input side electrolyser). The high degree of linguistic operating states made this application particularly suitable for fuzzy control. Jeong *et al.* (2005) deal with the fuzzy control of a fuel cell hybrid battery combination. The controller is used to determine the power split between the power fed to the load and the power fed to battery storage. Again, the control objectives relating to these split pathways was better represented by linguistic rules than could be achieved by conventional control techniques. Jurado and Valverde (2005) considered the use of a multiobjective function genetic algorithm to tune the rulebase and membership functions of a fuzzy logic controller applied to determine the required switched states of a pulse-width modulation inverter on the output of a solid oxide fuel cell. Though applications of fuzzy methods to the control of renewable energy systems is limited at present, the applications that have been attempted show that there is substantial potential for these methods in this sector.

3.3. Expert Systems

Fuzzy expert systems are applicable to decision making problems in which multiple evaluation criteria (inputs) are available to assist in the decision making process but the evaluation criteria may consist of a mixture of qualitative (linguistic) and quantitative (crisp) data and some of the criteria may be imprecise or ill-defined.

Several examples of fuzzy expert systems have been developed to assist decision-making in energy-related fields. Some of the earliest work is due to Dexter and Trewhella (1990) who applied 5 fuzzy input sets to the performance assessment of a hot water fan coil unit under PI control. The fuzzy input sets cover maintenance, energy use, control error, valve travel and occupant dissatisfaction. In a fault detection and diagnosis application, Barelli and Bidini (2005) use an auto-associative artificial neural network to examine the consistency of data measured from a 1MW lean burn gas engine cogeneration plant the results of which are used in fault detection. A fuzzy inference system is then used to evaluate the state-of-health of the plant (based on the results generated by the auto-associative model) and determine the diagnosis of any fault that may be evident from the data. Mamlook *et al.* (2001b) attempt to rank the merits of five different solar energy systems for energy planning prioritisation in Jordan. Two fuzzy weights are generated for each option – one based on the perceived benefits of the option and the other based on the option cost. The fuzzy weights are then translated into benefit-to-cost ratios from which a priority ranking of the various options follows. In related work they use a neuro-fuzzy network to generate a similar ranking set applied to power generating plant options (Mamlook *et al.*, 2001a). Kaminaris *et al.* (2006) conduct an analysis of renewables-to-electricity options using a three-step methodology involving the identification of fuzzy input variables, membership function setting and fuzzy inference to a "fuzzy project priority index" for each option. The input variables are initially grouped into life-cycle and development cost categories (43 inputs in all) and then reduced to a normalised fuzzy input variable for each category. In an optimisation application, Jana and Chattopadhyay (2004) attach both fuzzy and crisp constraints to input criteria for a multiobjective problem of selecting domestic lighting systems for rural settlements in India.

Three objectives are defined: minimised cost of direct energy; minimised non-local energy and maximised efficiency. Jaber *et al.* (2005) investigated energy conservation awareness among the general public using a fuzzy expert system by applying a fuzzy set to each of six knowledge areas. They score responses to a variety of energy-related questions which are interpreted by a fuzzy inference system into degrees of knowledge about the area that the question addressed. The results could be used by energy policy-makers to target training and awareness campaigns for improving public understanding about energy and, hence, its use. Beccali *et al.* (1998) compare a fuzzy approach with an established multi-criterion method for the ranking of fourteen renewable energy system options. Twelve evaluative criteria are used as initial inputs and these inputs are first split into qualitative and quantitative sets after which the latter set is fuzzified. Fuzzy evaluation weights are generated which form the basis of a decision-making strength for each option.

3.4. Data Modelling

Data modelling represents the most recent branch of fuzzy systems research, development and application. Recall that a fuzzy system is, of essence, merely a method for mapping an input variable onto an output variable through the use of linguistic rules. Input/output data sets whose relationship can be described qualitatively but for which a relationship using deterministic or stochastic methods has proved elusive represent situations that might be treated using one of the various fuzzy modelling approaches. Modelling methods are summarised in Table 1.

Considerable attention has been given to the application of fuzzy models to climate data forecasting, particularly solar radiation and much of the reported work involves comparing fuzzy models with other modelling methods. Gomez and Casanovas (2002) propose a solar radiation model based on both crisp and fuzzy inputs using fuzzy c-means clustering to define sky categories based on fuzzy membership functions with reduced parameters. In related work they develop a fuzzy model for solar radiation on inclined surfaces based on the ratio of normal solar irradiance to extraterrestrial irradiance (clearness index) and a brightness index which defines cloud thickness and opacity as inputs (Gomez and Casanovas, 2003). Santigosa *et al.* (2003) compare three models for the calculation of daily UV radiation from inputs of daily global radiation on the horizontal plane. The models compared are linear and non-linear regression and a fuzzy model which uses ANFIS with clustering. Zarzalejo *et al.* (2005) compare an ANFIS model with an artificial neural network (ANN) based model and auto-regression (AR) based models for predicting hourly global solar radiation using satellite images as inputs.

Table 1. Summary of modelling methods

Method	Basis	Strengths	Limitations
Deterministic	Fundamental laws of the physical sciences.	General applicability. Highly intuitive.	Limited robustness. Modelling uncertainties.
Stochastic	Model-fitting to historical time series using autoregression method.	Robust. Accurate.	Limited applicability. Non-interpretive. Requires good data.
Adaptive networks	Artificial neural networks; adaptive neuro-fuzzy inference systems.	Reduced data. Improved accuracy.	Limited applicability. Non-interpretive.
Static networks	Fuzzy inference systems.	Suitable for uncertain/ill-defined data. Easy to set up.	Limited applicability.

They conclude that the two network models outperform the AR models (though the performance of all the models is similar). Sfetsos (2000) considers forecasting techniques for wind speed prediction comparing AR, ANN and ANFIS modelling approaches. A wide variety of ANN model architectures is considered in this work. Again the ANN and ANFIS methods perform similarly and both slightly outperform the AR method in terms of root-mean-square modelling errors.

A significant amount of work has been directed at the use of fuzzy inference systems for modelling and filtering, uncertainty. Michalik *et al.* (1997) coded the use of electrical appliances in houses into time-banded usage (two-state: 0 or 1) and degree-of-usage (five linguistic states) and expressed the result as a fuzzy filter to be applied as an enhancement measure in conventional energy prediction models. Lee and Dexter (2005) describe a fuzzy sensor for dealing with cross-sectional temperature variations in discharge ducts from mixing dampers in air handling plants using computational fluid dynamics (CFD) modelling to characterise the fuzzy inputs. The output from the fuzzy sensor is expressed as a set of temperature distribution possibilities that can be used to quantify uncertainty in a measured value. Narasimhan and Asgarpoor (2000) adopt fuzzy inference with clustering of hourly power demand loads to map uncertainties in power onto uncertainties in failure rate and repair time for power generating plants. Huang and Shih (2002) deal with the short term load forecasting of power system demand using a fuzzy model with a small number of rules. The fuzzy model uses weighted inputs which are generated using a novel "group method of data handling" (GMDH) technique based on an evolving regression model whose variable set adjusts through a learning technique until an optimum model is acquired. Hong *et al.* (2004) modelled the power losses in an electrical distribution system using fuzzy c-means clustering to partition daily loss patterns into segments from which a fuzzy multiple regression model is used to fit the energy losses to energy load data for each segment.

A certain amount of attention has also been given to the use of fuzzy modelling applied to thermal comfort and HVAC plant. Hamdi *et al.* (1999) developed a fuzzy thermal comfort index based on the classical "predicted mean vote" (PMV) that, unlike classical PMV calculations, does not require an iterative solution. He *et al.* (2005) use a Takagi-Sugeno-based fuzzy supervisor to generate fuzzy weights that are used in three partitioned lower level fuzzy models for the control of temperature in an air handling plant. The lower level models are generated with respect to clusters in air mass flow rate identified as low, medium and high and are linked using a fuzzy linear varying parameter. The partitioning of the low level models helps to deal with plant non-linearities. Lu *et al.* (2005) model pipe (and, potentially, duct) networks using an ANFIS method for obtaining pressure set points for the control of pumps and other network devices. Škrjanc *et al.* (2001) developed a fuzzy model with T-S inference to establish the relationship between rollerblind position and internal temperature in an environmental test chamber for practical implementation. Results are compared with a theoretical model. Ruusunen and Leiviskä (2004) also use a T-S-based fuzzy model to predict CO_2 emissions from combustion in a wood-burning boiler by mapping a crisp fuel mass value onto three combustion phases (each with a fuzzy membership function) – ignition; burning and charring.

4. Case Studies

The review has confirmed that the application of fuzzy systems to energy and renewable energy in buildings is broadly based across three main areas of interest in the fields of decision-making, data modelling and control. Three case studies follow which demonstrate possible approaches to the design of fuzzy inference systems for this range of applications. It is not the intention to provide full and closed solutions to the presented case studies but merely to illustrate the potential that a fuzzy systems approach might have for dealing with each of the problem areas described.

4.1. Case Study 1: Fuzzy Expert System

4.1.1. Problem Definition

Interest is increasing in the rural application of small-scale renewable energy systems as embedded generators in small buildings such as housing. Examples include photovoltaic cladding, biofuel micro-CHP, vertical axis and ducted wind turbines. In many developing countries the motivation for doing this is to address limitations in the extent of grid-based power and fuel infrastructures in rural areas whereas, in developed countries, the motivation is often related to reliability problems of existing grid-based infrastructures particularly during periods of severe weather. Figure 9 illustrates embedded renewable energy generation possibilities for part of a row of terraced housing in a remote UK location.

Features:

- Vertical-axis wind turbines replace redundant chimney pots
- Roof covering using photovoltaic slates
- Roof-mounted evacuated tube solar collectors

Inset:
Biofuel Stirling engine-based micro-CHP

Fig. 9. Embedded renewable energy generators

There are numerous possible solutions to local embedded power and heat generation and making choices of candidate technologies in a given applications will usually require a balanced judgement in which at least four groups of evaluation criteria might be applied:

- Availability, reliability and design issues
- Operating and maintenance issues
- Socioeconomic and sustainability issues
- Cost

A variety of criteria can be set against each evaluation group. Whilst it may be possible to quantify some of these criteria, others will be qualitative and some may be relative or loosely-defined. Forming a balanced judgement about which technologies might form good choices and which may not can be assisted through the use of a fuzzy expert system application with the following objectives:

- Identity evaluation groups (i.e., possible fuzzy inputs) and, hence, group evaluation criteria for small scale embedded renewable energy systems
- Develop and apply a suitable scoring method to the group evaluation criteria forming inputs to a fuzzy inference system
- Design a fuzzy inference system
- Apply the fuzzy inference system to a set of candidate technologies

4.1.2. Evaluation Criteria

The case study will consider the evaluation of four alternative technologies in a remote UK application:

PV – photovoltaic roof cladding
WT – vertical axis wind turbines
D-Bio – local micro-CHP modules fuelled by a biofuel network
C-Bio – centralised biofuel CHP with district heating and power

Evaluation criteria have been identified and listed in Table 2 for each group.

Table 2. Case Study 1: Evaluation criteria

Design and technology	Operating and maintenance	Socioeconomic and sustainability
DT1: Extent of commercial availability	OM1: Extent of life cycle	SS1: Encourage community
DT2: Potential for heat recovery	OM2: Parts availability	SS2: Conservation of cultural heritage
DT3: Power generating efficiency	OM3: Operating skills needs	SS3: Encourage sustainable lifestyles
DT4: Grid connection possibility	OM4: Maintenance skills needs	SS4: Promote local community action
DT5: Power conditioning needs	OM5: Availability of specialist expertise	SS5: Contribute to local employment/training
DT6: Dispatchability	OM6: Accessibility	SS6: Use of local suppliers/skills
DT7: Site sensitivity – wind	OM7: Ease of commissioning	SS7: Protection from crime
DT8: Site sensitivity – solar	OM8: Ability to monitor/troubleshoot	SS8: Use of previously-developed land
DT9: Integration into host building	OM9: Safety in use/repair	SS9: Air pollution at point of use
DT10: Aesthetics		SS10: Preservation of wildlife habitats
DT11: Acoustics		SS11: Embodied energy
		SS12: Emissions during manufacture
		SS13: Potential for component repair/reuse
		SS14: Used component disposal

The evaluation criteria are not listed in any particular order. A cost criterion has not been included at this stage but will be returned to later.

Using existing knowledge about the relative capability of each technology option it is possible to score each evaluation criterion using an arbitrary 3-point scale where 1 = lowest score and 3 = best score (Table 3).

A normalised single score for each option's performance in each evaluation group is now possible by adding all scores for each option/group and dividing by the maximum score possible for that group. The results are summarised in Table 4 the contents of which form inputs to a fuzzy inference system.

Table 3. Case Study 1: Evaluation criteria scores

Criterion	Score			
	PV	WT	D-Bio	C-Bio
DT1	2	2	1	2
DT2	2	1	3	3
DT3	1	2	2	2
DT4	2	3	3	3
DT5	1	3	3	3
DT6	1	1	3	3
DT7	3	1	3	3
DT8	1	3	2	2
DT9	1	1	2	3
DT10	2	1	2	2
DT11	3	2	1	1
OM1	3	2	1	1
OM2	2	2	2	2
OM3	3	3	1	1
OM4	3	2	1	1
OM5	1	2	1	2
OM6	1	1	3	3
OM7	2	3	1	1
OM8	3	3	1	2
OM9	2	2	1	1
SS1	1	2	3	3
SS2	1	1	2	2
SS3	2	2	2	2
SS4	2	2	3	3
SS5	1	2	3	3
SS6	1	2	3	3
SS7	3	3	2	1
SS8	3	3	2	2
SS9	3	3	1	2
SS10	2	1	3	3
SS11	3	2	1	1
SS12	2	3	1	1
SS13	2	2	2	2
SS14	1	2	2	3

Table 4. Case Study 1: Normalised evaluation group scores

Evaluation Group	Overall Group Score			
	PV	WT	D-Bio	C-Bio
Design and technology	0.576	0.606	0.788	0.818
Operating and maintenance	0.741	0.741	0.444	0.518
Socioeconomic and sustainability	0.643	0.714	0.714	0.738

4.1.3. Fuzzy Inference System

Figure 10 shows the initial fuzzy inference system that has been applied.

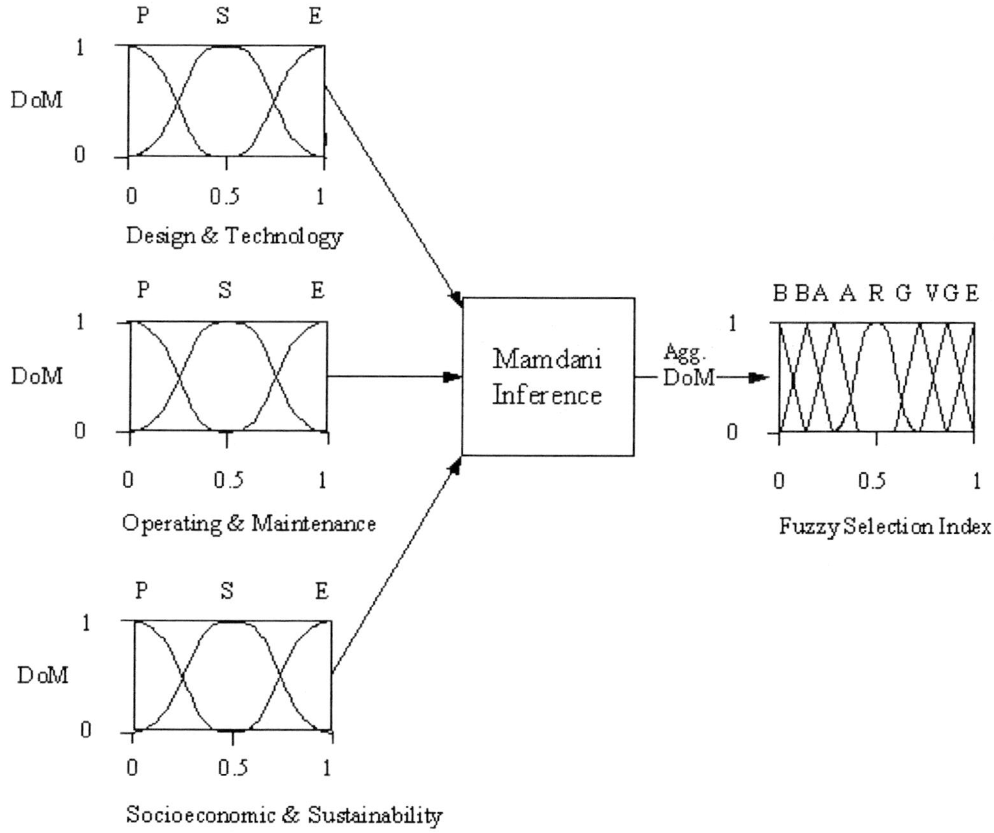

Fig. 10. Case Study 1: FIS

The membership functions chosen are, like any initial attempt at fuzzy system design, somewhat arbitrary. In general overlaps between membership functions will be chosen to permit smooth and continuous interpolation where needed but a certain amount of fine-tuning is inevitable after initial results have been obtained.

Three Gaussian membership functions were used for all three input variables each with a universe of discourse of 0 – 1 corresponding to the normalised evaluation group scores arrived at in the above. The three functions map scores which might be interpreted as “poor”

(P); "sound" (S) and "excellent" (E). By choosing a Gaussian set, continuous interpolation and a smooth transition between membership functions is expected. A max aggregation function was used between membership functions.

With three fuzzy inputs each with three membership functions then $3^3 = 27$ rules require to be set. Seven fuzzy output membership functions were applied to give a graded range of decisions (with some overlaps) among the 27 rules:

Bad	(B)
Barely adequate	(BA)
Adequate	(A)
Reasonable	(R)
Good	(G)
Very good	(VG)
Excellent	(E)

The rules adopted are given in Table 5.

Table 5. Case Study 1: Rulebase

Design and technology	Operating and maintenance	Socioeconomic and sustainability	Consequent
AND	AND	THEN	
P	P	P	B
S	P	P	BA
E	P	P	BA
P	S	P	BA
S	S	P	A
E	S	P	R
P	E	P	BA
S	E	P	A
E	E	P	R
P	P	S	B
S	P	S	A
E	P	S	A
P	S	S	BA
S	S	S	G
E	S	S	VG
P	E	S	BA
S	E	S	G
E	E	S	VG
P	P	E	B
S	P	E	A
E	P	E	R
P	S	E	BA
S	S	E	VG
E	S	E	VG
P	E	E	BA
S	E	E	VG
E	E	E	E

For the output, a crisp "fuzzy selection index" (FSI) variable was defined which would require to be determined by defuzzification. The FSI will take on a crisp value in the interval 0,1 indicating the strength of an option for design selection. Centroid defuzzification was used.

4.1.4. Application and Results

Initially, the fuzzy expert system model defined above was applied using the input evaluation group scores given in Table 3. No cost criterion has been used in this version of the model. This was done in order to obtain results based solely on technical and social criteria without the distorting affect of cost. The results are summarised in Table 6 revealing a slight lead by the centralised biofuel option.

Table 6. Case Study 1: Results (without cost criterion)

Option	Fuzzy Selection Index
PV	0.774
WT	0.797
D-Bio	0.771
C-Bio	0.814

To re-evaluate the selection procedure but this time with a cost criterion added to the input set, three adjustments to the model were made. First, to reduce the inputs, the socioeconomic/sustainability and operating and maintenance groups were combined by applying a max function to the two group evaluation scores. Second, a new "relative cost" input was created. The relative cost was expressed as a cost score of each option with reference to the most expensive of the options (zero implying highest relative cost, 1 implying least). This is valid for a comparison between a specific group of options but would require to be generalised for a larger group of options. Finally the rulebase was adjusted accordingly.

The cost input membership functions were set as shown in Fig. 11.

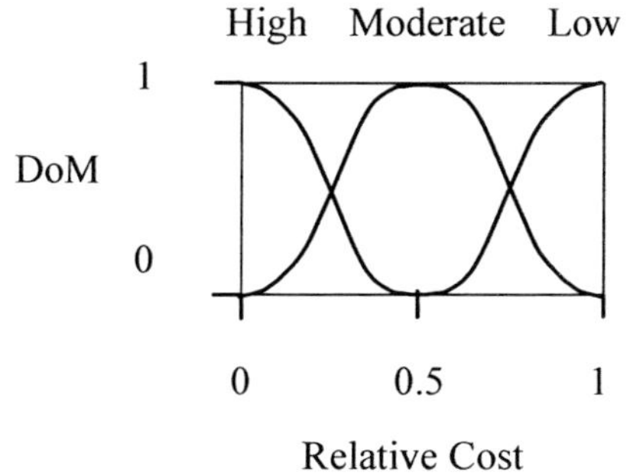

Fig. 11. Fuzzy cost input

Arbitrary relative cost scores were applied as follows, by assuming that the centralised biofuel CHP option would be the most expensive:

PV: 0.50

WT:	0.75
D-Bio:	0.25
C-Bio:	0.00

The modified rulebase is given in Table 7 and the results of this case are summarised in Table 8.

Table 7. Case Study 1: Rulebase with added cost input

Design and technology	Relative Cost	Max (socio/sustainability, operating and maintenance)	Consequent
AND	AND	THEN	
P	H	P	B
S	H	P	BA
E	H	P	BA
P	M	P	BA
S	M	P	A
E	M	P	R
P	L	P	A
S	L	P	R
E	L	P	G
P	H	S	B
S	H	S	BA
E	H	S	A
P	M	S	BA
S	M	S	R
E	M	S	G
P	L	S	A
S	L	S	G
E	L	S	VG
P	H	E	B
S	H	E	A
E	H	E	R
P	M	E	BA
S	M	E	G
E	M	E	G
P	L	E	A
S	L	E	VG
E	L	E	E

Table 8. Case Study 1: Results (with cost criterion)

Option	Fuzzy Selection Index
PV	0.583
WT	0.655
D-Bio	0.445
C-Bio	0.357

4.1.5. Conclusion

In this case study a fuzzy expert system for the selection of a renewable energy system as an embedded generator in a remote rural application has been evolved. The emphasis has been on demonstrating the potential of the method as opposed to arriving at a fully robust solution to the problem. Two interpretations of the problem have been explored. The first ignored the impact of cost and focused on technical and social factors only. The second approach added in a cost input.

Without a cost input, the fuzzy selection index for all options was fairly tightly bunched though the centralised biofuel CHP option was found to be just in front. When cost was introduced as an input the wind turbine option was significantly in front with the two biofuel CHP options lagging significantly behind both the wind turbine and photovoltaic options.

4.2. Case Study 2: Data Modelling

4.2.1. Problem Definition

The modelling of climate data has received significant attention in the past. The interest has usually stemmed from one of four avenues of interest:

- Forecasting
- Patching missing fields in measured data sets
- Expanding data to include details that may be difficult to measure
- Generation of weather data sets from synoptic data (e.g. for energy modelling)

This case study relates to the second of the above areas. Frequently in local climate data monitoring one or more key variables are either not measured or are measured at time intervals that are too high for the data to be of use in many instances. The measurement of relative humidity along with other main climate variables such as air dry bulb temperature and solar irradiance is an example of a variable that is often missing from a set of climate data observations on the grounds that it is of lower importance and yet it is a key variable for, amongst other things, the simulation modelling of energy use in buildings. This case study deals with the application of an adaptive neuro-fuzzy inference system (ANFIS) to the modelling of relative humidity when only air temperature is known.

The rational is quite simple. The relative humidity of air is dependent on the air dry bulb and wet bulb temperatures at a constant total air pressure. In temperate climates the difference between the dry bulb temperature and wet bulb temperature is typically low in cold (winter) weather but increases as the dry bulb temperature increases towards summer conditions. Thus the variation in relative humidity can be partly explained by the variation in dry bulb temperature for this type of climate. It therefore seems likely that some sort of fuzzy relationship between the two might exist.

The measured hourly dry bulb temperature and relative humidity taken in 1993 for Newcastle upon Tyne in the United Kingdom are shown in Fig. 12.

Though the peak hourly relative humidity in each day is always about the same (and likely to occur in the early hours of each morning) there is some evidence of an inverse relationship between the two variables but there is also considerable scatter and uncertainty.

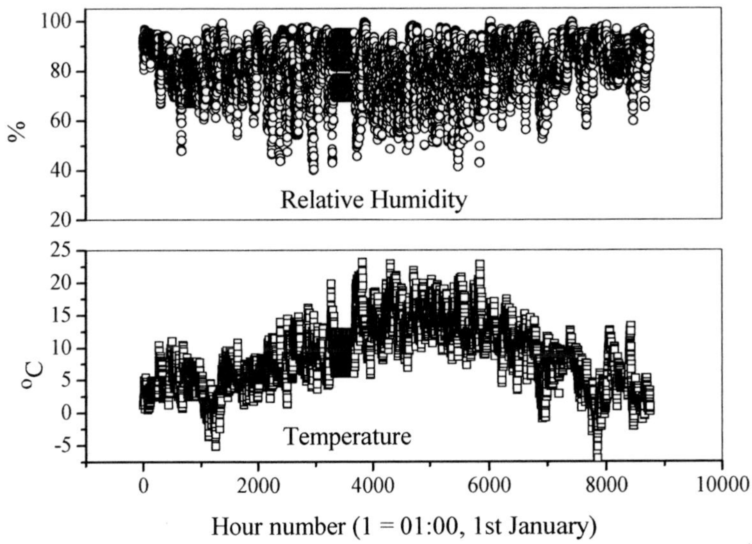

Fig. 12. Case Study 2: Temperature and relative humidities

The objectives of this case study therefore are:

- To generate an ANFIS model relating the relative humidity to the dry bulb temperature
- To establish how much of the variation in relative humidity can be described (i.e., how adequate the model is)
- To determine whether clustering the data might contribute to an improved model

4.2.2. ANFIS Model Development

The two annual hourly time series' (Fig. 12) were divided into two half sets of data pairs cut through the middle of the data after the complete removal of several patches of synthetic fill-in data (one of these is evident at approx. 3300 hours in Fig. 12). This resulted in two sets of 3876 data points. The data were then ranked according to the dry bulb temperature values. The first data set was used for ANFIS model training and the second set was used for model testing. The range of the training data set was $T \in (-8.55, 22.8)$ °C.

An initial model was generated using a conventional structure consisting of three Gaussian input membership functions acting on the entire training data set. With a single input single output structure just three rules are therefore generated and these map to linear Sugeno output membership functions. A product aggregation among input membership functions was applied and output aggregation was carried out using weighted averaging. ANFIS training was carried out using a hybrid (mixed least squares and backpropagation method). The model structure is as shown in Fig. 13.

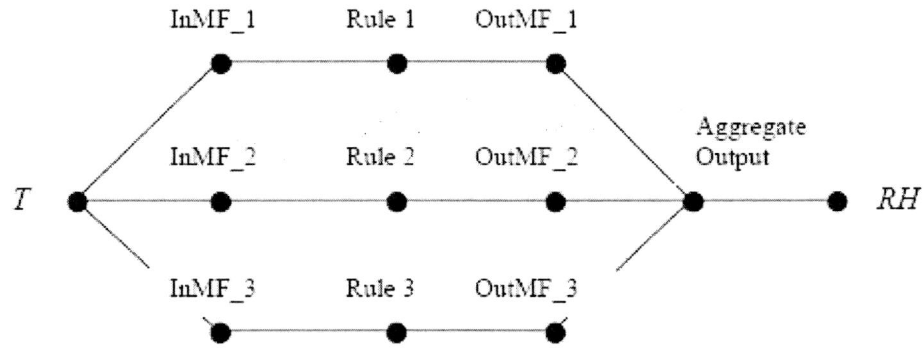

Fig. 13. Case Study 2: ANFIS model structure

A Gaussian fuzzy membership function ranging from zero base to unit height is given by:

$$\mu = \exp\left(\frac{-(x-x_c)^2}{2w^2}\right) \tag{12}$$

(where x and μ are the crisp input and degree of membership respectively, x_c is the crisp input value corresponding to the membership function's centre and w is the membership function width parameter). The membership functions were fitted with the following parameter sets at the conclusion of training which was completed with a small number of sweeps (10) through the network (see also Fig. 14).

$\mu_1(T, T\in(-8.55,22.8))$ (mid-range): $w = 6.653\text{K}$, $x_c = 7.119°\text{C}$
$\mu_2(T, T\in(-8.55,22.8))$ (low): $w = 6.673\text{K}$, $x_c = -8.537°\text{C}$
$\mu_3(T, T\in(-8.55,22.8))$ (high): $w = 6.638\text{K}$, $x_c = 22.81°\text{C}$

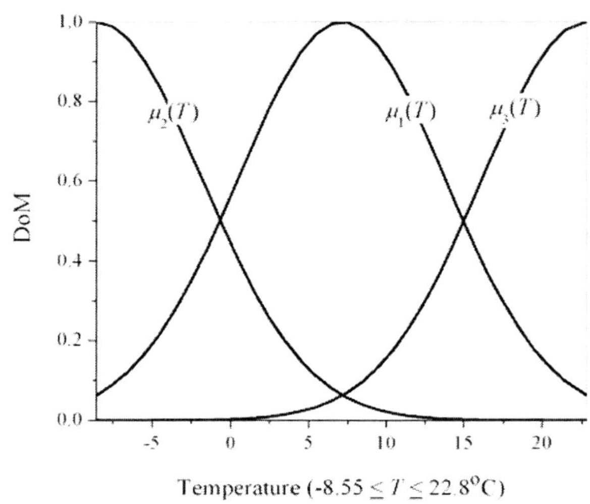

Fig. 14. Case Study 2: Fitted input membership functions

The following Sugeno output membership functions were obtained:

Output_1 = 3.483**T*+119.7
Output_2 = 2.104**T*+67.0
Output_3 = -1.229**T*+75.2

Results from the application of the above to the applied input, weighted by the extent to which input membership functions have fired, are averaged to give a crisp output result.

4.2.3. Application and Results

Applying the ANFIS fuzzy model described above to the temperature time series from the second (model testing) data set and plotting the results along with the measured relative humidities from the model testing set gave the results plotted in Fig. 15 (upper graph).

The general trend of the data is reasonably well reproduced by the ANFIS model though the model tends to overpredict the minimum relative humidity values that arise at the higher air dry bulb temperatures. A mean bias error (i.e., the average of all measured results minus corresponding model results) of –2.79% (relative humidity) confirms the tendency of the model to overpredict.

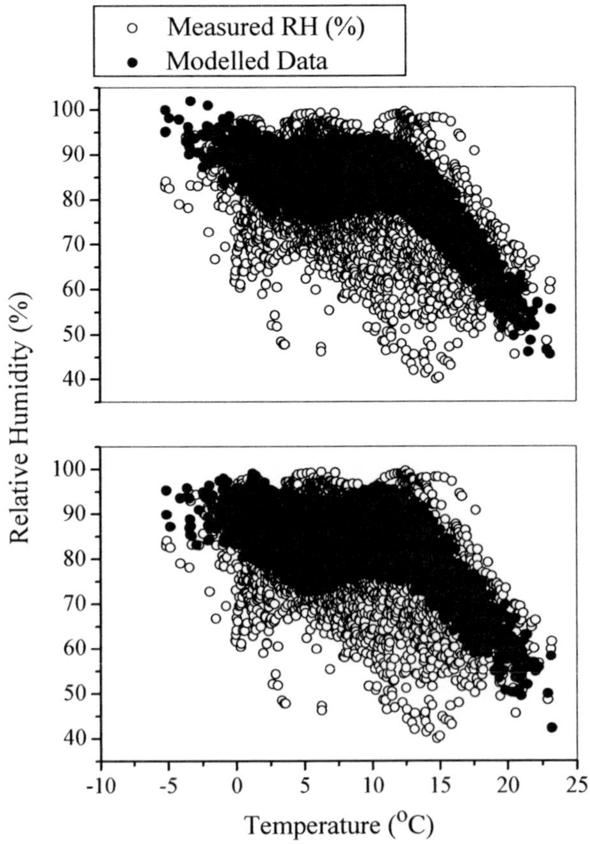

Fig. 15. Case Study 2: Results (Upper graph: Whole data model Lower graph: Clustered data model)

To consider the impact of clustering the data a second ANFIS model was generated by first clustering the training set. Three clusters were obtained using c-means clustering and they were found to be centred at the ($RH(T)$) pairings: 91.7(0.427); 85.5(9.432) and 69.5(15.5).

A Gaussian membership function was centred at each cluster centre for this case. Hence the ANFIS network for this case resembles Fig. 13 but this time the individual network pathways represent a fuzzy inference applied to a particular cluster. Though the net affect is the same number of input membership functions as was used in the whole data case, this technique can, in general, reduce the number of input membership functions required in ANFIS applications.

The new membership function set for this case is as follows and as shown in Fig. 16:

$\mu_{\text{Cluster1}}(T)$ (middle cluster):	$w = 4.408\text{K}, x_c = 9.432^\circ\text{C}$
$\mu_{\text{Cluster2}}(T)$ (low cluster):	$w = 4.430\text{K}, x_c = 0.427^\circ\text{C}$
$\mu_{\text{Cluster3}}(T)$ (high):	$w = 4.431\text{K}, x_c = 15.50^\circ\text{C}$

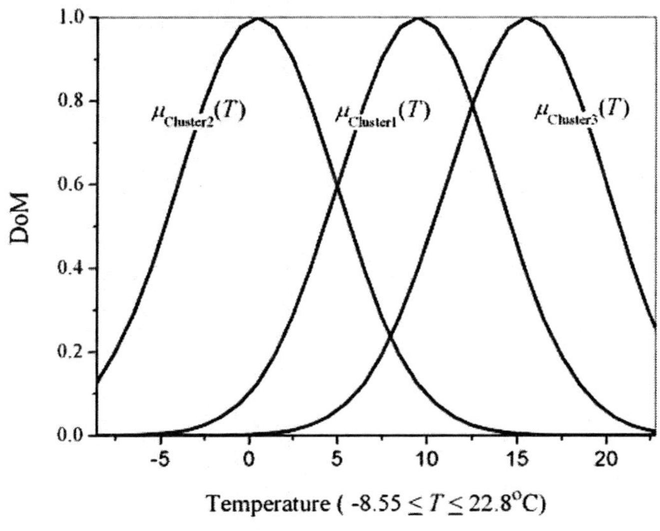

Fig. 16. Case Study 2: Fitted input membership functions (clustered data case)

The following Sugeno output membership functions were obtained for this case:

$Output\_1 = 4.385*T+49.3$
$Output\_2 = 0.519*T+94.2$
$Output\_3 = -1.660*T+85.2$

Results of an application of this version of the model to the test data set are given in Fig. 15 (lower graph). Some improvement in model performance is evident in terms of the spread of prediction. This is confirmed by a slight improvement in the mean bias error to –2.59% (relative humidity) for this case.

4.2.4. Conclusion

In this case study an adaptive neuro-fuzzy inference system has been used to model a relationship between relative humidity and dry bulb temperature from a typical weather data set. Two relatively simple alternative models have been developed, the first based on just three input membership functions applied to the whole training data set. A second model used a clustering technique to partition the data into three clusters and centred a membership function at each cluster centre. The emphasis has been on demonstrating the potential of the method as opposed to arriving at a fully robust solution to the problem.

The results demonstrated that a model could be generated which gave a reasonable prediction of the underlying trend in the data but had a tendency to overpredict. The clustered data approach led to a small improvement in the model.

4.3. Case Study 3: Fuzzy Logic Control

4.3.1. Problem Definition

The control of HVAC plant has received widespread attention by those engaged in the development of fuzzy logic controllers. There are three main reasons behind this:

- HVAC plant is typically highly nonlinear and contains significant transport delays which challenges stability in conventional controllers
- Conventional controllers require tuning to achieve stability and acceptable tracking
- The control objectives of many HVAC applications are difficult to define precisely but can easily be described intuitively

All of these issues can be addressed to some extent using fuzzy inference methods.

The control of the state of air leaving air handling equipment is a particularly common example which espouses all of the features listed above. Figure 17 shows the problem which invariably involves three sequenced processes; heating with a minimum input of fresh air at the damper set, free cooling by the damper set by mixing a variable proportion of fresh air and recirculating air and refrigeration-based cooling. For least energy use, the three processes are sequenced to ensure that only one process is active at any one time whilst the other two are at saturated states (i.e., fully off or at some pre-set minimum). Figure 18 illustrates the sequence.

This final case study illustrates the development of a set of fuzzy controllers for the sequence control of air temperature leaving an air handling plant and benchmarks this against a conventional control method using simulations. The case study objectives are:

- To develop a set of fuzzy controllers for sequence control of air handling plant outlet air temperature using linguistic rules
- To apply the controllers to a practical example using a simulation
- To set up and "adjust" a conventional strategy and compare the performances of the two methods

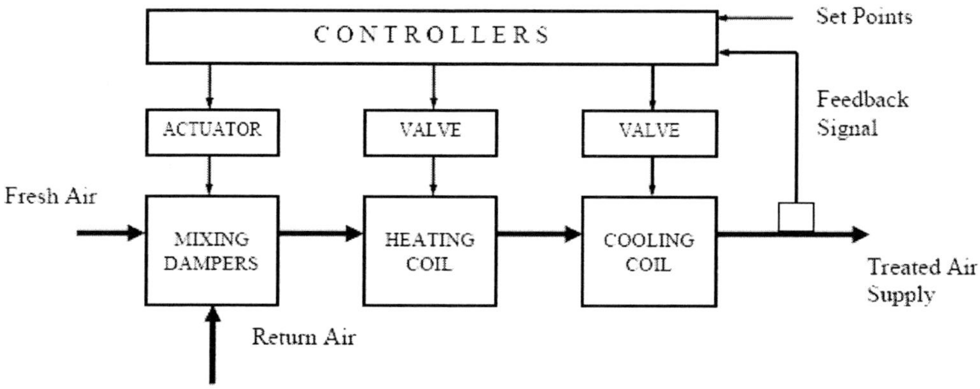

Fig. 17. Control of air temperature in air handling plant

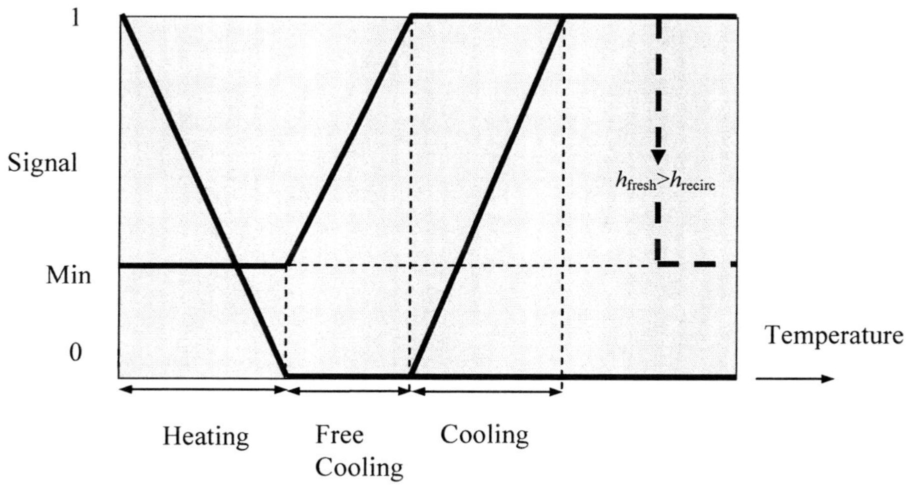

Fig. 18. Sequence control of air temperature

4.3.2. Fuzzy Inference System

A single fuzzy input set can be chosen to be applicable to both heating and cooling processes. This can be very simply structured around the idea that "if the air is warm then reduce the heating valve whilst the cooling valve is off" leading on to "if the air is very warm then increase the cooling valve whilst the heating valve is off". Somewhere in between these two statements free cooling by the damper set can also be applied. In principle, the same fuzzy set can be used for heating and cooling – all that needs to be changed is the direction of its outcome (i.e., increasing temperature implies *reduced* heating initially then *increased* cooling). The upper of the two fuzzy input sets in Fig. 19 defines a simple set that can be used for heating and cooling in which the term "TemperatureError" is the difference between the required temperature at the plant discharge and the actual value. Hence when "TemperatureError" is negative cooling by some degree is needed whilst, when positive, heating by some degree is needed. Other than that, it requires to be recognised that the damper's role is twofold. It can help with energy-efficient free cooling but it also requires to ensure that a critical minimum amount of fresh air input is never prejudiced in order to ensure good indoor air quality.

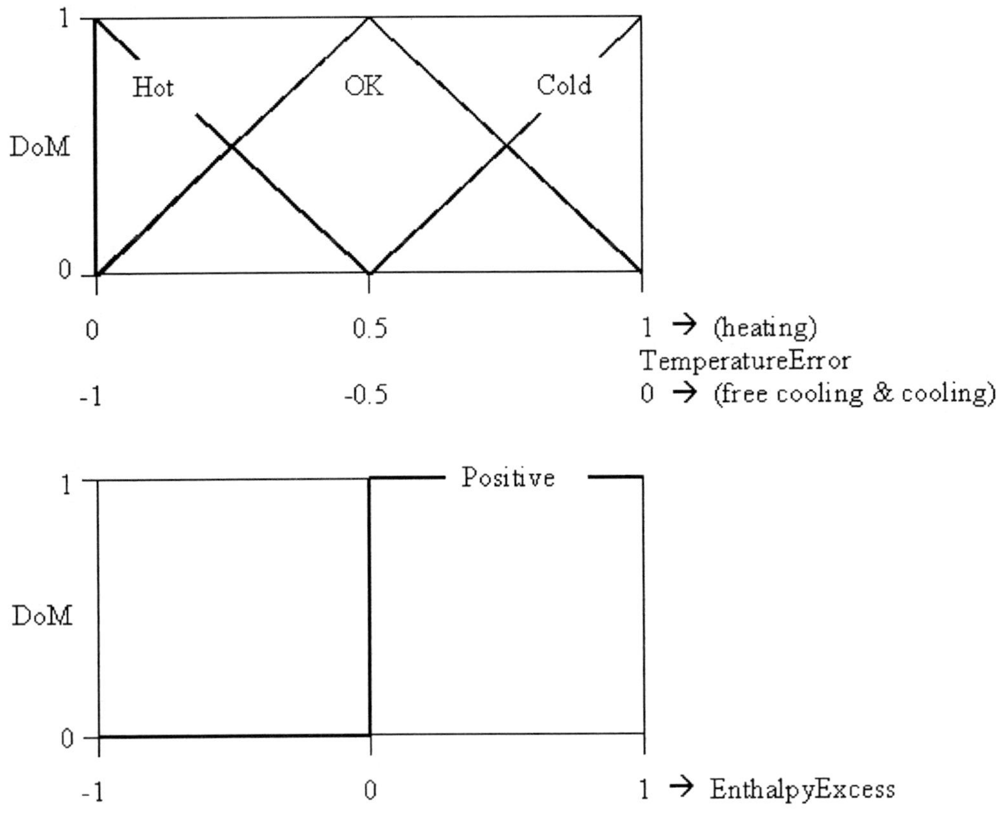

Fig. 19. Case Study 3: Fuzzy input sets

This can be achieved with a Boolean input which lets the damper function normally unless a limit to free cooling potential arises in which case the damper must be switched to a minimum fresh air input role only. The limit to free cooling potential arises when the external air enthalpy exceeds the return building air enthalpy (a condition which will be referred to here as "EnthalpyExcess"). The lower of the fuzzy input sets shown in Fig. 19 defines this switching state.

Three rule sets intuitively follow from the above, each for a controlled process. Table 9 gives the rules for the heating and cooling coil processes. For the damper set an AND operator is required to deal with the minimum fresh air reset function (Table 10).

Table 9. Case Study 3: Fuzzy rules for heating and cooling coils

IF	THEN (heating case)	THEN (cooling case)
TemperatureError is Hot	Valve is Shut	Valve is Open
TemperatureError is OK	Valve is Modulating	Valve is Modulating
TemperatureError is Cold	Valve is Open	Valve is Shut

Table 10. Case Study 3: Fuzzy rules for the damper set

IF	AND	THEN
TemperatureError is Hot	EnthalpyExcess is Positive	Damper is Minimum
TemperatureError is Hot	EnthalpyExcess is Not Positive	Damper is Open
TemperatureError is OK	EnthalpyExcess is Positive	Damper is Minimum
TemperatureError is OK	EnthalpyExcess is Not Positive	Damper is Modulating
TemperatureError is Cold	EnthalpyExcess is Positive	Damper is Minimum
TemperatureError is Cold	EnthalpyExcess is Not Positive	Damper is Minimum

Sugeno inference was used because of its appeal for control type problems. The advantages of this approach will become evident in the following. By applying the following output constants for a Sugeno inference of the input fuzzy sets then the valve position will simply follow a linear function of the fuzzy inputs:

"Shut" = 0
"Modulating" = 1
"Open" = 1

For instance when the crisp error (heating) is 0.25, the fuzzy error will be "Hot" and "OK" each to a degree of 0.5. Thus 0.5 of the "Shut" consequent will fire (i.e., 0) plus 0.5 of the "Modulating" consequent (i.e., 0.5) so the valve will be sent to 0.5 of its range. When the crisp error is 0.75 (heating), the fuzzy error will be "Cold" and "OK" each to a degree of 0.5 and so 0.5 of the "Open" consequent will fire (i.e., 0.5) plus 0.5 of the "Modulating" consequent (i.e., 0.5) so the valve will position at 1 (fully open). In effect, a very simple unity proportional gain control method is achieved with these settings.

A further advantage of this approach is that the control signal, u, on each channel (i.e., damper, heating valve or cooling valve) saturates precisely at 0 (shut) or 1 (fully open). To achieve this using Mamdani inference can require extensive manipulation of the membership functions.

4.3.3. Applications and Results

The resulting fuzzy controller set was applied to a nominal air handling plant with a cross section of 1m x 1m and an air throughput of $2m^3/s$. First, open loop step-response simulations of the three controlled processes were carried out and these were used to deduce tuning parameters for conventional PID controllers by using classical Ziegler-Nichols rules as would most likely be adopted in a practical implementation (Ziegler and Nichols, 1942). Results of the open loop response tests and the corresponding PID controller tuning parameters are given in Fig. 20 and Table 11 respectively (the derivative term in all cases was not considered necessary for this application and, therefore, not tuned).

A simulation was carried out with the plant alternatively controlled using the developed fuzzy controllers and by using the conventional PID controllers. A minimum damper position of 0.5 was used (i.e., a minimum fresh air rate of 50% of total air delivered) and target controlled temperature set points of 11°C, 12°C and 13°C were applied to the heating, free cooling and cooling control processes respectively. The return air state was fixed at 20°C and 50% relative humidity. The plant inlet air temperature was progressively stepped between values of -3°C, 12°C, 26°C, -3°C in order to test the response of each control channel in sequence and to investigate tracking against short term disturbances, a random number sequence with an amplitude of 2K was applied to the inlet air temperature. Results are shown in Figs. 21 and 22.

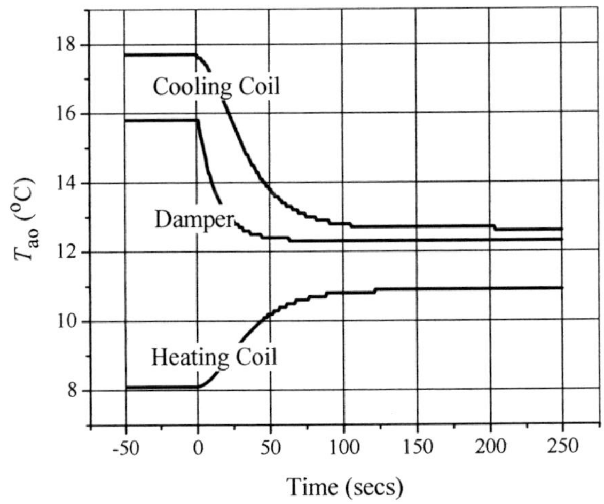

Fig. 20. Case Study 3: Plant open loop responses (T_{ao} – air outlet temperature)

Table 11. Case Study 3: PI controller parameters

Loop	K_c	K_i
Damper	-2.620	-3.490
Heating Coil	0.188	0.006
Cooling Coil	-0.108	-0.004

It is evident that the fuzzy controllers outperform the conventional PID controllers both in terms of stability and tracking. The PID controllers exhibit instability (heating channel) and excess integral wind-up on both the heating and cooling channels as evidenced by prolonged overshoot/undershoot immediately following a step change in inlet air temperature. Damper reset due to the "EnthalpyExcess" function on the fuzzy damper channel is evident in the damper control signal behaviour as the cooling coil becomes active as a result of high inlet air temperature (and, hence, enthalpy) (Fig. 22). Also evident is some "chatter" on the cooling channel of the fuzzy controller (Fig. 22) and also a small amount of offset on the fuzzy damper control channel (Fig. 21). In effect, the fuzzy implementation with Sugeno inference resembles a high gain proportional controller and so these conditions are to be expected. The introduction of a time-rate-of-change input can often reduce or eliminate these undesirable effects.

To put the results in perspective, it is only fair to acknowledge that with patient re-tuning and adjustment of the PID settings, the introduction of an improved integral wind-up facility and the introduction of a damper reset switch for high inlet air enthalpy, it is quite possible to end up with a PID control performance which is as good (or maybe better) than the fuzzy controller results presented here. But this is the very point about fuzzy control – the strategy adopted above is simple, intuitive and requires minimal fine tuning adjustment.

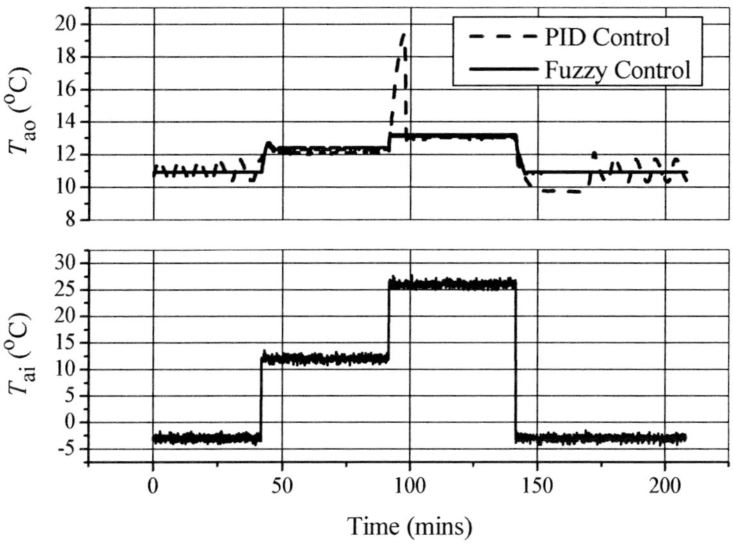

Fig. 21. Case Study 3: Simulated controller results (Top: Controlled outlet air temperature (T_{ao}) Bottom: Plant inlet air temperature (T_{ai}))

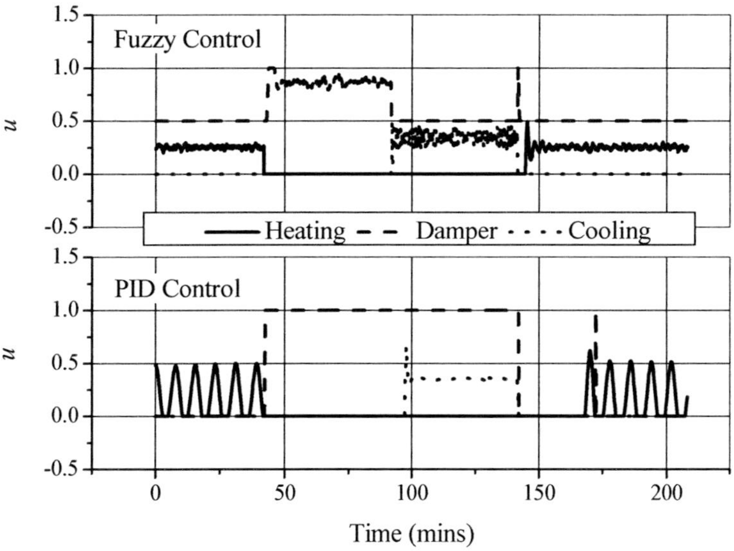

Fig. 22. Case Study 3: Simulated controller signals

4.3.4. Conclusion

This case study has considered the fuzzy control of air handling plant outlet air temperature. Three controllers were identified to independently work on damper, heating and cooling control processes thereby achieving sequence control. The fuzzy controllers with

Sugeno inference were "benchmarked" using conventional PID control. Results showed superior tracking and stability performance by the fuzzy controllers as well as the ease with which the fuzzy controllers were set up and implemented compared with the PID controllers.

5. Conclusions

This chapter has considered the application of fuzzy inference systems to energy and renewable energy systems in buildings. Fuzzy inference can be used to establish relationships between inputs and outputs using linguistic rules and can be especially productive in situations where the inputs are themselves uncertain, ill-defined, qualitative or consist of a mixture of qualitative and quantitative information. The range of fuzzy inference methods has expanded and matured during the past 30 years in the three main fields of decision-making, control and modeling. All three branches have found substantial application in the field of energy and renewable energy in buildings during the past 15 years with perhaps the greatest area of attention having been in the field of control system developments. Case studies have been used to demonstrate the application of fuzzy inference to these three areas and have demonstrated both acceptable results but also the relative ease and transparency with which the methods can be implemented. It would however be wrong to regard a fuzzy inference approach to problem solving as a panacea because of, rather than in spite of, the varied range of alternative methods that are available. It is often the case that solutions to a problem using fuzzy inference will be no better than could be achieved using another method though in many of these cases a fuzzy approach may well have been easier to define and implement.

References

Alcalá R., Casillas J., Cordón O., González A., Herrera F., 2005. A genetic rule weighting and selection process for fuzzy control of heating, ventilating and air conditioning systems. *Engineering Applications of Artificial Intelligence*, Vol. 18, pp. 279-296.

Aprea C., Mastrullo R., Renno C., 2004. Fuzzy control of the compressor speed in a refrigeration plant. *International Journal of Refrigeration*, Vol. 27, pp. 639-648.

Ardehali M.M., Saboori M., Teshnelab M., 2004. Numerical simulation and analysis of fuzzy PID and PSD control methodologies as dynamic energy efficiency measures. *Energy Conversion and Management*, Vol. 45, pp. 1981-1992.

Barelli L., Bidini G., 2005. Design of the measurements validation procedure and the expert system architecture for a cogeneration internal combustion engine. *Applied Thermal Engineering*, Vol. 25, pp. 2698-2714.

Beccali M., Cellura M., Ardente D., 1998. Decision making in energy planning: the ELECTRE multicriteria analysis approach compared to a fuzzy-sets methodology. *Energy Conservation Management*, Vol. 39, No. 16-18, pp. 1869-1881.

Bezdek J.C., 1981. *Pattern Recognition with Fuzzy Objective Function Algorithms,* New York: Plenum Press.

Bilodeau A., Agbossou K., 2005. Control analysis of renewable energy system with hydrogen storage for residential applications. *Journal of Power Sources,* in press, doi: 10.1016/jpowersour.2005.04.038.

Calvino F., Gennusa M.L., Rizzo G., Scaccianoce G., 2004. The control of indoor thermal comfort conditions: introducing a fuzzy adaptive controller. *Energy and Buildings*, Vol. 36, pp. 97-102.

Chu C.M., Jong T.L., Huang Y.W., 2005. Thermal comfort control on multi-room fan coil unit system using LEE-based fuzzy logic. *Energy Conversion and Management*, Vol. 46, pp. 1579-1593.

Dexter A.L., Trewhella D.W., 1990. Building control systems: fuzzy rule-based approach to performance assessment. *Building Services Engineering Research and Technology*, Vol. 11, No. 4, pp. 115-124.

Dounis A.I., Santamouris M.J., Lefas C.C., Argiriou A., 1995. Design of a fuzzy set environment control systems. *Energy and Buildings*, Vol. 22, pp. 81-87.

Gomez V., Casanovas A., 2002. Fuzzy logic and meteorological variables: a case study of solar irradiance. *Fuzzy Sets and Systems*, Vol. 126, pp. 121-128.

Gomez V., Casanovas A., 2003. Fuzzy modeling of solar irradiance on inclined surfaces. *Solar Energy*, Vol. 75, pp. 307-315.

Gouda M.M., Danaher S., Underwood C.P., 2001. Thermal comfort based fuzzy logic controller. *Building Services Engineering Research and Technology*, Vol. 22, No. 4, pp. 237-253.

Gouda M.M., Danaher S., Underwood C.P., 2005. Quasi-adaptive fuzzy heating control of solar buildings. *Building and Environment,* in press, doi: 10.1016/j.buildenv.2005.07.008.

Hamdi H., Lachiver G., Michaud F., 1999. A new predictive thermal sensation index of human response. *Energy and Buildings*, Vol. 29, pp. 167-178.

He M., Cai W.J., Li S.Y., 2005. Multiple fuzzy model-based temperature predictive control for HVAC systems. *Information Sciences*, Vol. 169, pp. 155-174.

Hong Y.Y., Chao Z.T., Yang M.S., 2004. A fuzzy multiple linear regression based loss formula in electric distribution systems. *Fuzzy Sets and Systems*, Vol. 142, pp. 293-306.

Huang S., Nelson R.M., 1991. A PID-law-combining fuzzy controller for HVAC applications. *ASHRAE Transactions*, Vol. 97, No. 2, pp. 768-774.

Huang S., Nelson R.M., 1994a. Rule development and adjustment strategies of a fuzzy logic controller for an HVAC system: part one – analysis. *ASHRAE Transactions*, Vol. 100, No. 1, pp. 841-850.

Huang S., Nelson R.M., 1994b. Rule development and adjustment strategies of a fuzzy logic controller for an HVAC system: part two – experiment. *ASHRAE Transactions*, Vol. 100, No. 1, pp. 851-856.

Huang S.J., Shih K.R., 2002. Application of a fuzzy model for short-term load forecast with group method of data handling enhancement. *Electrical Power and Energy Systems*, Vol. 24, pp. 631-638.

Jaber J.O., Mamlook R., Awad W., 2005. Evaluation of energy conservation programs in residential sector using fuzzy logic methodology. *Energy Policy*, Vol. 33, pp. 1329-1338.

Jana C., Chattopadhyay R.N., 2004. Block level energy planning for domestic lighting – a multi-objective fuzzy linear programming approach. *Energy*, Vol. 29, pp. 1819-1829.

Jeong K.S., Lee W.Y., Kim C.S., 2005. Energy management strategies of a fuel cell/battery hybrid system using fuzzy logics. *Journal of Power Sources*, Vol. 145, pp. 319-326.

Jurado F., Saenz J.R., 2002. Neuro-fuzzy control for autonomous wind-diesel systems using biomass. *Renewable Energy*, Vol. 27, pp. 39-56.

Jurado F., Valverde M., 2005. Enhancing the performance of a solid oxide fuel cell using multiobjective genetic algorithms. *Renewable Energy*, Vol. 30, pp. 881-902.

Kaminaris S.D., Tsoutsos T.D., Agoris D., Machias A.V., 2006. Assessing renewables-to-electricity systems a fuzzy expert system model. *Energy Policy*, Vol. 34, pp. 1357-1366.

Kolokotsa D., 2003. Comparison of the performance of fuzy controllers for the management of the indoor environment. *Building and Environment*, Vol. 38, pp. 1439-1450.

Kolokotsa D., Niachou K., Geros V., Kalaitzakis K., Stavrakakis G.S., Santamouris M., 2005. Implementation of an integrated inddor environment and energy management system. *Energy and Buildings*, Vol. 37, pp. 93-99.

Lah M.T., Zupančič B., Krainer A., 2005. Fuzzy control for the illumination and temperature comfort in a test chamber. *Building and Environment*, Vol. 40, pp. 1626-1637.

Leclercq L., Robyns B., Grave J.M., 2003. Control based on fuzzy logic of a flywheel energy storage system associated with wind and diesel generators. *Mathematics and Computers in Simulation*, Vol. 63, pp. 271-280.

Lee P.S., Dexter A.L., 2005. A fuzzy sensor for measuring the mixed air temperature in air-handling units. *Measurement*, Vol. 37, pp. 83-93.

Li S., Zhang W., Zhang R., Lv D., Huang Z., 2005. Cascade fuzzy control for gas engine driven heat pump. *Energy Conversion and Management*, Vol. 46, pp. 1757-1766.

Lu L., Cai W., Xie L., Li S., Soh Y.C., 2005. HVAC system optimization – in-building section. *Energy and Buildings*, Vol. 37, pp. 11-22.

Luk P.C.K., Low K.C., Sayiah A., 1999. GA-based fuzzy logic control of a solar power plant using distributed collector fields. *Renewable Energy*, Vol. 16, pp. 765-768.

Mamdani E.H., Assilian S., 1975. An experiment in linguistic synthesis with a fuzzy logic controller. *International Journal of Man-Machine Studies*, Vol. 7, No. 1, pp. 1-13.

Mamlook R., Akash B.A., Mohsen M.S., 2001a. A neuro-fuzzy program approach for evaluating electric power generation systems. *Energy*, Vol. 26, pp. 619-632.

Mamlook R., Akash B.A., Nijmeh S., 2001b. Fuzzy sets programming to perform evaluation of solar systems in Jordan. *Energy Conversion and Management*, Vol. 42, pp. 1717-1726.

Michalik G., Khan M.E., Bonwick W.J., Mielczarski W., 1997. Structural modeling of energy demand in the residential sector: 2. The use of linguistic variables to include uncertainty of customers' behaviour. *Energy*, Vol. 22, No. 10, pp. 949-958.

Miland H., Glöckner R., Taylor P., Aaberg R.J., Hagen G., 2006. Load control of a wind-hydrogen stand-alone power system. *International Journal of Hydrogen Energy*, Vol. 31, pp. 1215-1235.

Mohamed A.Z., Eskander M.N., Ghali F.A., 2001. Fuzzy logic control based maximum power tracking of a wind energy system. *Renewable Energy*, Vol. 23, pp. 235-245.

Narasimhan S., Asgarpoor S., 2000. A fuzzy-based approach for generation system reliability evaluation. *Electric Power Systems Research*, Vol. 53, pp. 133-138.

Patcharaprakiti N., Premrudeepreechacharn S., Sriuthaisiriwong Y., 2005. Maximum power point tracking using adaptive fuzzy logic control for grid-connected photovoltaic system. *Renewable Energy*, Vol. 30, pp. 1771-1788.

Ruusunen M., Leiviskä K., 2004. Fuzzy modeling of carbon dioxide in a burning process. *Control Engineering Practice*, Vol. 12, pp. 607-614.

Santigosa L.R., Martinez J.P., Lopez L.M., de Cardona Ortin M.S., Galvez J.B., 2003. Fuzzy inference systems applied to the daily ultraviolet radiation evaluation (295-385nm) from daily global radiation. *Solar Radiation*, Vol. 75, pp. 447-454.

Sfetsos A., 2000. A comparison of various forecasting techniques applied to mean hourly wind speed time series. *Renewable Energy*, Vol. 21, pp. 23-35.

Škrjanc I., Zupančič B., Furlan B., Krainer A., 2001. Theoretical and experimental fuzzy modelling of building thermal dynamic response. *Building and Environment*, Vol. 36, pp. 1023-1038.

So A.T.P., Chow T.T., Chan W.L., Tse W.L., 1994. Fuzzy air handling system controller. *Building Services Engineering Research and Technology*, Vol. 15, No. 2, pp. 95-105.

So A.T.P., Chan W.L., Tse W.L., 1997. Self-tuning fuzzy air handling system controller. *Building Services Engineering Research and Technology*, Vol. 18, No. 2, pp. 99-108.

Sözen A., Kurt M., Akçayol M.A., Özalp M., 2004. Performance prediction of a solar driven ejector-absorption cycle using fuzzy logic. *Renewable Energy*, Vol. 29, pp. 53-71.

Sugeno M., 1985. *Industrial Applications of Fuzzy Control,* New York: Elsevier Science.

Zadeh L.A., 1973. Outline of a new approach to the analysis of complex systems and decision making processes. *IEEE Transactions on Systems, Man and Cybernetics*, Vol. 3, No. 1, pp. 28-44.

Zarzalejo L.F., Ramirez L., Polo J., 2005. Artificial intelligence techniques applied to hourly global irradiance estimation from satellite-derived cloud index. *Energy*, Vol. 30, pp. 1685-1697.

Ziegler J.G., Nichols N.B., 1942. Optimum settings for automatic controllers. *Transactions of the American Society of Mechanical Engineers*, Vol. 64, pp. 759-768.

Zimmerman W., 1994. *Fuzzy Set Theory and its Applications,* Boston: Kluwer.

In: Artificial Intelligence in Energy and Renewable Energy Systems ISBN 1-60021-261-1
Editor: Soteris Kalogirou, pp. 309-338

Chapter 9

APPLICATION OF COMPUTATIONAL INTELLIGENCE TECHNIQUES TO ARCHITECTURAL AND BUILDING ACOUSTICS

Jingfeng Xu, Joseph Nannariello and Fergus Fricke
School of Architecture, Design Science and Planning,
University of Sydney, Australia

ABSTRACT

The theory and practice of acoustics, when applied to buildings, often give very different results. The reasons for this are many and include unknown properties, or variation in properties, of materials and constructions due to limited quality control, multiple design parameters, workmanship and limitations in measurement techniques. This chapter describes a series of investigations that show the potential importance of two computational intelligence (CI) techniques, artificial neural networks (ANNs) and evolutionary algorithms (EAs), in architectural and building acoustics applications and how these techniques achieve better outcomes than conventional analytical or numerical techniques.

In many cases designers of spaces want to predict the perceived acoustic quality, speech intelligibility or speech privacy of auditoria, concert halls, theatres, places of worship, hotel rooms or offices. In these situations theoretical approaches are not available and parametric investigations are dogged by subjective differences and non-linear relationships between preferences and combinations of acoustical parameters. In auditoria there is no theory linking acoustic quality with the many different measures used to describe the acoustics and the ANN method is the only known way of objectively predicting perceived acoustic performance. An ANN can also be used to predict acoustic properties of spaces using limited information before a detailed (and possibly unworkable) design has been completed.

Architects need guidelines to help them define the architectural design scheme for a listening space in the early stages of the design process. On the basis of the theories of subjective preference and sound field, genetic algorithms (GAs), a branch of EAs, can be applied to search for architectural design schemes that would produce high scale values of subjective preferences.

There are several instances where theoretical treatments do exist but the theory does not always agree well with results in practice or too much experimental work is required for the optimisation of performance. The examples described here involve the application of ANNs and EAs for the prediction and optimisation of sound insulation and absorption in buildings. This will include the prediction of wall sound insulation and the prediction and optimisation of airborne sound transmission of floor-ceiling constructions and the absorption performance of multi-layered acoustic absorbers using EA techniques.

Keywords: Computational Intelligence, Artificial neural networks, Evolutionary algorithms, Architectural and building acoustics

NOMENCLATURE

Af	Main floor slope of an auditorium (degrees)
A_W	Angle of side walls in main seating area of an auditorium (degrees)
C_{50}	Clarity factor assuming that the energy within the first 50 milliseconds is the early energy (dB)
C_{80}	Clarity factor assuming that the energy within the first 50 milliseconds is the early energy (dB)
CAS	Capacity number of seats of the auditorium
D_{mean}	Mean depth of an auditorium (mean distance from the front of platform to the rearmost wall) (m)
D_{MX}	Distance from front of podium to the rearmost wall (m)
E	Energy density (J/m^3)
EA_R	Enclosure's average rating coefficient (i.e. the sum of the rating coefficient of each surface/number of surfaces)
En	Entire data set
G	Strength factor (dB)
GEATbx	Genetic and evolutionary algorithm toolbox for use with Matlab
G_{low}	Strength factor at low frequencies: average of the values of G in the 125 and 250 Hz bands (dB)
G_{mid}	Strength factor at mid frequencies: average of the values of G in the 500 and 1000 Hz bands (dB)
H/W	Ratio of height to width
H_{mean}	Mean height of an auditorium (m)
$IACC_3$	Interaural cross-correlation coefficient averaged over three frequency bands (500, 1000 and 2000 Hz)
L/W	Ratio of length to width
L_{MX}	Maximum length of an auditorium (m)
N	Number of seats
R^2	Coefficient of determination
R_{OS}	Distance between the sound source and the sound receiver (m)
RT_{60}	Reverberation time with a sound decay of 60 dB (s)
$RT_{125\text{-}250}$	Reverberation time averaged over 125 and 250 Hz bands(s)
$RT_{500\text{-}1000}$	Reverberation time averaged over 500 and 1000 Hz bands(s)

RT_{mid}	Mid-band reverberation time with a sound decay of 60 dB (s)
$RT_{(bands)}$	Reverberation time (octave frequencies) (s)
$RT_{A[bands]}$	Reverberation time (octave frequencies) at listener positions (s)
R_X	X coordinate of the listener (m)
R_Y	Y coordinate of the listener (m)
R_Z	Z coordinate of the listener (m)
S/N	Signal-to-noise ratio (dB)
S	Scale value
$SP_{[bands]}$	Sound propagation in octave frequency bands (dB)
S_T	Total acoustical floor area (m^2)
S_X	X coordinate of the sound source (m)
S_Y	Y coordinate of the sound source (m)
S_Z	Z coordinate of the sound source (m)
Te	Test data set
T_I	Initial time-delay gap
Tr	Training data set
TR	Tube ratio (m^{-1})
TSA	Total surface area of the auditorium (m^2)
T_{sub}	Subsequent reverberation time (s)
V	Volume (m^3)
Ve	Verification data set
W_{mean}	Mean width of an auditorium (m)
W_{MX}	Maximum width of an auditorium (m)

ABBREVIATIONS

AI	Articulation Index
ANN(s)	Artificial Neural Network(s)
AQI	Acoustic Quality Index
BCE	Before the Common Era
BP	Back Propagation
BR	Bass Ratio
CI	Computational Intelligence
EA(s)	Evolutionary Algorithms
EDT	Early Decay Time
FEM	Finite Element Model
GA(s)	Genetic Algorithms
GRNN	General Regression Neural Network
IACC	Interaural Cross-Correlation Coefficient
IIC	Impact Insulation Class
LL	Listen Level
LF	Lateral Fraction
MLP	Multilayer Perceptron
MOGA	Multi-Objective Genetic Algorithm

MOP	Multi-objective Optimisation Problem
RBF	Radial Basis Function
SDI	Surface Diffusion Index
SD	Standard Deviation
SDR	Standard Deviation Ratio
SII	Speech Intelligibility Index
SP	Sound Propagation
STC	Sound Transmission Class
TL	Transmission Loss
UBC	University of British Columbia

1. INTRODUCTION

Architectural and building acoustics is all too often not given much attention in the design of buildings. It is usually a secondary consideration when issues such as structural integrity, architectural form, aesthetics, fire rating, thermal performance, views and light are concerned and yet for some buildings, such as residential apartments and auditoria, acoustics/noise control are major sources of conversation, concern and complaints.

Part of the reason for the lack of consideration of acoustics is that architects are generally "visual" people and find it difficult to understand and deal with sound. This is also partly because:

- acoustics is not a life-threatening issue (although it might be to the life of a building such as a concert hall as it may be unusable if the acoustics are too bad);
- the desired acoustics are often difficult to define, specify, predict and measure;
- compared with other building services and environmental considerations, there are minute amounts of energy involved (if all the acoustic energy from a person shouting was transformed into heat and transferred without loss into a perfectly insulated kettle with a kilogram of water in it, it would take approximately ten years of continuous shouting to boil the water!); and
- the range of acoustic conditions that people find enjoyable or acceptable is enormous and depends on an individual's personality, hearing acuity and previous experience, what activity is being undertaken, what the characteristics of the sound are, what associations the person has with the sound, what "information" content the sound has, who or what is responsible for making the sound, when and where it occurs and so on.

The comfort of seats, the aesthetics of the space, the lighting and thermal conditions can also play a part in acoustic quality judgements. It is therefore no wonder that it is often said that acoustics is an art rather than a science. To categorize all aspects of architectural and building acoustics as "art" rather than "science" would be a mistake however. Since the time of Pythagoras, who 2500 years ago applied scientific method to acoustics (Hunt, 1992), scientific acoustics has however usually been the poor relation of acoustics in architecture, music and theatre.

The acoustic design of auditoria has been of concern for millennia. Greek amphitheatre acoustic design (Cremer, 1975) was probably refined over hundreds of years. Acoustic design of Roman buildings was obviously of concern as indicated in the writings of Vitruvius (Morgan, 1960) in the first century BCE. While there were writings about acoustics in the 16th, 17th and 18th centuries it was not until the 19th century that acoustic design was taken seriously in the design and treatment of buildings such as:

- the British Houses of Parliament (Reid, 1853). Reid wrote in 1835 that, "any difficulty in the communication of sound in large rooms arises generally from the interruption of sound produced by prolonged reverberation", when he advised on the acoustic treatment for the House of Commons,
- the Paris Opera House (Garnier, 1880) which opened in 1876. Charles Garnier, the architect, wrote about acoustics, "I had read diligently in my books, and conferred industriously with philosophers – nowhere did I find a positive rule of action to guide me; on the contrary nothing but contradictory statements" and, ".... I leave success or failure to chance alone.", and
- the infamous acoustic failings in 1895 of the lecture theatre in the Fogg Art Museum at Harvard University which prompted a serious study of the physical behaviour of sound in large spaces (Sabine, 1922) and led to the now famous Sabine equation for reverberation time.

For a while the use of an appropriate reverberation time seemed to be sufficient to ensure satisfactory acoustics but it became apparent that there was much more to the design of concert halls than reverberation time. The shape of a concert hall was found to be important as was the background noise level, the stage design, the ceiling, the type and arrangement of seats, the surface finishes of the room and so on and on. There were far too many "architectural" parameters to consider to use as design criteria and when they were used such parameters were often found to be contradictory as, for any successful hall that contained one or more features (that were considered to be essential for a good hall) there was another successful hall which did not contain these features. Subsequently, many acoustic measures were developed such as:

- the rate of decay of a sound pulse in a room and variants on it (e.g. reverberation time, early decay time and reverberation time at different frequencies),
- various measures of early to late acoustic energy ratio (e.g. C_{50} and C_{80}),
- the amount of acoustic energy reflected from side walls (e.g. lateral energy fraction),
- the ratio of acoustic conditions on stage to that in the audience area (e.g. the reverberation time on the stage to that in the audience area),
- the delay time between the direct sound and the first reflected sound to reach an average audience member,
- many different measures of background noise level,
- the difference between the sound at the right and left ear of a member of the audience (interaural cross-correlation coefficient in its many forms),
- the intensity of sound at one or more seats, and

- the absence of reflections, which might lead to the sound being perceived as distorted or be heard as an echo.

Conventional numerical techniques used for architectural acoustic design purposes, such as "ray-tracing" and "image-source" methods require too much detailed information about the size, shape and surface finishes of a room to be of any use in the early design stage of a building. However, often it is at this initial design stage that irrevocable decisions are made about the auditorium that will probably account for 70 or 80 percent of its acoustic quality. Artificial neural networks (ANNs) offer the possibility of making informed decisions, at the sketch design stage of an auditorium, based on precedents and without the need for the detailed information needed for other types of numerical modelling. ANN techniques are shown to predict acoustic properties of spaces such as reverberation time (RT_{60}), strength factor (G), clarity factor (C_{80}), lateral fraction (LF) and early interaural cross-correlation coefficient ($IACC_3$) using limited information. The results obtained using ANNs have been shown to be more accurate predictors of acoustic quantities than the alternative analytical or numerical methods (see section 2.1).

Many different measures of acoustic quality have been proposed and used but few, other than reverberation time and sound level and variants of them, have remained in use for more than a couple of decades. One of the difficulties has been that most room acoustical parameters have been impossible to calculate with sufficient accuracy and the measurement of some parameters is not sufficiently reproducible to be useful. Another of the difficulties has been that even if acoustical parameters, rather than architectural ones, were used for design purposes there would be too many independent variables[1] and too few available measurements of these parameters. Subjective assessments of the acoustic quality of concert halls, the dependent variable, are too few and too unreliable. Beranek did an intensive study of concert hall acoustics in the early 1960s (Beranek, 1962) and more recently he has revised his prediction method (Beranek, 2004). His prediction method, which requires six acoustic input parameters, was not substantiated by statistical analysis. With so many inputs and so few cases to base the prediction method on there is obviously the possibility of significant errors and subjectivity involved, but the ANN provides the means for assessing the efficacy of Beranek's non-linear prediction method and assessing the probability of a correct prediction (see sections 2.2 and 2.3).

"Noisy data" are inevitable because of the way judgements on acoustic quality are made. For example, is the acoustic reputation of a concert hall made at the conductor's rostrum or the best seat, the worst seat, the average seat or the median seat in the auditorium? Also, how does one separate the acoustics of a hall from the performance heard by the person making the judgement? Certainly the stage acoustics are very important but it is not at all clear whether a hall that has "good" acoustics throughout is judged in the same way as one that has very good acoustics in some seats and very poor acoustics in other seats. Again ANNs provide a method of dealing with such data and other computational intelligence (CI) techniques, such as genetic algorithms (GAs), can potentially be used to optimise the design of a concert hall (see section 2.4).

---

[1] The difference between "art" and "science" has been said to be the number of variables involved and when there are more than about six, art takes over from science.

Another important aspect of architectural acoustics is speech intelligibility and speech privacy. Speech intelligibility is important in auditoria, live theatres, class rooms and conference rooms for instance. Speech privacy is important in situations such as doctors' waiting rooms and surgeries, between apartments in the same building, meeting rooms or offices where security, political, delicate, or confidential business or legal discussions take place. While methods of predicting speech intelligibility and privacy are well documented (for example, Cavanaugh *et al.*, 1962) there has been little work on "speech security" which is a more stringent requirement than "speech privacy" (Gover and Bradley, 2004). An ANN has provided better predictions of the thresholds of audibility than conventional techniques (see sections 2.5 and 2.6).

There are several instances where conventional theoretical treatments exist but the theory does not always agree well with results in practice or too much experimental work is required. The transmission of sound through a wall or floor/ceiling construction can be accurately predicted in simple cases but becomes very difficult as the complexity of the wall and floor/ceiling construction and that of the attached building structure increases. Such complexities are common in contemporary buildings where the cost of the materials, labour, structure and space to accommodate traditional constructions such as cavity brick walls are prohibitive and so light-weight, dry-wall constructions incorporating vibration isolation and acoustic damping are employed. The performance of such constructions is difficult to predict analytically and even by conventional numerical methods employing modal analysis or statistical energy analysis [see for example (Pierce, 1997; Manning, 1997; Craggs, 1997)]. This chapter describes the effective application of an ANN and EA to the prediction and optimisation of the sound insulation of walls and floor/ceiling constructions (see section 2.7).

How to develop better sound absorbing materials has been the subject of numerous studies such as (Zwickker and Kosten, 1949) but, like walls and floor/ceiling constructions, the useful theory is largely limited to simple configurations of materials with uniform properties. Considerable improvements can be obtained if two or three layers of porous absorbing materials, with different properties, are used rather than a single layer of material (Dunn and Davern, 1986). However, the experimental work required is prohibitive and multiple criteria optimisation is essential. Hence, evolutionary algorithms (EAs) must be employed to optimise the combination of commercially available materials to achieve the desired result without excessive trial and error testing is undertaken (see section 2.8).

2. Application Cases

In the past fifteen years, ANNs and EAs have been applied in the research of building and architectural acoustics to develop workable, useful and accurate tools with the prospect of being commercialised. This section introduces some of their successful applications in the areas of room acoustics, sound isolation in buildings and sound absorption of acoustic absorbent materials.

2.1. Predicting Acoustical Parameters of Concert Halls

Sabine's work led to the widespread use of reverberation time (RT) and for many years this was the only acoustical parameter used in the design of auditoria. However, the many anomalies inherent in the Sabine's equation and other related theoretical formulae are responsible for the often inaccurate prediction of reverberation times. These predictions are often not within 5%, the subjective difference limen for reverberation time. This reason and the absence of simple and accurate rules of thumb suitable for use at the early conceptual design stage led to the investigation of an alternative method of predicting reverberation time (Nannariello and Fricke, 1999). Neural networks were trained using constructional and acoustical data of auditoria as input variables. Importantly, the input variables associated with the absorption coefficients were replaced by simple rating coefficients in terms of the absorptiveness of materials.

On the basis of the results drawn from the above work (Nannariello and Fricke, 1999), networks with a reduced number of input variables (a network "dimensionality" reduction) have been developed to predict ($RT_{125\text{-}250}$) and ($RT_{500\text{-}1000}$) for auditoria (Nannariello and Fricke, 2002a). The concept was further extended to developing some basic relationships and rules of thumb on how simple geometric parameters affect reverberation time.

The network architecture used was a *Multilayer Perceptron* (*MLP*) with one hidden layer (as most problems seem to yield to a single hidden layer). A general schematic of a single hidden layer MLP is shown in Fig. 1. The topology (the number of hidden neurons) for the best trained NN model consisted of three hidden neurons. The set-up function for the neural network analyses for ($\mathrm{RT}_{60[125\text{-}250]}$) and ($RT_{60[500\text{-}1000]}$) reverberation time predictions was:

$$\mathrm{RT}_{60} = \mathrm{f}\,(\mathrm{V},\ \mathrm{W}_{\mathrm{MX}},\ \mathrm{H}_{\mathrm{MX}},\ \mathrm{TR},\ \mathrm{CAS},\ \mathrm{TSA},\ \mathrm{AFA},\ \mathrm{EA}_{\mathrm{R}}) \quad (1)$$

where:

V = Volume of the auditorium (m^3)
W_{MX} = Maximum width of the auditorium (m)
H_{MX} = Maximum height of the auditorium (m)
CAS= Capacity number of seats of the auditorium
TSA = Total surface area of the auditorium (m^2)
AFA= Audience floor area of the auditorium (m^2)
TR= Tube ratio of the auditorium (m^{-1}) [=Maximum length of auditorium/(maximum width*maximum height) (m^{-1})]
EA_{R}= Enclosure's average rating coefficient [= The sum of the rating coefficient of each surface/number of surfaces]
RT_{60} = Reverberation time of the auditorium (seconds)
$RT_{[\mathrm{bands}]}$ = Reverberation time of the auditorium (octave frequencies) (seconds)

The network parameters were varied for each *training* run and *training* for each network continued until no further improvement in the root mean square (*RMS*) error was experienced.

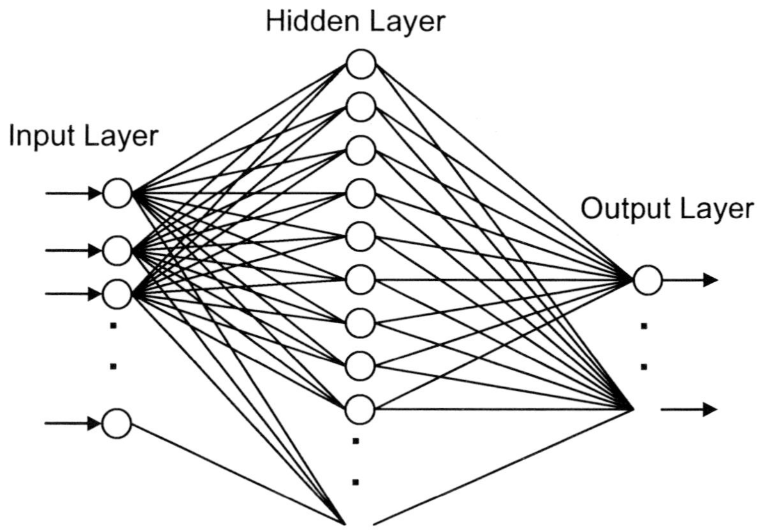

Fig. 1. General schematic of a single hidden layer Multilayer Perceptron (MLP) neural network

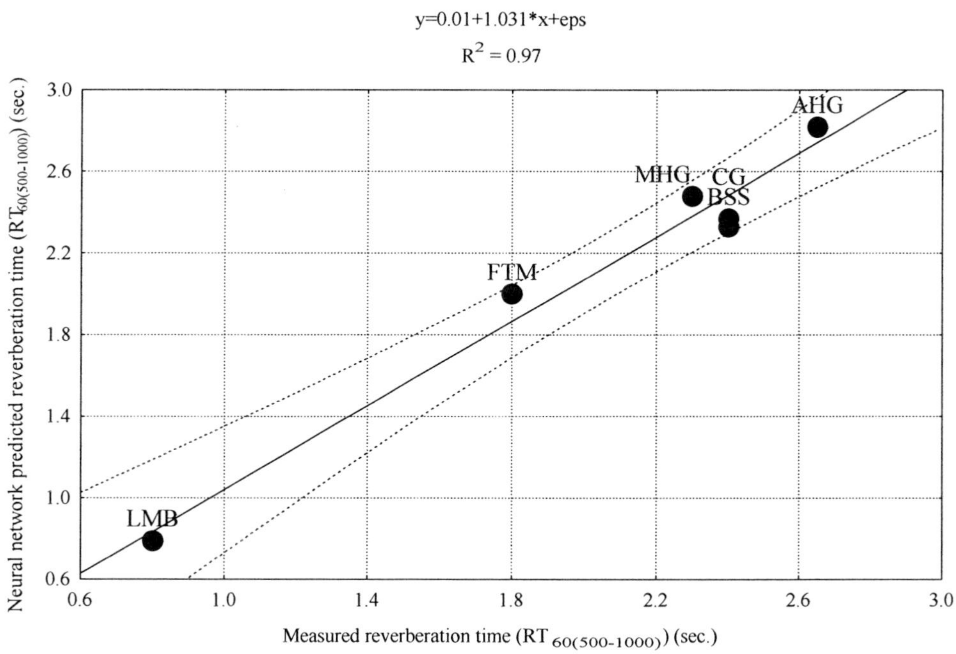

Fig. 2. Linear regression analysis for predicted versus measured low-band reverberation times ($RT_{60(125\text{-}250)}$) utilizing optimal neural network trained with 8 input variables (Nannariello and Fricke, 2002a). In the figure, AHG, BSS, MHG, SSM, TI and ZGS are the abbreviations of rooms, the dotted lines represent the 95% confidence limits and the solid line represents the best fit. In the fitted function equation eps equals the random error associated with measurements

The result of this work provided evidence that neural networks can be used to make predictions of reverberation times at low and mid frequencies for auditoria and that these predictions are as good as or better than existing methods. Furthermore, and more importantly, the results showed excellent strength of association and high percentage agreement between measured and predicted reverberation times (see Figs 2 and 3). The results were repeatable and were within the range of the subjective difference limen of 5%.

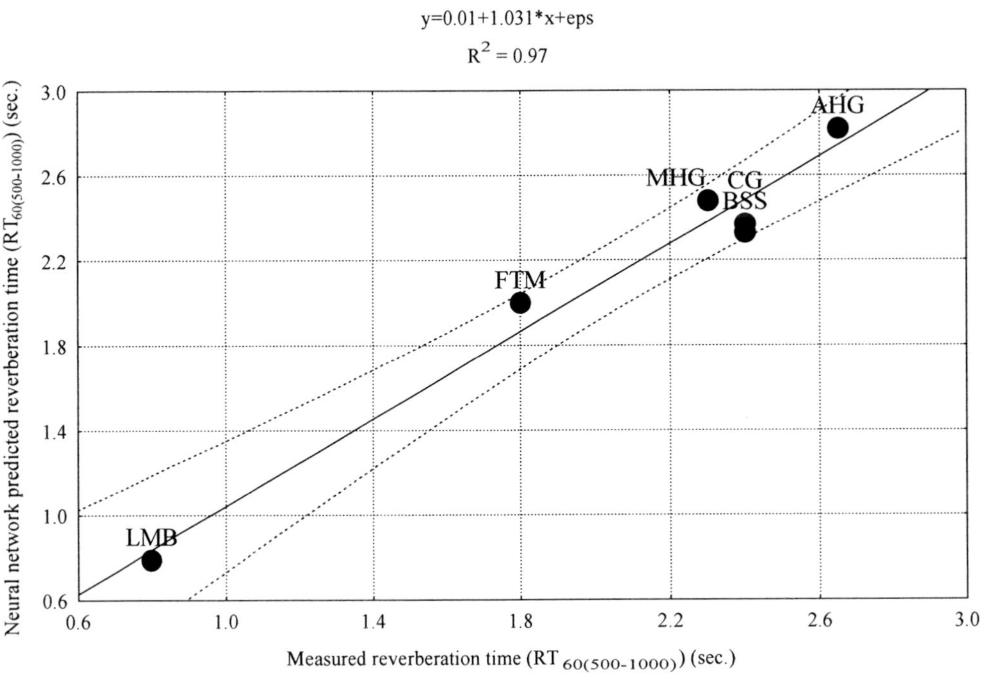

Fig. 3. Linear regression analysis for predicted versus measured mid-band reverberation times ($RT_{60(500-1000)}$) utilizing optimal neural network trained with 8 input variables variables (Nannariello and Fricke, 2002a). In the figure, AHG, BSS, MHG, SSM, TI and ZGS are the abbreviations of rooms, the dotted lines represent the 95% confidence limits and the solid line represents the best fit. In the fitted function equation eps equals the random error associated with measurements

It has long been realized that there is more to auditorium acoustics than reverberation time. Over the last 30 years or so, a number of objective acoustical parameters (related to the subjective assessment of the acoustical characteristics of auditoria) have emerged to aid the design of auditoria. Consequently a number of methods have been developed (Baron and Lee, 1988; Gade, 1991; Bradley, 1994; Volander,1995) to predict parameters such as the strength factor (*G*, a measure of the average loudness of sound in the auditorium), the clarity factor (C_{80}, a measure of the clarity of the sound), lateral energy fraction (*LF*, a measure of the envelopment of the sound) and interaural cross-correlation coefficient (*IACC*, the difference of the sound at each ear), but these methods have their limitations.

A method of predicting *G*, C_{80}, *LF* and (1-$IACC_{E3}$) values in auditoria using neural networks was therefore investigated and developed (Nannariello and Fricke, 2001b). As a trial of this concept and because well-documented measured data from halls is a rarity, neural networks were trained using ODEON room acoustics program (Lynge, 1998) predictions (Nannariello and Fricke, 2001a). One of the important outcomes of this work was that there was a good basis for carrying out further investigations using "noisy" and poorly distributed measured data to train neural networks to predict G values and possibly values of other acoustical parameters.

Subsequent work (Nannariello and Fricke, 2001b) provided evidence that non-linear models, such as neural networks trained with geometrical and measured acoustical data, could make predictions of the G, C_{80} and LF values in concert halls. These simple predictions were as accurate as those calculated using existing, more complex, models (Gade, 1991). The study demonstrated that neural networks could be trained with a handful of simple and available input variables such as the volume, maximum length, total floor area, reverberation time and tube ratio. Between five and eight input variables were used to train networks to predict seat-averaged G, G_{low}, G_{mid}, C_{80} and LF.

The network architecture used was a *MLP* with one hidden layer. The set-up functions for the neural network analyses were (the number of hidden neurons for the best trained neural network models are shown in square brackets at the end of the set-up function):

$$G = f(V, L_{MX}, D_{mean}/(W_{mean}*H_{mean}), S_T, RT_{500\text{-}1000})\ [2] \quad (2)$$

$$G_{low[125\text{-}250]} = f(V, L_{MX}, D_{mean}/(W_{mean}*H_{mean}), S_T, RT_{125\text{-}250})\ [2] \quad (3)$$

$$G_{mid[500\text{-}1000]} = f(V, L_{MX}, D_{mean}/(W_{mean}*H_{mean}), S_T, RT_{500\text{-}1000})\ [2] \quad (4)$$

$$C_{80} = f(V, L_{MX}, W_{MX}, D_{mean}/(W_{mean}*H_{mean}), S_T, Af, A_W, RT_{500\text{-}1000})\ [3] \quad (5)$$

$$LF = f(V, L_{MX}, W_{MX}, D_{mean}/(W_{mean}*H_{mean}), S_T, Af, A_W, RT_{500\text{-}1000})\ [3] \quad (6)$$

where:

S_T = Total acoustical floor area of the auditorium (m^2)
D_{mean} = Mean depth of the auditorium – mean distance from front of platform to rearmost wall (m)
$D_{mean}/W_{mean}*H_{mean}$ = Mean depth / (Mean width*Mean height of the auditorium) (=tube ratio, *TR*) (m^{-1})
H_{mean} = Mean height of the auditorium (m)
L_{MX} = Maximum length of the auditorium (m)
W_{mean} = Mean width of the auditorium (m)
G = Strength factor of the auditorium (dB)
G_{low} = Strength factor at low frequencies: average of the values of G in the 125 and 250 Hz bands (dB)
G_{mid} = Strength factor at mid frequencies: average of the values of G in the 500 and 1000 Hz bands (dB)
Af = Main floor slope of the auditorium (degrees)
A_W = Angle of side walls in main seating area of the auditorium (degrees)

Six input variables (one additional input variable, R_{OS}, distance between the sound source and receiver) were used to train networks to predict position-dependent G values. It was demonstrated that the exact positions of seats in a hall were not required to accurately predict the average parameters. For the 126 receiver positions, in the 8 auditoria tested, the neural network analysis produced excellent results.

In most cases, the absolute average errors and root mean squared errors for G, G_{low}, G_{mid}, were within their subjective limen of ±1 dB and for C_{80} and LF within their subjective limen of ±0.5 dB.

The idea of using neural networks to make predictions of auditorium attributes was further extended to using neural networks to develop some basic knowledge and rules of thumb on how simple geometric parameters affect the attributes of an auditorium, such as G (Nannariello and Fricke, 2001c). The use of acoustical parameters, such as reverberation time, as an input variable, was deliberately avoided. The results showed that neural networks trained with 5 simple geometric input variables – the hall volume, V; the maximum length, L_{MX}; maximum width, W_{MX}; the total acoustical floor area, S_T and the tube ratio, $D_{mean}/(W_{mean} x H_{mean})$ – gave accurate predictions of G. The prediction errors were well below the subjective difference limen for G. The other attendant benefit was that the neural network models produced relationships which, in most cases, agreed with the published literature (Gade, 1991; 1996; Beranek, 1996). Significantly, however, the work also showed that determining the cubic volume and number of seats is not sufficient as a "rule" in modern auditorium design. The optimization of G is dependent on a combination of geometrical parameters including the shape of the hall, which is represented by the tube ratio and the maximum dimensions (see Fig. 4).

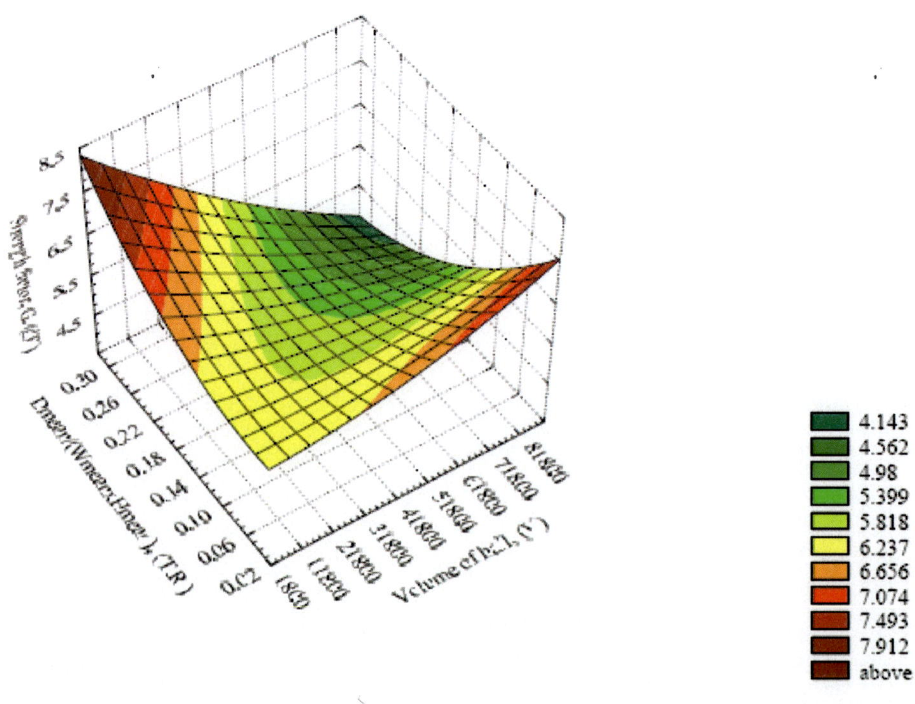

Fig. 4. Quadratic smooth surface plot (V,TR,G) showing relationship between averaged acoustic parameter G (dB), hall volume, V (m^3) and tube ratio(TR), $D_{mean}/W_{mean} x H_{mean}$ (m^{-1}) (Nannariello and Fricke, 2001c)

A neural-computation approach for predicting the early interaraul cross-correlation coefficient, ($IACC_{E3}$), in unoccupied auditoria was also investigated (Nannariello and Fricke, 2002b). Data from thirty-six auditoria were used in the neural network analysis. A multilayer perceptron, fully connected, three layer feedforward network architecture, based on the supervised learning procedure was used to build the neural networks. Seven input variables were used in the first layer and three hidden neurons in the middle layer. The set-up function for the neural network analyses was:

$$(1\text{-}IACC_{E3}) = f(V,\ L_{MX},\ W_{MX},\ D_{mean}/W_{mean}\text{x}H_{mean},\ S_T,\ A_w,\ RT_{mid}) \quad (7)$$

where the symbols specify quantities previously defined and RT_{mid} is the mid frequency reverberation time.

Results of the investigations showed that the neural network model could predict $IACC_{E3}$ values within the subjective difference limen, which is 0.075 ± 0.008. Five auditoria were used to validate the neural network analysis method and the errors between measured and predicted ($1\text{-}IACC_{E3}$) ranged from –0.05 to 0.02 (see Fig. 5). The neural network model used to make ($1\text{-}IACC_{E3}$) predictions was imbedded in an Excel spreadsheet so that designers and researchers, without assess to specialized neural network software, could use the results of the work.

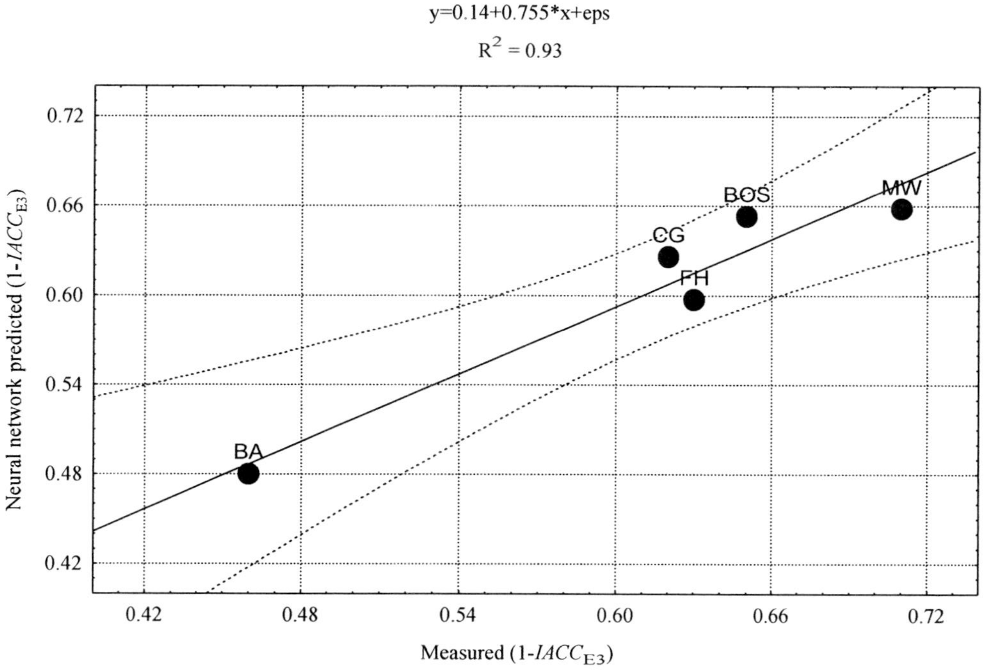

Fig. 5. Scatterplot showing linear regression analysis for neural network predictions of measured values of ($1\text{-}IACC_{E3}$) versus neural network predicted values of ($1\text{-}IACC_{E3}$) (Nannariello and Fricke, 2002b). In the figure, BA, BOS, CG, FH and MW are the abbreviations of rooms, the dotted lines represent the 95% confidence limits and the solid line represents the best fit. In the fitted function equation eps equals the random error associated with measurements

2.2. Predicting Acoustical Quality of Concert Halls

Architects and designers, when designing concert halls, still make use of precedents especially at the sketch design stage. This technique, in most cases, has not guaranteed good acoustics. A neural network analysis was undertaken to relate the acoustic quality index (*AQI*) of halls, as judged by conductors and musicians, to hall parameters. In the first of these (Fricke and Haan, 1995) the ten input parameters used were hall volume, surface area, number of seats, length, width, height, mean rake angle of the seats, SDI (surface diffusion index, visually assessed) A_w (angle of side walls) and hall shape. This was an important study in that it was the first to gather data on the opinions of musicians about the acoustic qualities of many different halls. It was also the first study to apply ANN analysis to an architectural acoustics problem. However the study was flawed in that the *MLP* network used was over-trained, the ANN software used could not be used to produce any statistical data on the reliability of the results and the hall shape could not be adequately described in the four categories used (rectangular, horse-shoe, fan and geometrical).

Thus a second study (Fricke and Han, 1999) was undertaken in an attempt to rectify the limitations of the first study. Two inputs were discarded (A_w and hall shape) and were replaced by the stage height and the degree of stage enclosure (visually assessed). Thus there were ten input variables for the *MLP* network with two neurons in the hidden layer and one output, the acoustic quality index (AQI).

The work demonstrated that neural networks offered the opportunity to study the non-linear interactions of the many variables involved in the acoustic performance of concert halls although the standard deviation ratio (the ratio of the standard deviation of the prediction error to that of the original output data) achieved was approximately 0.90 which left a lot to be desired. A standard deviation ratio (SDR) of 0.1 is considered an excellent fit of the data and an SDR of 1.0 is a poor fit (Statsoft, 1999). However, a modified *MLP* network using five inputs (number of seats, ratio of length to width, ratio of height to the cube root of volume, mean rake angle and the ratio of surface diffusivity index to stage enclosure), one hidden layer with four neurons and two classifications of acoustic quality ("good" and "poor") as the output, a 90% success rate was achieved.

In a related study Fricke (2000a), found that subjective assessments of SDI were not reliable and so one of the objectives of further work was to produce an ANN model which was independent of the subjectively assessed SDI factor that had been shown to be questionable. Other objects were to provide some statistical basis for Beranek's model (Beranek, 1996) of preferred values of acoustic quantities in good concert halls and to produce a simpler (less inputs) ANN model with a better prediction record using both "acoustical" and "architectural" inputs.

As indicated in Section 2.1 above, many measures of the acoustics of concert halls have been proposed. None of these measures is now considered sufficient in itself for the successful design or evaluation of a hall. Unfortunately multi-criteria issues are difficult to deal with and there does not seem to be a consensus on what combination of acoustic parameters creates good acoustics in a concert hall. Beranek, one of the founders of the modern architectural acoustics, opts for six orthogonal parameters; the visually assessed surface diffusion index (*SDI*), the early interaural correlation ($IACC_{E3}$), the time delay between the direct and first reflected sound at the centre of the main seating area (T_I), the early decay time (*EDT*), the measure of the average sound level in a hall at mid-frequency

(G_{mid}) and the bass ratio (*BR*) (Beranek, 1996). Beranek also shows how each parameter contributes to the overall acoustic quality of a hall and gives each a preferred value. However, he does not give the details of how he determined these relationships. Beranek's method also assumes a linear relationship between the *AQI* (Fricke and Haan, 1995) and each of the six parameters, which are given weightings to obtain the *AQI*, but the basis for these weightings is not given.

Using Beranek's six parameters as input variables, a neural network was trained to predict the *AQI* of halls (Fricke, 2000a) The results of the neural network analysis were used firstly to investigate the importance of surface diffusion (Fricke, 2000a) and then the veracity of Beranek's assessment method. Modified networks with acoustical as well as architectural inputs were then used to see whether better fits could be obtained to existing concert hall data than could be obtained using Beranek's six inputs (Fricke, 2000b).

From the results of the neural network analysis, it was postulated that Beranek's approach to the prediction of the acoustic performance of concert halls was reasonable. A *MLP* network with Beranek's six inputs and three hidden neurons gave a SDR of 0.4 (see ANN 1 in Table 1 below). Surprisingly, with only four inputs, ANN 2 (Table 1) has an even better SDR (0.37) though the difference in SDR is probably not significant. Tellingly, the worst SDR achieved, when SDI was not used as an input, was 0.87 (ANN 3). On this basis it is suggested that a better way of predicting the acoustic performance of concert halls may be to use trained neural networks because ANN has the ability of modelling the non-linear relationship between the inputs and outputs and to optimise the use of available information.

Table 1. Standard Deviation Ratios (SDRs) for different combinations of Beranek's parameters, together with N and V, for calculating AQI of concert halls (Fricke, 2000b)

ANN	IACC	T_I	EDT	G_{mid}	BR	SDI	N	V	SDR
1	*	*	*	*	*	*			0.40
2	*		*	*		*			0.37
3	*	*	*	*					0.87
4	*	*	*	*		*	*		0.25
5	*	*	*	*		*		*	0.51
6	*	*	*	*			*		0.58
7	*	*	*			*	*		0.57
8	*	*	*				*		0.64
9	*	*	*			*	*	*	0.51
10	*		*	*		*	*		0.47
11	*		*	*	*	*	*		0.33
12	*			*		*	*		0.42

Other neural networks were trained (ANNs 4 to 12, Table 1) with combinations of Beranek's input variables together with the number of seats (*N*) and the volume of the hall

(V). The results of the neural network analyses were presented as SDRs (see Table 1). They show that using some of Beranek's inputs plus the number of seats, N, can result in lower SDRs; the best SDR achieved being 0.25 by ANN 4.

The effect of changing the input values in ANN 4 on the acoustic quality of the Concertgebouw, the main concert hall in Amsterdam, was then investigated. This investigation showed the non-linear relationships between the output (AQI) and the different inputs (Fricke, 2000b).

Finally, because of the difficulty in obtaining reliable subjective acoustic quality ratings of concert halls and acoustic data on them (only 26 cases were available to train the networks presented in Table 1) it is more realistic and practical to use three acoustic quality categories (best, middle and worst) rather than a finer scale, or rank ordering of preferences. It is also more practical and useful, especially at the initial design stage, to use geometrical inputs rather than acoustic ones, or acoustic inputs that Nannariello (see section 2.1) has shown can be accurately obtained using a limited number of room geometrical inputs. Thus a study was undertaken using *MLP* networks with up to six inputs with different combinations of acoustic (EDT, IACC, G_{mid}, T_I and BR) parameters and architectural [V, V/N, N, H/W, L/W and shape (rectangular or non-rectangular)] parameters to predict the acoustic quality category of concert halls (Choi, 2004).

Using 32 cases for training and testing a simple *MLP* network, with three hidden neurons, a prediction accuracy of up to 90% was achieved. Using inputs of EDT, G_{mid}, T_I, V/N and shape a 90% success was achieved in predicting acoustic quality category (A, B or C) while an 80% success rate was achieved using only four inputs (EDT, G_{mid}, V, shape). Besides showing that ANNs for calculating acoustic quality can be simplified, this work also demonstrated that ANNs are capable of handling a problem that involves categorisation, which is impossible for some traditional analysis methods, such as multivariate regression analysis. It also indicated that ANNs with "geometrical" inputs only, which are much more useful for design purposes than "acoustic" inputs, could be developed.

2.3. Predicting Acoustical Quality of Small Music Rooms

A large amount of research has been undertaken on acoustics of auditoria for the performance of live music and for speech, but there has been very little research carried out on the acoustics of smaller rooms used for music practice and music teaching. There are several reasons for this but the main ones are that small rooms are rarely prestigious, poor acoustics would not result in financial disaster and classical acoustical prediction methods are not applicable to small rooms because of the errors in ignoring diffraction effects in such spaces. With data collected and measured using binaural recordings made in small music rooms, a method of predicting the acoustic quality of small music rooms, using a neural network, was developed (Osman and Fricke, 1996). The 36 rooms used in the investigations were parallelepipedic with volumes ranging from 24 to 427 cubic metres.

A combination of simple input variables (room dimensions, reverberation time and distribution of absorption) for four musical instruments (cello, saxophone, trumpet and guitar) was used to build a number of *MLP* networks with up to six hidden neurons. The neural network models were used to predict the acoustic quality of six small music rooms. From the results of the investigations it was postulated that neural network models can be used to predict acoustic performance of small music rooms and that specific acoustical and architectural parameters are required to predict the acoustic quality of rooms for specific musical instruments.

2.4. Optimising the Design of Classical Shoebox Type Concert Halls

In Section 2.2, the acoustic quality of concert halls was predicted using six "average" acoustic properties of the halls. The acoustic conditions in each seat will however be different and members of the audiences in seats at the front of a concert hall will experience markedly different acoustics to those at the rear for instance. This raises the issue of whether the reputation of concert halls is determined on the basis of the "best" seats, the "worst" seats, the average or median seats or whether it is based on the acoustics of the conductor's podium.

The theory of subjective preference allows a sound field to be evaluated in terms of the following four orthogonal acoustical parameters (Ando, 1998); the listening level (LL), the initial time-delay gap between the direct sound and the first reflection (T_1), the subsequent reverberation time (T_{sub}) and the magnitude of the interaural cross-correlation function (IACC) for a particular seat in a concert hall. Several experiments have been carried out to examine the independent effects of the four orthogonal parameters on subjective preference (Ando, 1998). Linear scale values of preference were obtained by using the law of comparative judgment. The units derived from experiments with different sound sources and different subjects were almost constant, so the scale values may be added to obtain (Sato et al, 2002):

$$S \approx S_1 + S_2 + S_3 + S_4 \tag{8}$$

where S_i ($i = 1, 2, 3, 4$) are the scale values for the respective orthogonal parameters.

On the basis of this theory, it was proposed that for designing sound fields in a concert hall, the ultimate goal is to maximize the scale values of preference and subsequently a maximum overall S (Sato *et al.*, 2002). A GA, a branch of evolutionary algorithms, was applied to search architectural design schemes that would produce high scale values of subjective preferences. At the start of a GA search, a shoebox shape was defined as the initial scheme for the hall. Using the image method, the orthogonal parameters were then calculated for the initial shape and various shapes generated by the GA. The scale values of subjective preference were employed as fitness functions. Those hall shapes that produced higher scale values were selected as parent chromosomes, i.e., binary strings. To create a new generation, the room shapes were modified and the corresponding movement of the vertices of the walls was encoded in chromosomes. After GA operations that included crossover and mutation, new offspring were created. The fitness of the offspring was then evaluated in terms of the scale values of subjective preference. This process was repeated until the end condition (2000 generations) had been satisfied.

Two models were investigated in the work (Sato *et al.*, 2002). In the first model, the geometry proportions of the shoebox were optimised and in the second model the floor plan was optimised with the results for the first model as the starting point of search. The optimisation for the first model produced a shoebox shape with the length/width ratio almost the same as that of the Grosser Musikvereinsaal, in Vienna, one of the wildly recognised excellent concert halls. However, conflicting height/width ratios were required for the maximum S_1 and maximum S_4 due to the different reflection requirements of the ceiling and walls to achieve the optimum LL and optimum IACC. The height/width ratio of the Grosser Musikvereinsaal is intermediate between those for the maximum S_1 and maximum S_4. The optimisation for the second model indicates that a leaf shaped plan was optimum with regard to the sidewalls. However, again due to the different reflection requirements, to maximise S_1 and S_4 the front wall and rear walls took on opposite characteristics, one requiring concave shape and the other requiring convex shape.

Although the work concluded that the maximisation of S_1 might take priority over that of S_4, a more useful study may be the use of multi-objective optimisation evolutionary algorithms to optimise S_1 and S_4 at the same time and find a "compromised" solution, such as that achieved in the Grosser Musikvereinsaal.

2.5. Predicting Sound Propagation in University Classrooms

Classrooms are primarily auditory-verbal learning environments in which students must have access to acoustic signals (speech) in order to understand and learn. University classrooms are designed and purpose built for this sole reason, that is, for speech communication from a source to a number of listeners. It would be very useful at the schematic design stage of a classroom to have an expeditious and accurate method of predicting the distribution of sound levels (speech levels).

There are a number of theoretical models that can be used to calculate sound levels in rooms. The most widely used is the Hopkins-Stryker equation (Davis and Davis, 1991). Using this equation, the sound level at a receiver can be predicted from the source sound level and the characteristics of the room. Unfortunately, this equation is only suitable for predicting the sound level from a steady sound source in a fully diffuse field and research and listening experiences have shown that the results from the traditional equation can be very inaccurate and usually produce predictions which are larger than measured values (Hodgson, 1996). The sound level in the room can also be predicted based on the statistical theory of room acoustics, such as Barron's revised theory (Barron and Lee, 1988). However, it requires further validation to use statistical models for predicting speech levels at the schematic design stage of a classroom. In another work Hodgson *et al.* (1999) showed that the statistical models for predicting speech level and student-activity noise level only explains less than 70% of the variation in data, therefore, predictions when using the models may be poor. It is for these and other reasons, such as the lack of success in applying traditional techniques and the extended time required to conduct calculations, that a method of using ANNs to predict the Sound Propagation (*SP*) in university classrooms was investigated and developed (Nannariello *et al.*, 2001).

The *SP* is the variation of sound pressure level, normalized to the source power level, with distance from an omnidirectional source. Constructional and acoustical data for 34 randomly chosen unoccupied University of British Columbia (UBC) classrooms were used as inputs for the neural network analyses.

The network architecture used was a *MLP* with one hidden layer. The topology varied accordingly with the set-up function. The set-up function for the neural network analyses are (the number of hidden neurons in the middle layer is shown in square brackets at the end of set-up function):

$$SP_{\text{r-p}[125]} = f(S\text{-Type}, V, L_{MX}, D_{\text{MX}}, W_{\text{MX}}, H_{\text{MX}}, S_{\text{T}}, S_{\text{x}}, S_{\text{y}}, S_{\text{z}}, R_{\text{x}}, R_{\text{y}}, R_{\text{z}}, R_{\text{OS}},, RT_{\text{A}[125]})\ [4] \qquad (9)$$

$$SP_{\text{r-p}[250]} = f(S\text{-Type}, V, L_{MX}, D_{\text{MX}}, W_{\text{MX}}, H_{\text{MX}}, S_{\text{T}}, S_{\text{x}}, S_{\text{y}}, S_{\text{z}}, R_{\text{x}}, R_{\text{y}}, R_{\text{z}}, R_{\text{OS}},, RT_{\text{A}[250]})\ [5] \qquad (10)$$

$$SP_{\text{r-p}[500]} = f(S\text{-}Type, V, L_{MX}, D_{\text{MX}}, W_{\text{MX}}, H_{\text{MX}}, S_{\text{T}}, S_{\text{x}}, S_{\text{y}}, S_{\text{z}}, R_{\text{x}}, R_{\text{y}}, R_{\text{z}}, R_{\text{OS}},, RT_{\text{A}[500]})\ [4] \qquad (11)$$

$$SP_{\text{r-p}[1000]} = f(S\text{-}Type, V, L_{MX}, D_{\text{MX}}, W_{\text{MX}}, H_{\text{MX}}, S_{\text{T}}, S_{\text{x}}, S_{\text{y}}, S_{\text{z}}, R_{\text{x}}, R_{\text{y}}, R_{\text{z}}, R_{\text{OS}},, RT_{\text{A}[1000]})\ [4] \qquad (12)$$

$$SP_{\text{r-p}[2000]} = f(S\text{-}Type, V, L_{MX}, D_{\text{MX}}, W_{\text{MX}}, H_{\text{MX}}, S_{\text{T}}, S_{\text{x}}, S_{\text{y}}, S_{\text{z}}, R_{\text{x}}, R_{\text{y}}, R_{\text{z}}, R_{\text{OS}},, RT_{\text{A}[2000]})\ [5] \qquad (13)$$

where:

D_{MX} = Distance from front of podium to rearmost wall (m)
S_{X}, S_{Y}, S_{Z} = Source coordinates (m)
R_{X}, R_{Y}, R_{Z} = Listener coordinates (m)
R_{OS} = Source-listener distance (m)
$RT_{\text{A}[125]}$ to $RT_{\text{A}[2000]}$ = Octave frequency band reverberation times at listener positions (seconds)
$SP_{[125]}$ to $SP_{[2000]}$ = Sound propagation in octave frequency bands (Hz).

The results of this work showed that neural networks trained with variables that have a causal relationship to the acoustical quality of the UBC classrooms produce reliable and accurate predictions. *RMS* errors for *SP* in each of the frequency bands, were within the subjective difference limen for steady-state sound pressure levels, which is about 1 dB (i.e. $\Delta E/E = 0.26$, where E is the energy density). Furthermore, results showed that *SP* predictions for classrooms were in better agreement with measured values than those obtained using Barron's revised theory or the Hopkins-Stryker equation.

The good "fit" between measured and predicted *SP* values for the four classrooms tested was highlighted by the high correlation, $R^2 = 0.97$ (see Fig. 6) . The average (standard deviation) of the variations between measured and predicted *SP* in the octave bands 125 to 2000 Hz ranged between –0.72 (0.35) and +0.69 (1.05) dB. These results indicate that there is a good basis for using geometrical and acoustical variables to train ANNs to predict SP in university classrooms.

2.6. Predicting Architectural Speech Security

Offices and meeting rooms are often intended for confidential discussions where eavesdroppers and others outside the room should not be able to hear what is being said. The acoustical privacy of such rooms can be referred to as architectural speech security. Improved speech security is achieved by constructing the boundaries of the room to have higher sound transmission losses. When speech security is of concern, it is desirable to be able to predict the degree to which the "transmitted" speech is audible or intelligible at positions outside the room.

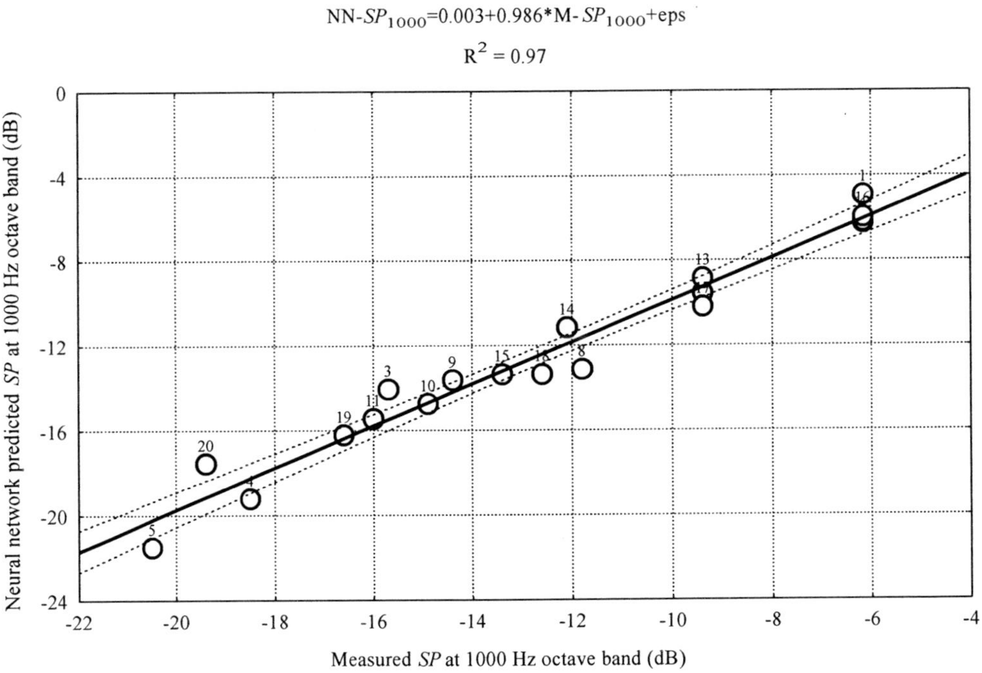

Fig. 6. Typical scatterplot showing a global linear regression analysis for neural network predictions of SP for 20 listener positions in the 1000 Hz octave band in 4 classrooms (Nannariello *et al.*, 2001). In the figure the dotted lines represent the 95% confidence limits and the solid line represents the best fit. In the fitted function equation eps equals the random error associated with measurements

The current method for predicting the speech intelligibility, privacy and security is to first develop a certain measure and then relate the measure to the subjective scores using a transfer function (Gover and Bradley, 2004; ANSI, 1969; 1997). Speech privacy and speech security have been related to signal-to-noise (S/N) type measures, where the signal is the speech from the adjacent space. The simplest measure is the difference of A-weighted speech and noise levels. More sophisticated measures such as the Articulation Index (AI) (ANSI, 1969) and its more recent replacement the Speech Intelligibility Index (SII) (ANSI, 1997) are known to be better related to speech intelligibility within rooms. However, research (Gover and Bradley, 2004) has shown that SII and AI cannot be used to describe conditions for high levels of speech security which would correspond to acoustical conditions below SII=0 or AI=0, where SII and AI are not defined. The difference in A-weighted levels is not limited in this way but it is much less accurately related to intelligibility scores. A more successful measure is the SII-weighted S/N ratio, which is a weighted sum of one-third-octave band S/N ratios using the same frequency weightings as the SII measure. Although the SII-weighted S/N ratio predicts the intelligibility score and threshold reasonably well, the S/N loudness ratio provides a more accurate estimation of the thresholds of cadence and audibility (Gover and Bradley, 2004).

An approach of applying ANNs to directly represent the functional relationship between the octave band (250 Hz – 8 kHz) S/N ratios and the speech intelligibility score and security was investigated (Xu *et al.*, 2005a). The objectives of the ANNs are to predict:

a) the speech intelligibility score, namely the percentage of words correctly identified by each individual,
b) the intelligibility threshold, namely the percentage of listeners able to correctly identify at least one word,
c) the cadence threshold, namely the percentage of listeners able to detect the cadence of the speech, and
d) the audibility threshold, namely the percentage of listeners able to hear the presence of the speech.

A total of 9500 individual intelligibility score test cases were used as the database for predicting the intelligibility score and 500 intelligibility threshold test cases were used as the database for predicting the intelligibility threshold. For predicting the cadence threshold and the audibility threshold 160 test cases were used. In all prediction situations, a three layer *MLP* network architecture (six input neurons, three hidden neurons and one output neuron), based on the supervised learning procedure was applied.

Table 2 summarises the statistical analysis results for ANN predictions. For comparison, in the previous work (Gover and Bradley, 2004), the SII-weighted S/N measure has an R^2 of 0.762 for predicting the intelligibility scores and an R^2 of 0.919 for predicting the intelligibility threshold. The S/N loudness ratio has an R^2 of 0.956 for predicting the cadence threshold and an R^2 of 0.899 for predicting the audibility threshold.

Table 2. Statistical analysis results for ANN predictions (Xu *et al*, 2005a)

	ANN model for predicting intelligibility scores				ANN model for predicting intelligibility threshold				ANN model for predicting cadence threshold				ANN model for predicting audibility threshold			
	Tr[a]	Ve[b]	Te[c]	En[d]	Tr[a]	Ve[b]	Te[c]	En[d]	Tr[a]	Ve[b]	Te[c]	En[d]	Tr[a]	Ve[b]	Te[c]	En[d]
rms error	0.205	0.208	0.208	0.206	0.116	0.098	0.104	0.111	0.079	0.056	0.072	0.076	0.096	0.089	0.033	0.091
Error S.D.[e]	0.205	0.208	0.208	0.206	0.117	0.099	0.105	0.111	0.079	0.049	0.067	0.077	0.096	0.088	0.034	0.091
S.D.[f] ratio	0.477	0.476	0.475	0.476	0.280	0.246	0.244	0.267	0.192	0.115	0.139	0.183	0.254	0.211	0.096	0.240
R^2	0.773	0.773	0.774	0.773	0.921	0.940	0.941	0.929	0.963	0.987	0.982	0.967	0.936	0.958	0.992	0.942

[a]Tr = training data set.
[b]Ve = verification data set.
[c]Te = test data set.
[d]En = entire data set.
[e]Error S.D. = standard deviation of the prediction error.
[f]SDR = standard deviation ratio.

The investigation (Xu *et al.*, 2005a) indicated that the ANN approach can use the S/N ratio information to directly predict the subjective speech intelligibility score and security thresholds. Compared with the previous work (Gover and Bradley, 2004) that used one-third-octave band S/N ratios, the ANN approach produced comparable, or better prediction results in terms of the audibility threshold, using only the “octave band” S/N ratios.

2.7. Predicting and Optimizing the Sound Insulation of Walls and Floor-Ceilings

The acoustic performance of various types of wall and floor-ceiling construction has become more and more important in the field of architectural acoustics as residential densities and occupant amenity expectations increase and sound systems become more powerful and produce bass sound more effectively. Building codes and occupant satisfaction require the architects to be fully informed as to the sound insulation performance of walls and floor-ceilings. Manufactures require that their products are tested in standard test facilities to ensure compliance with building codes. However, in the product development stage, there are vast ranges of building materials and construction assemblies available for manufacturers to select from. Therefore, numerical analysis, as a complement to the expensive and time consuming laboratory tests, is needed to predict the sound insulation performance of constructions that have not been measured.

2.7.1. Predicting the Sound Insulation of Walls

Using the results from acoustic laboratory tests of known steel and timber stud drywall constructions, an investigation of the application of ANNs for predicting the transmission loss (TL) at specific frequencies and the sound transmission class (STC) was carried out (Coomes and Fricke, 2001). Basic parameters (stud frame type and size, mass of wall construction, type and thickness of any cavity absorption, overall partition width, minimum sheet lining thickness and the difference in sheet lining from one side to the other as well as the inclusion of lining vibration isolation), were used as inputs for the neural network analysis (See Fig. 7, for instance). A three-layer *MLP* network architecture (seven input neurons, three hidden neurons and one output neuron) was applied. In the training of each series of data approximately 400 networks with different combinations of inputs and hidden neurons were looked and the best ten retained.

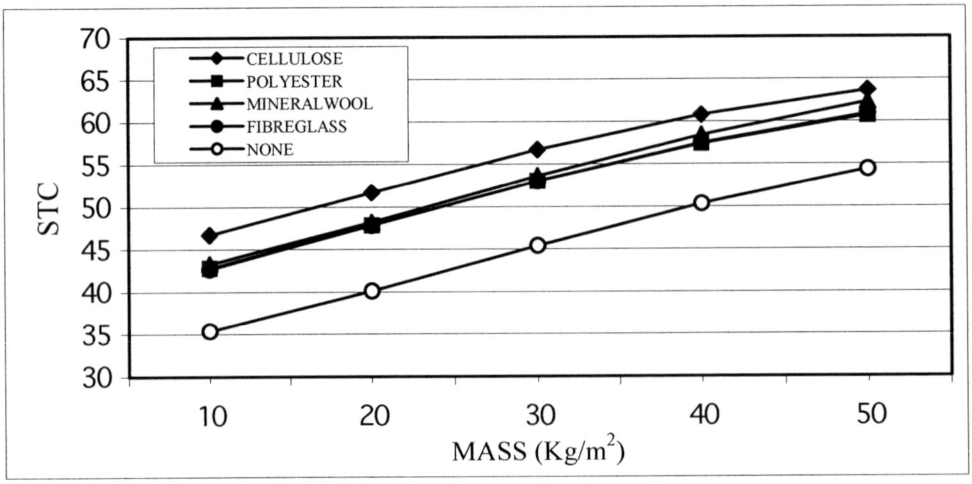

Fig. 7. Predicted STC values as a function of the mass per unit area of steel stud partitions with isolated fixings, an overall thickness of 170 mm and different cavity absorbers (Coomes and Fricke, 2001)

The results obtained were highly encouraging with neural network designs achieving predictions for STC values within a similar range to those determined by a number of acoustic laboratories for comparable wall constructions. For instance, using data from the National Research Council of Canada on all types of dry wall constructions, the prediction of STC using ANN achieved an *RMS* error of 2.01, which is less than the just noticeable difference in sound level that is easily detected by the majority of people. It was also suggested that neural networks analysis could potentially overcome the problem of obvious different test results in different "standard" laboratories (Coomes and Fricke, 2001).

In particular, the neural network analysis offers the possibility of obtaining wall sound transmission ratings in a "virtual" laboratory. That is, the transmission loss results for any construction in any laboratory could be used to predict the transmission loss for the same construction in any other laboratory, virtual or real, without the need to undertake extensive "round-robin" testing of a particular construction.

ANNs were also applied to establish the room and wall parameters that influence sound level difference between rooms at low frequencies (40 – 100 Hz) (Fora-Moncada and Gibbs, 2002). The inputs and outputs of the network were generated by a Finite Element Model (FEM) (Maluski, 1999). The accuracy of the FEM was such that the eigenfrequencies in each room were processed within an error of 10%. The data set was separated into two parts, the training set and the verification set. The training set was formed by 375 cases and the verification set by the remaining 15. Three neural network architectures, Backpropagation (BP), Radial Basis Function (RBF) and General Regression Neural Netwok (GRNN), were tested for each model and test results showed that in the frequency range of 40 – 100 Hz, RBF and GRNN performed better than BP. Moreover, GRNN exhibited rapid learning and prediction. The input variables that yielded the best prediction of sound level difference between rooms were; volume of the source room, volume of the receiving room, ratio of the source room length to the wavelength, ratio of receiver room length to wavelength, wavelength at the third octave band frequency and edge fixing conditions of the partition. The cross-validation average prediction error was 4 dB, with all cases included, or 3 dB, if equal room configurations were excluded. Fifteen room pairs (unequal room configurations and with larger volumes) had correlation coefficients greater than 0.8 and the average error was usually less than 3 dB and often less than 2 dB. The remaining cases had coefficients less than 0.8. Therefore, the best results were obtained when predicting for bigger and unequal room volumes.

2.7.2. Predicting and Optimising the Sound Insulation of Floor-Ceilings

Using results from laboratory tests on floor-ceilings, the STC and the impact insulation class (IIC) of typical floor-ceiling constructions were predicted using ANNs (Xu and Fricke, 2003). Two categories of data were considered:

***Category 1*:** All cavity floors, including solid wood joist type, wood I-joist type, wood truss-joist type and steel joist type floors, with resilient channels directly attached to the joists and with sound absorbing material in the cavity; and

***Category 2*:** Solid wood joist floors with resilient channels directly attached to the joists and with sound absorbing material in the cavity.

Statistical parameters of ANN predictions of Category 1 data were compared with predictions using conventional multivariate regression analysis (Warnock and Birta, 1998). The correlation coefficients, R^2, from multivariate regression analysis are close to those of the test dataset of the ANNs. However the advantage of neural network analysis is that nominal values such as joist types, timber or other materials, can be easily included. Also, a sound absorbing material type or code which contains both the information about the density and thickness of the material can be used as one of inputs of the NN model.

As there were limited cases available for an accurate and meaningful parametric study of floors with wood I-Joist, wood truss and steel joist, only wood solid joist floors (Category 2) were used for parametric study of the constructions of floor-ceilings (Xu and Fricke, 2003). For this category, the predictions using the multivariate regression equations (Warnock and Birta, 1998) had R^2s of 0.97 for STC and 0.79 for IIC. ANNs showed a similar R^2 for STC (0.98 for the verification dataset and 0.95 for the test dataset) and a better R^2 for IIC (0.94 for verification dataset and 0.83 for the test dataset). In addition, using the sensitivity analysis (Statsoft, 1999) of the neural network, it was found that the length of the joist had no effect on the acoustical performance of floor-ceilings, which confirmed the conclusion from the laboratory experiment (Warnock and Birta, 1998).

The parametric study indicated that the most important parameter that determines STC performance of floor-ceiling constructions is the mass of the floor and ceiling layers. However, other parameters also have an effect on the overall STC performance of floor-ceiling systems. Hence, there is a need to resolve – once there is a NN model that accurately predicts the STC of a certain floor-ceiling construction – how one can optimize the STC performance of floor-ceiling constructions within the range of design parameters utilised in buildings. For instance, an architect may wish to know the construction to achieve an STC 55 rated floor-ceiling that uses a relatively low depth type of solid timber joists, say 235mm deep and the least mass possible. Consequently the question is posed as to which type of floor-ceiling construction can meet these requirements.

In the subsequent work (Xu *et al.*, 2004a), the accuracy of the *MLP* neural network model (six input neurons, three hidden neurons and one output neuron) for predicting the STC of floor-ceiling constructions was improved and then it was demonstrated how an evolutionary algorithm [Genetic and Evolutionary Algorithm Toolbox for use with Matlab (GEATbx) (Pohlheim, 2004)] could be successfully applied to search the multidimensional space provided by the neural network model and find an appropriate floor-ceiling construction which would meet the design requirements of the architect. The improved neural network model was used as the objective function. Real-value variable representations and multi-population searching were used. Initial values were randomly generated by the GEATbx within the ranges of input parameters at the start of the searching. Discrete recombination and real value mutation genetic operators were employed. As an example, it was found that to achieve an STC 55 floor-ceiling construction using 235mm deep solid timber joists, the following construction would be required:

- a 20 kg/m^2 minimum surface weight of the floor and ceiling layers,
- 235 mm thick cellulose fibre sound absorbing material, with a density of 58 kg/m^3,
- a 610 mm joist spacing, and
- a 610 mm resilient metal channel spacing.

The above EA search result was verified on the basis of the conclusions from the laboratory tests (Warnock, 1999).

2.8. Optimising the Acoustic Performance of Multi-Layered Absorbent Linings

Multi-layered absorbent linings composed of porous materials and/or air gaps are widely used to absorb the sound energy and control the reverberance in a room. An extreme case, however, is to use them to achieve an anechoic condition, namely a free field condition, in a room. If an absorbent lining has its normal absorption coefficient not less than 99 percent at any frequency above a certain frequency (cut-off frequency), this absorbent lining can be defined as an anechoic lining with a characteristic cut-off frequency (Beranek and Sleeper, 1946). Hence, the design of an anechoic lining can be considered as a multi-objective optimization problem (MOP).

The design of multi-layered absorbent linings was first studied experimentally (Davern, 1980). The approach used was to design the lining system by using conventional measurements in an impedance tube. However, the work proved to be tedious and time consuming because of the numerous trial and error measurements involved. The subsequent work undertaken proved that the overall impedance of a multi-layered lining system could be calculated by the repeated application of a single layer impedance equation (Dunn and Davern, 1986). Since the acoustic absorption of the lining can be easily calculated based on the impedance, the acoustic absorption of any combination of materials can be rapidly tested using a computer. The inputs to the equation were simply the bulk acoustic properties (characteristic impedance and propagation constant) and the thickness of each layer of material (Dunn and Davern, 1986).

On the basis of the above work an approach of employing a multi-objective evolutionary algorithm as an optimiser to aid and speed up the design of the multi-layered absorbent linings as anechoic linings was investigated (Xu *et al.*, 2004b). The multi-objective evolutionary algorithm used was "Multi-Objective Genetic Algorithm (MOGA) with goal attainment" (Fonseca and Fleming, 1993; 1998) which was embedded in GEATbx. Fully-reticulated (i.e., open-cell) polyurethane foams were investigated and the bulk acoustic properties of each material were calculated using the curve-fitting equations (Dunn and Davern, 1986; Cummings and Beadle, 1993). It was also assumed that any required thickness of the material was available. Hence, variables were represented as continuous real values and real valued evolutionary operators were applied accordingly. Four cases were studied. The optimisation object of the first three cases was to find a multi-layered lining that can achieve 100 Hz cut-off frequency with a minimum overall thickness. The difference of the first three cases was just the range of materials for consideration during the optimisation operations. For the fourth case, the object was to find a multi-layered lining that can achieve a cut-off frequency of 100 Hz with a fixed overall thickness. The fixed thickness was in fact a constraint to the search, which was tackled by using a penalty function.

Using the same multi-objective evolutionary algorithm, a thorough investigation on the application of multi-layered polyurethane foams as the anechoic linings was carried out (Xu *et al.*, 2006). The investigation includes aspects such as the efficacy of a single layer of material, the minimum number of layers of multi-layered lining to achieve the minimum overall thickness for low (100 Hz), mid (250 Hz) and high (500 Hz) cut-off frequencies, the three-layered lining composite for low to mid cut-off frequencies and the effect of air gaps.

Applying the method developed in the above work (Xu *et al.*, 2004b; 2006) a multi-layered anechoic lining for the anechoic room at the MARCS Auditory Laboratories, the

University of Western Sydney was designed (Xu *et al.*, 2005b). The design target was to find a multi-layered anechoic lining that has a 250 Hz cut-off frequency with an overall thickness less than 300 mm. Sixteen of the most commonly used commercial fibrous building insulation materials available in Australia were investigated and the bulk acoustic properties of materials were measured in the impedance tube (ISO, 1996) because there was no equation that can be used to calculate them. In order to apply GEATbx, the measured bulk properties and fixed thicknesses of the materials were organized into matrices with the index of each column of the matrix, ranging from one to sixteen, representing each material. Another variable required is the integer multiples of the fixed thickness of each material. During the EA searching process, the program would pick up the material of interest by pointing at the corresponding column of the matrix and vary the overall thickness of the material by assigning a certain integer multiple. In other words, the EA works on the integer value representation. Accordingly, the evolutionary operators that can work with integer value representation were applied in the GEATbx. Fourteen design options (i.e., material combinations) were found by the evolutionary algorithm. These options were then evaluated in accordance with their costs and measured acoustic absorption performances. Finally, the completed anechoic room, where the optimized design was applied, was qualified and the results showed that a large percentage (75% - 85%) of the distance between the sound source and the room boundaries, on the traverses made, were anechoic.

3. Conclusion

This chapter has presented evidence that computational intelligence (CI) techniques, ANNs and EAs in particular, have an important role to play in architectural and building acoustics practice and research. The reason is not hard to discern; ANN provides more robust and reliable models than the available analytical and numerical models and EA is an efficient optimisation tool capable of solving multi-object optimisation problems.

It is worth looking more deeply at why ANNs and EA are useful in the analysis of architectural and building acoustics issues. There appear to be a number of reasons:

a) The multi-factorial and multi-objective nature of many architectural acoustics issues.
b) Architectural issues can often be qualitative and complex to analyse when subjective opinions and assessments are involved.
c) CI can handle non-linear relationships.
d) CI can be used where data is in categories rather than on a continuous numerical scale.
e) ANN models can be based on field or laboratory data which is often very noisy.
f) EAs can be used to develop optimal solutions to difficult design issues where there are several and possibly conflicting design requirements.

There are also a number of limitations:

a) Inability to reliably model situations for which there are no precedents.

b) The lack of case data on which to base ANN.
c) Limited opportunities to rationalise the solutions provided by ANNs and EAs.

The future for CI techniques in architectural and building acoustics seems assured but, like many professions, there is a reluctance to embrace new methods even though they can be shown to be superior to existing ones. The exception to the rule appears to be in cases where there are clear financial incentives and that there are no "insurance" and other peripheral issues involved. Overcoming these will be difficult and it will probably only happen if CI techniques are used conjunctively with conventional methods until professional practitioners gain confidence in them. This however may take some time.

REFERENCES

Ando Y., 1998. *Architectural Acoustics – Blending Sound Sources, Sound Fields and Listeners.* New York: AIP Press/Springer Verlag.

ANSI, 1969. S3.5-1969 *American National Standard Methods for the Calculation of the Articulation Index.* New York, NY: American National Standards Institute.

ANSI, 1997. S3.5-1997 *American National Standard Methods for Calculation of the Speech Intelligibility Index.* New York, NY: American National Standards Institute.

Barron M., Lee L-J., 1988. Energy relation in concert auditoria. I. *Journal of the Acoustical Society of America,* Vol. 84, No. 2, pp. 618-628.

Beranek L.L., Sleeper H.P., 1946. The design and construction of anechoic sound chambers. *Journal of the Acoustical Society of America*, Vol. 18, No. 1, pp. 140-150.

Beranek, L.L., 1962. *Music Acoustics and Architecture*. New York: Wiley.

Beranek L.L., 1996. *Concert and Opera Halls: How They Sound.* Woodbury, NY: Acoustical Society of America.

Beranek L.L., 2004. *Concert Halls and Opera Houses*. New York: Springer-Verlag.

Bradley J.S., 1994. Comparison of concert hall measurements of spatial quality. *Journal of the Acoustical Society of America*, Vol. 96, No. 6, pp. 3525-3535.

Cavanaugh W.J., Farrell W.R., Hirtle P.W., Watters B.G., 1962. Speech privacy in buildings. *Journal of the Acoustical Society of America*, Vol. 34, pp. 475–492.

Choi Y.J., 2004. *Towards Better Predictions of Concert Hall Acoustic Quality.* PhD thesis. The University of Sydney: School of Architecture, Design Science and Planning.

Coomes J., Fricke F.R., 2001. The prediction of wall sound insulation using neural networks. *Journal of Building Acoustics*, Vol. 8, No. 2, pp. 103 – 119.

Craggs A., 1997. Acoustic modelling: Finite element method. In M. J. Crocker, editor. *Encyclopaedia of Acoustics*. New York: John Wiley and Sons, pp. 165-172.

Cremer L., 1975. Different distributions of the audience. In Mackenzie R., editor. *Auditorium Acoustics.* London: *Applied Science*, pp 145-159.

Cummings A., Beadle S.P., 1993. Acoustic properties of reticulated plastic foams. *Journal of Sound and Vibration*, Vol. 175, No. 1, pp. 115-133.

Davern W.A., 1980. Flat-walled graded density anechoic lining. *Proceedings of 10th International Congress on Acoustics*. Sydney, 1980.

Davis D., Davis C., 1991. Audio measurements. In: Ballou G., editor. *Handbook for Sound Engineers: The New Audio Cyclopaedia,* 2ne ed., Indiana: SAMS, pp. 1365 – 1412.

Dunn I.P., Davern W.A., 1986. Calculation of acoustic impedance of multi-layer absorbers. *Applied Acoustics*, Vol. 19, No. 5, pp. 321 – 334.

Fonseca C.M., Fleming P.J., 1993. Genetic algorithms for multiobjective optimization: formulation, discussion and generalization. In: Forrest S., editor. *Proceedings of the Fifth International Conference in Genetic Algorithms*. San Mateo, CA: Morgan Kaufmann, pp. 416 – 423.

Fonseca C.M., Fleming P.J., 1998. Multiobjective optimization and multiple constraint handling with evolutionary algorithms – part I: a unified formulation. *IEEE Transactions on Systems, Man and Cybernetics – Part A: System and Humans*, Vol. 28, No. 1, pp. 26 – 37.

Fora-Moncada A., Gibbs M., 2002. Prediction of sound insulation at low frequencies using artificial neural networks. *Journal of Building Acoustics*, Vol. 9, No. 1, pp. 49 – 71.

Fricke, F.R., Haan, C.H., 1995. Predicting the acoustics of concert halls using an artificial neural network. *Acoustics Australia*, Vol. 23, No. 3, pp. 87-95.

Fricke F.R., Han Y.G., 1999. A neural network analysis of concert hall acoustics. *Acustica*, Vol. 85, No. 1, pp. 113-120.

Fricke F.R., 2000a. Visual assessments of the surface diffusion properties of concert halls. *Applied Acoustics*, Vol. 60, No. 2, pp. 253-261.

Fricke F.R., 2000b. Concert hall acoustics: an alternative approach. *Journal of Building Acoustics*, Vol. 7, No. 4, pp. 233-246.

Hunt F.V., 1992. *The Origins of Acoustics*. New York: Acoustical Society of America.

Gade A.C., 1991. *Prediction of room acoustical parameters*. A paper presented on the 121st Meeting of the Acoustical Society of America, Baltimore.

Gade A.C., 1996. Room acoustic properties of concert halls: quantifying the influence of size, shape and absorption area. Honolulu: 3rd ASA/ASJ meeting, paper 5aAA1, pp. 1-6.

Garnier C., 1880. *Le nouveau Opera de Paris*. Paris: Ducher et Cie.

Gover B.N., Bradley J.S., 2004. Measures for assessing architectural speech security (privacy) of closed offices and meeting rooms. *Journal of Acoustical Society of America*, Vol. 116, No. 6, pp. 3480-3490.

Hodgson M., 1996. When is diffuse-field theory applicable. *Applied Acoustics*, Vol. 49, No. 3, pp. 197-207.

Hodgson M., Rempel R., Kennedy S., 1999. Measurement and prediction of typical speech and background noise levels in university classrooms during lectures. *Journal of the Acoustical Society of America*, Vol. 105, No. 1, pp. 226-233.

ISO, 1996. *ISO 10534-1:1996 "Acoustics – Determination of sound absorption coefficient and impedance in impedance tubes Part 1: Method using standing wave ratio"*. Geneva: International Organization for Standardization.

Lynge C., 1998. *ODEON Room Acoustical Program Version 3.1 Manual*. Technical University of Denmark: The Department of Acoustic Technology.

Maluski S., 1999. *Low Frequencies Sound Insulation in Dwellings*. PhD thesis. Sheffield, UK: Sheffield Hallam University.

Manning J.E., 1997. Statistical modelling of vibrating systems. In M. J. Crocker, editor. *Encyclopaedia of Acoustics*. New York: John Wiley and Sons, pp. 925-935.

Morgan M.H., 1960. *The ten books on architecture*. Translation of "De Architectura" from Vitruvius. New York: Dover Publications.

Nannariello J., Fricke F R., 1999. The prediction of reverberation time using neural network analysis. *Applied Acoustics*, Vol. 58, No. 3, pp. 305-325.

Nannariello J., Hodgson M., Fricke F.R., 2001. Neural network predictions of speech levels in university classrooms. *Applied Acoustics*, Vol. 62, No. 7, pp. 749-767.

Nannariello J., Fricke F.R., 2001a. The use of neural network analysis to predict the acoustic performance of large rooms: Part I Predictions of the parameter G utilizing numerical simulations. *Applied Acoustics*, Vol. 62, No. 8, pp. 917-950.

Nannariello J., Fricke F.R., 2001b. The use of neural network analysis to predict the acoustic performance of large rooms: Part II Predictions of the acoustical attributes of concert halls utilizing measured data. *Applied Acoustics*, Vol. 62, No. 8, pp. 951-977.

Nannariello J., Fricke F.R., 2001c. A neural network analysis of the effect of geometric variables on concert hall G values. *Applied Acoustics*, Vol. 62, No. 12, pp. 1397-1410.

Nannariello J., Fricke F.R., 2002a. The prediction of reverberation time using optimal neural networks. *Journal of Building Acoustics*, Vol. 9, No. 1, pp. 5-28.

Nannariello J., Fricke F.R., 2002b. A neural computation method of predicting the interaural cross-correlation coefficient. $IACC_{E3}$ for auditoria. *Applied Acoustics*, Vol. 63, No. 6, pp. 627-641.

Osman M.R., Fricke F.R., 1996. Predicting the acoustical quality of small music rooms. *Proceedings of 1996 Australian Acoustical Society Conference*, pp. 13-15.

Pierce A.D., 1997. Mathematical theory of wave propagation. In M. J. Crocker, editor, *Encyclopaedia of Acoustics*. New York: John Wiley and Sons, pp. 21-37.

Pohlheim H., 2004. *Genetic and Evolutionary Algorithm Toolbox for Use with MATLAB Version 3.5*, Available from: http://www.geatbx.com/.

Reid R.B., 1835. On the construction of public buildings in reference to the communication of sound. Transactions of the British Association.

Sabine W.C., 1922. *Collected Papers on Acoustics*. Harvard University Press, New York: Reprinted by Dover, 1964.

Sato S., Otori K., Takizawa A., Sakai H. Ando Y., Kawamura H., 2002. Applying genetic algorithms to the optimum design of a concert hall. *Journal of Sound and Vibration*, Vol. 258, No. 3, pp. 517 – 526.

StatSoft, 1999. *STATISTICA Neural Networks Version 4.0A Manual*. Tulsa, OK: StatSoft Inc.

Vorlander M., 1995. International round robin on room acoustical computer simulation. *Proceedings of 15th International Congress on Acoustics*, pp. 689-692.

Warnock A.C.C., Birta J.A., 1998. *Summary Report for Consortium on Fire Resistance and Sound Insulation of Floors: Sound Transmission Class and Impact Insulation Class Results, Internal Report IRC-IR-766*. Canada: Institute for Research in Construction, National Research Council Canada.

Warnock A.C.C., 1999. Controlling the transmission of airborne sound through floors. *Construction Technology Update*, No. 25, pp. 1 – 4.

Xu J., Fricke F.R., 2003. The prediction of airborne and impact sound transmission through floor-ceiling constructions using neural network analysis. *Proceedings of the 10th International Congress on Sound and Vibration*, pp. 4533-4540.

Xu J., Nannariello J., Fricke F.R., 2004a. Predicting and optimising the airborne sound transmission of floor-ceiling constructions using computational intelligence. *Applied Acoustics,* Vol. 65, No. 7, pp. 693-704.

Xu J., Nannariello J., Fricke F.R., 2004b. Optimising flat-walled multi-layered anechoic linings using evolutionary algorithms. *Applied Acoustics*, Vol. 65, No. 11, pp. 1009-1026.

Xu J., Bradley J.S., Gover B.N., 2005a. An artificial neural network approach for predicting architectural speech security. *Journal of Acoustical Society of America*, Vol. 117, No. 4, pp. 1709-1712.

Xu J., Bucholz J., Fricke F.R., 2005b. Flat-walled multi-layered anechoic linings: Optimization and application. *Journal of the Acoustical Society of America*, Vol. 118, No. 5, pp. 3104-3109.

Xu J., Bucholz J., Fricke F.R., 2006. Multi-layered polyurethane foams for flat-walled anechoic linings. *Applied Acoustics*, Vol. 67, No. 5, pp. 476-485.

Zwickker C., Kosten C.W., 1949. *Sound Absorbing Materials*. Amsterdam: Elsevier.

In: Artificial Intelligence in Energy and Renewable Energy Systems ISBN 1-60021-261-1
Editor: Soteris Kalogirou, pp. 339-362

Chapter 10

ACTIVE LEARNING STRATEGIES FOR THE NEURAL ESTIMATION OF ENGINE MAPS

Marta Neve and Giuseppe De Nicolao
Department of Computer Engineering and Systems Science
University of Pavia, Italy

ABSTRACT

The chapter deals with the sequential optimal design of experiments for the reconstruction of engine maps from static data by means of artificial neural networks. Since the collection of static data is expensive and time consuming, the problem of selecting the most informative set of experiments arises. In the neural network literature, this problem goes under the name of active learning and is solved by either optimizing the entropy or the variance of the estimation error. If Multi-Layer Perceptron (MLP) networks are used, the selection of the next experimental point is based on the linearization of the model around the current parameter values. Conversely, if linear-in-parameter models are employed, such as Radial Basis Function (RBF) networks with fixed basis functions, the entire sequence of optimal experiments can be computed in advance without knowing the results of the previous experiments. In the chapter, it is shown that for Bayesian RBF networks the optimal sequence of experiments can be designed in advance even when the regularization parameter has to be tuned. In fact, within an optimal sequential design scheme, it is seen that the so-called regularization parameter tends toward a constant so that it can be fixed from a certain step onward. The proposed active learning strategies are applied to the estimation of the map that gives air-flow as a function of manifold pressure and crankshaft speed.

Keywords: IC engine models, mean value models, engine maps, neural networks, bayesian estimation, regularization, active learning, optimal design of experiments.

NOMENCLATURE

D_k	Input-output training set
$f(x)$	Unknown function
$J_\gamma(\vartheta)$	Cost function
n	Crankshaft speed (rpm)
p	Manifold pressure (mbar)
$q(\gamma)$	Equivalent degrees of freedom
v	Measurement error
x_i	Input measurement
$\bar{x}^j$	j-th centre of the basis functions
X	Sampling region
X_u	Region of interest
y_i	Output measurement

Greek

γ	Regularization parameter
ϑ	Weight (parameter) vector
λ^2	Parameter variance
σ^2	Noise variance (Kg^2/h^2)
$\varphi^j(x)$	j-th basis function

ABBREVIATIONS

AIC	Akaike Information Criterion
ANN	Artificial Neural Network
DOE	Design Of Experiments
FPE	Final Prediction Error
GCV	Generalized Cross Validation
IC	Internal Combustion
ML	Maximum Likelihood
MLP	Multi Layer Perceptron
MSE	Mean Square Error
MVM	Mean Value Model
OCV	Ordinary Cross Validation
RBF	Radial Basis Function
RMSE	Root Mean Square Error

SALP1	Sequential Active Learning Problem 1
SALP2	Sequential Active Learning Problem 2
SSR	Sum of Squared Residuals

1. INTRODUCTION

One of the problems arising in the identification of engine models is the estimation of nonlinear engine maps. Such maps take the form of scalar multivariable functions and are typically reconstructed on the basis of static measurements collected during bench tests. This procedure is used, for instance, to identify the volumetric efficiency, the air flow as a function of manifold pressure and crankshaft speed, the mean indicated torque and the friction torque. Various approaches have been proposed in the literature (De Nicolao *et al.*, 1996) ranging from polynomial and spline models (Hastie and Tibshirani, 1990) to MLP (Multi Layer Perceptron) and RBF (Radial Basis Function) neural networks (Haykin, 1963; Lippmann, 1981; Poggio and Girosi, 1990). All these methods have to deal with the trade-off between the need for high precision and the availability of limited experimental data. In particular, the estimated map should ensure both good interpolation and extrapolation properties as there are regions where static data cannot be collected but must be covered by the model because they can be reached during dynamic transients. Moreover, static data are expensive in terms of test bench occupation and operator working time. Therefore, it is important to optimize the choice of the static experiments in order to maximize the information they convey.

The aforementioned considerations motivate the interest for methodologies borrowed from the theory of optimal Design of Experiment (DOE), (Fedorov, 1972) that in the neural network literature goes also under the name of Active Learning (MacKay, 1992; Cohn, 1994; Sollich, 1994; Sung and Niyogi, 1995; Fukumizu, 2000; Sugiyama and Ogawa, 2000). Examples of applications to engine modelling are also available in (Muller *et al.*, 2000; Arsie *et al.*, 2001). Arsie *et al.* (2001) used an active learning methodology to derive a procedure that selects the new experimental point maximizing the state of knowledge about the unknown map given the previous experimental data. The gain of knowledge is measured either by means of the entropy (MacKay, 1992) or in terms of variance of the generalization error (Cohn, 1994). Arsie and co-workers concluded that the techniques based on active learning guarantee good accuracy and generalization properties with a significant improvement with respect to heuristic selection of training data. An important consequence is that, with a proper design of the experiments, a reduction of the training set is possible without deteriorating the approximation quality. Conversely, with a given number of experiments the optimal design will yield better models.

The active learning scheme is an iterative one. For a given training set, the algorithm looks for the next experimental point (e.g. the pair speed-pressure) which is most informative. In order to compute the gain of knowledge due to the new datum (e.g. the air flow measurement), it is necessary to perform a linearization of the model based on the current training set. After having chosen the experiment that maximizes the gain of knowledge, a new training set is formed and the model is updated. It is clear that the procedure is intrinsically a step-by-step one. It should be noted that it is not possible to select in advance a sequence of experiments because the result of each experiment is needed in order to choose the subsequent one. The main reason for such a limitation is the need of linearizing the current model. If the model is linear with respect to its parameters, its linearization does not depend on the actual values of the parameters. Hence, an entire sequence of experiments can be optimally selected in advance without waiting for the outcomes of the experiments. For this reason, in the present chapter the map estimation problem is approached using RBF neural networks, because they can be handled as linear-in-parameter models. Moreover, this class of models has been recently employed with success in order to estimate engine maps (De Nicolao *et al.*, 1996; Neve *et al.*, 2003; 2004). Of course, the results presented in this chapter extend to any other type of linear-in-parameter model, e.g. polynomials and splines.

However, the development of an active learning scheme cannot ignore the issue of the choice of complexity of the model. In its simplest form this corresponds to deciding the order of the model (e.g. the order of the polynomial approximation) and is carried out by resorting to cross-validation or statistical criteria such as Generalized Cross Validation (GCV), Ordinary Cross Validation (OCV), Final Prediction Error (FPE), Akaike Information Criterion (AIC) and C_p (Akaike, 1974; Golub *et al.*, 1979; Ljung, 1987; Söderström and Stoica, 1989; Hastie *et al.*, 2001).

Alternatively, the flexibility of the model can be tuned within a Bayesian paradigm, as done in the present chapter. In the Bayesian approach the model parameters are regarded as random variables and are assigned a prior distribution. Usually, such a distribution depends on a positive scalar, the so-called regularization parameter, which has to be tuned in order to optimize the model flexibility (Tykhonov, 1963). Various tuning methods are possible, e.g. cross-validation, GCV, OCV, C_p and Maximum Likelihood (ML) (Hall and Titterington, 1987; Hastie *et al.*, 2001). A common feature of these techniques is that they call for an iterative algorithm, because the cost functions depend on the residuals so that the model has to be estimated for several values of the regularization parameter until the criterion is optimized. The need of re-computing the estimates several times entails a significant computational burden. Another important consequence is that the advance design of a whole sequence of experiments (that would be viable if the regularization parameter was fixed) is no more possible because each datum is needed for tuning the regularization parameter before the next experimental point is decided.

A possible solution to these drawbacks relies on the behaviour of the estimates of the regularization parameter. In fact, it is expected that these estimates converge to an asymptotic value as the number of training data tends to infinity. In view of this, a two-phase procedure is proposed. In the first phase, a certain number of data are collected one at a time readjusting the regularization parameter at each iteration until the value of the regularization parameter stabilizes. In the second phase, the regularization parameter is kept equal to the average of the last values calculated in the first phase and the subsequent experimental points can be decided all in one go before performing the corresponding experiments.

The two-phase procedure has been tested on a simulated benchmark problem. The data are obtained adding pseudo-random noise to the mathematical model of the air flow map identified in a previous study. In this context, the pairs (speed, pressure) that are most informative have to be chosen. The advantage of using simulated data is the possibility of assessing the quality of approximation by computing the difference between the true engine map and the estimated one. The results demonstrate that the two-phase procedure provides an effective approach to active learning of engine maps. In particular, after a certain number of data have been collected, the readjustment of the regularization parameter does not bring significant advantage in terms of model accuracy.

In this chapter the use of neural networks for engine map estimation is addressed. However, Artificial Neural Networks (ANNs) have a much wider range of applications in the field of Internal Combustion (IC) engines, see (Kalogirou, 2003) for a review. Their ability to learn and represent complex nonlinear functions of several variables renders them a valuable tool in a number of modelling problems. In particular, they can be exploited for engine diagnosis and control as well as for virtual sensor design. For an overview of combined control and diagnosis strategies that may involve ANNs the interested reader may refer to Kim *et al.* (1998) which focuses on the air fuel ratio control system. Among other possible control applications it is worth mentioning ignition timing control which aims at optimising either brake torque (Wang *et al.*, 2001) or combustion characteristics (Jorach *et al.*, 1997; Yang *et al.*, 2002). An interesting application in the field of virtual sensor design is the reconstruction of cylinder pressure from cylinder head vibration signals using RBF networks (Du *et al.*, 2001).

The chapter is organized as follows; Section 2 gives a brief overview of the main concepts at the basis of Bayesian learning techniques for the reconstruction of multi-dimensional functions. Section 3 presents an active learning procedure that minimizes the generalization error over a region of interest. Section 4 describes the RBF neural network approach used to estimate the air flow map as a function of pressure and speed and enters into details regarding the tuning of the regularization parameter by means of the so-called C_p statistics. Section 5 presents the benchmark used to assess the effectiveness of the proposed two-phase active learning approach and shows the results obtained when applying such a methodology to two different problems (the first considering a region of interest coinciding with the sampling one and the second with distinct regions). Finally, some conclusions are drawn highlighting the advantages and disadvantages of the proposed methodology and some future developments are presented.

2. Fundamentals of Bayesian Learning

In this section a concise review of the Bayesian approach for the reconstruction of a multi-dimensional function $f(x)$ is provided. The unknown function $f(x): R^n \mapsto R^1$, has to be estimated on the basis of k discrete and noisy observations:

$$y_i = f(x_i) + v_i, \quad i = 1, \dots, k \tag{1}$$

where the measurements errors v_i are independent and identically distributed with zero mean and variance $Var[v_i] = \sigma^2$. In the following, the set $D_k = \{(x_i \quad y_i), i = 1, \dots, k\}$ denotes the input-output training set. Concerning the unknown function, the following assumption is made:

$$f(x) = \varphi^T(x)\vartheta, \quad \vartheta \in R^m \tag{2}$$

$$\varphi(x) = \left[\varphi^1(x) \quad \varphi^2(x) \quad \ldots \quad \varphi^m(x)\right]^T \tag{3}$$

where $\varphi^j(x)$ are suitable basis functions, e.g. polynomials or trigonometric functions. If $\varphi^j(x) = \varphi^j\left(\left\|x - \overline{x}^j\right\|\right)$, that is φ^j is radially symmetric with respect to the centre $\overline{x}^j$, Eq. (2) is a Radial Basis Function (RBF) Neural Network, and ϑ is the weight vector. In the Bayesian approach, the parameter vector ϑ is modelled as a random vector, e.g. normally distributed with zero mean and variance matrix $Var[\vartheta] = \lambda^2 I$. Moreover, ϑ is assumed to be independent of the measurement errors v_i.

Bayesian estimation relies on the computation of the posterior distribution of ϑ given the training set D_k. It is well known (Hastie *et al.*, 2001), that such a posterior is still Gaussian with:

$$\vartheta_k := E[\vartheta | D_k] = A_k^{-1} \sum_{i=1}^{k} \varphi_i y_i \tag{4}$$

$$Var[\vartheta | D_k] = A_k^{-1} \tag{5}$$

$$A_k := \sum_{i=1}^{k} \varphi_i \varphi_i^T + \frac{\sigma^2}{\lambda^2} I \tag{6}$$

where the shorthand notation $\varphi_i := \varphi(x_i)$ is used. The ratio σ^2/λ^2 is also known as regularization parameter. The posterior expectation ϑ_k is the so-called Bayes estimate. Consequently, the estimate of the function $f(x)$ will be:

$$f_k(x) = \varphi^T(x)\vartheta_k \tag{7}$$

Remarkably, once the basis functions φ^j have been fixed, the Bayes estimate ϑ_k is obtained as the solution of a system of linear equations, see Eq. (4). This is a major advantage with respect to other type of networks, such as the multilayer perceptron ones, whose training calls for nonlinear optimization which may suffer from local minima and convergence problems.

3. Active Learning for Bayesian Models

In this section, the active learning problem for Bayesian estimation is discussed. In order to assess the quality of the estimated function f_k, one may refer to its generalization capability. If the problem of predicting the value of the function at a specific point x_u is considered, the generalization performance can be measured by the Mean Square Error:

$$MSE_k(x_u) := E\left[(f_k(x_u) - f(x_u))^2\right]. \tag{8}$$

It was proven by Sung and Niyogi (1995), that:

$$MSE_k(x_u) = \varphi_u{}^T A_k{}^{-1} \varphi_u \tag{9}$$

where $\varphi_u := \varphi(x_u)$. Given a training set D_k, the problem of choosing the next sampling point x_{k+1} so as to optimise the generalization performance of ϑ_{k+1} goes under the name of sequential active learning problem.

Definition 1. Sequential active learning problem 1 (SALP1): Let X be the region in the input space where additional measurements can be taken. Select the sampling point x_{k+1} so as to minimize $MSE_{k+1}(x_u)$.

Let $\Delta(x_{k+1}) := MSE_k(x_u) - MSE_{k+1}(x_u)$ be the variation of generalization performance due to the introduction of the additional point x_{k+1}. Recalling Eq. (9), it would seem that the evaluation of $\Delta(x_{k+1})$ requires the inversion of A_k, which may be burdensome when k is large. The next lemma shows that such an inversion can be avoided by a proper update formula.

Lemma 1, (Cohn, 1994):

$$\Delta(x_{k+1}) = \frac{\varphi_{k+1}{}^T A_k{}^{-1} \varphi_u \varphi_u{}^T A_k{}^{-1} \varphi_{k+1}}{\sigma^2 + \varphi_{k+1}{}^T A_k{}^{-1} \varphi_{k+1}} \tag{10}$$

$$A_k{}^{-1} = A_{k-1}{}^{-1} - \frac{A_{k-1}{}^{-1} \varphi_k \varphi_k{}^T A_{k-1}{}^{-1}}{\sigma^2 + \varphi_k{}^T A_{k-1}{}^{-1} \varphi_k} \tag{11}$$

Then, the solution of the SALP1 is obtained as:

$$x_{k+1}{}^{opt} = \arg\max_{x_{k+1}} \Delta(x_{k+1}) \tag{12}$$

In real world problems, one is interested with the generalization capability over a region of interest rather than on a single point x_u. In the following, the region of interest will be denoted as X_u. Notably, X_u does not necessarily coincide with the admissible sampling region X. In fact, one may be concerned with the generalization performance in regions where samples cannot be taken because of physical or economic constraints. The following index measures the expected generalization performance on a point x_u randomly chosen in X_u according to a probability density function p_u, known as environmental probability:

$$\overline{MSE}_k = \int_{X_u} MSE_k(x_u)p_u(x_u)dx_u \tag{13}$$

It can be seen that:

$$\overline{MSE}_k = tr\left(A_k^{-1}C\right) \tag{14}$$

$$C = \int_{X_u} \varphi_u\varphi_u^T p_u(x_u)dx_u \tag{15}$$

Definition 2. Sequential active learning problem 2 (SALP2): Let X be the region in the input space where additional measurements can be taken. Select the sampling point x_{k+1} so as to minimize $\overline{MSE}_{k+1}(x_u)$.

In analogy with Lemma 1, the following lemma can be exploited in order to solve SALP2 efficiently.

Lemma 2, (Cohn, 1994):

$$\overline{\Delta}(x_{k+1}) = \overline{MSE}_k - \overline{MSE}_{k+1} = \frac{\varphi_{k+1}^T A_k^{-1}CA_k^{-1}\varphi_{k+1}}{\sigma^2 + \varphi_{k+1}^T A_k^{-1}\varphi_{k+1}} \tag{16}$$

It is worth noting that, in view of Eqs. (9), (14) and (15), the solutions of both the SALP1 and SALP2 do not make use of the output measurements y_j, $j = 1,\ldots,k$. Consider the problem of selecting l points x_{k+i}, $i = 1,\ldots,l$, so as to optimize the generalization over X_u. By an iterative application of the solution of the SALP2, all the l points can be decided in advance before any output measurement y_{k+i} is taken. This would not be possible for nonlinear estimators because the selection of each new point is based on a linearization which depends on past output measurements.

As a matter of fact, such an "advance design" is possible only if all the parameters of the Bayesian model are known. Although the measurement error variance σ^2 may be known in many cases, the variance λ^2 is hardly known and has to be tuned on the basis of D_k. The possible tuning criteria include OCV, GCV, Maximum Likelihood and the C_p statistics (Hall and Titterington, 1987; Hastie *et al.*, 2001). It should be noted that whenever λ^2 is updated when passing from D_k to D_{k+1}, the numerically efficient formulas for computing A_k^{-1} without matrix inversion are no more applicable. In order to recover the possibility of an advance design, a two-step procedure is proposed. At first, the sequential active learning strategy is applied updating λ^2 at each iteration until the estimated λ^2 converges to a steady state. After that, λ^2 is fixed and the advance design can be carried out to obtain the desired number of further sampling points.

4. BAYESIAN NETWORKS FOR ENGINE MAP ESTIMATION

Mean Value Models (MVMs) are widespread tools for engine dynamics description and simulation in automotive industries and research centres. A major advantage of such kind of models is their simple structure together with their good predictive performance in terms of Mean Square Error between measured and simulated intake manifold pressure and engine speed (Hendricks and Sorenson, 1990; Heywood, 1992; Siviero *et al.*, 1995; Holliday and Davis, 1998; Guzzella and Onder, 2004). Hence, these models are used for control purposes (e.g. for an efficient fluid film compensation in idle control, or for an adequate fuel dosage mostly in rapid transient conditions), for diagnosis purposes (e.g. detecting a hole in the manifold wall or a drift in the pressure sensor) or for safety applications (engine shut off can be forced if a sudden fault in the Drive-by-Wire throttle valve is detected).

The construction of an MVM requires the identification of the nonlinear map that gives the air flow past the throttle plate as a function of manifold pressure (p) and crankshaft speed (n). Usually, the reconstruction of this three-dimensional surface relies on static data, namely measures of the air flow corresponding to several pairs (n_i, p_i) that should cover the region of interest. This is done using an engine bench (see Fig. 1), keeping the engine operating point at fixed values, thus making steady-state measurements possible.

Fig. 1. Engine bench used for collecting steady-state measurements

When reconstructing an engine map, one must face the difficulty of obtaining a large number of measures covering all regions of interest. In fact, every single static measure is expensive in terms of both test bench occupation and operator working time (Muller *et al.*, 2000). Moreover, there are points in the *(n,p)* plane that can be reached during engine transients, but where static data cannot be collected as they are not stability points for the engine. For these reasons, active learning techniques may play an important role in the engine map estimation procedure since they produce a sensible reduction in the amount of required data by properly selecting the most informative ones (Röpke, 2003).

Another issue to be considered is given by the choice of a suitable engine map estimator. A good technique should guarantee:

- flexibility of the surface in order to reproduce the profile of the true map;
- smoothness of the estimated map that should not suffer from overfitting;
- reliable extrapolation in regions where static data are not available;
- proper handling of boundary conditions in order to exploit the prior knowledge on the shape of the map.

All these requirements can be met by resorting to a Bayesian Basis Function neural network with a large number of neurons whose weights are adjusted during the training phase through a regularization method (Neve *et al.*, 2003; 2004). Bias reduction is accomplished by searching the model within a rich functional class whereas overfitting is avoided by using regularization rather than least-square regression for parameter estimation (Poggio and Girosi, 1990; MacKay, 1992; Sjoberg and Ljung, 1992). Finally, an explicit handling of boundary conditions together with the smoothness properties assured by the use of regularization lead to reliable extrapolations.

Assuming that the unknown map function is indicated by $f(x)$, with $x = [n \quad p]^T$, and recalling Eq. (2), the basis functions are defined as:

$$\varphi^j(x) = \int_0^n \int_0^p h\left(\left\|\xi - \bar{x}^j\right\|\right) \cdot d\xi_1 d\xi_2 \quad j = 1, \ldots, m \tag{17}$$

$$h(r) = e^{-\frac{r^2}{2c^2}} \tag{18}$$

where c is a design parameter that controls the width of the gaussian function *h(r)*. Note that $f(x)$ is modelled as a linear combination of gaussian functions, $h(r)$, integrated over the (n, p) space, that is $f(x)$ is modelled as the integral of an RBF neural network with gaussian kernel. Once the value of c and the centres $\bar{x}^j$ have been properly selected, the overall model is linear in the unknown parameter vector ϑ. The m centres of the network are located on a regular grid spanning the operating region in the (n, p) plane. As for the parameter c it can be either tuned following statistical criteria or assigned according to some rule-of-thumb (e.g. equal to half the maximum distance between the training inputs x_i).

A diagram of the ANN is shown in Fig. 2. The input is the two-dimensional vector $x = [n \quad p]^T$ and the scalar output $f(x)$ is the reconstructed map. The m neurons are in a one-to-one correspondence with the basis (activation) functions $\varphi^j(x)$. The weights of the network, i.e. the elements of the vector ϑ, are to be learnt using the training set D_k.

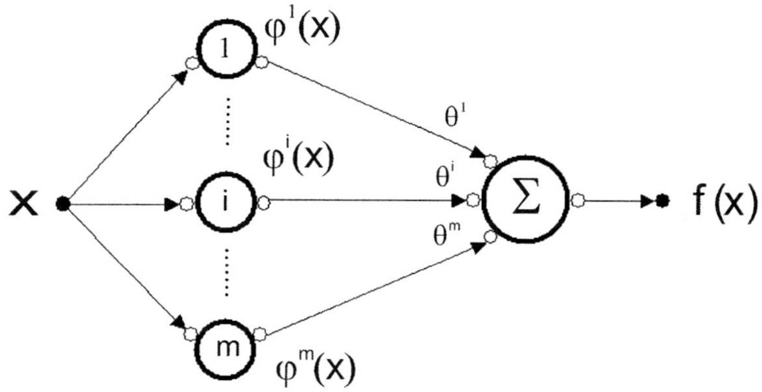

Fig. 2. Diagram of the ANN used for engine map reconstruction. The input is the two-dimensional vector $x = [n \quad p]^T$ and the scalar output $f(x)$ is the reconstructed map

According to the regularization approach, the estimated parameter vector is computed as:

$$\hat{\vartheta} = \arg\min_{\vartheta} J_\gamma(\vartheta) \tag{19}$$

$$J_\gamma(\vartheta) = SSR(\vartheta) + \gamma\|\vartheta\|^2 \tag{20}$$

$$SSR(\vartheta) := \sum_{i=1}^{k} \left(y_i - \varphi^T(x_i)\vartheta\right)^2 \tag{21}$$

where SSR is the usual sum of squared residuals and γ is the so-called regularization parameter. If γ is "small" the surface $f(x)$ will tend to be wiggly and, in the limit as $\gamma \to 0$, data interpolation will be obtained. Conversely, larger values of γ will yield smoother surfaces and in the limit, a null constant function. Once a suitable value of γ is chosen, the calculation reduces to the solution of a system of linear equations:

$$\hat{\vartheta} = \left(\Phi^T\Phi + \gamma I\right)^{-1}\Phi^T y \tag{22}$$

where $\Phi = [\varphi(x_1) \quad \ldots \quad \varphi(x_k)]^T$ is the regression matrix and N represents the number of available data. Note that Eq. (22) is strictly equivalent to Eq. (4) if $\gamma = \sigma^2/\lambda^2$, so that the Bayesian estimate of the unknown parameters coincides with the one obtained through the regularization approach. As already mentioned, in most cases either σ^2 or λ^2 (or both) are unknown so that γ must be properly tuned. A possible solution is looking for the value of γ minimizing the C_p statistics (Hastie *et al.*, 2001):

$$C_p = SSR + 2q(\gamma)\sigma^2 \tag{23}$$

$$q(\gamma) = tr\left[\Phi\left(\Phi^T\Phi + \gamma I\right)^{-1}\Phi^T\right]. \tag{24}$$

The real number $q(\gamma)$ represents the so-called equivalent degrees of freedom, a number that ranges from 0 (the degrees of freedom of the null constant function) to $\min(m, N)$ as γ varies from ∞ to zero (Wahba, 1990).

5. Active Learning of the Air-Flow Map

The effectiveness of the proposed active learning strategy has been tested on a simulated benchmark. The data are obtained by adding pseudo-random noise to a mathematical model of the air flow map identified from experimental data as discussed in Neve *et al.* (2004). The advantage of resorting to simulated data is twofold. First of all, this allows the comparison of the reconstructed map with the "true" one to assess the quality of the estimation. In this simulated framework the comparison is more objective than performing validation using experimental data divided into a training and a validation set. Moreover, it is possible to emulate arbitrary sampling schedules with respect to the number and location of training data. The variance of the pseudo-random noise is $\sigma^2 = 10^4$ (Kg^2/h^2). It has been assumed that static measurements can be collected in the rectangular region $400 \le p \le 1000$ (mbar); $1000 \le n \le 7500$ (rpm). For computational reasons, the sampling region has been discretized using a 10x10 grid (the potential sampling points are the knots of the grid). The centres of the neural network are located on a regular 10x10 grid spanning the region $0 \le p \le 1000$; $0 \le n \le 7500$. The training has been performed on normalized inputs $-1 \le x \le 1$; $-1 \le y \le 1$, where x and y correspond to p and n, respectively. In such a normalized setting the width parameter c was set equal to 0.3, which guarantees a reasonable smoothness of the approximating surface.

Two problems have been considered that differ with respect to the choice of the region of interest. In both cases the environmental probability p_u is assumed constant over a region of interest. In the first problem, such a region coincides with the sampling one. The active learning procedure has been applied using the four corners of the sampling region as starting points. Initially, active learning has been used for selecting 50 sampling points performing the tuning of λ^2 via the C_p statistics at each iteration. The values of the estimated λ^2 are plotted against the iteration number in Fig. 3. The asymptotic convergence towards a steady state value is apparent. This observation justifies the use of the following two-phase procedure. In the first phase, the sampling points are evaluated iteratively performing the tuning of λ^2 at each step, until the estimated values become reasonably stable.

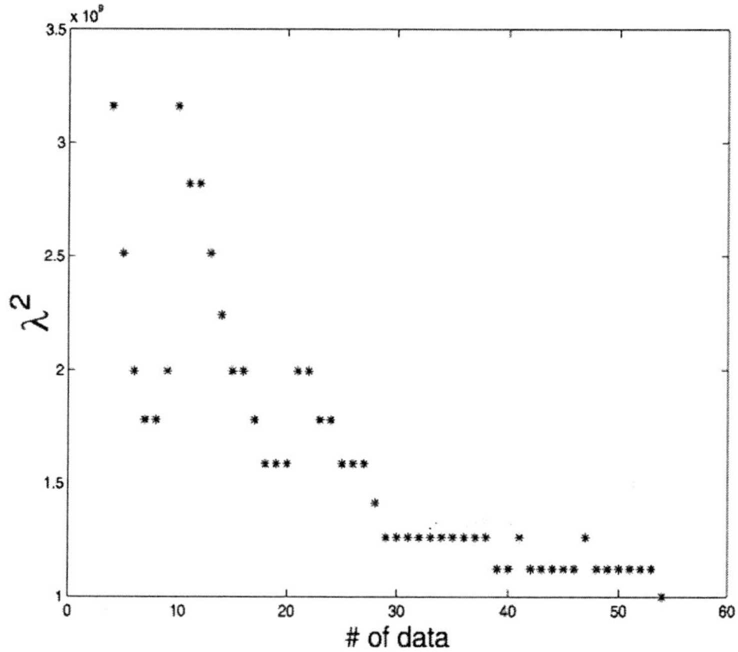

Fig. 3. Problem 1. Estimated λ^2 as a function of the number of data (asterisks)

In practice convergence is assessed by checking whether $\left|\ln\left(\lambda^2{}_{k+i}\right)-\ln\left(\lambda^2{}_k\right)\right|<0.35$, $i=1,\ldots,10$, where $\lambda^2{}_k$ denotes the estimate of λ^2 computed at the k-th step. If the condition is satisfied, in the second phase λ^2 is kept fixed and equal to the average of the last ten values. It is worth mentioning that, as it is well-known in regularization problems, the estimate is relatively robust with respect to the regularization parameter provided that changes within the same order of magnitude are considered.

As already mentioned at the end of Section 3, using a constant λ^2 is computationally advantageous and allows an advance design of the sampling points before collecting the data. In the considered example, the two-phase procedure has been applied collecting 20 data in phase I and other 30 data in phase II. The sampling points selected in the two phases are plotted in the left panel of Fig. 4. It is worth noting that there are locations that are chosen more than once. In the right panel of the same figure the first ten points are plotted together with their order of choice. In Fig. 5, the performance of the two-phase procedure is compared with that of the original approach that tunes λ^2 at each step. It can be seen that applying the two-phase procedure entails only a slight degradation in terms of RMSE between the estimated and true engine map evaluated on the region of interest. The precision of the obtained estimates is reasonably good if one considers that the RMSE of the map estimated using all the 100 points in the grid of the sampling region is 22.7 $[(Kg/h)^{1/2}]$ ($\lambda^2 = 1.8 \cdot 10^9$).

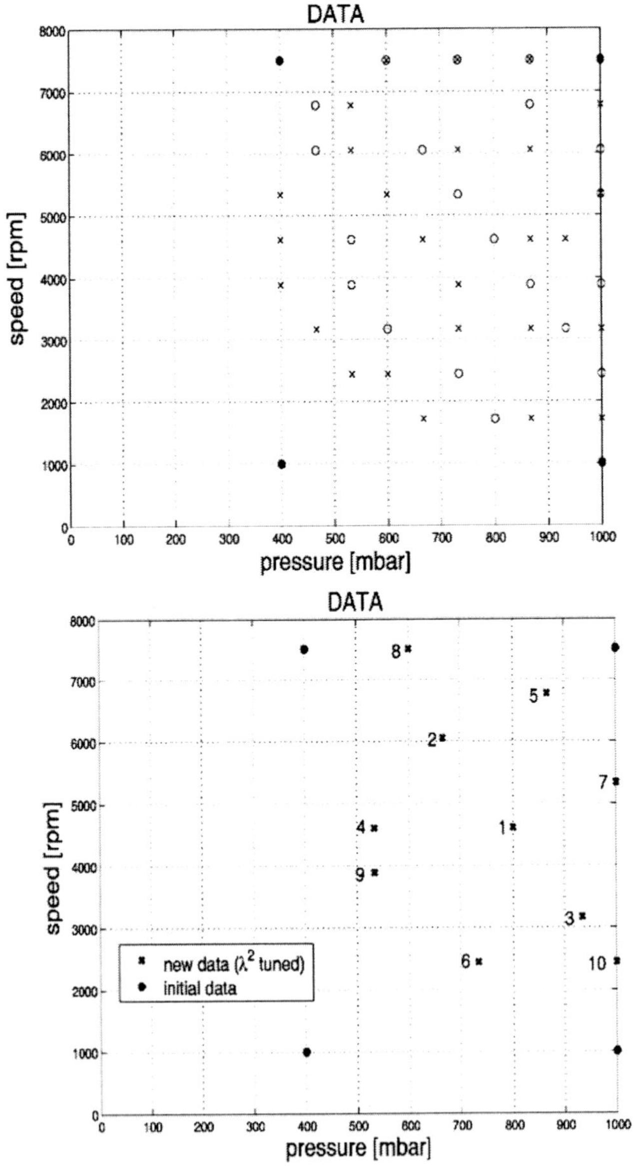

Fig. 4. Problem 1. Top panel: data selected in phase I by the active learning procedure while tuning λ^2 (circles) and data selected in phase II with λ^2 fixed to its steady state value (crosses). The initial data are also reported (full circles). Bottom panel: the first ten points numbered according to the order of choice

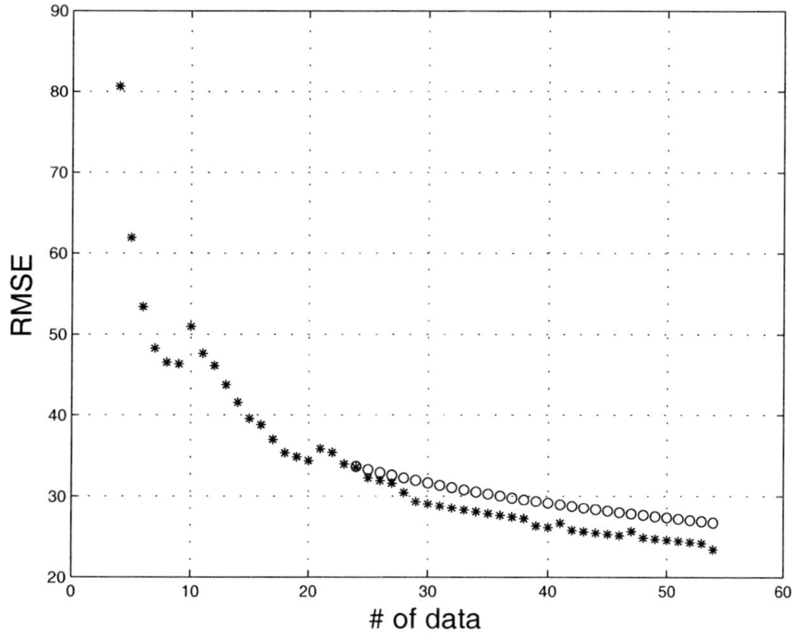

Fig. 5. Problem 1. Root Mean Square Error (RMSE) between the estimated map and the true one. Comparison between the RMSE obtained tuning λ^2 for each of the 50 data collected (asterisks) and the RMSE obtained in phase II fixing λ^2 to its steady state value after the first 20 data (circles)

Finally, in Fig. 6 the true air flow map and the three maps estimated using the active learning procedure (with 10, 20 and 20+30 points) are plotted. The results obtained using the 50 data selected by the two-phase scheme are completely satisfactory and comparable to the results obtained in Neve *et al.* (2003) and Neve *et al.* (2004) using 92 experimental data.

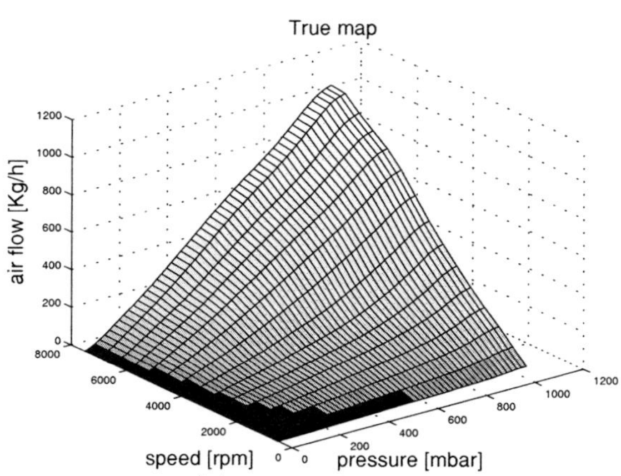

Fig. 6. (Continued on the next page)

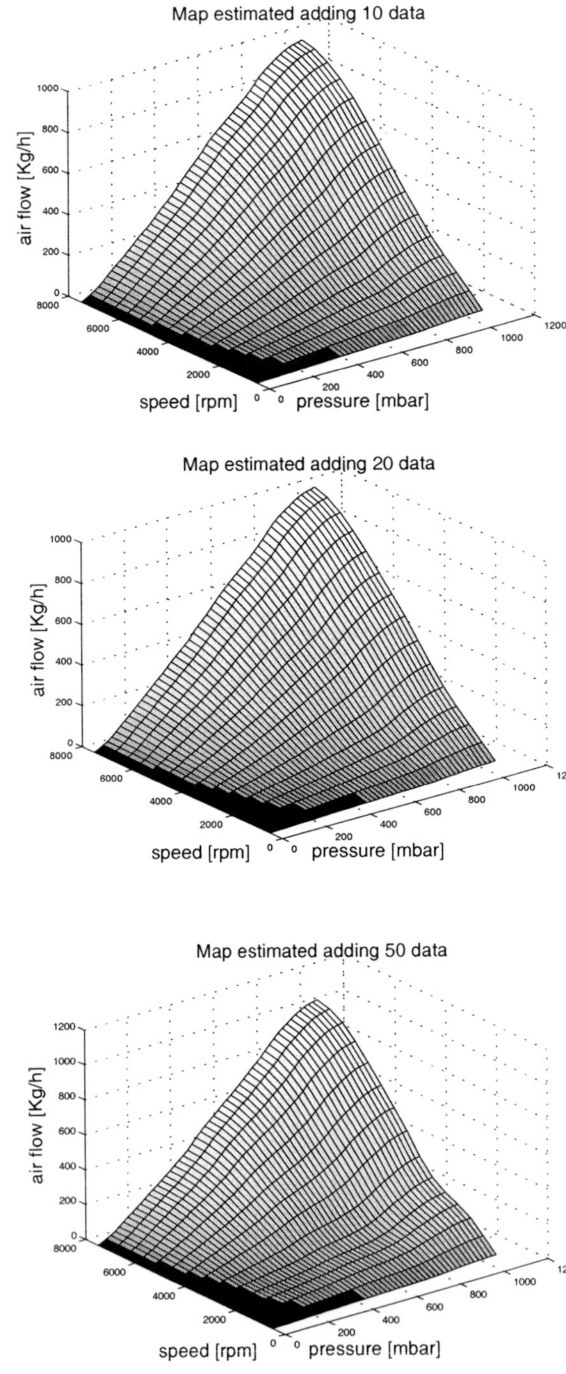

Fig. 7. Problem 1. The true map and the air flow maps estimated on the basis of 10, 20 and 20+30 data collected by means of the proposed active learning procedure

In the second problem the region of interest is $200 \leq p \leq 700$ (mbar); $1000 \leq p \leq 6000$ (rpm) and does not coincide with the sampling one, see Fig. 7. The choice of this region could be motivated by the need of developing a model for the design of the idle-speed controller. The region of interest is not included in the sampling region because some points are outside the engine stability region where static measurements can be collected. On the other hand, the map model must cover also these points because they can be reached during dynamic transients. Again, the starting points are the corners of the sampling region.

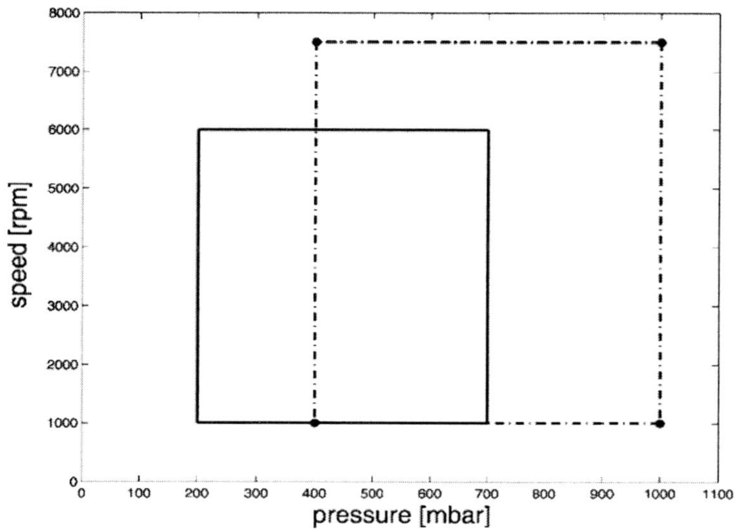

Fig. 8. Problem 2. Sampling region (dash-dot) and region of interest (solid) with respect to the air flow map estimation problem. The initial data are also reported (full circles)

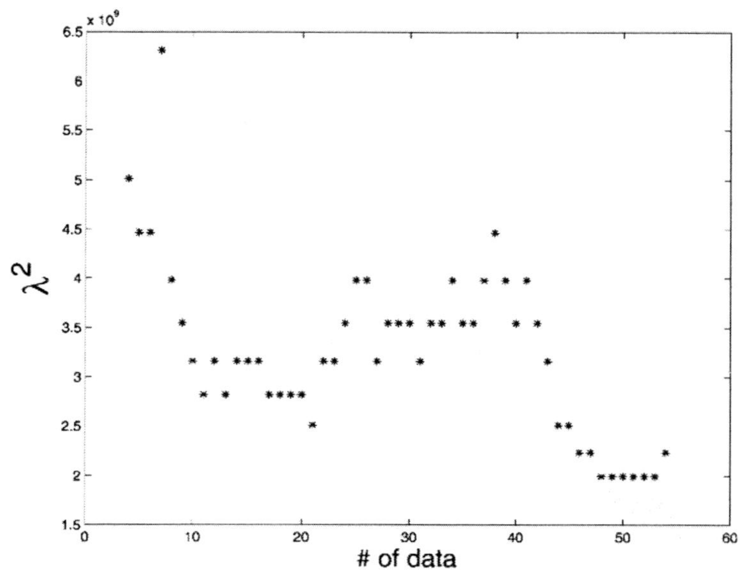

Fig 9. Problem 2. Estimated λ^2 as a function of the number of data (asterisks)

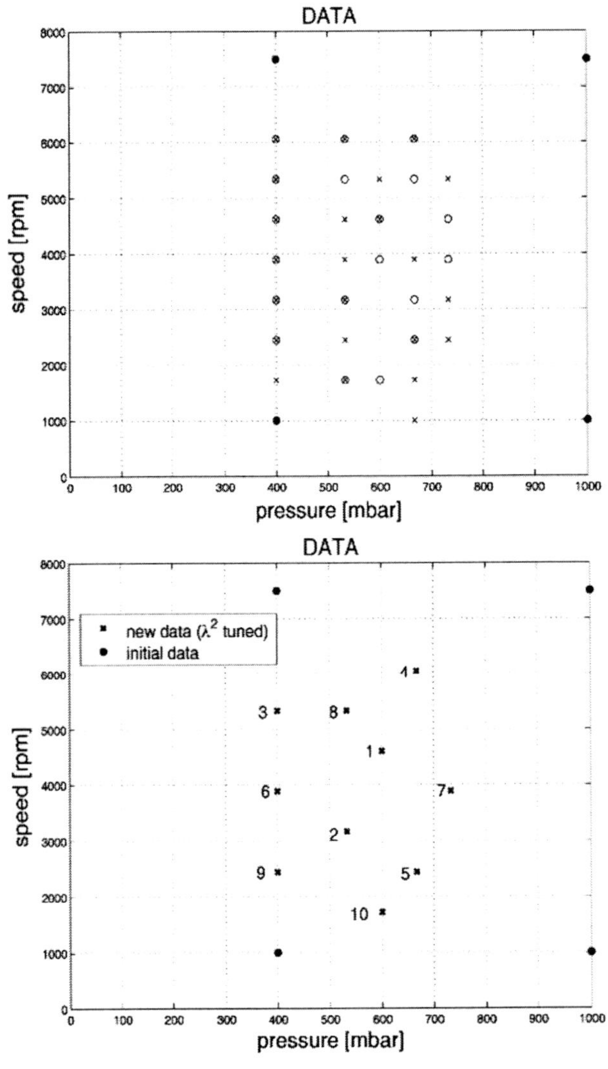

Fig. 10. Problem 2. Top panel: data selected in phase I by the active learning procedure while tuning λ^2 (circles) and data selected in phase II with λ^2 fixed to its steady state value (crosses). The initial data are also reported (full circles). Bottom panel: the first ten points numbered according to the order of choice

When 50 sampling points are selected tuning λ^2 at each iteration, the values of λ^2 reported in Fig. 8 are obtained. The same criterion of convergence has been applied as that used in Problem 1. In the problem at hand, 20 data have been collected in phase I and other 30 in phase II. The 50 sampling points and the details of the first ten are reported in Fig. 9. A comparison between Fig. 4 and Fig. 9 highlights the different spatial distribution of the selected points that, in the latter case, are more concentrated towards the lower left corner which corresponds to the intersection of the sampling region with the region of interest.

In Fig. 10, the performance of the two-phase procedure is compared with that of the original approach that tunes λ^2 at each step. Again, the use of a fixed λ^2 appears to be more than satisfactory. For the sake of comparison, the RMSE of the map estimated using all the 100 points in the grid of the sampling region is $34.89\,(Kg/h)^{1/2}$] ($\lambda^2 = 2.99 \cdot 10^9$). In Fig. 11, the true air flow map is compared with the maps estimated using 10, 20 and 20+30 points. The global quality of approximation is worse than that of Problem 1, see Fig. 6, but this is not surprising because the region of interest is smaller and shifted.

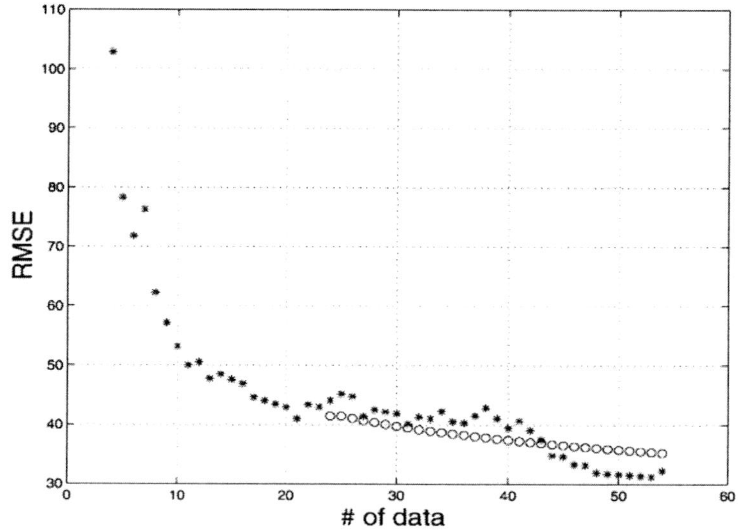

Fig. 11. Problem 2. Root Mean Square Error (RMSE) between the estimated map and the true one. Comparison between the RMSE obtained tuning λ^2 for each of the 50 data collected (asterisks) and the RMSE obtained in phase II fixing λ^2 to its steady state value after the first 20 data (circles)

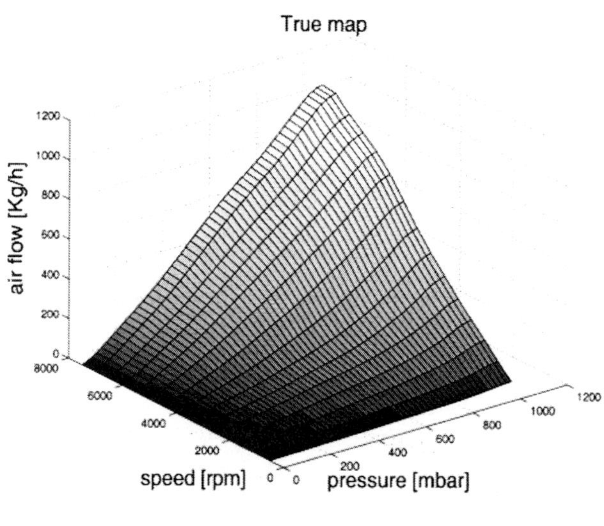

Fig. 12. (Continued on next page)

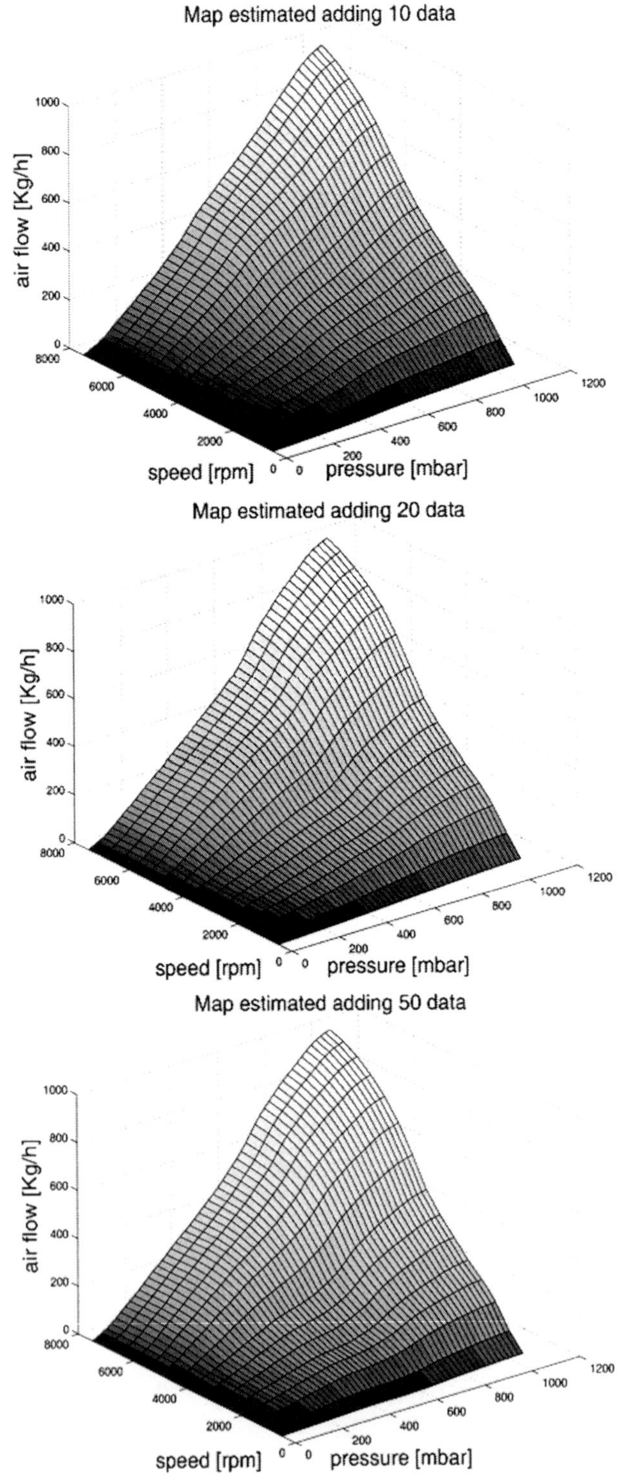

Fig. 13. Problem 2. The true map and the air flow maps estimated on the basis of 10, 20 and 20+30 data collected by means of the proposed active learning procedure

6. Conclusion

In this chapter, the problem of selecting the most informative static experiments for engine map reconstruction has been considered. Typically, such a problem is solved by means of iterative active learning schemes which choose the next experimental point on the basis of the available training set. This means that, in general, an advance design is not possible. However, if the engine map is described by a linear-in-parameter model the choice of the next point does not depend on the previous experimental outcome. Hence the interest for linear models such as Bayesian RBF neural networks. Nevertheless, recalling that Bayesian methods use the available training set to tune the so-called regularization parameter, it would seem that an advance design is still not feasible. To obviate this drawback, in the present chapter a two-phase procedure has been proposed. In the first phase, the regularization parameter is tuned at each step until its estimate has reached, at least approximately, a steady-state value. Then, in the second phase, the regularization parameter is kept fixed and the advance design of the whole sequence of future experiments is made possible. As a consequence in the second phase the training of the ANN is faster because after a new datum is added the weights can be computed directly according to Eq. (4).

The proposed procedure has been tested on a simulated benchmark. The results are satisfactory and demonstrate the usefulness of active learning methodologies in order to improve accuracy or reduce the number of experiments. From the practical point of view, one could consider the following strategy. First an initial set of experiments is decided on heuristic grounds. Assuming that the regularization parameter is reliably estimated on the basis of this training set, an optimal advance design is performed in order to choose all the experiments that will be conducted in a second and final experimental session.

For what concerns the future developments three main directions can be envisaged. First of all, other types of Bayesian estimators may be considered. These may include classical models such as polynomial ones, but with Bayesian priors, or other methods such as Gaussian Processes (MacKay, 1998; Williams, 1999; Smola and Schölkopf, 2003) which are the basis of Regularization Networks models (Poggio and Girosi, 1990).

A second development is the extension of the two-phase approach to parametric models estimated via least squares methods. In this case, the advance design of the experiments is hindered by the need of choosing the model complexity (e.g. the polynomial order). Again, a possible solution would be to estimate the model order on a reduced set of experiments (possibly heuristically determined) and then design the future experiments all in one go.

Finally, it may be worth exploring the use of data collected during dynamic onboard tests for engine map estimation. For instance, dynamic data could be processed to obtain a preliminary model on the basis of which few but informative static experiments can be chosen.

Acknowledgements

This research has been partially supported by the Italian Ministry of University and Research through the FIRB Project "Learning theory and application" and the PRIN Project

"New methods and algorithms for identication and adaptive control of technological systems".

References

Akaike H., 1974. A new look at the statistical model identification. *IEEE Transactions on Automatic Control*, Vol. AC-19, No. 6, pp. 716-723.

Arsie I., Marotta F., Pianese C., Rizzo G., 2001. Information based selection of neural networks training data for S.I. engine mapping. *SAE Technical Paper*, No. 01P23-196.

Cohn D.A., 1994. Neural network exploration using optimal experiment design. *A.I. Memo, Artificial Intelligence Laboratory, MIT*, Vol. 1491.

De Nicolao G., Scattolini R., Siviero C., 1996. Modelling the volumetric efficiency of I.C. engines: parametric, nonparametric and neural techniques. *Control Engineering Practice*, Vol. 4, No. 10, pp. 1405-1415.

Du H., Zhang L., Shi X., 2001. Reconstructing cylinder pressure from vibration signals based on radial basis function networks. *Proceedings of the Institution of Mechanical Engineers, Part D: Journal of Automobile Engineering*, Vol. 215, No. D6, pp. 761-767.

Fedorov V. V., 1972. *Theory of Optimal Experiments*. Academic Press.

Fukumizu K., 2000. Statistical active learning in multiplayer perceptron. *IEEE Transactions on Neural Networks*, Vol. 11, pp. 17-26.

Ghahranami Z., Jordan M. I., Cohn D. A., 1995. Active learning with statistical models. *Advances in Neural Information Processing Systems, MIT*, Vol. 7.

Golub G. H., Health M., Wahba G., 1979. Generalized cross-validation as a method for choosing a good ridge parameter. *Technometrics*, Vol. 21, pp. 215-224.

Guzzella L., Onder C.H., 2004. *Introduction to Modeling and Control of Internal Combustion Engine Systems*. Germany: Springer-Verlag.

Hall P., Titterington D. M., 1987. Common structure of techniques for choosing smoothing parameters in regression problems. *Journal of the Royal Statistical Society*, Vol. 49, pp. 184-198.

Hastie T. J., Tibshirani R. J., 1990. *Generalized Additive Models*. London: Chapman and Hall.

Hastie T. J., Tibshirani R. J., Freedman J., 2001. *The Elements of Statistical Learning*. NY: Springer-Verlag.

Haykin S., 1963. *Neural Networks a Comprehensive Foundation*. NY: MacMillan College Publishing Company.

Hendricks E., Sorenson S. C., 1990. Mean value modelling of spark ignition engines. *SAE Technical Paper*, No. 900616.

Heywood J.B., 1992. *Internal Combustion Engine Fundamentals*. NY: McGraw-Hill.

Holliday T., Davis T. P., 1998. Engine mapping experiments: A two-stage regression approach. *Technometrics*, Vol. 40, pp. 120-126.

Kalogirou S. A., 2003. Artificial intelligence for the modelling and control of combustion processes: a review. *Progress in Energy and Combustion Science*, Vol. 29, No. 6, pp. 515-566.

Kim Y. W., Rizzoni G., Utkin V., 1998. Automotive engine diagnosis and control via nonlinear estimation. *IEEE Control Systems Magazine*, Vol. 18, No. 5, pp. 84-89.

Jorach R., Enderle C., Decker R., 1997. Development of a low NO_x truck hydrogen engine with high specific power output. *International Journal of Hydrogen Energy*, Vol. 22, No. 4, pp. 423-427.

Lippmann R., 1987. An introduction to computing with neural nets. *IEEE Acoustics, Speech, and Signal Processing Magazine*, Vol. 4, pp. 4-22.

Ljung L., 1987. *System Identification: Theory for the User*. Englewood Cliffs, New Jersey: Prentice-Hall.

Mackay D. J., 1992. Bayesian interpolation. *Neural Computation*, Vol. 4, pp. 415-447.

MacKay D. J., 1998. Introduction to Gaussian processes. In: Bishop C. M., editor. *Neural Networks and Machine Learning*. Springer-Verlag.

Müller N., Hafner M., Isermann R., 2000. A neuro-fuzzy based method for the design of combustion engine dynamometer experiments. *SAE Technical Paper*, No. 2000-00P-198.

Neve M., De Nicolao G., Prodi G., Siviero C., 2003. Nonparametric neural identification of the air flow map. *Proceedings of the 6th International Conference on Engine for Automobile on Cd ROM.*

Neve M., De Nicolao G., Prodi G., Siviero C., 2004. Nonparametric estimation of engine maps using regularized basis-function networks. *Proceedings of the First IFAC Symposium on Advances in Automotive Control on Cd ROM.*

Poggio T., Girosi F., 1990. Networks for approximation and learning. *Proceedings IEEE*, Vol. 78, pp. 1481-1497.

Röpke K., 2003. *Design of Experiments (DOE) in der Motorenentwicklung*. Expert Verlag.

Siviero C., Scattolini R., Gelmetti A., Poggio L., Serra G., 1995. Analysis and validation of mean value models for SI-engines. *Proceedings of the First Workshop on Advances in Automotive Control*, pp. 1-6.

Sjöberg J., Ljung L., 1992. Overtraining, regularization and searching for minimum in neural networks. *Proceedings of the 4th IFAC Symposium on Adaptive Systems in Control and Signal Processing*, pp. 669-674.

Smola A. J., Schölkopf, B., 2003. Bayesian kernel methods. In: Mendelson S., Smola A. J., editors. *Machine Learning, Proceedings of the Summer School, Australian National University*. Springer, pp. 65-117.

Söderström T., Stoica P., 1989. *System Identification*. London: Prentice Hall International.

Sollich P., 1994. Query construction, entropy, and generalization in neural-network models. *Physical Review E*, Vol. 49, No. 5, pp. 4637-4651.

Sugiyama M., Ogawa H., 2000. Incremental active learning for optimal generalization. *Neural Computation*, Vol. 12, No. 12, pp. 2909-2940.

Sung K. K., Niyogi P., 1995. Active learning the weights of a RBF network. *Proceedings of IEEE Workshop on Neural Networks for Signal Processing*, Vol. 49.

Tykhonov A. N., 1963. Solutions of incorrectly formulated problems and the regularization method. *Soviet Mathematics Doklady*, Vol. 4, No. 4, pp. 1624.

Wahba G., 1990. *Spline Models for Observational Data*. Philadelphia: Society for Industrial and Applied Mathematics.

Wang W., Chirwa E. C., Zhou E., Holmes K., Nwagboso C., 2001. Fuzzy neural ignition timing control for a natural gas fuelled spark ignition engine. *Proceedings of the*

Institution of Mechanical Engineers, Part D: Journal of Automobile Engineering, Vol. 215, No. D12, pp. 1311-1323.

Williams C. K. I., 1999. Prediction with Gaussian processes: from linear regression to linear prediction and beyond. In: Jordan M., editor. *Learning and Inference in Graphical Models*. Boston: MIT Press, pp. 599-621.

Yang Z. Z., Wei J. Q., Fang Z. Y., Li J. D., 2002. An investigation of optimum control of ignition timing and injection system in an in-cylinder injection type hydrogen fuel engine. *International Journal of Hydrogen Energy*, Vol. 27, No. 2, pp. 213-217.

In: Artificial Intelligence in Energy and Renewable Energy Systems ISBN 1-60021-261-1
Editor: Soteris Kalogirou, pp. 363-394

Chapter 11

THE APPLICATION OF ARTIFICIAL INTELLIGENCE TECHNIQUES IN COMBUSTION SYSTEMS

John Ward, Chee-Keong Tan, Alex Z. S. Chong and Steven Wilcox
Intelligent Systems Research Centre
School of Technology, University of Glamorgan
United Kingdom

ABSTRACT

All combustion systems are inherently non-linear to some extent in their behaviour so that an exact analysis of the system is only possible under certain assumptions. Without these assumptions, the analysis of a real application requires a large number of non-linear equations, whose solution is either almost impossible or takes too much computational time and effort. While conventional mathematical modelling approaches have provided the fundamentals to better understand combustion processes, their applications for real time diagnostic, modelling, control and optimisation have been limited. In particular, the inability to capture imprecision with the required level of tolerance, and the in-flexibility against distortions in the operating variables, make them less attractive. Consequently, new Artificial Intelligence (AI) based techniques such as; Artificial Neural Networks (ANNs), Fuzzy Logic (FL) and Genetic Algorithms (GAs) are finding increasing application in these areas. Above all, the ability to capture the non-linearities of a real system accurately and the versatility in being able to accommodate with ease, the various conventional and advanced strategies within their structures, have made these AI techniques much more attractive. The primary objective of this chapter is to review work at the University of Glamorgan and elsewhere in this area with a particular emphasis placed on combustion related problems ranging from coal and gas fired furnaces and boilers, gas turbine combustion and internal combustion engines. Overall, this chapter is organised into three sections, each of which describe some of the applications of the three AI techniques (i.e., ANNs, FL and GAs) in various areas such as condition monitoring, fault diagnosis, modelling and prediction, control and optimisation.

Keywords: Artificial intelligence, Neural networks, Fuzzy logic, Genetic algorithms, Combustion systems.

NOMENCLATURE

A	Total surface area (m^2)
A_f	Throttle valve opening (degree)
Al_2O_3	Aluminum oxide
CaO	Calcium oxide
CF	Fixed carbon content of coal (% by weight, dry-ash-free)
CO	Carbon monoxide
CO_2	Carbon dioxide
D	Diameter of burner quarl (m)
E	Exhaust valve fault
F	Fuel injector fault
Fe_2O_3	Iron oxide (hematite)
H	Hydrogen
HC	HydroCarbon
I	Diagnostic index
J	Cost function
K_2O	Potassium oxide
$\dot{m}_{fi}$	Fuel injection mass flow rate (kg/s)
m_{gas}	Natural gas flow rate (kg/s)
MgO	Magnesium oxide
$m_{pa}(t)$	Mass of the fuel layer at time t (kg)
$m_{pa}(t_0)$	Fuel mass at the start-up (kg)
msp	Model set-point air-fuel ratio (dimensionless)
N	Engine rotational speed (RPM)
Na_2O	Sodium oxide
NO_x	Nitrogen oxides
O_2	Oxygen
P	Pressure (in Eqs 3-6) (mbar)
P_2O_5	Phosphorus pentoxide
P_{el}	Electric power produced (kW)
$p_{i.man.}$	Intake manifold pressure (mbar)
R_0	Vitrinite reflectance (%)
SiO_2	Silicon dioxide
T	Combustion temperature (K)
T	Mixture temperature (in Eqs 3-6) (K)
T_{cj}	Exhaust temperature of j-th cylinder (K)
TiO_2	Titanium dioxide (IV)
T_r	Maximum coolant temperature (K)
u	Throttle angle (degree)
wf	Weighting factor, from 0 to 1

$x_1(t)$	Proportional value of the fuel mass
$\hat{y}$	Actual air–fuel ratio (dimensionless)
$\boldsymbol{Y}_{NOx}$	Predicted NO_x by ANN model (ppm)
$\boldsymbol{Y}_{CO}$	Predicted CO by ANN model (ppm)
$\boldsymbol{Y}_{UBC}$	100 - Predicted CBO by ANN model (% by weight, dry-ash-free)

Greek

λ	Control weighting factor which penalises excessive changes of the control input
	Air to fuel ratio
ϕ	The objective value

Subscripts

d	Nominal operation condition
i	Inlet condition
o	Outlet condition

ABBREVIATIONS

AFR	Air-Fuel Ratio
AI	Artificial Intelligence
ANN(s)	Artificial Neural Network(s)
ARX	Auto-Regressive with eXogenous variable
ARMAX	Auto-Regressive, Moving Average with eXogeneous variable
ASTM	American Standard for Testing and Materials
BP	Back Propagation
CBO	Carbon BurnOut
CCD	Charge Coupled Device
CCSEM	Computer Controlled Scanning Electron Microscopy
CEMS	Continuous Emissions Monitoring Systems
CFD	Computational Fluid Dynamics
CPU	Central Processing Unit
CV	Calorific Value
EDTGA	ElectroDynamic ThermoGravimetric Analyzer
FAR	Fuel-to-Air Ratio
FDAS	Fuzzy Diagnosis and Advice System
FIR	Finite Impulse Response
FL	Fuzzy Logic
FLC	Fuzzy Logic Controller
FLT	Fuzzy Logic Tuner

GA(s)	Genetic Algorithm(s)
ICAO	International Civil Aviation Organisation
ICE	Internal Combustion Engine
IFRF	International Flame Research Foundation
IR	Infrared Radiation
LTO	Landing and Take-Off
MAPE	Mean Absolute Percentage Error
MBPC	Model Based Predictive Control
MLP	Multi Layered Perceptron
MPC	Model Predictive Controller
N	Normal
NNBC	Neural Network Based Controller
NPV	Net Present Value
PEMS	Predictive Emissions Monitoring Systems
pf	Pulverised Fuel
PI	Proportional-Integral
PID	Proportional-Integral-Derivative
ppm	Parts Per Million
RAS	Random Amplitude Signals
RBF	Radial Basis Function
RPM	Revolutions Per Minute
SOC	State Of Charge
SOM	Self Organising Map
TGA	ThermoGravimetric Analyser
THC	Total HydroCarbon
TWC	Three-Way Catalytic converter
UBC	UnBurned Carbon in ash
UHC	Unburned HydroCarbon

1. INTRODUCTION

Despite the contributions made by nuclear, hydraulic, solar and wind energy sources, almost 90% of the world's energy used currently for transportation, power production and heating is produced by combustion of some form of the hydrocarbon fuels. Thus the study of combustion is of continuing importance especially if we are to use these fuels efficiently and economically to conserve resources and reduce air pollution in a world of increasing population and increasing energy needs. Nowadays, it is true that a large amount of combustion-related diagnostic research and development are motivated by environmental concerns and by the necessity to meet regulations for air quality, while simultaneously maintaining or improving combustion efficiency. Thus, faced by an increasingly complicated existence, new scientific tools and techniques for combustion diagnostics are continuously being developed. This is largely because there remains a large class of problems related to combustion processes, which are very difficult to solve by conventional mathematical modelling approaches. These problems are frequently characterised by the following features:

(1) It is not always possible to develop a mathematical model for the problem, which adequately reflects the actual physical process.
(2) The solution methodology employed by a human expert is not capable of expression in conventional algorithmic or a mathematical form.
(3) The complexity and size of the problem is such that complete computational solutions cannot be obtained in a reasonable time.

Thus a large number of AI techniques have been increasingly employed in these systems. AI techniques have been regarded as the new breed of knowledge-based systems that cannot be merely described by the traditional mathematical representations. In a broad perspective, these knowledge-based approaches underlie what is called "soft computing". These methods include artificial neural networks (ANNs), fuzzy logic (FL) and genetic algorithms (GAs). Especially, ANNs with their unique leaning capability have been successfully applied to solve some long-standing problems where conventional approaches have proven to be difficult or have not achieved the desired speed, accuracy or efficiency. In addition, these methodologies in the most part are complimentary rather than competitive. Increasingly, these approaches are being utilised in combination and are commonly referred to as "hybrid systems". It should also be appreciated that no single technique alone will be ideal for every application problem, rather it is always a complement of different techniques, which make the system more useful in a practical sense, and this is best illustrated in Fig. 1.

For the last decade or so, there has been a plethora of journals and conferences devoted to various engineering applications of AI in combustion systems (Kalogirou, 2003). In this respect, it is timely to review some of the aspects in this book. In this chapter, a variety of applications are covered including; process condition monitoring, fault diagnosis, system modelling, control and optimisation. They all represent important issues in the operation and management of modern combustion systems. In particular, it is very encouraging to see the economic values and benefits that have resulted from the employment of AI based techniques.

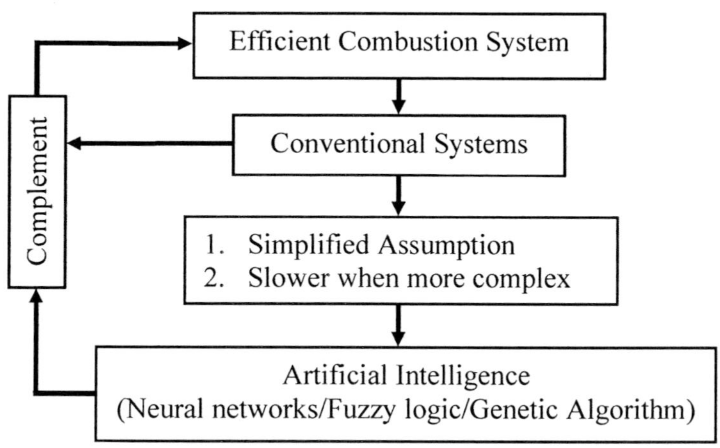

Fig. 1. The role of artificial intelligence

This chapter is divided into sections in which the perspective of the authors on the potential of different AI techniques, such as ANNs, FL and GAs, are given by highlighting selected application examples and where possible by guiding the reader to most recent literature.

2. Application of Artificial Neural Networks

Artificial Neural Networks (ANNs) are by far the most widely used AI technique for solving the majority of the problems related to combustion, which are generally ill defined and non-linear in nature. In particular, the feedforward multilayer perceptron (MLP) networks are already well known for their capability as a universal non-linear function approximator, which has found diverse applications in the modelling and control of combustion processes. Another class of ANNs, often known as the Kohonen Self-Organising Map (SOM), have also found numerous applications especially in problems relating to pattern recognition such as those found in condition monitoring and fault identification.

2.1. Modelling of the Thermo-Chemical Properties of Solid Fuels

Fuels, in the form of solid, gaseous or liquid, represent the primary input to any combustion system, which produces energy (in the form of heat) upon reaction with oxygen. The thermo-chemical properties of a solid fuel, for instance, are important factors, which can affect both the conversion efficiency of the fuel, and safe operation of the combustors in which they are burned or consumed. There are standard laboratory equipments available for characterising these properties but often they are both tedious and expensive to perform. Consequently, ANNs have been used increasingly to estimate the thermo-chemical properties for new fuels based on the heuristic knowledge from historical fuel databases.

2.1.1. Predictions of the Heating Values of Fuels

In almost any combustion system, the heating value of fuels represents an important design parameter, which defines the energy content of the fuel. Estimation of the heating value from the composition of the fuel is one of the basic steps in performance modelling and evaluation in combustion systems. While the heating value of a natural gas can be calculated unambiguously from the composition of its constituents (Kenneth and James, 1995), accurate determination of the heating value for coal and biomass is much more complicated due to the thermal effects of ash formed during the combustion. Currently, the heating values of solid fuels are commonly determined by means of a bomb calorimeter in accordance to the ASTM D2015 standard. However this method requires tedious laboratory procedures and is relatively costly. Thus, numerous correlations are available in the literature, which have been developed to predict the heating values using information either from the proximate or ultimate analysis of the fuel (Parikh *et al.*, 2005; Channiwala and Parikh, 2002). These correlations are often derived using multiple linear regression methods, which may not fully capture the actual non-linear function underlying the data. Alternatively, with their inherent ability to learn from complex non-linear data, back-propagation (BP) neural networks can be used instead to predict the heating values of fuels. Given sufficient training data and hidden neurons, it has been demonstrated that BP neural networks can outperform the linear regression models over a wide range of fuel compositions (Dong *et al.*, 2003; Tan, 2003).

2.1.2. Prediction of Ash Deposition Tendency of Coal

ANN techniques have also been used to predict the ash fusion temperature of coal blends, which is an important parameter characterising the slagging propensity of coal (Yin *et al.*, 1998). At present, the experimental research indicates that while the ultimate analysis and the mineral matter in blended coal can be approximated by the weighted average of its component coals, the fusibility of the coal blend is in general a non-linear function of its blend composition. Many techniques, such as the ternary equilibrium phase diagrams and multiple linear regression models, have been proposed in the past to predict the ash fusion temperature but with limited success. This was largely due to the fact that the complex interactions of ash-forming species within coal blends were, in general, not precisely known. However, based on the generic understanding that the chemical composition of coal ash dictates its fusion temperature, a two-layer MLP neural network with 8 hidden-layer neurons could be used to learn the underlying non-linear relationship between the ash compositions and its corresponding fusion temperature. Using the eight major oxides of coal ash, the authors proposed an overall architecture for the ANN-based predictor as illustrated in Fig. 2. The results suggested that the ANN-based predictor consistently outperformed the conventional linear regression models.

On the other hand, the knowledge of certain ash-forming species in coal is of paramount importance to determine various performance indices for assessing the ash deposition tendency in pulverised coal fired boilers (Salehfar and Benson, 1998). Currently, they can only be determined using the Computer-Controlled Scanning Electron Microscopy (CCSEM) methods, which are rather time-consuming and expensive. Thus the authors have developed an ANN based predictor which uses the bulk ash composition to accurately and quickly determine the CCSEM-based analysis results. The structure of the BP neural network consisted of one hidden layer with 32 neurons. The inputs to the ANN were the weight, as a percentage, of Na_2O, MgO, Al_2O_3, SiO_2, P_2O_5, K_2O, CaO, TiO_2 and Fe_2O_3 oxides and these were then used to predict the evolution of Quartz, Kaolinite, Montmorillonite, K Al-Silicate (illite) and pyrite. The developed ANN model thus allowed the plant operator to quickly change or modify the operating conditions and coal blends to alleviate the ash related problems in the boiler.

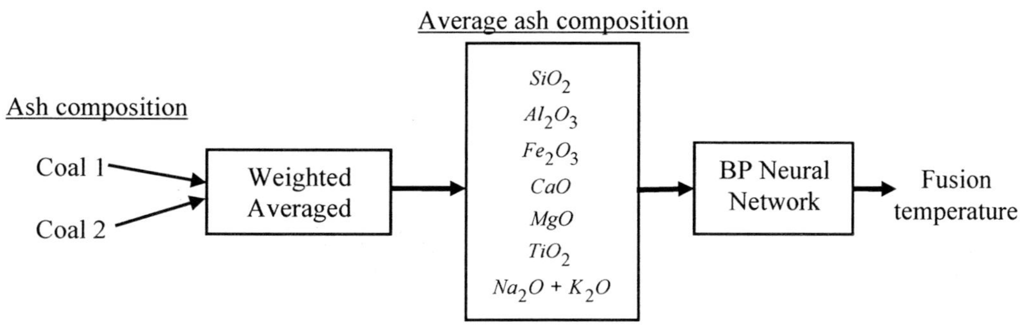

Fig. 2. Schematics of the ash fusion temperature predictor

2.1.3. Modelling of Coal Devolatilisation and Coal/Char Combustion Rates

An accurate and efficient model of solid fuel devolatilisation is essential for incorporation into Computational Fluid Dynamics (CFD) models suited for combustion modelling. Existing devolatilisation models in CFD codes used for industrial combustion applications typically correlate volatile yield with particle temperature, or specify the devolatilisation rate through single or, very seldom, multi-step Arrhenius reaction schemes. These models cannot simulate all the parallel consecutive reactions taking place during devolatilisation, and for the most part, their constants require re-calibration for each new fuel, which is an expensive experimental process. Although more advanced and accurate devolatilisation models, such as FLASHCHAIN, CPD and FG-DCV are available; their implementation in CFD codes has been hindered due to the shear additional computational burden. To overcome these problems, an ANN-based devolatilisation model was developed using a database, which contain more than 100 different types of fuels including coal/blends, biomasses and waste materials (sawdust and sewage sludge) from a diverse set of experimental conditions (heating rates varying between 10^3 to 10^6 K/s) and apparatus (drop-tube furnace, TGA, and EDTGA). The inputs used to train the ANN model included temperature, atomic ratios (hydrogen/carbon content, oxygen/carbon content), fixed carbon, ash and volatile content of the fuel. These input parameters directly affect the softening and swelling, fluidity, bond breaking and cross-linking of the coal's organic matrix ignition and thus can be used to directly predict the weight loss of particles during devolatilisation. A single-hidden-layer BP neural network having one hidden layer with five neurons and one output-layer neuron was used in the modelling work. The results showed that with sufficient training data, the predictions made by the ANN model were in good agreement with experimental results and comparatively better than the single-step reaction model. Above all, a massive saving of 64% in CPU time relative to the single-step model was reported and clearly the savings would be much greater with respect to the more complex devolatilisation models (Abbas *et al.*, 2003).

In the design and operation of coal-fired burners, it is important to know the combustion rate of coal/char to ensure flame stability and high efficiency in carbon conversion (Zhu *et al.*, 1999). In modelling the coal/char combustion rate, numerous mathematical models have been developed but with limited practical application. Some models involve complex formulae using parameters that are not readily available, whilst others work well only with a particular coal type. In an attempt to generalise such a model (for low temperature regimes) to different coal types, the authors adopted an ANN technique. It was generally known that the coal/char combustion rate depends on a number of factors such as combustion temperature (T), the rank of the coal, the extent of coal pyrolysis and the total surface area (A) of the char. In order to quantify the rank of a coal, both the vitrinite reflectance (R_0) and fixed carbon content (CF) of the coal were considered separately. The extent of coal pyrolysis was quantified as the ratio of char to coal hydrogen content (H_{char}/H_{coal}). A database containing the combustion rate reactivity of 55 chars derived from 26 coals covering a wide range of rank and geographic origin was established to train and test the neural networks. The heat treatment temperature of the chars ranged from 1000 to 1500°C and the combustion rate reactivity of the chars were measured using thermo-gravimetric analysis in a temperature range of 420–600°C. The following three sets of inputs were used, respectively, to train the neural network:

(1) R_0, H_{char}/H_{coal}, A,T;
(2) CF, H_{char}/H_{coal}, A, T;

(3) CF, H_{char}/H_{coal}, T;

The results showed that when sufficient amount of training data are available, a neural network model could be developed to predict the combustion rates of coal chars with good accuracy and robustness for all three sets of input data shown above. The "best" neural model in this case consisted of one hidden layer with 4 tan-sigmoid neurons (sum-square-error < 0.1) when using fixed carbon content, H_{char}/H_{coal}, total surface area of the char and combustion temperature as inputs.

2.2. Condition Monitoring and Fault Diagnosis

For the improvement of reliability, safety and efficiency, advanced methods of supervision, fault-detection and fault diagnosis become increasingly important for many technical processes. For combustion systems, often there are many process/sensor measurements available, which can be directly related and monitored for fault detection and diagnosis. However the interpretation of these measurements for automatic fault identification can be an overwhelming task for human experts due to the shear amount and complexity in the data. While the classical limit-value based fault detection methods are simple and easy to apply, they are however, only able to react after a relatively large change of a feature, i.e., after either a large sudden fault or a long-lasting gradually increasing fault. In addition, an in-depth fault diagnosis is usually not possible (Isermann, 2005). Therefore advanced methods of supervision and fault diagnosis, which can provide early detection of small faults with abrupt or incipient time behaviour, are highly desirable. The idea of using ANNs for fault diagnosis is not new as there has been an increasing number of successful applications reported in the literature. There are generally two ANN methods for condition monitoring and fault diagnosis applications; the first method is based on training the ANN to recognise specifically known faults whilst the second method is based on assessing the ANN model for the departures from normal operating conditions (i.e., novelty detection). Both methods are essentially complementary with the novelty detection being particularly useful in cases where fewer examples of faulty data are available.

2.2.1. Detection of Near Burner Slag Formation in Pulverised Coal Burner

Tan *et al.* (2003a) have developed an intelligent monitoring system to detect the presence of so-called "burner eyebrow" on pulverised coal fired burner by measuring both the optical and acoustic characteristics of the combustion flame. The study essentially involved a series of experiments with a wide range of coals over a range of burner conditions using a 150kW pulverised coal burner fitted with simulated eyebrows of different sizes. Results obtained from monitoring the infrared (IR) radiation and sounds emitted by the flame were processed into a number of features, which were then used to train a hybrid system which was based on the SOM and radial basis function (RBF) neural networks as shown in Fig. 3. In the approach, a two-dimensional (15x15 neurons) SOM network first learns the eyebrow-class probability density function and this is then used by the RBF network to determine the degree of membership, as a fuzzy vector, of the input pattern to each of these classes. Subsequently the results are defuzzified to return the class identity of the input pattern, as either class 1, 2 or 3. In a series of tests with two coals the system was able to distinguish the different eyebrows with a high degree of accuracy if representative data sets were used to train the network for each coal.

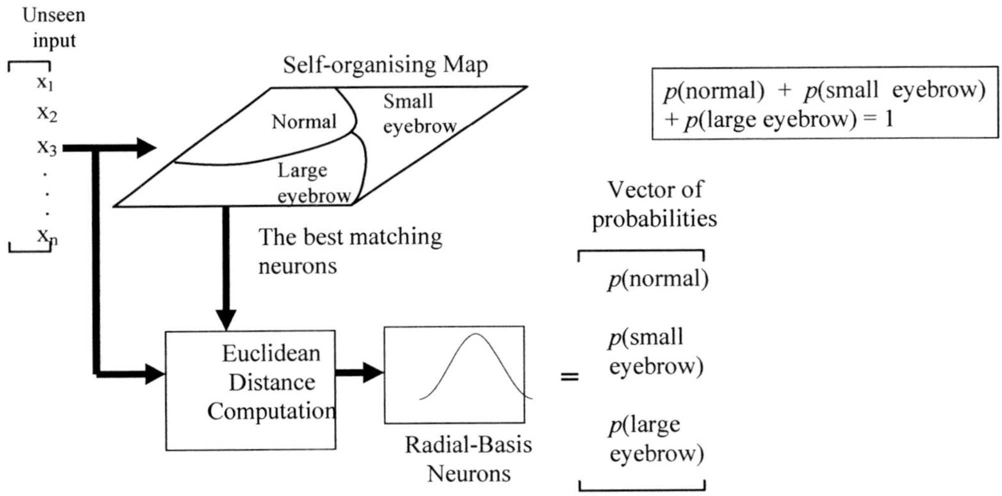

Fig. 3. Schematics of the hybrid classifier for burner slagging

In further tests with a range of six different coals the system was also able to distinguish between a clean burner and one fitted with a particular sized eyebrow. In this case it proved to be possible to use only the features from three of the coals in the training process and the data from the remaining fuels for validation. The monitoring system therefore appears to be relatively independent of changes to the coal quality if it has been trained with a representative range of coals.

The authors also presented a possible method to detect burner eyebrows via the evaluation of so-called "eyebrow indices" using a SOM which has been trained solely using data obtained from clean burner operation (Tan *et al.*, 2005). The approach is attractive in practice, since it will be very difficult to obtain training data for a range of eyebrows since it will not be possible to determine the geometry and dimensions, or even the presence, of a slag deposit by visual inspection whilst the boiler is in operation. Therefore it will be necessary to develop a technique to distinguish the presence of at least relatively large eyebrows based solely on training the SOM with data gathered over a period of "clean" burner operation for a range of coals. Any subsequent data from the sensors during further burner operation can then be related to the SOM by calculating the Euclidean distance from the new input to the "nearest" neuron on the SOM. Providing that the SOM has been adequately trained there is a substantial probability that the patterns of any further clean burner data will be closely related so that the average "distance" will be relatively small. Conversely the input patterns from the sensors in the presence of a significantly large eyebrow will be less well matched so that the distances from the best matching neuron should be substantially greater. This distance can then be used to define an "eyebrow index" and a suitable threshold value should be able to be employed to indicate the need for remedial action, or a requirement for further investigation of the presence of severe slag deposits.

This approach was simulated by using 50% of the data, obtained by operating the burner without the eyebrows, to train the SOM network. This proportion of the data was randomly selected but included results for all the six coals over the whole range of excess air levels so that the SOM was representative of a wide range of "normal" clean burner operation. The eyebrow indices for the remainder of the clean burner data as well as for all other data representing the eyebrows of different sizes studied in the project were then determined by calculating the distances of each of the patterns with respect to its "nearest" neuron on the SOM. The resultant average eyebrow indices for the training data and the various categories of test data are presented in Fig. 4.

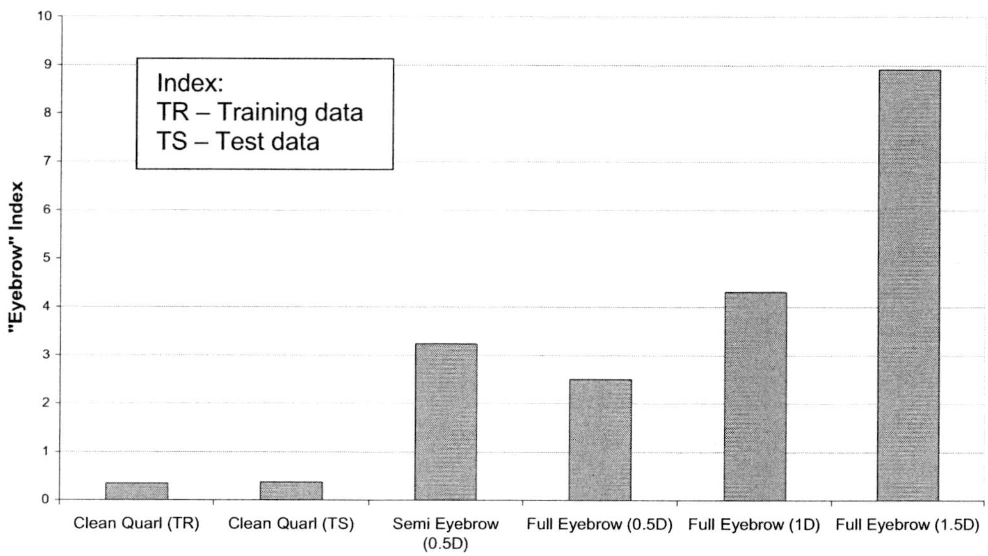

Fig. 4. Test results of the novelty detector for burner slagging

The index for the clean burner test data is very similar to the training index whereas much higher values are apparent for all the different eyebrows. Moreover the numerical value of this eyebrow index increases significantly as the thickness of the artificial eyebrow increases so that it should be possible to identify the presence of a severe slag deposit by defining an appropriate threshold value for the index. It is also noted that, for the same eyebrow thickness of 0.5D, the index of the semi-annular eyebrow appears to be higher than that of the full-annular eyebrow, probably due to more severe flame distortion. Consequently the suggested technique by comparing data from in-service sensors with a SOM trained solely with clean burner measurements appears to offer a promising means of detecting the formation of burner eyebrows. It should also be noted that the basic SOM could be gradually updated during boiler operation as long as the burner can be considered to be relatively "slag free".

2.2.2. Fault Diagnosis of a Diesel Engine

By measuring the pressure in the combustion chamber, an ANN based fault diagnosis system has been designed to provide an early warning for leaking exhaust valves and fuel injector nozzles in 4-stroke 2 cylinder air cooled diesel engines (Sharkey *et al.*, 2000). These faults were physically induced in the engine by replacing the exhaust valve with one that had been cut across its seating face and the fuel injector nozzle, which was not atomising properly. A total of 3600 operating conditions were simulated (from 1800 RPM to 2800

RPM) from both faulty conditions as well as during normal operation. For each of these operating conditions, 7200 pressure measurements were acquired which corresponds to 10 measurement points per degree of crank angle for one complete engine cycle (equal to two complete rotations of the crankshaft). However, only 200 pressure measurements were considered as the possible input variables to the neural networks due to the fact that these faults should affect the characteristics of combustion during the combustion stroke (corresponds to measurement points from 1801 to 2000) and were therefore sufficient for identifying the faults. These 200 input variables were subsequently reduced to form two 100-dimension input vectors of equally spaced variables. These could be further reduced to two additional 50-dimension input vectors of equally spaced variables by discarding every other variable from the two 100-dimension input vectors. From the 3600 operating conditions, 1200 were reserved for testing and validating the trained networks. The overall architecture of the system consisted of three modules, as shown in Fig. 5, each containing three MLP networks (see Table 1) which were trained using either the entire training set containing examples of all three classes, i.e., Normal (N), Exhaust valve fault (E) and Fuel injector fault (F), or subsets of the data each containing one of the three pair wise combinations of the three classes (i.e., NE, NF and EF). Upon training, any new input vector was presented first to the highest-level NEF module that determined to which lower level module (NE, NF and EF) the input pattern should be directed. The output from each module was essentially computed by taking a majority vote on the output of the three sub networks 1, 2 and 3. The performance of the system was then tested against a decision tree algorithm (Quinlan's C5.0) and a human expert (an engineer with 15 years of experience) and it was demonstrated than the ANN-based fault diagnostic system outperformed them consistently.

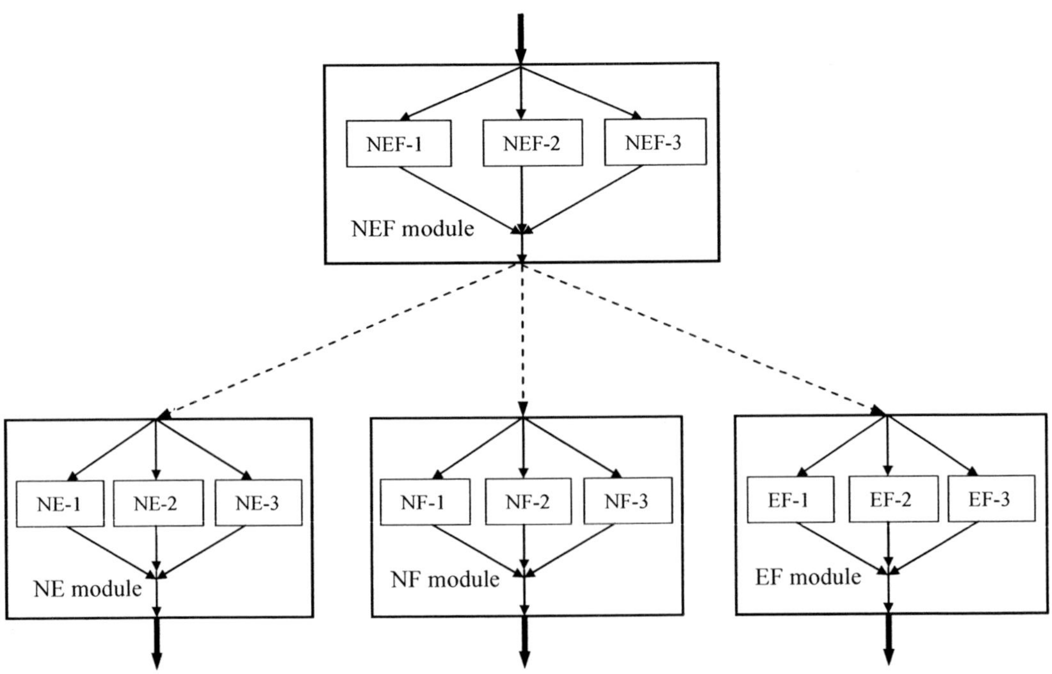

Fig. 5. Structure of the ANN based engine diagnostic system

Table 1. Architecture of the ANNs used in the engine diagnostic system

ANN Modules	Dimension of Inputs	Hidden neurons	Output neurons	Output classes
NEF-1	100	80	3	
NEF-2	50	20	3	N, E or F
NEF-3	50	20	3	
NE-1	100	80	2	
NE-2	50	20	2	N or E
NE-3	50	20	2	
NF-1	100	80	2	
NF-2	50	20	2	N or F
NF-3	50	30	2	
EF-1	50	40	2	
EF-2	50	20	2	E or F
EF-3	50	20	2	

2.3. System Identification and Control

A reliable and accurate process model is necessary for the design of any process controller. Many process modelling and control approaches are based upon a representation of the dynamic relationship between the cause and effect variables. In such approaches, the process is typically approximated via some form of linear dynamic model, such as finite impulse response (FIR), auto-regressive with exogenous variable (ARX) and auto-regressive, moving average with exogeneous variable (ARMAX) models (Ljung, 1999). Once determined the dynamic process model of the system can be integrated within a variety of process monitoring, control and optimisation algorithms. In process control, for example, the model can be incorporated within a model based predictive control (MBPC) algorithm, such as Generalised Predictive Control (Clarke *et al.*, 1997). Unfortunately, these models tend to be linear in nature and while they can provide acceptable performance for many systems, they may be unsuitable in the presence of significant non-linearities such as those encountered in combustion processes. For such processes it may be beneficial to employ a model that reflects the non-linear relationship between the cause and effect variables. In this sense, with their inherent ability to approximate any nonlinear continuous function without requiring any prior knowledge, ANNs could provide a more generic, non-linear solution for such processes. On the other hand, inverse model control represents another area where ANNs have found increasingly applications. An inverse modelling technique essentially involves prediction of the inputs to the plant given the past data of the plant's inputs and the outputs together with the desired future outputs (Ashhab, 2004). The following sections detail some of the work on how ANNs have been applied successfully to the identification and control of various combustion processes.

2.3.1. Control of Air-Fuel Ratio in Spark-Ignition Engine

Control of automotive internal combustion engine represents one of the most complex control problems due to the highly non-linear dynamic behaviour of the air and fuel injection system. Consequently, Wang *et al.* (2006) investigated the feasibility of using ANNs to control the air-fuel ratio (AFR) of a simulated spark-ignition engine. First of all, the system dynamics of the engine were derived using a well known and widely used benchmark engine model (Hendricks *et al.*, 1996) and subsequently used to train and test the ANN-based controller. In order to analyse the performance of the controller for a wide range of practical driving conditions, the sine-wave throttle angle bounded between 20° and 40° with a maximum of 0.5° uncertainty was introduced. A set of random amplitude signals (RAS) was then simulated for the fuel injection to obtain a representative set of input–output data. The range of fuel mass flow rates for this excitation signal was between 0.0014 and 0.0079 kg/s. The duration of each square wave was set to be 1 second, which was equal to 10 times the sampling frequency of the simulation system. A set of 7000 data samples of the AFR was collected and then divided equally for training and testing of the neural network model. The AFR was then modelled by a two-layer RBF neural network with a set of selected input variables as fuel injection ($\dot{m}_{fi}$), throttle angle (u) and air–fuel ratio ($\hat{y}$). By trial and error, the best architecture of the neural network, which gave the lowest prediction error, was consisted of 12 hidden neurons with a second-order inputs as shown in Fig. 6, which was trained using the online recursive training algorithm. The structure of the model predictive controller (MPC) is shown in Fig. 7. The nonlinear optimiser minimises the errors between the set point and the engine output by using the cost function:

$$\mathbf{J} = \sum_{i=t+N_1}^{t+N_2} \left[msp(i) - \hat{y}(i)\right]^2 + \lambda \sum_{i=t}^{t+N_u} \left[m_{fi}(i) - m_{fi}(i-1)\right]^2 \tag{1}$$

In Eq. (1), N_1 and N_2 define the prediction horizon whilst N_u defines the control horizon. Furthermore, λ represents a control weighting factor which penalises excessive changes of the control input, i.e., the fuel injection, $\dot{m}_{fi}$. Hence, the control problem reduces to searching a set of optimal control output such that the cost function **J** is minimised. By using $N_1 = 1$, $N_2 = 6$, $N_u = 0$ and $\lambda = 0.6$, the performance of the MPC was compared against traditional proportional plus integral (PI) control algorithm that is currently used in production electronic control units. The control experiments were simulated by changing the throttle angle from 25° to 30° with the objective of maintaining the AFR within ±1% of the stoichiometric AFR (i.e., 14.7) and the test results showed that although the steady state response of the PI controller is acceptable, its transient response tends to exhibit large overshoot and long settling time. The MPC controller, on the other hand, reduces the maximum overshoot above +1% bound (AFR = 15.2) and below −1% (AFR = 14) respectively. Furthermore, the corresponding settling times to steady state AFR were reduced by 25% to 50% when compared to the PI controller. Similar improvement using the MPC approach has also been reported by Mu *et al.* (2005) in controlling the aircraft gas turbine engine against a conventional PID controller.

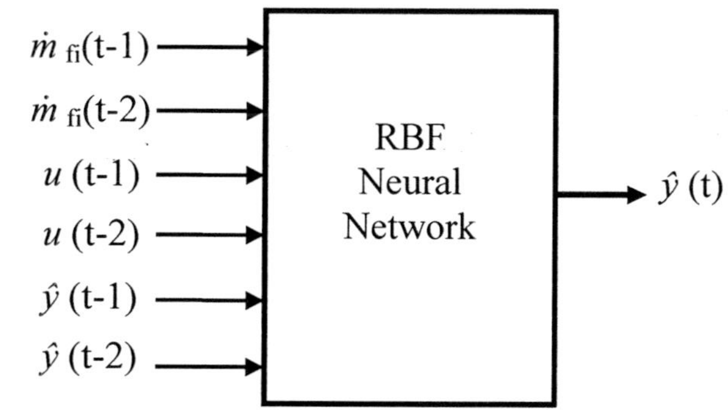

Fig. 6. Input structure of the ANN predictive model

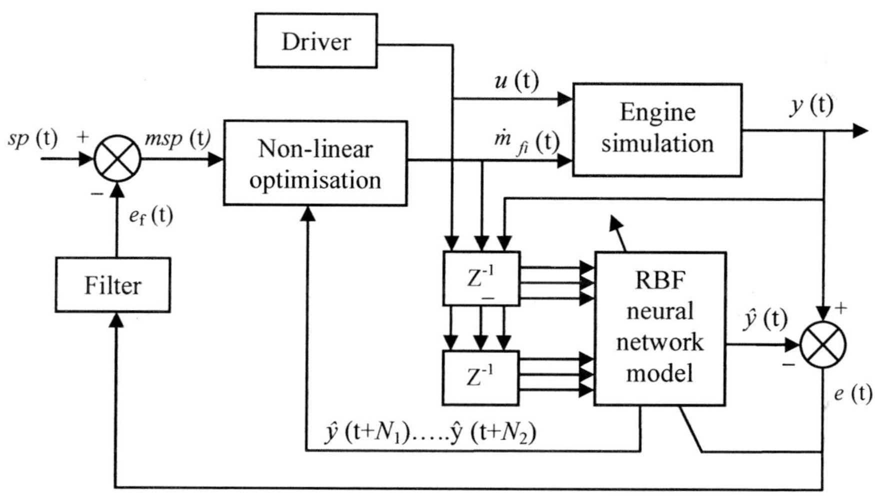

Fig. 7. Overall controller strategies for engine air/fuel ratio

2.3.2. Neural network based controller for chain grate stoker boiler

A novel Neural Network Based Controller (NNBC) has been developed based on a series of comprehensive tests conducted on a pilot-scale stoker test facility and was subsequently implemented on an industrial chain grate stoker fired boiler (Chong *et al.*, 2000). In this work, data collected during 300 hours worth of operation of a 0.75 MW chain grate stocker boiler were used to train two neural networks. The overall structure of the NNBC is shown in Fig. 8. The primary aim is to supply the minimum amount of combustion air required without incurring unacceptably high CO emissions and carbon in ash losses, thus improving the combustion efficiency. In principle, the system first detects changes in the load required and subsequently delivers a near optimum coal feed and airflow. This was accomplished by training a two-layer direct-inverse neural network model (so-called setting network), which consisted of 30 hidden neurons, to predict the desired 6 controller inputs (grate and rotary valve speeds and four primary air flow rates) based on the required load demand.

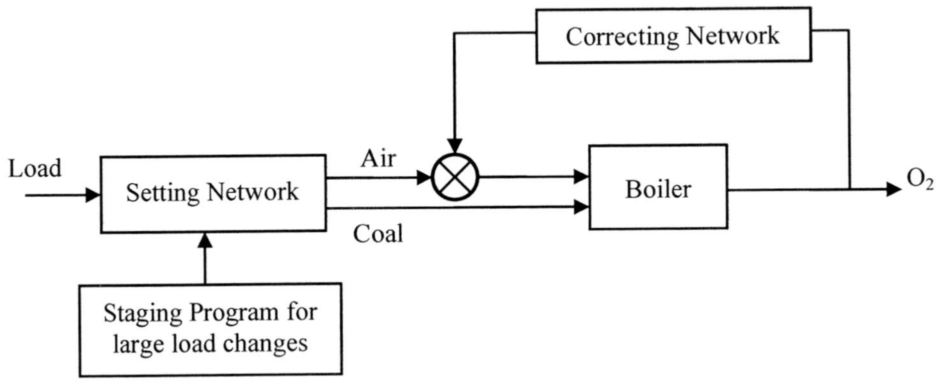

Fig. 8. Structure of the ANN based controller for chain grate stoker boiler

Hence, the setting network essentially mimics the human operator when operating the boiler after start-up. If the magnitude of the load change is higher than a pre-set limit, then an optimum staging profile of coal and air feed is delivered based on past experience. After quasi-steady-state conditions are reached the second phase of the neural network based controller is activated to fine tune the primary airflow rate in order to optimise the combustion process on the grate. This is achieved by monitoring the concentration of oxygen in the flue gas on-line and feeding the error into the corrective network, which then delivers the amount of the necessary adjustment in order to consistently maintain an optimum air/fuel ratio during steady-state combustion. The neural network based decision maker has one input (sample oxygen reading), two outputs (adjustment of the primary air flow to the middle two sections of the grate) and a hidden layer with 25 neurons. The function of this network is to deliver a decision as to how much primary air to the main combustion zone requires adjustment, during steady-state, in order to maintain the optimum level of flue gas oxygen concentration at the stack.

2.3.3. Neural Network Based Controller for Pulverised Coal Fired Burner

A controller strategy has been developed that utilised low cost sensors and AI techniques to control the combustion of a 150 kW pulverised fuel burner as shown in Fig. 9. The system consisted of three neural network models, which represent the excess oxygen (O_2) and emissions of nitrogen oxides (NO_x) and carbon monoxide (CO). These ANN models were derived from data collected from a series of experiments conducted on a pilot scale test furnace using a single IFRF variable swirl, low NO_x, pulverised fuel (pf) burner with a gross thermal capacity of 150kW. During these tests, appropriate sensors were employed to measure the infrared radiation and combustion noise generated by the flames together with the gaseous emissions in the flue. Standard signal processing techniques were then used to extract a range of "features" from the sensor signals and these features were subsequently used to train the ANNs to predict the excess O_2 level and pollutant emissions such as NO_x and CO. The control system then incorporates the ANNs, which were trained with the best features from the sensor signals, and these ANNs were used to estimate the NO_x and CO emissions based only on data from the sensors.

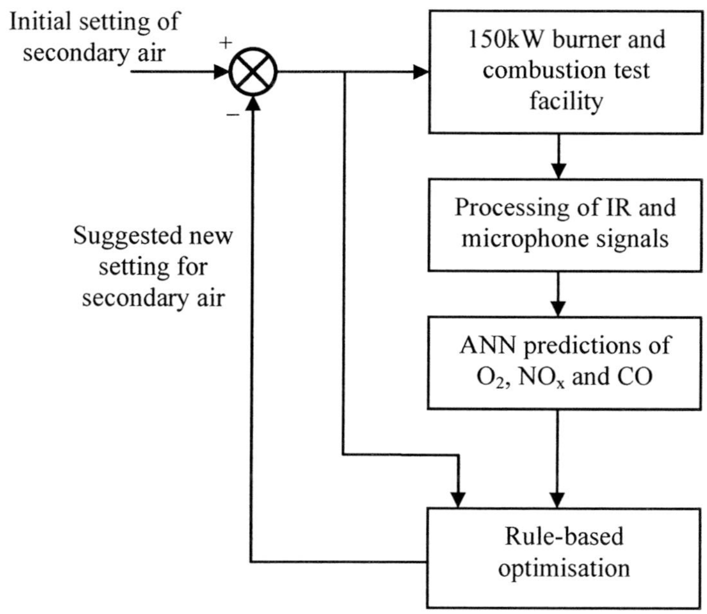

Fig. 9. Overall strategies for controlling pulverised coal burner

These estimates were then used to make corrections to the excess air to reduce the NO_x whilst maintaining the CO emissions within acceptable limits by means of a set of rules which were generated from observation of the previously obtained experimental data under the various test conditions as were the magnitudes of the adjustments which were required for the combustion air flow rate. The aim of the system was therefore to recognise and classify the state of the burner and supply an appropriate airflow to maximise the combustion efficiency. The results suggested that although lacking of the desired speed of response, the system was able to reduce both NO_x and CO emissions to "low" values. Moreover, the approach offers a promising opportunity to optimise individual burner performance in large-scale multi-burner arrangement boiler (Tan *et al.*, 2003b).

2.3.4. Predictive Emission Monitoring Systems

The technology for *Continuous Emissions Monitoring Systems* (CEMS) for use in combustion systems has been around for many years. Traditionally, monitoring of emissions such as NO_x, CO and particulates is done with physical sensors that are expensive to maintain. On the other hand, *Predictive Emissions Monitoring Systems* (PEMS), provide both cost savings as well as more robust monitoring of emissions and therefore have found a wide range of application in combustion engines and furnaces (Kibrya and Botros, 2002; Berry *et al.*, 2004; Heitz *et al.*, 1997). Depending on the type of combustion system, PEMS work by accurately predicting emissions based upon more easily measurable variables, such as air flow into the combustion system, temperature of combustion air, fuel flow and exit oxygen concentration, for example. First, conventional Continuous Emissions Monitoring System (CEMS) measurements were performed, providing historical data over a wide range of operation. Then an empirical predictive model was created from the historical data, which can then put online to assess the emissions. For example Heiltz *et al.* (1997), has investigated the use of ANNs to develop predictive models for the total hydrocarbon (THC), NO_x and CO emissions from six multiple hearth incinerators burning sewage waste. A series of experiments were first conducted to simulate a wide range of operating conditions while collecting a range essential operating and process variables. The input variables used to train the neural networks includes temperatures, pressure and oxygen concentration measurements at different locations in the incinerator, supplied air and solid fuel feed rates and other auxiliary measurements which could immediately affect the emissions of THC, NO_x and CO. All data were acquired by field sensors and CEMS analysers and subsequently stored in computer for off-line analysis and training of the neural network models. The structure of the neural networks consisted of three hidden layers with six, four and two neurons respectively. After training, the neural network models were subsequently put online and validated against the CEMS measurements and the results showed that the maximum error was between 7 and 9 percent, which were well within the regulatory requirements. The authors concluded that PEMS could help reduce emissions without the large capital expenditure for new CEMS type pollutant control equipment.

The concept of predictive emission monitoring can be extended for the purpose of combustion optimisation in which a set of optimum settings of the boilers could be determined so as to minimise the pollutant emissions or maximise combustion efficiency. A system known as GNOCIS (Slatsky *et al.*, 1997) has been developed by PowerGen Technology Centre in collaboration with Tessella, for optimising the emissions and combustion efficiency of coal or oil fired power stations. It receives approximately 100 operating parameters, such as the boiler load, coal feed rate, amount of combustion air and the combustion air temperature, and deliver advice that the operators can apply to improve various measures of the power stations performance, such as NO_x and CO emissions and unburned carbon in ash. At the heart of the system is a neural network model of the power station trained on long-term historical data of these operating parameters. Subsequently, a constrained nonlinear optimisation algorithm was applied to identify the best set points for the plant. Several field trials were conducted at Gaston and Kingsnorth power station in the 90's and the results suggested that with proper configuration the GNOCIS could offer an improvement of as much as 0.5% and 15% in the combustion efficiency and NO_x emission respectively.

3. Application of Fuzzy Logic

Although fuzzy logic systems are very different from that of neural networks, they can be used to control nonlinear systems to a degree not possible with conventional linear control system such as the PID controller. Neural network and fuzzy logic systems are fundamentally different despite the fact that both perform mappings. On the one hand, neural networks tend to map a measured input into a measured output by typically minimising the least square error. The fact that this so-called training error is not zero implies some uncertainties in the data, the degree of which is dependent on the amount and quality of the data used in the training phase. On the other hand, fuzzy logic systems tend to map a known input into unknown output using a known but fuzzy inference procedure. Consequently, the imprecision or uncertainty in the data is recognised and dealt with in the fuzzy logic system whilst for the neural network systems the degree of precision and generalisation in the input/output relationship are retained in the networks upon training. Perhaps the greatest attribute of fuzzy logic systems is their ability to describe linguistically and to perform decision-making and control actions without having to explicitly represent the process behaviour using complex mathematical equations. Furthermore, the fuzzy relationships (i.e., fuzzy rules) are readily available for examination and can be used to explain the cause and effect of the decision or actions taken, thus rendering the model more transparent. The following sections aim to describe some of the applications of fuzzy logic systems in combustion related processes.

3.1. Fuzzy Logic Controller

In control applications fuzzy logic systems have been used as stand-alone controllers or in combination with conventional PID-type controller. In order to use a fuzzy logic system as a stand-alone controller, a front-end "fuzzifier" and a rear-end "defuzzifier" are added to the usual input-output mapping. A fuzzy logic controller commonly consists of four sections; rules, fuzzifier, inference engine, and defuzzifier. Once the rule has been established, the controller can be considered as a nonlinear mapping from the input to the output. On the other hand, in conventional PID-type controller, fuzzy logic has been utilised mainly to adapt the controller parameters to improve overall system robustness while maintaining stability in the control.

3.1.1. Incinerator Temperature Control

There are many factors affecting the combustion process in incinerators. Among these, adequate control of the incinerator temperature represents one of the most important step to reduce emissions. Generally, there are two key factors affecting the incineration temperature, namely, the calorific value (CV) and feed-rate of the feedstock. In practice, variability in the CV of the waste feedstock can be significant and thus impose non-linearity in the system. To overcome this problem a fuzzy logic controller was developed by Shen *et al.* (2005), with the primary objective of stabilising the incinerator temperature to an optimum set point temperature of about 850-950°C. The fuzzy system consisted of two input variables, namely, the temperature deviation from the set point and the rate of change of the temperature. These inputs were used to control the feeding duration and the feed stopping time. A set of so-called adaptive fuzzy rules was then derived which take into account the weighting factor of the input variables at different temperature regimes. Below a temperature of 900°C the effect of the adaptive rules was to optimise the increase in the temperature while at temperatures above 900°C the effect was to stabilise the temperature. For comparison, a classical fuzzy system based on fixed rules was also designed and applied to several waste incinerators. The results suggested that the control quality of the fuzzy adaptive controller is superior to that of the classical fuzzy controller in terms of robustness and stability.

3.1.2. An Adaptive Fuzzy Tuned PID Controller for Utility Boiler

Currently, PID controllers represent the most widely used control strategic in large-scale pulverised coal fired utility boiler. However, despite their simple structure, reliability and robustness in performance at fixed load conditions, they are generally not suitable in situations where the load condition fluctuates within a large operating range. Moreover, this type of controller needs frequent tuning, which can be difficult and time consuming. Consequently, Rangaswamy *et al.* (2005), proposed the design of adaptive fuzzy tuned PID controllers to regulate both the fuel and air inputs to the boiler while maintaining the desired steam pressure at the turbine inlet, regardless of the change in the steam demand. In their current approach, the two fuzzy logic tuners (FLT) utilised the PID controller outputs and the online adaptive error for fuel and air respectively (i.e., the differences between the actual and required fuel and air into the boiler) as inputs and used these to modify the gains of the PID controllers for fuel and air accordingly. The adaptive fuzzy tuned PID controllers were tested in a laboratory scale test facility and compared with conventional PID controller. The results of the tests highlighted the robustness of the FLT scheme for both positive and negative step changes in load and the adaptive PID controllers consistently delivered less overshoot and faster settling time following the load changes.

3.1.3. Control of a Premixed Gas Burner

An application of fuzzy logic for optimising the combustion flame of a pilot scale premixed gas burner based on two-dimensional flame imaging technique has been demonstrated (Tuntrakoon and Kuntanapreeda, 2003). The objective of the controller was to ensure good characteristics of the combustion flame by optimising the size of blue region of the flame. In general, a good combustion should exhibit solely blue flame while orange flame indicated insufficient excess air. The flame images were captured using a CCD camera and were subsequently pre-processed to extract two input information for the fuzzy logic controller, namely, the relative proportion of the blue and orange parts of the flame. The controller then processed these inputs and delivered the required air flow-rate so as to minimise the proportion of the orange part of the flame.

3.1.4. An Efficient Energy Management Strategy for Hybrid Vehicle

A generalized fuzzy logic controller (FLC) has been used to optimize the fuel economy and reduce the emissions of hybrid vehicles with parallel configuration (Naim *et al.*, 2004). Using the driver input, the state of charge (SOC) of the energy storage, the motor/generator speed, the current gear ratio and vehicle speed, a set of 44 rules have been developed, in a fuzzy controller, to effectively determine the power split between the electric machine and the internal combustion engine (ICE). The underlying theme of the fuzzy controller is to optimize the fuel flow and reduce NO_x emissions. The parameters in the fuzzy rules can be adjusted to trade-off the fuel economy and the NO_x emissions of the vehicle. The authors claimed that the controller has the potential to improve the composite urban/highway fuel economy by more than 10% (when tuned for fuel economy) over traditional strategies that optimize the engine operation only. It also has the potential of reducing the NO_x emission by more than 25% over the composite urban/highway cycle (when tuned for NO_x emission reduction) over traditional strategies that optimize engine operation.

3.2. Process Modelling

The basic architecture of the type of fuzzy logic systems that have been introduced so far uses fixed membership functions, which have been somewhat arbitrarily chosen. Furthermore, the rules used in the systems were essentially pre-determined by the users based on the interpretation of the behaviour of the processes under investigation. However, there will be some modelling situations in which it is impossible to deduce the appropriate membership functions. Rather than choosing the parameters associated with a given membership function arbitrarily, these parameters could be chosen so as to tailor the membership function to the input/output data in order to account for these type of variation in the data values. This is where the so-called adaptive neuro-fuzzy learning techniques can be useful. This learning method works similarly to that of neural networks. In this sense, the adaptive parameters of the fuzzy membership function represent the weights whilst the fuzzy rules represent the neurons activation.

3.2.1. Modelling of CO_2 Emission from Wood Burning Combustion System

For example, Ruusunen and Leiviskä (2004) explored the use of this type of adaptive fuzzy model to mimic the CO_2 emission from a small-scale wood burning combustion system. In general, burning of wood is a series of thermal reactions in which three main stages can be specified: ignition, burning and char-combustion. The combustion process exhibits significant dynamical changes due to the characteristic of each burning stage. To enable control towards higher efficiency and lower emissions, real-time information about combustion quality such as CO_2 emission is essential. Consequently, the authors explored the use of an adaptive fuzzy algorithm to identify a predictive CO_2 emission model with the objective of reducing the operating costs and overcoming the lack of real-time performance of the conventional flue gas analysers. The predictive emission model was derived from the knowledge of the measured combustion temperature along with the so-called proportional value of the fuel mass defined as:

$$x_1(t) = \frac{m_{pa}(t)}{m_{pa}(t_0)} \tag{2}$$

Where $m_{pa}(t)$ is the mass of the fuel layer at time t, and $m_{pa}(t_0)$ is the fuel mass at the start-up or just after fuel feeding. To estimate the ongoing burning stage using the calculated value in the above equation, the local operating regimes are defined by ignition, burning and char-combustion fuzzy sets as shown in Fig. 10. Similarly, two fuzzy sets for low and high combustion temperatures were defined. The membership functions for the temperature fuzzy sets had the same form to that of ignition and char-combustion and were spaced uniformly between 0K and 900K. The model parameters were identified using the 720 data points collected from the first experiment and validation was subsequently carried out using the remaining 4885 data points collected from another five separate experiments. The best model performance delivered an overall mean absolute percentage error (MAPE) of 14%, which was 1% lower than the error criterion for the gas analyser. The validation results also suggested that the performance of the fuzzy model was 200% and 40% more accurate than that of linear and neural network models respectively.

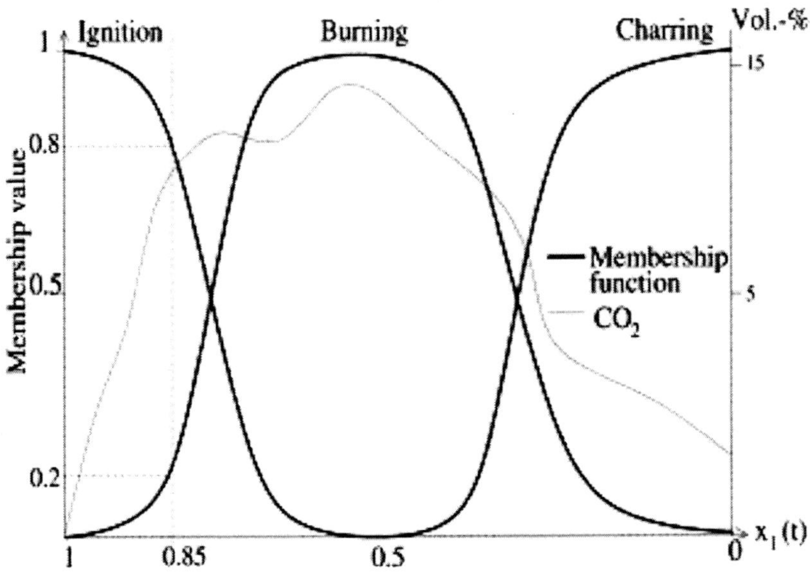

Fig. 10. The fuzzy sets for modelling CO_2 emissions

3.2.2. Engine Load Prediction for Off-Road Vehicles

Engines of large off-road vehicles such as dozers, harvesters or mobile equipment for road construction, have to produce not only enough power for moving their own weight, but are also loaded with some additional, time-varying tasks that depend on the particular type of equipment. A combine harvester, for example, has to move through the field, but has also the task of cutting, processing and storing the crop. The latter set of tasks can consume more than 70% of the total engine power. Consequently, the engine load can approach its maximum allowable limit in some operating regimes and, therefore, precautions must be taken to avoid engine overload, especially under automatic ground speed control of the vehicle. An accurate model to predict the engine load is therefore essential to design a model predictive control algorithm. Maertens *et al.* (2004) thus proposed a scheme based on adaptive fuzzy model to accurately predict the one-step-ahead value of the engine load based on the past history of the vehicle weight and ground speed, engine load and crop feed-rate. The validation results showed that the fuzzy model could be potentially useful for implementation into a model predictive control scheme.

3.3. Fault Diagnosis

Fuzzy logic can be particularly useful in fault diagnosis when the basic relationships between the faults and symptoms are at least partially known. To perform a diagnosis, this qualitative knowledge can now be expressed in the form of rules: IF <*antecedent*> THEN <*consequent*>. The antecedent (or premise) part contains facts in the form of symptoms S_i as inputs, and the consequent (or conclusion) part includes the corresponding faults F_j as a logical cause of the symptoms. If several symptoms indicate a fault, then they are associated by AND and OR connectives, leading to rules in the form: IF < S_1 AND S_2 > THEN <F_1>.

3.3.1. Diagnosis of Emissions and Fuel Consumption of Transport Vehicles

The emission from and fuel consumption of transport vehicles are important factors to be considered due the increasingly more stringent environmental regulation. Nowadays, periodic examination of vehicles emission is compulsory in many countries. This has motivated the development of a diagnostic system that can be used to detect and subsequently identify the possible causes of high emissions during the routine check. Consequently, Kilagiz *et al.* (2005) have developed a fuzzy expert system which could identify possible fuel system faults, ignition system faults, in-take valve and exhaust valve faults and suggest a solution advice for these faults, based on measurements of CO, HC, CO_2, O_2 and air/fuel ratio (λ). The structure of the fuzzy diagnosis and advice system (FDAS, shown in Fig. 11) has been specifically designed for Fiat, Renault, and Ford brand cars that were the mostly used cars in Turkey, but can also be customised for other brands of car. Firstly, input values were fuzzified in the fuzzification subsystem. The fuzzified values were then passed to the first inference mechanism where the first rule base was evaluated for the calculation of the fuel consumption. The results from the first inference mechanism were also passed to the second inference system where the second rule base applied. This second inference system decides if there were fuel system fault, ignition system fault, exhaust and intake valves fault or other faults caused by engine wearing, and subsequently suggested advice to repair these faults. The FDAS was tested for 832 cars, which were repaired according to expert system advice. The results were encouraging that 98% of these cars complied again with the emission limits after being diagnosed.

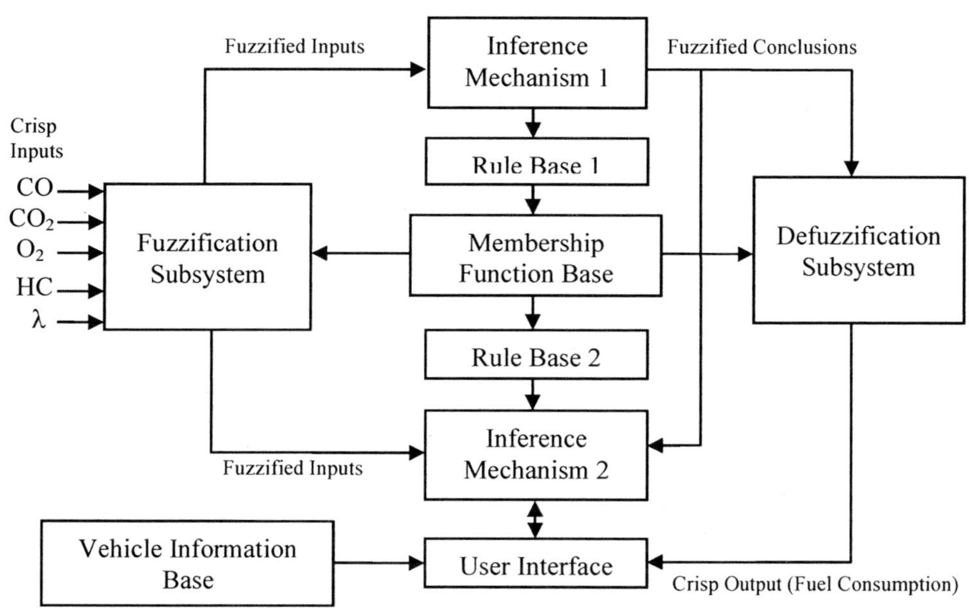

Fig. 11. Overall structure of the fuzzy emission and fuel consumption diagnostic system

3.3.2. Fault Diagnosis for a Cogeneration Internal Combustion Engine

Barelli and Bidini (2005) proposed a diagnostic procedure for internal combustion engines in a cogeneration plant. The plant under study is based on a Caterpillar 3516 alternative internal combustion engine of 1 MW nominal electric power, adapted to natural gas fuel and turbocharged with after-cooler. The engine can be subdivided into several modules and then analysed to extract the main process characteristics that characterise its "health state". A set of diagnostic indices (I) were then derived as:

Aspiration module: $$I_1 = \frac{(P_o - P_i)}{(P_o - P_i)_d} \tag{3}$$

Compressor module: $$I_2 = \frac{(T_o - T_i)}{(T_o - T_i)_d} \qquad I_3 = \frac{(P_o - P_i)}{(P_o - P_i)_d} \tag{4}$$

After-cooler module: $$I_4 = \frac{(T_i - T_o)}{(T_i - T_o)_d} \qquad I_5 = \frac{(P_i - P_o)}{(P_i - P_o)_d} \tag{5}$$

Turbine module: $$I_6 = \frac{(T_i - T_o)}{(T_i - T_o)_d} \tag{6}$$

Engine blocks:

$$I_7 = \frac{A_r}{A_{rd}} \qquad I_8 = \frac{N}{N_d} \qquad I_9 = \frac{T_r}{T_{rd}} \qquad I_{10} = \frac{P_{i,man}}{P_{i,man,d}}$$
$$I_{11} = \frac{(\dot{m}_{gas} / P_{el})}{(\dot{m}_{gas} / P_{el})_d} \qquad I_{cj} = \frac{T_{cj}}{T_{cjd}} \quad j = 1,\ldots,16 \tag{7}$$

where

T_{cj}	Exhaust temperature of j-th cylinder (K),
A_f	Throttle valve opening (degree),
N	Engine rotational speed (RPM),
T_r	Maximum coolant temperature (K),
$p_{i.man.}$	Intake manifold pressure (mbar),
m_{gas}	Natural gas flow rate (kg/s),
P_{el}	Electric power produced (kW).

Subscripts i and o refer to the inlet and outlet conditions, while d indicates nominal operation conditions. Base on the knowledge of the operation of the engine, a series of cause-effect relationships were then derived. For example, clogging of the air filters increases the pressure drop across the aspiration unit (I_1 ↑). This consequent reduction in the actual air intake causes a reduction in the combustion efficiency and hence the power output. This in turn causes the auxiliary throttle valve to open (to compensate for the reduction in the air intake) under the effort of the control system that attempt to maintain the target load (I_7 ↑). In the compressor module, fouling is indicated by an increase in both the compression ratio (I_3 ↑) and the outlet gas temperature (I_2 ↑) at constant outlet pressure. In any case, the occurrence of one of these symptoms means compressor malfunction. Moreover, an increase in the outlet pressure (I_3 ↑), at constant inlet pressure causes reduction in mass flow to the compressor. Under this situation, the control system regulates the opening of the butterfly valve (I_7 ↑). Similar diagnostic reasoning were then formulated for the after-cooler, expansion and the engine block modules thus deriving a set of diagnostic rules for the fuzzy inference system. However, the proposed system has not been tested at that time which the author envisaged was vital to evaluate its actual potential in real implementation.

3.3.3. Gas Turbine Fault Diagnosis

Ogaji *et al.* (2005) investigated the use of fuzzy logic to construct a diagnostic tool for military turbofan gas engine. Generally, the causes of gas-path impairment such as fouling, erosion, tip-clearance changes, corrosion, and ingested-object damage lead to performance deteriorations of the engine, thus resulting in alterations in its heath-component parameters, efficiencies and flow capacities. Turbomatch (simulation software developed at Cranfield University) was used to generate the data used to set up and test the accuracy of the proposed fuzzy-system. According to the authors, this method of producing the rules, using an engine-behaviour model, preserves non-linearities in the gas-path diagnostics. The complete fuzzy system has nine inputs measurements (i.e., mass flow-rate at station-2, low-pressure shaft speed, high-pressure shaft speed, total pressure at station-3, 25 and 26, temperature at station-3 and 25, and fuel flow-rate) and 12 outputs (i.e., two health parameters for each of the six components). In the current study, the usefulness of the system was demonstrated by considering the steady-state cruising condition for the aircraft while focusing only on one of the gas-path components, the intermediate-pressure compressor, whose health can be characterised by two parameter variations, the compressor-efficiency change (devEFFY) and the compressor-flow capacity change (devFC). Two groups of 20 test cases were then generated to assess the effectiveness and accuracy of the developed fuzzy-process and the results showed that the average percentage error for the examined 20 test cases is about 7.5%. The authors also claimed that the complete system could cope with multiple gas-path fault isolation problems that result from the degradation of more than a single engine-component simultaneously.

4. Applications of Genetic Algorithms

Genetic Algorithms (GAs) have been mainly used to solve optimisation problems. Its greatest strength lies in the fact that the analytical function of the process behaviour is not required since they do not require information on the gradient of the solution space. This makes them robust to deploy in highly non-linear and ill-defined problems typically encountered in combustion furnaces and engines.

4.1. Optimisation of Pulverised Coal Burner for Waste Co-Firing

The optimisation of burner operation in conventional pulverised coal fired boilers for co-combustion applications represents a significant challenge. In particular, there are currently no reliable analytical models or techniques which would allow an operator of a large utility boiler to predict whether a particular type of biomass material or waste can be successfully co-fired without causing serious problems such as excessive pollutant emissions, corrosion, ash deposition and degraded combustion efficiency. Consequently, Tan *et al.* (2006) have proposed an intelligent system, which made use of neural networks and genetic algorithms as a possible means to overcome such problems. The effectiveness of the proposed system was demonstrated by a case study that simulates the co-combustion of coal with sewage sludge in a 500kW pilot-scale combustion rig equipped with a swirl stabilised low-NO_x burner. The combustion rig was simulated using commercially available Computational Fluid Dynamics (CFD) software which generated data of different burner operating conditions and then used to train several ANNs to predict the co-combustion performance, such as NO_x, CO and unburned carbon in ash (UBC). Once trained, the ANNs were able to make estimations of unseen situations in a fraction of the time taken by the CFD simulation. Consequently, the networks were capable of representing the underlying physics of the CFD models and can be executed efficiently for a large number of iterations in the optimisation loop. For the present process of co-combustion of coal with sewage sludge in the test rig, there were 4 critical operating variables for the burner, namely, the secondary and tertiary air swirl-angles and flow-rates. The objective of optimising the combustion process can be viewed as the determination of the optimal settings for these operating variables so as to ensure good carbon burnout and low CO and NO_x emissions at the furnace exit. In the present study, two methods were proposed to describe this multi-objective function. In the first method, a linear objective function was defined such that:

$$\phi = wf * \boldsymbol{Y}_{NOx} + (1 - wf)*(\boldsymbol{Y}_{CO} + \boldsymbol{Y}_{UBC}) \tag{8}$$

where

ϕ	The objective value
wf	Weighting factor, from 0 to 1
$\boldsymbol{Y}_{NOx}$	Predicted NO_x by ANN model (ppm)
$\boldsymbol{Y}_{CO}$	Predicted CO by ANN model (ppm)
$\boldsymbol{Y}_{UBC}$	100 - Predicted CBO by ANN model (% by weight)

In the above objective function $\boldsymbol{Y}_{NOx}$ represents the "penalty" of NO_x emissions while the combined terms of $\boldsymbol{Y}_{CO}$ and $\boldsymbol{Y}_{UBC}$ represents the "penalty" of reduced combustion efficiency. It should be mentioned that $\boldsymbol{Y}_{NOx}$, $\boldsymbol{Y}_{CO}$ and $\boldsymbol{Y}_{UBC}$ were scaled so that they always fall within the range 0 to 1 so as to minimise thc undue effects of large magnitudes. Furthermore, the values of wf determine the relative importance of NO_x emissions and the combustion efficiency in the multi-dimension search space and need to be determined *a priori*. In the second method, fuzzy logic was used to encode the multi-objective fitness function using a set of linguistic rules as follows:

(1) IF (NO_x is high) AND (CO is high) AND (CBO is low) THEN (Penalty cost is high)
(2) IF (NO_x is low) AND (CO is low) AND (CBO is High) THEN (Penalty cost is low)
(3) IF (NO_x is medium) AND (CO is medium) AND (CBO is medium) THEN (Penalty cost is medium)

Five operating cases were selected to demonstrate the usefulness of the proposed system. These cases were identified to have exhibited non-optimum burner operation. In particular, they exhibited nominally high NO_x emissions at a relatively low to medium excess oxygen levels (3.1% to 4.6%) while some exhibited unacceptably high CO emission and relatively low carbon burnout (CBO) at a nominally high excess oxygen level (6.2%). The results suggested that this type of approach could be very useful for optimising actual boiler operation where plenty of plant measurements are readily available for the development of this type of system. The paper also concluded that, although both types of objective function have achieved similar results, the fuzzy objective function has the advantage that it eliminates the need to assume a predefined form of the objective function and its ability to adapt to the different weighting factors required to achieve optimum solutions for different burner operating conditions.

4.2. Optimization of Reaction Rate Parameters for Chemical Kinetic Modelling of Combustion

A general inverse procedure for determining the optimum rate coefficients for chemical kinetic schemes based upon limited net species production data is presented by Harris *et al.* (2000). The objective of the optimization process is to derive rate parameters such that, the given net species production rates at various conditions are simultaneously achieved by searching the parameter space of the rate coefficients in the generalized Arrhenius form of the reaction rate mechanisms. Thus, the goal is to match the given net species production rates and subsequently to ensure the accurate prediction of net species production rates over a wide range of conditions. The reaction rate data were retrieved using an inversion technique whose minimization process is based on GAs. The results presented are based upon the recovery of reaction rate coefficients for hydrogen/nitrogen/oxygen flames. The successful identification of the reaction rate parameters which correspond to product species measurement data from a sequence of such experiments clearly suggests that the use of other chemical kinetic schemes and the optimization of higher-order hydrocarbon schemes can now be realized. The results of this study therefore demonstrate that the GA inversion process promises the ability to assess combustion behaviour for fuels where the reaction rate coefficients are not known with any confidence and, subsequently, accurately predict emission characteristics, stable species concentrations and flame characterization. Such predictive capabilities are of paramount importance in a wide variety of industries.

4.3. Optimization of a Distributed Multi-Generation Energy System

A simultaneous optimisation of the thermodynamic, economic and pollutant emissions of a distributed combined heating, cooling and power generation (tri-generation) system in an urban residential area in Beijing has been presented by Li *et al.* (2006). In the current study, gas turbine, internal combustion engine, absorption chiller and gas boiler (centralized and exclusively for hot water demand) were included in the heating/cooling/hot water and power generation plant superstructure. The typical energy supply system currently used for the Beijing residential building sector consisted of centralised natural gas boilers for space heating and hot water supply, electrical chillers for cooling and all of the power demand satisfied by electricity imported from the grid. Furthermore, the units' thermal performance and cost function are derived based on commercial available data. In the analysis, the system Net Present Value (NPV) was set as the objective function to be maximised. It was calculated based on the principles of engineering project techno-economic evaluation and the economic environment of Beijing. A GAs optimiser subsequently solved this resultant mixed integer and non-linear programming problem. This method thus provide the opportunity either to the policy makers to establish the best regulation solution to achieve a certain emission reduction or to the project developers to find out the best energy supply system to invest in order to reach the maximum profit under different economic and environmental legislation contexts.

4.4. Controller Design for Aero-Engines

The design of an optimum fuel distribution in staged combustion of an aero-engine was presented by Breikin *et al.* (2006). Staged combustion represents one of the most effective ways of reducing NO_x while maintaining the other pollutants at low levels. This is achieved by splitting the fuel load among several fuel and air stages, thus enabling the control of the local fuel-to-air ratios (FAR) to minimise NO_x formation rates as the power increases. These optimum patterns for the fuel splits for different engine operating conditions are important to enable an appropriate control strategy to be devised for fuel distribution control and therefore need to be known *a priori*. This steady-state fuel split schedule optimisation over the engine operating range is a multi-objective optimisation problem. It requires finding an optimal fuel split between the pilot and main zones, which would result in the minimisation of NO_x, CO and Unburned Hydrocarbons (UHC) emissions over the International Civil Aviation Organisation (ICAO) reference emissions for landing and take-off (LTO) cycle. Furthermore, the solution should achieve low emissions whilst simultaneously ensuring against both weak and rich extinction and incomplete combustion, so a set of non-linear design criteria needs to be satisfied simultaneously. In this multi-objective optimisation, the GA approach demonstrated advantages over conventional optimisation approaches with the capability of handling function discontinuity, multi-modality and employing a mixture of decision variable types. Furthermore, its population-based search approach offers the opportunity to evolve a family of solutions.

4.5. Model of Three-Way Catalytic Converters During the Warm-Up Phase

New regulations for emission control require the improvement of the system composed by spark ignition ICE and three-way catalytic converter (TWC). In particular, an important problem is to minimize harmful emissions during the transient warm-up phase where the TWC is not working yet and, hence, a large amount of pollutants are emitted in the air.

Toward this goal a dynamical thermo-chemical TWC model simple enough for the design and test of warm-up control strategies is presented by Glielmo and Santini (2001). The model is obtained through art asymptotic approximation of a more detailed model, i.e., by letting the adsorption coefficient between gas and substrate tends to infinity. Further, the authors presented a fast integration algorithm based partly on a method of linear space-discretisation and partly on the method of characteristics for quasi-linear hyperbolic partial differential equations; the separation being allowed by a two-time scale analysis of the system. The model has been identified, through a purposely designed GA and validated on experimental data.

5. Conclusions

Over the last few decades, artificial intelligence has produced a number of powerful tools. This chapter has reviewed three of those tools, namely, neural networks, fuzzy logic and genetic algorithms. From the various applications presented in this chapter, it can be realised that the solutions to practical combustion system have grown out of conventional approaches as it became clear that more advanced techniques were required to analyse and discover knowledge from these highly complex and non-linear processes. Applications of these AI tools have become more widespread due to the power and affordability of present-day computers. Most of the presented work shows that artificial intelligence techniques do not only provide solution to problems, but also serve as powerful knowledge management tools, which can be used to better describe the actual behaviour in real-world combustion applications. Among these techniques neural networks have an edge over the others, mainly because of their ability to learn from large amounts of available data and subsequently produce some inferential/predictive models that can be very useful in many situations. Overall, the work presented in this chapter suggested that the use of artificial intelligence techniques has helped to solve many combustion related problems, which have not been possible by conventional approaches and/or resulted in the improvement of system performances.

References

Abbas T., Awais M.M., Lockwood F.C., 2003. An artificial intelligence treatment of devolatilization for pulverized coal and biomass in co-fired flames. *Combustion and Flame*, Vol. 132, No. 3, pp. 305-318.

Ashhab S., 2004. A combination of neural net modelling and constrained optimization towards inverse control. *Proceedings of the Fourth IASTED International Conference on Modelling, Simulation and Optimization*, August 17-19 2004, Kauai, HI, United States, pp. 66-71.

Barelli L., Bidini G., 2005. Design of the measurements validation procedure and the expert system architecture for a cogeneration internal combustion engine. *Applied Thermal Engineering,* Vol. 25, No. 17-18, pp. 2698-2714.

Berry P., Kukla P., Fengshou G., Ball A.D., 2004. Diesel engine pollutant prediction and remote vehicle monitoring Part I: The prediction of diesel engine smoke emission using neural networks. *International Journal of COMADEM*, Vol. 7, No. 4, pp. 24-30.

Breikin T.V., Herbert I.D., Kim S.K., Regunath S., Hargrave S.M., Thompson H.A., Fleming P.J., 2006. Staged combustion control design for aero engines. *Control Engineering Practice*, Vol. 14, No. 4, pp. 387-396.

Channiwala S.A., Parikh P.P., 2002. A unified correlation for estimating HHV of solid, liquid and gaseous fuels. *Fuel*, Vol. 81, No. 8, pp. 1051-1063.

Chong A.Z.S, Wilcox S.J., Ward J., 2000. The development of a neural network based system for the optimal control of chain-grate stoker-fired boilers. *Proceedings of ASME Heat Transfer Division*, Vol. 363, No. 3, pp. 103–109.

Clarke D.W., Mohtadi C., Tuffs P.S., 1997. Generalised predictive control. Part 1: the basic algorithm and part 2: extensions and interpretations. *Automatica*, Vol. 23, No. 2, pp. 137–160.

Dong C., Jin B., Li D., 2003. Predicting the heating value of MSW with a feed forward neural network. *Waste Management*, Vol. 23, No. 2, pp. 103-106.

Glielmo L., Santini S., 2001. A two-time-scale infinite-adsorption model of three way catalytic converters during the warm-up phase. *Journal of Dynamic System Measurement and Control,* Transaction of ASME, Vol. 123, No. 1, pp. 62–70.

Harris S.D., Elliott L., Ingham D.B, Pourkashanian M., Wilson C.W., 2000. The optimization of reaction rate parameters for chemical kinetic modeling of combustion using genetic algorithms. *Computer Methods in Applied Mechanics and Engineering*, Vol. 190, No. 8-10, pp. 1065-1090.

Heitz M.W., George B., Welp J.E., 1997. Using neural networks to predict incinerator emissions: A case study. *Proceedings of the 1997 Air and Waste Management Association's 90th Annual Meeting and Exhibition on CD ROM*, Paper No: 97-MP11.01, Jun 8-13 1997, Toronto, Canada.

Hendricks E., Chevalier A., Jensen A., Sorenson S.C., 1996. Modelling of the intake manifold filling dynamics, SAE paper 960037.

Isermann R., 2005. Model-based fault-detection and diagnosis – status and applications. *Annual Reviews in Control*, Vol. 29, No. 1, pp. 71-85.

Kalogirou S., 2003. Artificial intelligence for the modelling and control of combustion processes: A review. *Progress in Energy and Combustion Science*, Vol. 29, No. 6, pp. 515-566.

Kenneth R.H, James C.H., 1995. Heating value, relative density and compression factor for dry or wet natural gas from composition. *Flow Measurement and Instrumentation*, Vol. 6, No. 2, pp. 121-125.

Kibrya G., Botros K.K., 2002. A neural network based predictive emission monitoring model for NO_x emission from a gas turbine combustor. American Society of Mechanical Engineers, *Petroleum Division (Publication) PD*, Vol. 1, pp. 137-145.

Kilagiz Y., Baran A., Yildiz Z., Çetin M., 2005. A fuzzy diagnosis and advice system for optimization of emissions and fuel consumption. *Expert Systems with Applications*, Vol. 28, No. 2, pp. 305-311.

Li H., Nalim R., Haldi, P.A., 2006. Thermal-economic optimization of a distributed multi-generation energy system-A case study of Beijing. *Applied Thermal Engineering*, Vol. 26, No. 7, pp. 709-719.

Ljung L., 1999. *System identification - Theory for the user*, Prentice Hall. , Englewood Cliffs, New Jersey.

Maertens K., Johansen T.A., Babuka R., 2004. Engine load prediction in off-road vehicles using multi-objective nonlinear identification. *Control Engineering Practice*, Vol. 12, No. 5, pp. 615-624.

Mu J., Rees D., Liu G.P., 2005. Advanced controller design for aircraft gas turbine engines. *Control Engineering Practice*, Vol. 13, No. 8, pp. 1001-1015.

Naim A.K., Mutasim A.S., Niels J.S., 2004. Emissions and fuel economy trade-off for hybrid vehicles using fuzzy logic. *Mathematics and Computers in Simulation*, Vol. 66, No. 2-3, pp. 155-172.

Ogaji S.O.T., Marinai L., Sampath S., Singh R., Probert S.D., 2005. Gas-turbine fault diagnostics: a fuzzy-logic approach. *Applied Energy*, Vol. 82, No. 1, pp. 81-89.

Parikh J., Channiwala S.A., Ghosal G.K., 2005. A correlation for calculating HHV from proximate analysis of solid fuels. *Fuel*, Vol. 84, No. 5, pp. 487-494.

Rangaswamy T.R., Shanmugam J., Mohammed, K.P., 2005. Adaptive fuzzy tuned PID controller for combustion of utility boiler. *Control and Intelligent Systems*, Vol. 33, No. 1, pp. 63-71.

Ruusunen M., Leiviskä K., 2004. Fuzzy modelling of carbon dioxide in a burning process. *Control Engineering Practice*, Vol. 12, No. 5, pp. 607-614.

Salehfar H., Benson S.A., 1998. Electric utility coal quality analysis using artificial neural network techniques. *Neurocomputing*, Vol. 23, No. 1-3, pp. 195-206.

Sharkey A.J.C., Chandroth G.O., Sharkey N.E., 2000. A multi-net system for the fault diagnosis of a diesel engine. *Neural Computing and Application*, Vol. 9, pp. 152-160.

Shen K., Lu J., Li Z., Liu G., 2005. An adaptive fuzzy approach for the incineration temperature control process. *Fuel*, Vol. 84, No. 9, pp. 1144-1150.

Slatsky M.D., Allison J.S., Page G.C., Noblett, J.G., Stallings J.W., 1997. Advanced on-line optimization tool for electric utility boilers - an update. *Proceedings of the 7th Annual ISA POWID/EPRI Controls and Instrumentation Conference*, June 16-18, 1997, Knoxville, TN, USA, Vol. 40, pp. 221-232.

Tan C.K., 2003. How do I Use Artificial neural network to predict the calorific value of solid fuels. *Online Combustion Handbook*, International Flame Research Foundation (IFRF), File no. CF237, 6pp.

Tan C.K., Wilcox S.J., Ward J., Lewitt M., 2003a. Monitoring near burner slag deposition with a hybrid neural network system. *Measurement Science and Technology*, Vol. 14, No. 7, pp. 1137-1145.

Tan O.H., Wilcox S.J., Ward J., Lewitt M., 2003b. The development of a monitoring and control system for pulverised coal flames using neural networks. *Proceedings of ASME 2003 International Congress*, Paper No: IMECE2003-41409, pp. 91-98.

Tan C.K., Wilcox S.J., Ward J., Lewitt M., 2005. An intelligent monitoring system for the detection of slag deposition on a pulverized coal fired burner. Proceedings of the Institution of Mechanical Engineers, Part A: *Journal of Power and Energy*, Vol. 219, No. 5, pp. 383-394.

Tan C.K., Wilcox S.J., Ward J., 2006. Use of artificial intelligence techniques for optimisation of co-combustion of coal with biomass. *Journal of the Energy Institute*, Vol. 79, No. 1, pp. 1-7.

Tuntrakoon A., Kuntanapreeda S., 2003. Image-based flame control of a premixed gas burner using fuzzy logics. *14th International Symposium, ISMIS 2003, Lecture Notes in Artificial Intelligence*, Vol. 2871, pp. 673-677.

Wang S.W., Yu D.L., Gomm J.B, Page G.F., Douglas S.S., 2006. Adaptive neural network model based predictive control for air–fuel ratio of SI engines. *Engineering Applications of Artificial Intelligence*, Vol. 19, No. 2, pp. 189-200.

Yin C., Luo Z., Ni M., Cen K., 1998. Predicting coal ash fusion temperature with a back-propagation neural network model. *Fuel*, Vol. 77, No. 15, pp. 1777-1782.

Zhu Q., Jones J.M., Williams A., Thomas K.M., 1999. The predictions of coal/char combustion rate using an artificial neural network approach. *Fuel*, Vol. 78, No. 14, pp. 1755-1762.

In: Artificial Intelligence in Energy and Renewable Energy Systems ISBN 1-60021-261-1
Editor: Soteris Kalogirou, pp. 395-438

Chapter 12

LEARNING CONTROL OF FLUIDIZED-BED COMBUSTION PROCESSES FOR POWER PLANTS

Enso Ikonen and Jenő Kovács
Systems Engineering Laboratory
Department of Process and Environmental Engineering
University of Oulu, Finland

ABSTRACT

Applications of learning control in fluidized bed combustion (FBC) are considered. This chapter begins with a brief discussion on learning control of complex systems. The basic control structures in typical FBC power plants are described, followed by discussion on applications of advanced techniques of modelling, identification and control in fluidized bed power plants. Emphasis is put on adaptivity and learning in the applications. Studies on three artificial intelligence (AI) approaches are then considered in more detail.

First, modelling using distributed logic processors (DLP) is considered. Experiences in identification of steady-state FBC flue gas emissions using these fuzzy relational modelling techniques are discussed. The importance of transparency of the modelling approach is emphasized.

Second, on-line identification of non-linear time-varying process models is considered. Focus is on the collection of an information rich set of process data, based on clustering of the incoming measurement data. To illustrate the approach, modelling of multi-fuel FBC emissions is considered, using data collected from a 25 MW district heating thermal power plant.

In a final application example, genealogical decision trees in tracking control of a fluidized bed combustion plant are considered. This novel population-based optimization technique, closely related to techniques of particle filtering, is applicable to a wide scope of optimization problems with a dynamic character. The approach is illustrated considering a multivariable tracking problem, using a dynamic model for a FBC plant.

AI-inspired techniques have a lot to offer when developing methods for advanced identification, monitoring, control and optimization. The techniques can be easily extended for applications in industrial process environments other than power plants.

Keywords: boilers, control, fuzzy systems, identification, modelling, neural nets, optimisation, population-based systems, power plants, thermal plants.

NOMENCLATURE

A	Input covariance/weighting matrix
B	Output covariance/weighting matrix
c	Concentration (Nm^3/ Nm^3)
f	Some nonlinear function
F	Air flow (Nm^3/s)
	Some nonlinear function, state equation (in Section 4)
h	Some nonlinear function, measurement equation
J	Cost
k	Sampling index
K	Number of samples
N	Number of particles
P	Power (MW)
Q	Fuel feed (Q_C) (kg/s),
	Number of logic processors (in Section 2)
	Fuel type (in Section 3)
	Dimension of system output (in Section 4)
t	Time (s)
T	Horizon length
u	Controller output
u, U, v	State-space system input / controller output
y	Model output
y, Y	State-space system output
x, X	System state
W	Fuel inventory (kg)
z	Wiener model intermediate variable
μ	Stoichiometric ratio (mol/mol)
φ, ϕ	Input vector
θ	Parameter vector

Subscripts

1	Primary
2	Secondary
c	Recirculation
C	Char
Hz	In Hertz
kg	In kilograms
n	Sampling index

NOx	NO_x
O2	Oxygen
r	Residence time

Superscripts

^ (hat)	Estimate
ref	Reference
T	Matrix transpose

ABBREVIATIONS

AI	Artificial Intelligence
ARMAX	Auto-Regressie Moving Average eXogenous input
BI	Bounded Input
CFBC	Circulated Fluidized Bed Combustion
CHP	Combined Heat and Power
CN	Combined Network
CPC	Combustion Power Control
DLP	Distributed Logic Processor
DWLP	Distributed Wiener Logic Processor
FBC	Fluidized Bed Combustion
GPC	Generalized Predictive Control
IMC	Internal Model Control
LP	Logic Processor
LTI	Linear Time-Invariant
MIMO	Multiple-Input Multiple-Output
MISO	Multiple-Input Single-Output
MRA	Model Reference Adaptive
MRMSE	Mean RMSE
MSE	Mean Squared Error
NARX	Nonlinear Auto-Regressive eXogenous input
NOE	Nonlinear Output-Error
NARMAX	Nonlinear ARMAX
PI	Proportional-Integral
PIP	Proportional-Integral-Plus
RLS	Recursive Least Squares
RMSE	Root Mean Squared Error
RPE	Recursive Prediction Error
SIMC	Skogestad IMC
SIR	Sampling Importance Resampling
SISO	Single-Input Single-Output
SNN	Sigmoid Neural Network

SOM	Self-Organizing Map
WLP	Wiener Logic Processor

1. INTRODUCTION

Most engineering approaches to *artificial intelligence* (AI) are characterized by two fundamental properties: the ability to learn from various sources and the ability to deal with plant complexity. Learning systems that are able to operate in uncertain environments based on incomplete information are commonly referred to as being intelligent. There exists a number of other systems characterized by these properties, that are not easily categorized as AI-systems. Therefore, in this first section, the role of the more general category of *advanced* techniques in the field of power plant automation is considered. In particular, the aim is to present and discuss views to advanced learning control of fluidized bed combustion (FBC), and thereby to sketch the potential role of AI-methods in this field. The discussion is by no means complete, and does not attempt to provide a review on this vast topic. On the contrary, a subjective approach is adopted, reflecting the experiences of the authors.

First, learning control of complex systems is discussed. This is followed by a brief introduction to the fluidized bed combustion (FBC) process and its control. Methods of artificial intelligence have found most applications in advanced process control, where the intelligent and learning abilities of these systems can be fully exploited. Discussion of some sample elements of an advanced FBC control system ends the first section, highlighting the learning aspects in the sample applications.

1.1. Learning Control of Complex Systems

Let us start by giving a loose characterisation of learning control of complex systems. A *complex system* is a possibly large-scale system that is significantly non-linear (*e.g.*, due to constraints or process nonlinearities), multivariable (with severe interactions between process variables), uncertain (*e.g.*, probabilistic or fuzzy), and/or time varying (requiring repeated tuning or on-line learning). Even one of the aforementioned characteristics – when significant enough – may lead to a complex system. A system may also be considered to be a complex one due to other reasons than discussed above (forms of available information, conflicting objectives, speed requirements, *etc.*). Single-input single-output (SISO) linear time-invariant (LTI) systems, or systems that can be approximated to a sufficient accuracy with SISO LTI systems, are not complex systems. As a consequent, techniques based on the SISO LTI assumption are not sufficient for controlling complex systems.

Advanced control methods (adaptive, predictive, multivariable, robust, *etc.*) are based on the availability of a model of the process to be controlled. Hence, *identification* of process models may be a key issue. Identification includes, among other things, structure and parameter estimation. In non-linear structures, iterative estimation techniques are common; at each iteration the system is progressively adapted towards better performance. This leads to the use of adaptation and learning techniques.

To adapt is to change a behaviour to conform to new circumstances (Ikonen and Najim, 2002; Brown and Harris, 1994). An *adaptive system* is able to adjust to the current environment. It is memoryless in the sense that all new information adjusts the old one. A *learning system*, instead, is able to recognise the current environment (state) and to recall previously learned associated information from its memory: A learning system is an adaptive system with memory, and one may expect that a learning system improves its behaviour with time. Learning systems are particularly useful whenever complete knowledge about the environment is *a priori* unknown, expensive to obtain, or impossible to quantify.

When learning systems are synthesised with modelling techniques, *learning models* emerge. Furthermore, when learning models are applied in advanced control methods, this results in *learning control* systems. A typical learning control system concerns a learning technique applied for updating a process model, which in turn is used in the controller design. From this point of view, particularly close connections exist between indirect adaptive control and learning control.

Learning control of complex systems can include other types of systems as well; learning may involve some other part of the control system (the controller parameters directly, control objectives, actuator constraints, adjustment rates, *etc.*).

1.2. Fluidized-Bed Combustion Process

In fluidized bed combustion (FBC) (see, *e.g.*, Hupa and Boström, 1991; Kunii and Levenspiel, 1991), the combustion chamber contains a large quantity of finely-divided particles (called bed material) such as sand. The primary combustion air, entering from below through a distribution plate, lifts these particles until they form a turbulent bed, which behaves like a boiling fluid. Start-up burners heat the bed to some 500°C, at that temperature any combustible material fed into the bed ignites almost immediately. The fuel is added to the bed and the mixture of sand and fuel is kept in constant movement by the primary air. The heat released as the material burns maintains the bed temperature, and the turbulence keeps the temperature uniform through the bed. The heat capacity of the solid bed particles gives the system thermal stability, which makes variations in fuel properties less critical than with many other combustion systems. The normal operating temperature of the bed is between 750-950°C; at these relatively low temperatures the ash and moist fuels do not melt or sinter and the formation of NO emission is relatively low, compared for example to pulverised combustion. Fuel properties like ash content, particle size and moisture, are of less importance. However, agglomeration may occur when using high alkali content biomass fuels.

Fluidized bed combustion (FBC) is distinguished by:

- low average combustion temperatures (~850°C)
- high combustion efficiency ~95%
- flexibility to fuel quality
- low NO_x emission
- easy SO_2 reduction
- high excess air levels (30%),

- intermediate particle sizes (1--3 mm),
- long residence times of fuel particles (several minutes) and
- vigorous particle motion (dominates heat transfer and reaction processes.)

There are many reasons for using fluidized bed combustion in energy production. Low flue gas emission levels can be achieved and many types of solid fuels can be used, either separately or mixed with each other. Fluidized bed combustors burn a wide variety of fuels including coals (ranging from anthracite to lignite), peat, wood chips and saw dust, bark, sludge, municipal waste, recycled fuels (REF) and refuse derived fuel (RDF). Fuels are burned alone or in combination. Low ash fusion temperature coals and peat can be burned without slagging or sintering, because the boiler operates at temperatures well below the ash melting point. High ash fuels are ideal since the inherent ash forms the inert bed material. RDF can be burned successfully because good mixing prevents formation of stratified pockets of incomplete combustion. High moisture fuels are burned without difficulty since there is a large reservoir of heat to dry the wet incoming fuel. Fuel treatment is simple since the combustion is fairy insensitive to fuel particle size, in the sense that large pieces of fuel automatically remain suspended in a fluidized bed for a longer period of time.

There exists a variety of different combustors: bubbling, circulated, pressurized:

- In bubbling FBC (BFBC), the velocity of the raising air stream through the bed is about 1-3 m/s. Bubbling FBC is characterised by a dense bed in the bottom of the reactor with a clear phase above, the freeboard. Bed is usually 0.7-1.1 m deep and most of the combustion takes place in the bed (depending on the fuel). The hot combustion gases burst through the surface of the bed, and pass through the freeboard to the heat-transfer surfaces. Secondary and tertiary air is injected at the freeboard to provide complete combustion and optimization of emissions reduction.
- Circulated FBC (CFBC) uses high gas velocities, at most 6-10 m/s, and the rising air carries the bed solids along with it. Thus, no upper surface is defined. In order to avoid emptying the bed, it is then necessary to separate the particles that have been removed from the bed and return them, thus giving circulation of the bed material. Recirculation provides also longer residence time for the fuel particles for complete combustion. The high velocity gas stream gives good mixing of sand and burning fuel particles. There is, thereof, a high level of combustion and good transfer of heat to the walls of the boiler.
- Pressurized fluidized-bed combustion (PFBC) technology uses a combustion process similar to that of atmospheric FBC (CFBC and BFBC), but the boiler operates at higher pressures (from 0.5 to 2 MPa). The gas is cleaned downstream from the PFBC boiler, and expanded in a gas turbine. In this combined-cycle (steam-gas) arrangement, a 40-45 % overall efficiency of power generation can be achieved.

The fluidized bed boiler is particularly suitable when it is intended to use widely differing fuels with varying heat values. This is the case when using fuel with low heat values (coal as support fuel), coal with high sulphur content, fuel which contains a high proportion of fines, *etc.* The lighter fractions of the fuel are returned to the reactor through grille and cyclone separators. Multi-fuel capability is also due to the ability to alter the heat transfer coefficient by shifting bed inventory from the lower to the upper furnace. The bubbling bed is ideal for chemical reaction and drying applications, but is inferior to CFBC in the flexibility required by multi-fuel operation.

The plant units range from small units of less than 3 MW up to over 550 MW for biomass fuels and 1000 MW using coal; some examples include:

- Alholmens Kraft power plant in Pietarsaari, Finland, burns peat, bark, REF, forest residues, sawdust, coal and oil. This 550_{th} MW circulated FBC is the world's largest CHP plant;
- Rauhalahti power plant in Kuopio, Finland, is a 295 MW_{th} bubbling FBC CHP, burning peat and wood;
- Lagisza coal power plant in Poland is a circulating FBC with 970 MW_{th} and 460 MW_{e}.

1.3. Basic Control Loops

The main control loops (see, *e.g.*, Klefenz, 1986) of the combustion and steam generation processes in a typical constant pressure boiler-following control scheme FBC plant are as follows:

- steam pressure - fuel feed and corresponding primary air flow
- oxygen trim - secondary air
- steam temperature - spray water
- drum level - feed water
- furnace pressure - flue gas induced draught fan
- bed pressure - bottom ash screw
- SO_2 capture - lime feed
- furnace temperatures, fluidization - auxiliary fuel feeds, air distribution

The main focus of FBC control is between the relationships of fuel, air and steam flows. To make it simple, the main loop in the control of an industrial fluidized bed boiler is the control of steam pressure by fuel feed. A three-point controller can take into account the non-minimum phase phenomena of the two phase system in drum level control. The oxygen concentration in flue gases is a demonstrative measure of the combustion conditions, and is controlled by the secondary air flow (oxygen trim). Together with furnace oxygen content, the bed temperatures are a key element in controlling the flue gas emissions. Bed temperatures can be controlled by fuel flow (supporting fuel), primary air flow and flue gas recirculation, but are usually not under automatic control.

A coal-fired plant is relatively stable to control, but has the problem of fuel inventory. The control strategy of peat fired FBC differs from that of the high-grade fuels, as peat burns fast in the furnace and the variations in fuel quality are unpredictable and significant. Often, the construction and combustion strategy of a peat-fired FBC is close to the multi-fuel boilers (coal used as a support fuel) where the basic principle is to control (limit) excessive burning in the bed. This can be done by reducing the oxygen concentration in the bed using lower primary air flows (staging) or by lowering the oxygen concentration in the air flows (flue gas recirculation).

1.4. Advanced Control of FBC

In the control of power plants, it is helpful to categorize the tasks of automated operations:

(1) Stabilisation level: local regulators and controllers
(2) Supervisory control level: selection of optimal set points and trajectories
(3) Production planning level: plant optimisation

In principle, the level 1 needs simple and robust methods; complicated structures should be avoided. Level 2 is a good candidate for methods of learning control, such as model-based techniques. This level is also characterized by the need to interact with operators, in parallel with performing automatic control. Plant optimisation (level 3) is probably too important (and contains too many uncertain factors) to be left to a fully automated controller. However, means for aiding human decision making can find use of the learning and optimisation methods. Applications in level 2 require dynamic models, whereas static descriptions are usually sufficient for level 3 tasks.

The steam side is relatively well understood and can be solved using conventional approaches, whereas the open control problems in a FBC largely relate to the manipulation of combustion conditions. These include large variety of fuels and their quality, inhomogeneous and stochastic combustion conditions, complexity of the emission formation, and lack of proper measurements from within the furnace. In our opinion this part of power plant control may greatly benefit from the application of the artificial intelligence techniques.

The control of a multi-fuel FBC process can be posed as a hierarchical control problem (Ikonen and Kortela, 1994), as shown in Fig. 1. The upper stage computes optimal set-point values for the lower stage, which tries to track the given set points in the presence of deterministic and stochastic disturbances and process constraints. In long term optimisation (Henttonen *et al.*, 1992), the emissions have to be taken into account when calculating optimal set point values for the process controllers. In combustion power control, the short term disturbances in fuel feed are compensated.

Reported applications of AI-innovated techniques in power plant monitoring, control and optimisation include load forecasting, coal quality monitoring and furnace modelling, sensor interpretation and life extension, as well as fuzzy versions of lower level air and fuel control.

In the remaining part of this section, a closer look is taken at fuel inventory control, combustion power control, optimisation of emissions, oxygen modelling and control, and drum control.

The topics of combustion power control and optimisation of emissions are further separated, as the combustion control is rather concentrating on the combustion stabilisation eliminating the disturbances from the fuel feed (varying properties of fuel, fuel feeder discrepancy) and the emission optimisation is dealing with airflow control of the freeboard ensuring the optimal combustion conditions. Both of these problems are related to the flue gas oxygen content, as it appears as an intermediate measured variable in the former, but it is the controlled variable in the latter. The role of learning control in these applications is discussed and some case examples related to FBC are briefly presented. A more thorough examination of three case studies is given in Sections 2-4.

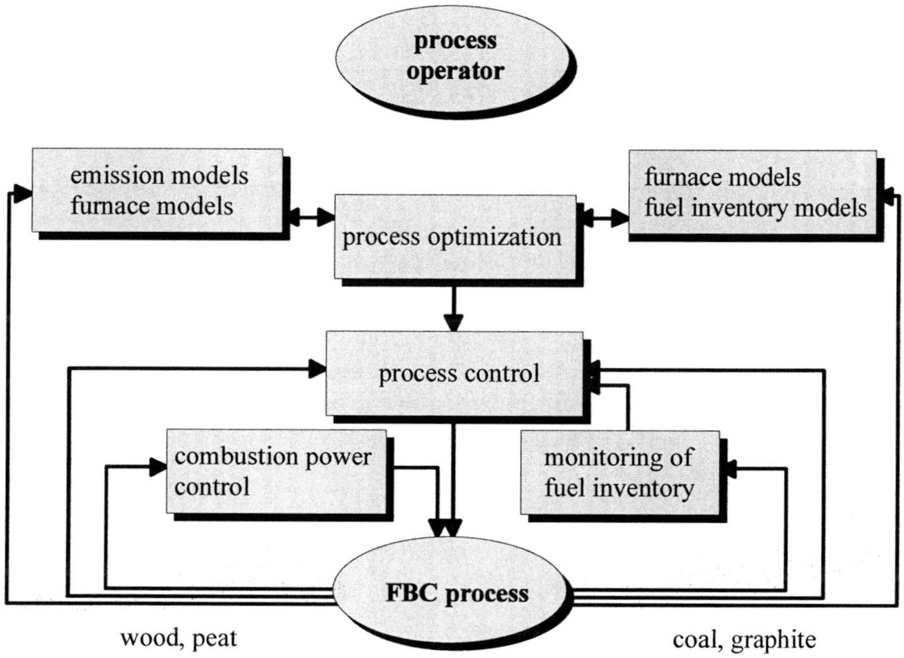

Fig. 1. Control concept for a multi-fuel fluidized bed combustor

1.4.1. Fuel Inventory Modelling

Bed fuel inventory, also called bed carbon load or carbon hold-up, can be understood as the amount of fuel in the bed that has not burned yet. It is one of the key variables when burning fuels such as coal in the bubbling FBC process. It describes the potential amount of heat available in the bed and has an important role when power and emission controls are being developed.

Formation of fuel inventory can be a very long process and is difficult to capture from common process measurements. Serious problems, such as bed smelting and sintering, may occur if the fuel inventory gets too high and more air is introduced to the bed, thus leading to an excessive rise in bed temperatures. Insufficiently low inventory levels may cause difficulties in controlling the combustion power and limit fast responses in boiler power during load changes. Higher oxygen content in the bed may also increase nitrogen emissions in the flue gas.

A furnace model was suggested in Ikonen and Kortela (1994) consisting of differential equations which describe the dynamic behaviour of bed fuel inventory, bed emulsion phase oxygen content, flue gas oxygen content, bed temperature, freeboard temperature, power output and drum pressure. Fuel inventory calculations were based on a steady-state shrinking core model for the burning of char particles (Donsi *et al.*, 1986). Flue gas oxygen content was calculated by making a molar balance over the freeboard; bed and freeboard temperatures by making an energy balance over the bed and freeboard, respectively. Power output and drum pressure were also considered in the model to make the usability of it better for control simulation purposes. The model was adapted to measurements from a 25MW bubbling FBC, and relatively accurate simulations could be obtained.

Iteration of the steady-state model turned out to be time-consuming, however. This problem could be overcome to some extent by training a sigmoid neural network (SNN) model to mimic the behaviour of the steady-state model (char combustion rate). Within a restricted operating area (fixed boiler design, fixed particle size distribution, *etc.*), the results were good giving computing speeds suitable for on-line simulations.

Due to the integrating nature of the bed fuel inventory, it is difficult to capture the amount of inventory using only steady-state information. Due to lack of direct measurements, it is difficult to find out the slow long-term changes due to, *e.g.*, fuel quality changes because bias and noise are always present in the plant measurements. However, an on-line indication of the fuel inventory would be very useful for adaptive control purposes. For this reason, some attempts (see, *e.g.*, Ikonen and Kotajärvi, 1993) were made to measure the most important input and output variables of the process, estimate the parameters of a time-series model recursively, and to correlate them with the measured bed fuel inventory. The essence of this approach is in on-line learning; adaptation of linear models based on noisy measurements. A functional relationship between changes in the model parameters and the fuel inventory was found; however, obtaining experimental information about the fuel inventories turned out to be impractical and the results were difficult to extend to other fuels and plants.

By the end of the century, the industry and vendors became more interested in circulated FBC's where the control of fuel inventory is less critical. In Karppanen (2000), interest was restricted to the case of temporal dramatic increases in char content in a circulated FBC. This was done since a rapid change in fuel inventory was seen as being the main control problem, due to the threat of sintering. The approach was based on fuzzy decision making. A number of process inputs (feed changes, state of bed and flue gas, *etc.*) were fuzzified into suitable classes and used in rule-based decision making with simple rules. This approach indicated possibilities to include 'experience-based' information in a more convenient way. Notice, that plant operators typically have a set of their own rules-of-thumb for estimating the state of the bed. These are often heuristic and subjective (even intuitive), and based on information that is difficult to measure (*e.g.*, looking at the flames inside the furnace, feeling the vibrations of the plant, *etc.*) The system was applied in a pilot-scale CFBC thermal plant with some success.

1.4.2. Combustion Power Control

The simplest way to improve thermal efficiency and reduce nitrogen emissions (at a given temperature) is to decrease the excess air ratio. In order to be able to do this, more stable burning conditions are required, as even short drops in the bed oxygen concentration may produce CO up to the amount of being explosive.

Combustion power control (CPC) aims at stabilising the furnace conditions when burning inhomogeneous fuels. In the basic FBC control configuration, the main loop is the control of steam pressure by fuel feed, complemented by the oxygen trim. When a disturbance in the fuel feed occurs, such as a change in the effective heat value, the control system compensates the fuel feed as soon as the disturbances are detected in the drum pressure. This means that the disturbance will propagate though the furnace. The oxygen trim reacts faster, but acts on the secondary air flows with minor effect in maintaining the power level. The idea of combustion power control (Kortela and Lautala, 1981) is based on fast estimation of the combustion power from oxygen consumption in the furnace, and a direct application of this information in regulation by compensating the fuel feed accordingly. In this way the disturbances in fuel power can be detected and compensated before they affect the drum pressure. When compensating the fuel feed, the heat transfer and fluidization are not affected as they are when acting on the air supply, thus leading to more stable burning conditions. In principle, the idea of combustion power control is applicable when burning fuels that have high reactivity and which don't accumulate in the bed.

CPC has been successfully applied in pulverized plants. Also numerous applications to fluidized bed combustors have been considered and implemented. In Karppanen (1998) the optimisation of fuel feeds in a multi-fuel CFBC is considered, with the estimation of fuel heat value based on a fuzzy combustion power estimator. Fuzzy implementation of the steam pressure PI-controller was also considered. In this type of lower level control systems, however, the additional value brought by using AI-inspired techniques over more conventional methods can be difficult to quantify.

1.4.3. Flue-Gas Emission Optimization

Harmful flue gas emissions such as carbon monoxide and dioxide (CO, CO_2), nitrogen oxides (NO_x, N_2O), and sulphur oxides (SO_2) are a result of the complex burning phenomena and the individual construction of the plant in question. Various national emission taxation procedures and emission trading, such as the ones resulting from the Kyoto agreement, have made their minimisation a profitable task. In addition to the developments in the plant construction and flue gas cleaners, the process operating conditions are also an important and cost-effective way to affect these emissions.

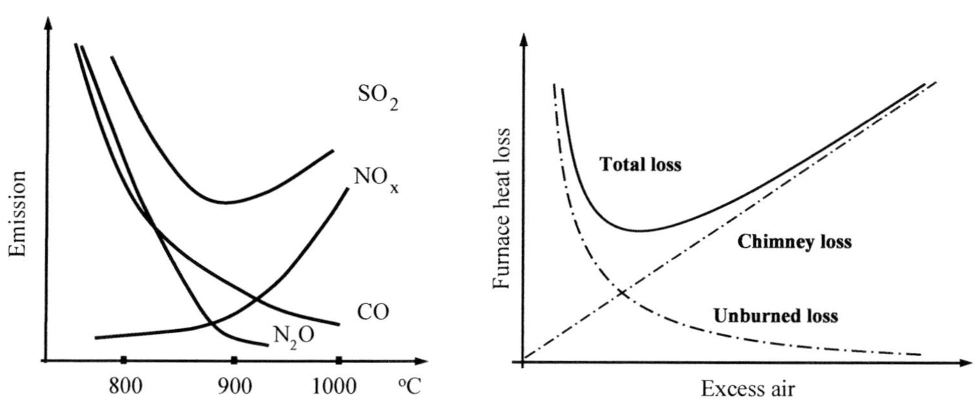

Fig. 2. Illustration of the dependencies of emissions vs. temperatures, and heat loss vs. excess air ratio

The properties of the combustion (solids residence time, bed temperature, excess air, air staging, bed pressure and calcium addition) strongly effect the production of different flue gas components. The complexity of the problem lies on the fact that neither of the mentioned properties modifies all the components in the same manner, but rather decreasing a component results in increasing another one. Figure 2 illustrates the temperature dependency of emissions (see Hupa, 2005; Lindsley, 2000; Leckner, 1997). The optimization of flue-gas emissions becomes therefore an important factor in the maximization of plant profitability.

The flue-gas oxygen content is an important indicator of the level of emission formation and the thermal efficiency of a FBC boiler. The optimal level of oxygen content is determined by the flue-gas thermal losses, the CO losses and the emissions. The optimal set point depends also on the load level, ranging typically between 3% and 6%. If the oxygen content in the furnace gets too low, a plant shutdown occurs due to safety reasons. Due to inevitable fluctuations, the set point then needs to be set higher than the optimal one. Applying advance control, the fluctuations of the oxygen content can be decreased, thereby allowing lower mean levels and smaller thermal losses.

In high-level emission optimisation tasks of practical importance, one has to make a compromise between several different objectives. In a "simple" case, the target of the optimization is to find optimal long term set values for process control system with respect to harmful emissions and thermal efficiency. In order to be able to optimise the plant operation, models for the variables of the overall cost function are required. In Ikonen and Kotajarvi (1993) and Henttonen *et al.* (1992), optimisation using linear models was considered. These models can also be used in expert systems giving information, help and suggestions to the plant operators. In Leppäkoski (1999), an expert system was developed and installed to a bubbling fluidized bed boiler to give information about flue gas emissions and to advice plant operators when NO_x-, SO_2-, and CO-emission limits were exceeded. When used for training, the system suits well both on-line and off-line use, as the user can immediately see the effects of his/her actions.

Since FBC boilers accept a large variety of fuels, the fuel mixture can be used for optimization. For example, in multi-fuel biomass combustion a typical combination is that of peat and wood. The emission formation can be greatly influenced by the ratio of the fuels, due to the lower sulphur content in wood. Increasing the amount of wood results in lower bed temperatures, which may lower NO_x emissions, but may also result in incomplete combustion and losses. A rule-based instruction system was suggested by Mononen *et al.* (1996) to help the operators select the optimal peat/wood ratio.

1.4.4. Flue Gas Oxygen Modelling

Properties of biomass fuels (heat value, moisture content, homogeneity, density and mix ability) may vary in a large range. The fluctuations cause agitated combustion conditions, even if a constant fuel feed is maintained, leading to increased emission levels and variations in the generated heat flow. With fuels such as peat, this problem is particularly pronounced. These property variations are difficult to predict or measure on-line. Only their effects on combustion and steam generation can be observed, by measuring the flue gas O_2 content and the steam pressure.

Figure 3 gives a schematic drawing of a hybrid combustion model, developed by Hímer *et al.* (2004). The model predicts the generated power and the oxygen content in the flue gases, as a function of the fuel feed (screw feeder rotation speed, Q_{Hz}) and the air flows (primary - F_1, secondary - F_2 and recirculation air flow - F_c).

As an intermediate variable, the unmeasurable dry fuel feed rate (Q_{kg}) is calculated. The model consists of two parts; a fuel flow model estimates the amount of dry fuel (Q_{kg}) fed to the system, the combustion model describes the chemical reactions providing the flue gas components and the combustion power. The stoichiometric equations were based on physical and chemical relations. The fuel flow model was learned from data.

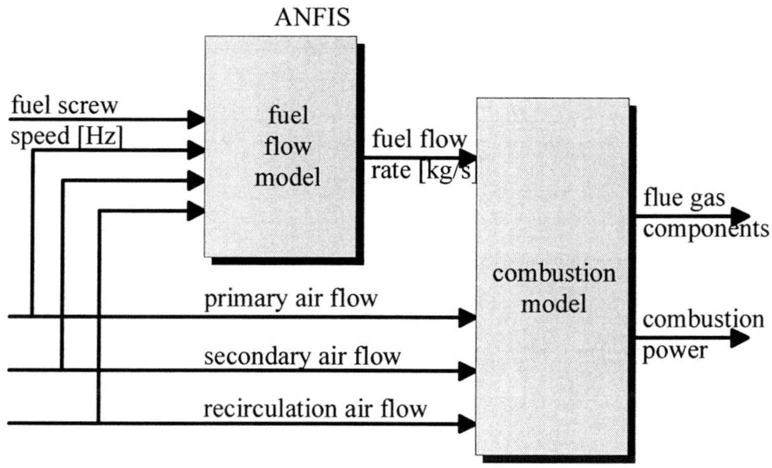

Fig. 3. Hybrid combustion model

In order to learn the model from data, the model input-output signals need to be available. The air flows and the screw speed (F_1, F_2, F_c, Q_{Hz}) can be measured on-line. However, the output variable (Q_{kg}) is not measurable. Therefore, it was generated by using the inverse of the stoichiometric model based on F_1, F_2, F_c, and O_2 measurements. The resulting data patterns (F_1, F_2, F_c, Q_{Hz}, Q_{kg}) could then be used to estimate the model parameters.

The fuel flow model was implemented using the ANFIS (Adaptive Neuro-Fuzzy Inference System) model structure (Jang *et al.*, 1997); where the antecedent and consequent parameters were estimated from data. The input variables were selected using a tree search algorithm, by incrementally growing the input variable set until no further improvement in MSE criterion could be obtained. The approach was experimented using data from a 100 MW CFBC power plant, with success.

1.4.5. Flue Gas Oxygen Control

At the lower, stabilisation level, the flue-gas oxygen can be maintained by adjusting the combustion conditions by manipulating the secondary air flows, *i.e.*, the oxygen trim. Figure 4 gives an example of such a control system (Benyó *et al.,* 2005a).

The quality variations of a fuel and its feeding system (heating value, moisture content, feeder discrepancy, *etc.*) cause the main disturbances for the boiler control, especially when considering biomass fuels. The quality changes in fuel appear both in the flue-gas oxygen content and in the steam pressure. Since the changes in flue-gas oxygen content appear much faster then in the steam pressure, controlling the O_2 content provides a fast action for stabilising the combustion. Compensating the disturbances in the fuel feed by the air flow leads to fluctuations in the released thermal power. However, the control action is only of short term and around the operation point.

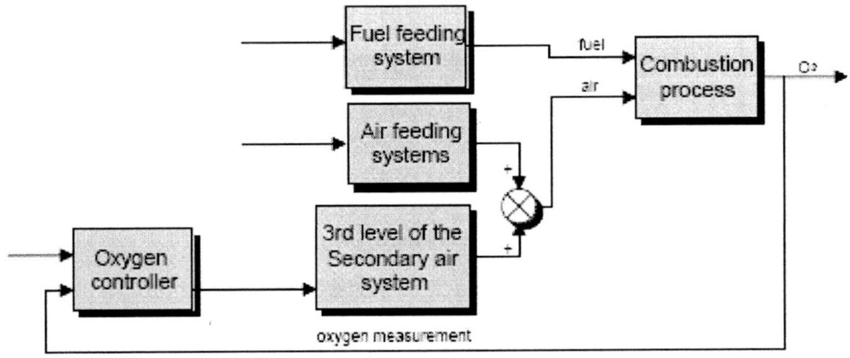

Fig. 4. Schematic diagram of the oxygen content control

Advanced model-based controllers take advantage of process models. The design of an oxygen content controller is challenged by the nonlinear, multivariable nature of the process. The main features of the process can be captured by a model taking advantage of nonlinear identification techniques. For a convenient handling of dynamics, a Wiener/Hammerstein approach has shown to be useful (Leppäkoski, 2004; Benyó *et al.*, 2005b).

Two major challenges in the design of flue-gas oxygen content control are the long time-delay of the fuel feed system, and the significant noise associated with the measurements. In the following, two model-based approaches are briefly described.

- The Smith-predictor is a straightforward solution to delay compensation. However, few of the process variables can be determined exactly; unmodelled dynamics and the unmeasurable variation of fuel properties must be also taken into account. The Smith-predictor extended with a robust filter (see Fig. 5), as proposed for the Internal Model Control (IMC) structure by Normey-Rico *et al.* (1997), can successfully address this problem (Paloranta *et al.*, 2004).

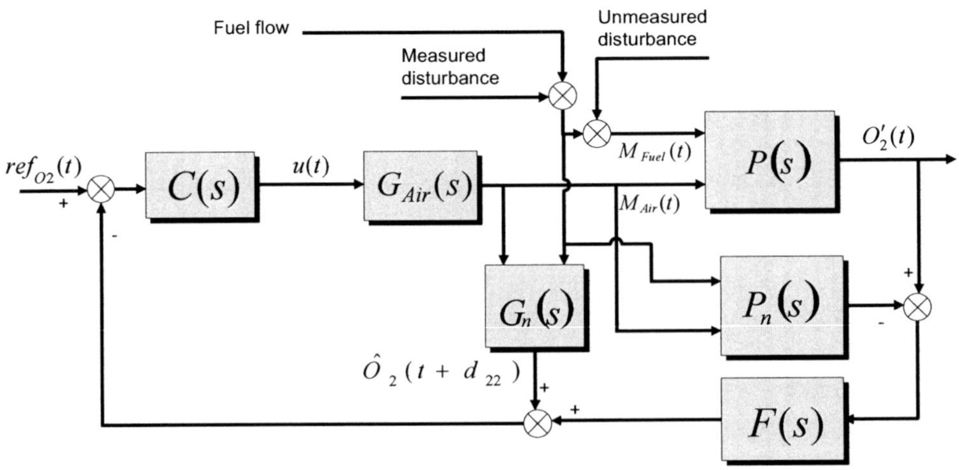

Fig. 5. Smith-predictor control of O_2 content

- The generalized predictive controller (GPC) approach offers a more advanced solution. In a cascade GPC structure (Benyó *et al.*, 2006), the outer loop is the oxygen controller calculating the set point for the 3rd level secondary airflow (see Fig. 6) and the inner loop generates the required air flow. The new cascade structure allows the inner and outer control loop into one GPC controller. The great advantage of this structure – in contrast to applying a GPC in the outer and another GPC in the inner loop – is the possibility to tune the robustness of the two loops separately from each other.

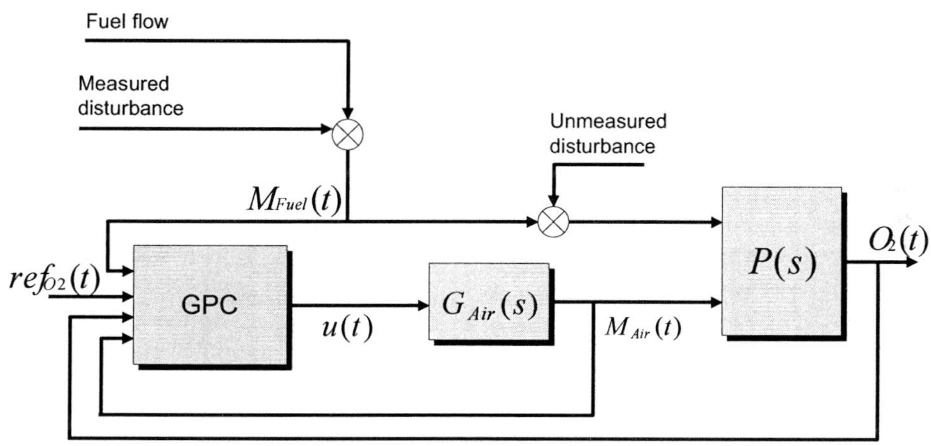

Fig. 6. GPC oxygen content controller

The Smith predictor and the GPC are based on linear plant models, but can take advantage of more complex process descriptions, *e.g.*, via linearization or development of equivalent strategies for nonlinear models.

1.4.6. Drum Control

The steam generation system of a boiler (see, *e.g.*, Åström and Bell, 2000) consists of a drum system (drum with water tubes) and a superheater system (superheaters with attemperators). The drum is fed by the feed water generated by the water supply system. The water is evaporated by the heat released in combustion. The fresh steam out of the drum goes through the superheaters. Since a turbine requires a constant steam temperature, water sprays are used to control the temperature of the superheated steam. Steam flow is controlled by a throttle valve.

A first-principle-based mathematical model of the complete steam generation process and a dynamic model of a drum was developed in Kovács (1998) for a circulated FBC with natural circulation drum. The model was based on preliminary assumptions resulting in a simple form, but still provided the main features of the process, such as the well-known "shrink-and-swell" effect. The model required only a few parameters (geometrical and metallurgical properties and water-steam parameters.) From a set of differential equations, a dynamic Auto-Regressive Moving-Average eXogenous (ARMAX) operating point model was developed; the output was the change in the drum level, with feed water, steam and heat flows as inputs. The parameters were calculated from the first-principle model, with a possibility to fine-tune (estimate) them using plant data. Compared with measurement data from a 300MW circulated FBC plant, the main characteristics of the plant performance were provided by the model.

The main goal of boiler feed water flow control is to supply as much water as the evaporation rate. At the same time, the drum water level must be maintained around a certain set-point. The feed water flow is controlled by a control valve in the case of small disturbances; a bigger change in the feed water flow demand requires the control of the speed of the water pump. The third requirement for the feed water flow is to maintain the pressure. One part of the feed water flow is sprayed to the attemperator to control the temperature of the superheated steam. To ensure the demanded water flow into the superheated steam flow, the attemperator flow must have a certain pressure controlled by a pump.

Proportional-Integral-Plus (PIP) control is an approach to state-variable feedback pole assignment for linear, discrete-time systems. In Kovács (1998), the Model Reference Adaptive (MRA) PIP control was applied in simulations to the control of the boiler *drum water level*. The evaporation system (drum with water tubes and water supply) is a complex system with hard rate and level constraints. Therefore a specific extension of PIP-control was considered, and adaptive model reference techniques applied to the extended (non-adaptive) PIP control. Considering the "shrink-and-swell" effect, after simplification one can say that at every operating point the steam (acting as a disturbance) and the feed water (acting as a control input) flows must be equal. The solution was to generate the control input from the steam flow measurement (rough control) and the PIP control output (fine control), which can be considered as a three-point controller. Simulation results showed fast regulation and tracking properties, even in the case of rate constraints.

1.4.7. Superheater Modelling

From modelling point of view, the steam *superheating* process can be seen as a combination of two sub processes; the hydrodynamic and the thermodynamic process. The dynamic time constants of these are significantly different; the hydrodynamic changes take place in a tenth of seconds while the time constants of the thermal process are in the magnitude of tens of seconds. Therefore these processes can usually be modelled separately. When emphasis is put on the identification aspects for the model, a Wiener/Hammerstein approach can be very useful. In several studies, a Multiple-Input Single-Output (MISO) Wiener/Hammerstein model describing the thermal process of a single steam superheater stage has been considered (Benyó *et al.*, 2005b; Mononen and Ikonen, 2005). The non-linear static part is based on the energy balance of the media involved in the process, while the linear part involves the main time dependencies of the process. The derivatives of the model can be easily obtained, which allows the application of efficient second order methods for identification.

As an example, in Benyó *et al.* (2005b), the nonlinear static part is based on a cascade Wiener-Hammerstein model of the superheater process (see Fig. 7). The static part expresses the outlet steam temperature drawn based on the energy balance equations at the wall, steam, and flue gas, and based on convective and radiative heat transfer. The path of dynamic effects of the input variables and the dynamics of output temperatures are captured by the linear transfer functions. Thus, advantages of Wiener-Hammerstein modelling – fusion of physical and experimental information – are exploited. The physical phenomena are captured by the static model and the dynamical behaviour is identified based on measurement data. Separation of the model blocks makes the identification simple and feasible, and greatly reduces the complexity in validation.

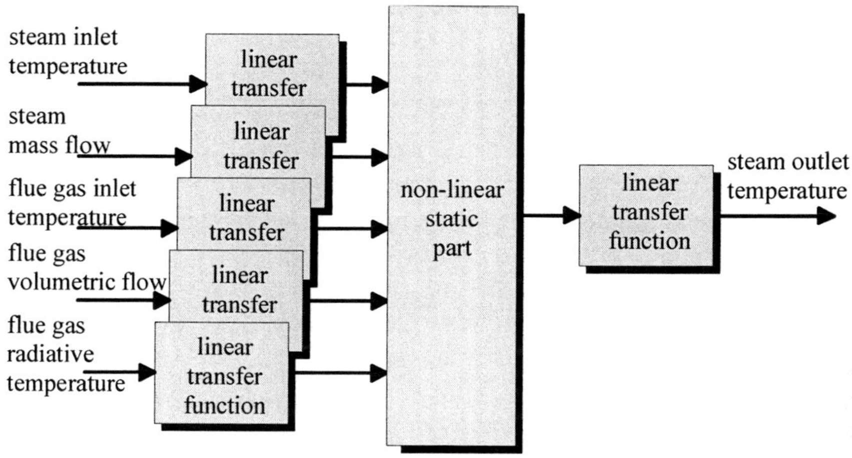

Fig. 7. A cascade Wiener-Hammerstein model for a superheater

1.5. On the Role of AI Techniques in FBC

In this section we aimed at making a brief overview on the role of advanced identification and control approaches in power plant monitoring, regulation, control and optimization. A subjective view was taken, so as to propagate our own experiences in this field. Needless to say, much significant research has been and is currently being conducted in many universities and research centres world-wide. Reviewing this work was not within our scope.

Needs for advanced techniques are typically characterized by the desire to properly handle plant non-linearities, the multivariable nature of the problems, and the necessity to adapt to changing plant conditions. Clearly, the areas of application of 'advanced' and 'learning' methods greatly overlap with those of artificial intelligence. In the field of fluidized bed combustion control, the many promising applications arise from the uncertainties and complexity of the combustion process, and the difficulties in obtaining reliable measurements from the changing furnace conditions.

In the next three sections, three case studies are described in more detail:

- modelling of FBC flue-gas emissions from data using fuzzy relational models,
- on-line identification of multi-fuel FBC NO_x emissions using adaptive prototypes, and
- multivariable FBC trajectory following control using genealogical decision trees.

In these applications, a variety of AI-inspired techniques are examined, including sigmoid neural networks, Kohonen self-organizing maps, fuzzy relational models, population based optimization schemes, and their hybrid combinations; together with comparisons and combinations with more conventional techniques. From function approximation point of view, many of the AI-inspired model structures can be classified as a modification of a basis function network approach (Ikonen and Najim, 2002). Often, the related learning problems can be efficiently solved using 'conventional' gradient-based techniques. On the other hand, population-based stochastic techniques provide extremely flexible tools for solving efficiently both difficult and also more simple optimization problems, both for 'conventional' and AI-inspired model structures. In the case examples, we describe only briefly the associated algorithms, ample references are given for the interested reader. Instead, emphasis is put on application of the considered techniques and on discussion of future trends in power plant automation. By our selection of case studies we wish to emphasize the importance of fusion of process knowledge with intelligent learning and interpretation of the results as the key to the successful development of feasible techniques and profitable applications.

2. Modelling of FBC Flue-Gas Emissions from Data Using Fuzzy Relational Models

Distributed logic processor (DLP) models are fuzzy relational models, the parameters of which can be estimated using standard gradient based techniques. A major difference compared to many of the other neuro-fuzzy modelling approaches (ANFIS, LOLIMOT, *etc.*), is that the DLP approach primarily adjusts the logical mappings (rule-bases), instead of the input-output partition (fuzzy sets). Since the concepts modelled via fuzzy sets are not altered in the adaptation, the familiarity of these concepts is not degenerated. Instead, emphasis is placed in finding the logical relationships between the concepts. A number of studies have been dedicated to DLP's (Pedrycz *et al.*, 1995; Najim and Ikonen, 1999; Ikonen *et al.*, 2000; Ikonen and Najim, 1997; 2004). This section starts with a brief look at the model structure, followed by an extension to modelling of dynamic systems using a Wiener LP system structure. Finally, the approach is illustrated with a case study on modelling NO_x emissions.

2.1. Process Identification Using DLP's

Logic processors (LP) (Pedrycz, 1993; 1994; Ikonen, 1996) are aimed at carrying out approximations of logical relationships, fuzzy functions. The LP input interface translates the H-dimensional input data coming from the modelling environment into a format that fits the conceptual level of the logical part. The interface is formed via a collection of H fuzzy sets and relevant matching mechanisms. An LP involves AND nodes situated at the 'hidden layer', followed by a single OR node at the output layer. The OR node is given by:

$$\begin{aligned} \mathrm{OR}(\mathbf{\Psi},\mathbf{\Omega}) &= \mathrm{OR}(\psi_1 \,\mathrm{AND}\, \omega_1; \psi_2 \,\mathrm{AND}\, \omega_2; \ldots; \psi_R \,\mathrm{AND}\, \omega_R) \\ &= \mathrm{S}(\mathbf{\Psi}\, \mathrm{t}\, \mathbf{\Omega}) \\ &= 1 - \prod_{r=1}^{R} (1 - \psi_r \omega_r) \end{aligned} \tag{1}$$

where column vectors $\mathbf{\Psi} = [\psi_1, \psi_2, ..., \psi_R]$, $\psi_r \in [0,1]$ and $\mathbf{\Omega} = [\omega_1, \omega_2, ..., \omega_R]$, $\omega_r \in [0,1]$ ($r = 1,2,...,R$) are the OR node input and parameter vectors, respectively; R is the number of inputs to the OR node. S denotes an s-norm (t-conorm), t denotes a t-norm. The computable form (bottom line) is obtained by choosing the t-norm as a product ($a \mathrm{t} b = ab$) and the s-norm as the probabilistic sum ($a \mathrm{s} b = a + b - ab$).

Similarly, the output of the r'th AND node is given by:

$$\begin{aligned} \mathrm{AND}_r(\mathbf{\Phi}, \mathbf{\Theta}_r) &= \mathrm{AND}(\varphi_1 \text{ OR } \theta_{r,1}; \varphi_2 \text{ OR } \theta_{r,2}; ...; \varphi_H \text{ OR } \theta_{r,H}) \\ &= \mathrm{T}(\mathbf{\Phi} \,\mathrm{s}\, \mathbf{\Theta}_r) \\ &= \prod_{h=1}^{H} (\varphi_h + \theta_{r,h} - \varphi_h \theta_{r,h}) \end{aligned} \tag{2}$$

where $\mathbf{\Phi} = [\varphi_1, \varphi_2, ..., \varphi_H]$, $\varphi_h \in [0,1]$ and $\mathbf{\Theta}_r = [\theta_{r,1}, \theta_{r,2}, ..., \theta_{r,H}]$, $\theta_{r,h} \in [0,1]$ (h=1,2,…,H) denote the input and parameter vectors of the r'th AND node, respectively; H denotes the number of inputs to each of the R AND nodes. T denotes a t-norm and s an s-norm. Again, the computable form is obtained by choosing the t- and s-norms as a product and probabilistic sum, respectively.

An LP network and the notation are illustrated in Fig. 8. A nonlinear element, such as a parameterised sigmoid function, may follow the OR-node, in order to enhance the nonlinear mapping capabilities of the structure.

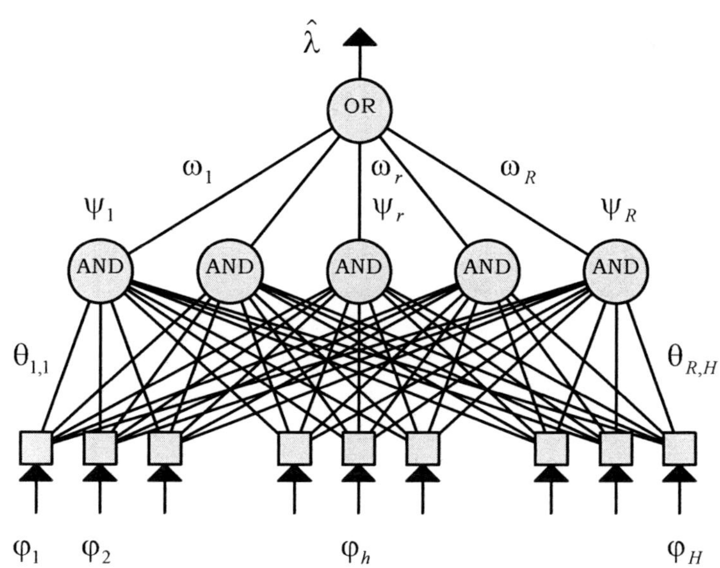

Fig. 8. Logic processor network consists of a 'hidden' layer of AND-nodes, combined by an OR-node. Each node is associated with a set of parameters with a logical interpretation

In *distributed LP's* (DLP's) (Pedrycz *et al.*, 1995), the LP's are organised in parallel and the inputs and outputs of LP's are associated with the elements of the input-output interface (*i.e.*, fuzzy partitions, modelling landmarks). If the outputs are associated with singletons a 0-order Sugeno type of model is obtained. The outputs of individual MISO LP's can be combined to a crisp singleton, $\hat{y} \in \Re$, by a node computing the weighted average,

$$\hat{y} = \frac{\sum_{q=1}^{Q} \tau_q \hat{\lambda}_q}{\sum_{q=1}^{Q} \hat{\lambda}_q} \tag{3}$$

where $\tau_q \in \Re$ represents the centre of the q'th output class (q=1,2,...,Q). $\hat{\lambda}_q \in [0,1]$ is the output of the corresponding logic processor. Fig. 9 shows a DLP model configuration partitioning the output space into Q classes. For simplicity, the figure illustrates a separate lattice partitioning for each input $x_p \in \Re$, resulting in an input interface consisting of H inputs $\varphi_h \in [0,1]$.

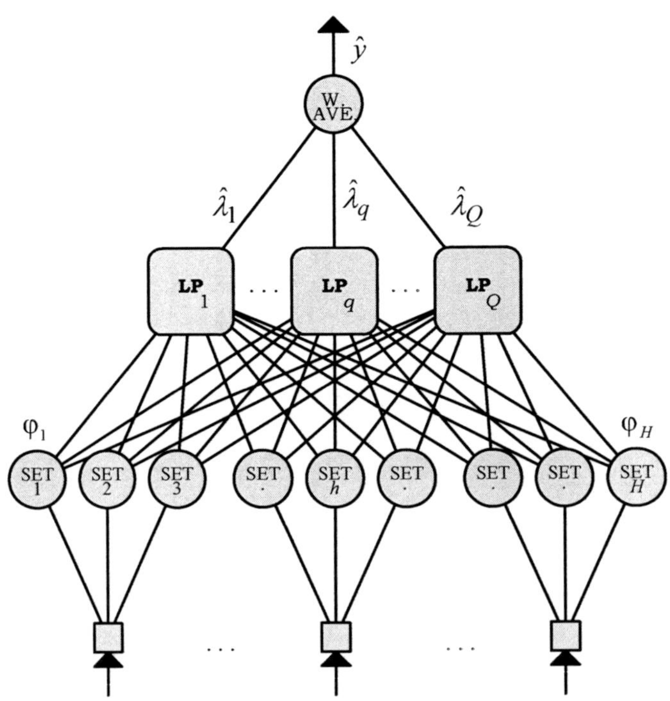

Fig. 9. A distributed logic processor network can be constructed by placing a number of LP's in parallel. The logical results (beliefs) are translated into a crisp number by taking a weighted average

In DLP's, the rule base of each LP is parameterized. The input-output partitions (fuzzy sets) are designed so as to fit the conceptual level of the problem's logical part. The target of learning (*parameter estimation*) is to find out the weightings (parameters in $\boldsymbol{\Omega}$ and $\boldsymbol{\Theta}$) within the rule base. An algorithm for training a DLP model can be derived by estimating the parameters of each LP separately, and formulating the training task of an LP as the minimisation of some quadratic cost function. Gradient-based training methods can be used, since the chain-rule can be applied for computing the required gradients. Several training methods have been investigated: Simple gradient descent with a constant or adaptive step size (Pedrycz *et al.*, 1995); recursive prediction error method and its modifications (Ikonen, 1996), Levenberg-Marquardt method; random search based on games of stochastic learning automata (Ikonen and Najim, 1997), and optimisation under constraints (Najim and Ikonen, 1999). In practice, standard gradient methods work well, and also an on-line implementation would seem possible. The computational burden is non-negligible, but since explicit expressions for the model gradients can be obtained, the parameter estimation task is not excessive.

In most approaches to adjusting fuzzy models, the minimization of a prediction error criterion or alike is accomplished via tuning the input/output fuzzy sets. In view of the minimization criterion, very efficient algorithms can be derived. However, the problem with such a tuning is that the training procedure changes the *concepts* (modelled by fuzzy sets), instead of revealing the *logical relationships* between them. Consequently, the modified concepts may no longer express the constructive structure adopted by the users, and much of the *transparency* of the model – relying on the use of familiar concepts – will be lost. Notice, that the benefits of efficient training, simple structures and accurate predictions may be of little importance, if the result of the training is a set of concepts that is not naturally (or even subjectively) related to the problem at hand. In DLP's, however, the parameterization affects the logical mapping between the concepts (the rule-base), thereby estimating the logical mapping between user-defined concepts. Using simple threshold techniques, it is straightforward to automatically express the contents of the LP rule bases as a set of rules. This duality in presentation between numerical and rule-based formats is a major benefit of the DLP modelling approach, providing transparency and ease-of-use.

2.1.1. Distributed Wiener Logic Processors

The identification of *dynamic* systems using DLP's is straightforward using the *time-series* approach (Pedrycz *et al.*, 1995). In this approach, the model input vector (regression vector) is chosen to consist of a set of delayed input-output data, resulting in structures such as Nonlinear Auto-Regressive eXogenous (NARX), Nonlinear Output-Error (NOE), Nonlinear Auto-Regressive Moving-Average eXogenous (NARMAX), *etc.* (Ikonen and Najim, 2002). However, the dimension of the nonlinear mapping can become excessive, prohibiting practical applications: In particular, the parameter estimation becomes time and computing power consuming, and the analysis of the model resulting from the identification becomes cumbersome. Ikonen and Najim (2004) attempted to overcome these problems via the *Wiener* and *Hammerstein* system paradigms. In the following the distributed Wiener logic processors (DWLP) are briefly outlined.

A *Wiener logic processor* (WLP) is a mapping from an *I*-dimensional real space to the unit interval, $\Re^I \to [0,1]$. A Wiener logic processor consists of a linear dynamic part, an input interface, a logical mapping and a nonlinear element. Figure 10 shows a schematic diagram of a WLP.

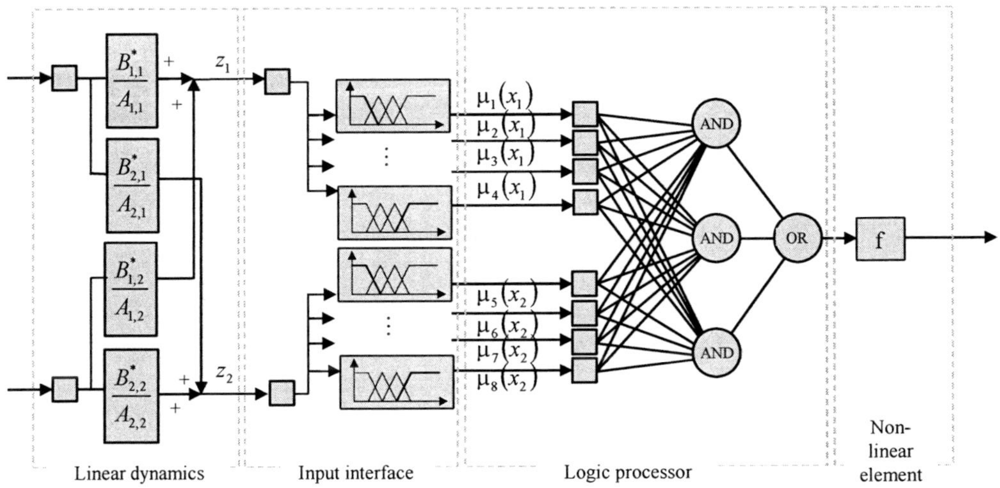

Fig. 10. In a Wiener logic processor the system inputs are preceded by a set of linear filters

The heart of a Wiener logic processor is a Logic Processor (LP). The system composed of the LP fuzzy function and the input interface is a static one. A system of linear filters precedes the static logical part, see Fig. 10. The parameters of the dynamic filters can be estimated simultaneously with those of the logical structures, using gradient based techniques similar to general Wiener-modelling approaches (see, *e.g.*, Ikonen and Najim, 2001; 2002). The *distributed* Wiener LP model (DWLP) is composed of the Wiener LP's organised in parallel.

The main benefit of the DWLP approach is in the *transparency* of the obtained dynamic nonlinear model. The static part can be logically decomposed into a set of simple rules, and examined vs. steady-state prior knowledge, for example. The dynamics can be described and analyzed using well-known and efficient techniques from linear systems and control theory, such as techniques in the frequency domain or based on the poles and zeros of the system. Transparent handling of dynamics in nonlinear systems is a highly nontrivial task, and this type of local model approaches provide one of few alternatives. Since the distributed model structure consists of Wiener systems in parallel, it is also possible to capture restricted nonlinear dynamics, provided that the dynamics do not exhibit output-multiplicity.

2.2. DLP Modelling of FBC Flue Gas Emissions

In this subsection, the performance of the DLP approach is briefly illustrated in modelling FBC flue gas emissions. For details and extensions, see the series of work reported in Ikonen (1996); Ikonen and Najim (1997; 2004); Ikonen *et al.* (2000), and Najim and Ikonen (1999).

Measurement data were collected during process experiments at a 25 MW_t semi-circulated FBC thermal power plant. During a five day test period, 47 continuous signals were measured from the plant with a sampling interval of 4 seconds. The test period consisted of two days of operation using saw-dust and three days using peat as fuel. Day-time test programme consisted of open and closed-loop experiments using different primary/secondary air ratios, excess air ratios, and power levels. During the nights, the plant was operated using saw dust supported by coal. The thermal plant was in full operation (connected to the district heating system of a small city in northern Sweden) during the process experiments.

Peat NO_x emissions, c_{NOx}, [ppm] were identified using the DLP approach, and results were compared with identification using Sigmoid Neural Networks (SNN) and linear structures. The oxygen concentration, c_{O2}, in flue gases [%/100], flue gas residence time in the furnace [s] and primary air stoichiometry [mol/mol] signals were chosen as inputs, based on *a priori* knowledge and rough modelling by linear regression. NO_x and O_2 were measured after the throat and before stack, respectively. Gas residence time, t_r, was estimated based on filtered air flows, furnace temperatures and furnace geometry; primary air stoichiometry, μ_1, was based on the measured primary air flow and fuel feed rate (Ikonen, 1996). Signals were scaled between -1 and 1. Training and test data sets were constructed by taking one and five minute averages, resulting in 515 and 164 data patterns, respectively.

Figures 11-13 illustrate the simulations (Ikonen, 1996). The solid lines show the model prediction, the dots show the corresponding test data. The simulations consider a 13 hour period of plant operation; data from the 5 last hours were not used for parameter estimation. The bottom left corner in each plot shows the Root-Mean Squared Error (RMSE) criterion for both training and test data.

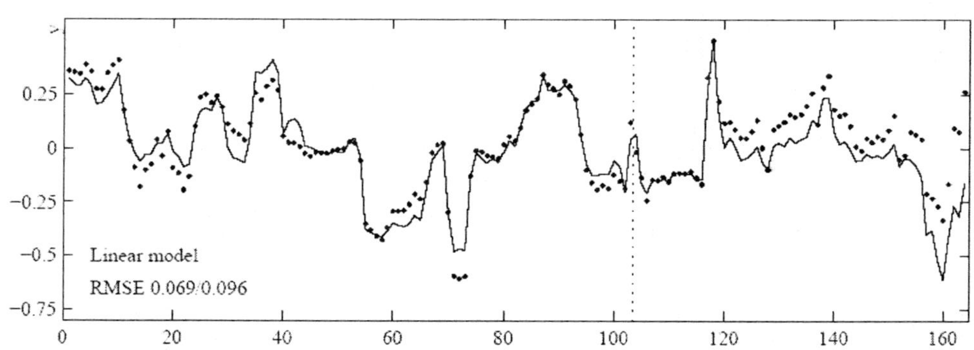

Fig. 11. Prediction on test data (linear model)

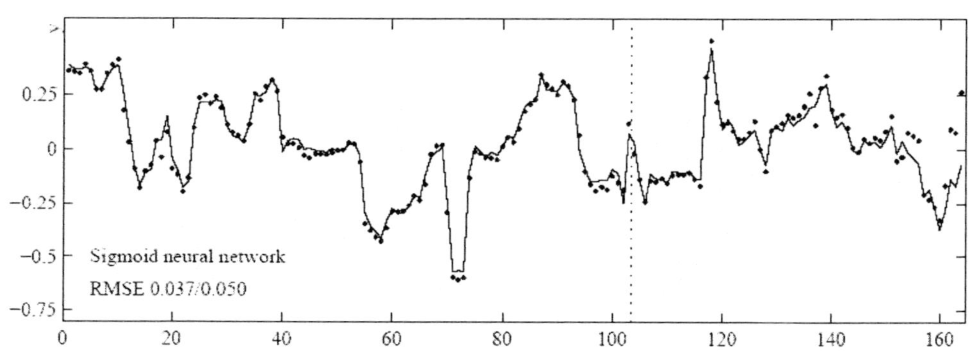

Fig. 12. Prediction on test data (sigmoid neural network model)

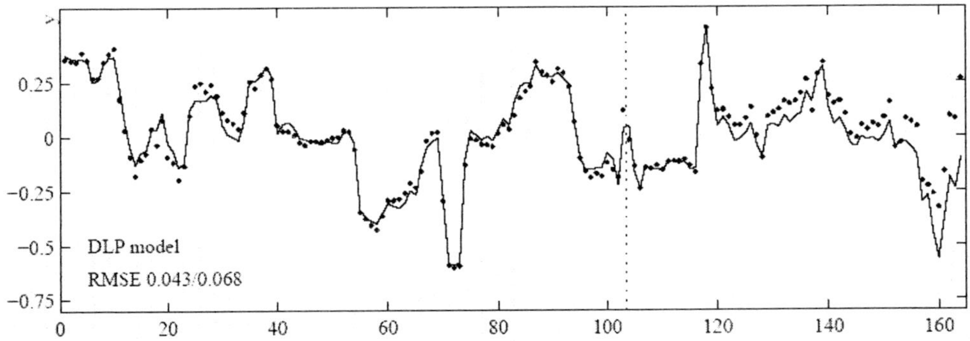

Fig. 13. Prediction on test data (DLP model)

The main conclusion from the experiments was that the DLP models for the FBC NO_x emissions were capable of competing with linear and non-linear (SNN) regression methods. Compared to linear regression models, the number of parameters in DLP's was large and the training times were long. Neither were the prediction accuracies of SNN models achieved. However, the explanatory status of a DLP model complemented these shortcomings. The resulting models were transparent and clear presentations of the main features of the behaviour of a non-linear system.

Conversion from DLP weights to rules (Pedrycz, 1993; Ikonen and Najim, 2004) is illustrated in Table 1. Numerical values of the LP weights are shown in the table, values passing the threshold have been underlined. Examining the table, the OR node 'O' can be further decomposed into "A3 or A5" and AND nodes 'A3' and 'A5' further translated into a rule format. These are shown in the rightmost column. The rule interpretation of the network is shown at the bottom of the Table, and the contents of the LP can be simplified into if-then rules. Obviously the model accuracy suffers from this type of simplification. However, the conditions for a high NO_x content become clear – flue gas oxygen is medium or high AND residence time is short or medium – and can be easily grasped after the translation from numerical weights into a rule format. For purposes requiring a more refined model, the unsimplified numerical description can be used.

The basic application intended for the DLP emission models is in on-line process optimisation. For this application the availability of up-to-date models is essential (remember that the variety of fuels in a FBC may be very large, and change often). The DLP structure allows several ways for interaction (numerical simulation, numerical optimization, linguistic what-if analysis) and learning (from 'experience-based' rules, from measured sampled process data). The DWLP approach further extends these very appealing properties into identification of dynamic processes. From numerical point of view, there are no technical problems in parameter estimation. However, data-based approaches to structure identification and identification of dynamic models are more difficult, as the general possibilities for obtaining experimental data are limited. Recall that typically the time constants in power plants of this magnitude are long (in the scale of minutes), in addition the possibilities for experimenting are constrained by economical and safety issues.

Table 1. Numerical to logical conversion of an LP

α=0.1		A1	A2	A3	A4	A5	IF O THEN c_{NOx} IS HIGH
O	c_{NOx}	0.23	0.00	1.00	0.57	1.00	A3 OR A5

		LOW	NOT LOW	MED	NOT MED	HIGH	NOT HIGH	
A1	c_{O2}	0.08	0.31	0.50	0.21	0.00	0.66	
	t_r	0.05	0.28	0.32	0.82	0.19	0.77	
	μ_I	0.57	0.40	0.32	0.65	0.31	0.11	
A2	c_{O2}	0.48	0.47	0.47	0.82	0.76	0.97	
	t_r	0.98	0.47	0.55	0.32	0.86	0.00	
	μ_I	0.14	0.56	0.51	0.69	0.05	0.11	
A3	c_{O2}	1.00	0.36	1.00	1.00	0.00	1.00	c_{O2} IS HIGH
	t_r	0.63	1.00	1.00	1.00	1.00	0.00	AND t_r IS NOT LONG
	μ_I	1.00	1.00	1.00	1.00	1.00	1.00	
A4	c_{O2}	0.00	0.74	0.77	0.21	0.29	0.98	
	t_r	0.61	0.06	0.94	0.67	0.00	0.54	
	μ_I	0.13	0.22	0.12	0.95	0.55	0.84	
A5	c_{O2}	1.00	0.00	0.00	1.00	0.81	1.00	c_{O2} IS NOT LOW AND c_{O2} IS MED
	t_r	0.12	1.00	1.00	1.00	1.00	0.00	AND t_r IS NOT LONG
	μ_I	1.00	0.36	1.00	1.00	1.00	1.00	

IF			
			c_{O2} IS HIGH
		AND	t_r IS NOT LONG
	OR		c_{O2} IS NOT LOW
		AND	c_{O2} IS MED
		AND	t_r IS NOT LONG
THEN			
			c_{NOx} IS HIGH

In high-level optimisation tasks, such as the minimisation of harmful NO_x emissions in FBC flue-gases, there are a large number of more or less vague criteria that need to be fulfilled, minimised, or maximised. In a power plant, these range from concerns of MW-demand, safety, economy, other emissions, maintenance, *etc.* down to the closeness of the end of a shift. Full automation of such optimisation seems like a very difficult problem, and the decisions are usually left to the operating personnel. In addition, no matter how clever the automated optimisation would be, the responsibility of taken actions rests on the shoulders of the personnel. What can be done, however, is to develop means of extracting knowledge of the plant behaviour. This extracted information needs to be simple enough so that it can be used with the other pieces of knowledge available. If accurate simulations can also be made with the same models, then both advanced optimisation routines and human supervision can be based on a same description of a process. In this view, the distributed logic processor approach is a very attractive tool for modelling.

3. On-Line Identification of Multi-Fuel FBC NO_X Emissions Using Adaptive Prototypes

In the field of process engineering, modelling is the major area of application of function approximation tasks. Typically, function approximation is conducted using sigmoid neural networks (SNN) or alike, whereas clustering tasks take advantage of unsupervised learning schemes such as Kohonen's (1990) self-organizing map (SOM). Also in applications of process control, fault diagnosis, *etc*, sigmoid neural networks often serve as models of process and process inverse. It has frequently been pointed out that the ability of neural networks to represent non-linear mappings and hence to model non-linear systems is the most significant.

If the process to be modelled is time-varying, adaptive or learning methods are required and the model needs to be updated continuously using on-line measurements. In on-line adaptation with real industrial processes, recursive algorithms may run into trouble due to the poor information content in the measurement data (*e.g.* due to long periods of functioning in the same operation point). In this section, a modelling approach is presented where the tasks of collecting a proper set of training data and creating a functional mapping between the input/output training patterns are explicitly separated. A SOM is used to provide adaptive partitioning of the model input space. Each SOM node is associated with an adaptive prototype that represents the up-to-date process behaviour in the corresponding operating region. The prototypes are used as training data in the on-line adaptation of a process model. The process model then learns the process characteristics and can be used for applications such as monitoring, control or optimisation. In the following, this configuration is referred to as the combined network.

In the reminder of this section, the combined network is introduced and a learning algorithm is given, followed by experimental results using data from a full-scale fluidized-bed combustion power plant.

3.1. Recursive Identification

In many continuous (as opposed to batch) industrial processes, significant changes of operating point are less frequent. In a thermal power plant, for example, the load typically has two or three common levels (*e.g.*, high load in the morning and afternoon, medium load during the day-time and low load in the night). During these periods, the energy production level is kept constant.

In the modelling of such processes, two mainstreams of approaches can be distinguished: adaptive and learning. In both approaches, the quality of the data is of great importance. In an *adaptive* approach, a simple model (*e.g.*, linear) is adjusted so as to fit the model to the behaviour of the plant around its current operation point (*e.g.*, load level). When the operation point is changed, a new set of parameters is recursively identified (Ljung and Söderström, 1983). Recursive identification presupposes that a parameter estimation algorithm can follow the properties of the process independently of the time passed from the initialisation of the estimation procedure. The abilities of the algorithm to follow the process can be enhanced by discounting old information. One assumes often that the estimated parameters change slowly over some time horizon. A natural way to do this is to consider a criterion where older values are discounted. With an exponential forgetting profile, the computations can be represented in a simple recursive form using a constant forgetting factor (exponential forgetting algorithm). Also time varying forgetting factors can be considered (constant trace method, *etc.*), forgetting factor can be assumed to be a matrix (selective forgetting, *etc.*), or the forgetting may be directed according to some criterion. In a *learning* approach, a global nonlinear model of the plant is constructed, covering the complete operating space of the plant. In this context, learning then refers to gaining initial process knowledge and adaptation to time-varying changes in the process (wearing, *etc.*).

The power and flexibility of basis function networks (*e.g.*, sigmoid neural networks) in tasks of function approximation is due to the low number of assumptions made about the shape of the function to be approximated. For this reason, rich process information that describes the behaviour of the process in a wide operating area is needed for parameter estimation in non-linear black-box structures. Recursive identification in on-line situations may lead to problems when the information provided by the measurement data is poor, which results to computational problems in the parameter estimation (covariance blow-up, parameter drift, *etc.*). These problems are emphasized when identifying global models using uniform function approximators.

3.2. Adaptive Clustering of the Input Space

In the basic scheme of recursive estimation, the model parameters are updated with the latest sample measured from the process. As often the optimization step size needs to be set small, an apparent increase in the speed of learning, vs. sampling rate, is obtained by a repeated (iterative) use of measured data patterns. In the delay coordinate (or moving window), a set of lagged process measurement patterns is used at each sample time for adjusting the model. In this approach, the training data reflect well the current characteristics of the process. However, if the process is operated in a steady state, the information content in this kind of training data set may be poor for the purposes of identifying a global model.

An alternative to the delay coordinate approach is to partition the process operating space into regions and represent the behaviour within each region by a sample, *i.e.*, a prototype. When new measured information arrives, only the prototype of the region matching with the current measurement is adapted. This allows the collection of data from a wide operation area, without excessive storage requirements. However, since an industrial process can rarely be operated on the whole input-output space (in the sense that the physical variables are dependent), a more efficient approach would base the partitioning of the input space only on the space that can be attained by the process. Further, if this clustering on the operating region is time-varying, the resulting prototype set will reflect the time varying spatial distribution of the latest measurements.

The preceding discussion is illustrated in Fig. 14, where training data are collected in a simple one-dimensional case. The input space data consist of noisy measurements of five steady state sequences, Fig. 14(a). In the delay coordinate approach, Fig. 14(b), the four training patterns consist of the current measurement and its three lagged delay coordinates. The training data are up-to-date, but reflect only the short term history of the plant behaviour. Using the non-adaptive input space clustering, Fig. 14(c), latest information on the whole input space is collected. However, the spatial distribution is not reflected and one of the prototype patterns is never updated. Using the adaptive grid for clustering, Fig. 14(d), the Voronoi regions adapt to the plant information. Information from the whole operating region is collected, and all the prototypes are used efficiently.

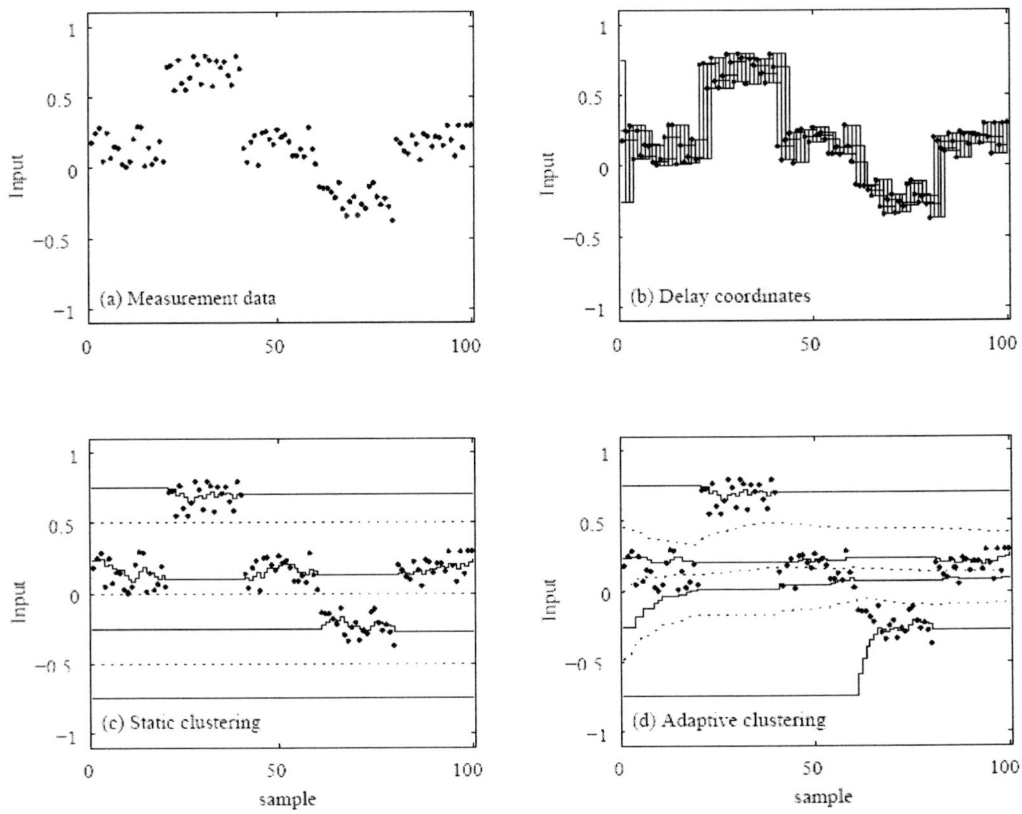

Fig. 14. Illustration of collecting a training data set of four samples in a one-dimensional case. Training patterns are collected from measurement data, see plot (a), using the delay coordinates (b), static clustering with filtered prototypes (c), and adaptive clustering with filtered prototypes (d). The dashed lines separate the grid regions in plots (c)-(d), solid lines show the time evolution of the four prototype patterns

The preceding discussion leads to propose an identification approach which allows the input space partitioning to be adaptive during the identification, and uses the resulting prototypes as training data for the non-linear model. Hence the steps of collecting process information and function approximation (curve fitting) are explicitly separated. The first step ensures that a representative data set is available, while the second step ensures that a suitable generalization over the prototype data is made. Successful completion of these tasks is highly nontrivial, and requires both flexible algorithms and good engineering. In the next subsection, the SOM and SNN are used for the mechanization of these two tasks, respectively.

3.3. Combined Network Algorithm

The structure of the combined network is shown in Fig. 15. Let us denote the input measurement vector at sampling instant k by $\varphi(k)$, and the corresponding measurement of the output by $y(k)$. The best matching node (the winner), $c(k)$, among the SOM nodes, $\hat{\varphi}_n(k-1)$, $n = 1,2,...,N$, is found and the map is updated so that the SOM internal topology is preserved, according to a self-organisation scheme. For an efficient algorithm, see *e.g.*, the Recursive Least Squares (RLS) method (Ruoppila *et al.*, 1993) with probability weighing (Hyötyniemi, 1994). To define a SOM network, the network topology and the learning algorithm need to be specified. In the optimal case, the internal topology should coincide with the "real topology" of the process. However, the real topology of the process is usually unknown. Note, however, that even if the space spanned by the measurements has a large dimension, the real degrees of freedom of industrial processes are almost always much less. With a suitable (non-linear) projection, the plant could be described with a sufficient accuracy using a model of reduced dimension. In practice, a simple 2-dimensional grid topology is a common choice as the resulting map can be easily used for graphing.

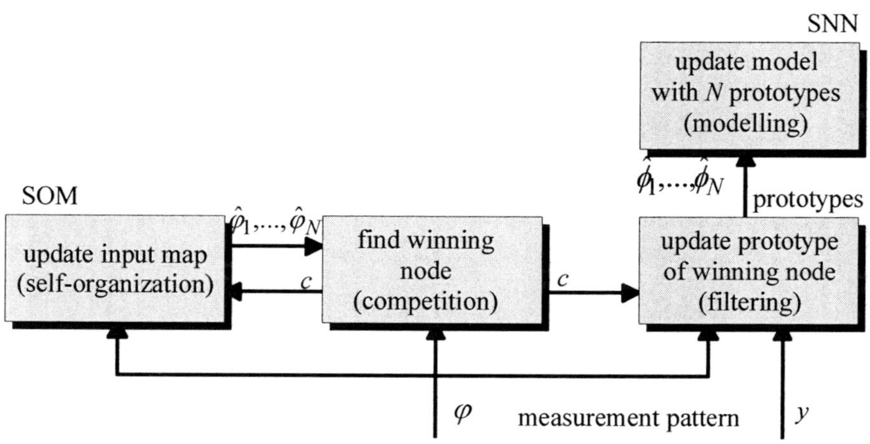

Fig. 15. Signals and computation flow in the combined network. The arriving measurement is clustered according to the input space map (SOM). The input map and the prototypes are updated recursively. The updated prototype set is used for updating the nonlinear model (SNN)

Each SOM node, $n = 1,2,...,N$, is associated with a prototype vector, denoted by $\hat{\phi}_n(k)$. Based on the clustering given by the SOM, the prototype vector associated with the winning unit is adapted to the corresponding input-output measurement at sampling instant k, $\phi(k) = \left[y(k)\ \varphi^T(k)\right]^T$. For the winning node, the update rule is simply:

$$\hat{\phi}_{c(k)}(k) = \hat{\phi}_{c(k)}(k-1) + \eta(k)\left(\phi(k) - \hat{\phi}_{c(k)}(k-1)\right) \tag{4}$$

where $n = 1,2,...,N$, N is the number of SOM/prototype nodes. The prototype filter coefficient $\eta(k) \in [0,1]$, should be tuned faster than the SOM organisation and its main function is to act as a low-pass filter of the measurement noise.

When a new measurement pattern at sampling instant k has arrived and the corresponding SOM node and the prototype vector have been updated, the SNN parameters $\hat{\theta}$,

$$\hat{y} = \mathrm{f}(\varphi, \hat{\theta}), \tag{5}$$

are updated using all N prototype vectors. The SNN architecture is selected taking into consideration the bias-variance balance of the problem. The parameters of the SNN parameter estimation algorithm should, in principle, be tuned fast enough so that the changes in the prototype patterns are reflected by the model. The Recursive Prediction Error (RPE) method (Ljung and Söderström, 1983) with the Bounded Information (BI) (or matrix regularisation) modification provides an efficient recursive parameter estimation scheme. A common batch-alternative is the Levenberg-Marquardt method (see, *.e.g.*, Ikonen and Najim, 2002).

If so desired, the SOM and SNN and the related parameter estimation techniques can be replaced with other types of schemes (see, *e.g.*, Ikonen and Najim, 2002). Alternative clustering methods (fixed lattice, expectation maximization, fuzzy c-means, *etc.*) will result in alternative grouping of data; different function approximation schemes (power series, radial basis functions, CMAC, *etc.*) will provide different generalization behaviour. If *a priori* knowledge about the process behaviour is available, parameterized models based on first-principle knowledge could be considered; if the explanatory status of the model is of prior importance, fuzzy modelling could be examined.

3.4. Application of the Combined Network in Fluidized-Bed Combustion

In this section, the combined network modelling approach is illustrated using data from an industrial fluidized bed combustor (FBC). In the following, simulations of an on-line identification of a three-input one-output static model of NO_x-emissions of an industrial fluidized-bed combustion power plant are described. In Ikonen and Kortela (1995), large dimensional FBC data were examined.

Five-day measurement data were collected during process experiments at a 25 MW_t semi-circulated FBC thermal power plant. For the NO_x-emissions [ppm], fuel type, oxygen concentration in flue gases [%/100], and flue gas residence time in the furnace [s] measurements were chosen as model inputs, based on *a priori* knowledge and rough modelling by linear regression. Fuel type (peat or saw dust) was observed from the SO_2 measurement, filtered and hard-cut so that a rough, flag-like behaviour was obtained. The signals were scaled so that a reasonable area would lay in the unit interval. A pseudo three-week data set (3×5-days) was constructed by repeating twice the measured five-day data set. The resulting measurement data used for parameter estimation consisted then of $K = 118968$ data patterns $\phi(k)$ ($k = 1,2,...,K$) of three inputs and one output, sampled with a rate of 0.1 Hz. A test data set of 319 patterns was constructed by taking measurements from the five-day data set with a sampling rate of 0.001 Hz.

A total of 11 simulation runs were conducted, as shown in Table 2. In the identification experiments using the combined network (CN I – CN VII), different Self-Organizing Map (SOM) topologies and algorithm parameters were experimented. For comparison, models were identified also using the delay-coordinate approach (simulations DC I – DC IV).

Table 2. Simulation results. See text for more explanations

Experi-ment name	Parameters of the training algorithms				Evaluation of SNN performance		
	Topo-logy	radius σ	BI a_{min}	BI a_{max}	MRMSE 3×5d	RMSE after 5d	MRMSE last 5d
CN I	4×4×4	0.15	10^{-5}	0.01	0.14	0.12	0.12
CN II	4×4×4	0.25	10^{-5}	0.01	0.15	0.13	0.13
CN III	4×4×4	0.25	10^{-3}	0.5	0.16	0.13	0.12
CN IV	8×8	0.25	10^{-5}	0.01	0.15	0.12	0.12
CN V	8×8	0.15	10^{-5}	0.01	0.15	0.12	0.13
CN VI	8×8	0.25	10^{-3}	0.5	0.16	0.14	0.13
CN VII	8×8	0.15	10^{-3}	0.5	0.17	0.13	0.17
DC I	64 delay coordinates		10^{-7}	0.001	0.22	0.37	0.19
DC II	64 delay coordinates		10^{-6}	0.01	0.20	0.25	0.19
DC III	64 delay coordinates		10^{-5}	0.1	0.22	0.37	0.21
DC IV	64 delay coordinates		10^{-4}	0.3	0.23	0.30	0.23

In the classification of input data, two dimensional and three dimensional SOM/prototype-layer grid (lattice) topologies were experimented. The three-dimensional topology (4x4x4) coincides with the dimension of the input data, whereas dimension reduction is performed by the network by the two-dimensional topology (8x8). Two values for the SOM neighbourhood radius were experimented for controlling the strength of the neighbourhood around the winning node; recall that each signal was scaled to unit interval. The length of the moving window (64) was selected so that the number of prototype patterns equals that of the SOM-based approaches.

A simple SNN network of 5 hidden sigmoid nodes and a linear output node was used throughout the simulations. For the smooth mapping needed in NO_x emission modelling, a small five-hidden-node network judged sufficient. The RPE-BI algorithm (Ikonen, 1996) has two parameters for controlling the 'forgetting' of old data: a_{min} and a_{max}, which give a lower and upper bound for the parameter covariance matrix, respectively. In practice, the upper bound (a_{max}) is the dominant parameter, and was chosen based on a compromise between the speed of the adaptation and accuracy of the search. Two sets of values for a_{min} and a_{max} were used in the experiments with the combined network (see Table 2). With the delay-coordinate approach, a small search was performed in order to find a good set of parameters of the BI algorithm.

Table 2 summarises also some results from the simulations. In order to evaluate the performance of the identified model, three root-mean-squared errors (RMSE) were computed from the SNN model predictions on test set data. During the simulations, the RMSE was computed after each 100 iterations, k = 100, 200, ..., 118900, as well as at the end of simulations, K = 118 968. From these evaluations, two mean RMSE (MRMSE) were computed: the mean RMSE of all evaluations (3×5 days), and the mean RMSE during the last third of evaluations (performance during presentation of the last 5-day data). In addition,

Table 2 gives the RMSE for the model after the presentation of first third (5-days) of the data set.

Based on Table 2, we can conclude that the selection of the dimensions of the SOM/prototype network is less crucial. The smallest RMSE's were obtained in experiment CN I, where the 3-dimensional 4×4×4 topology was used. However, the performance of the 2-dimensional SOM/prototype network was almost as good. Instead, the neighbourhood parameter plays an important role in the self-organisation of the SOM. With a larger radius, the network neighbourhood is stronger, pulling the neighbouring nodes closer together. In the three-dimensional topology, the effect of this parameter is emphasized, since there are only four elements to each direction.

Figure 16 illustrates the mapping provided by the network, after 118 900 measurement samples. For clarity, only fuel type and residence time are shown. The SOM code vectors (left) cluster the incoming measurement samples (right), roughly equalizing their probability distribution. This distribution is also reflected by the prototype vectors (middle) containing the data for further on-line identification of a SNN process model. Note, how some prototype patterns lay relatively far from the associated SOM nodes. Since prototypes contain no neighbourhood, it is possible that a SOM node associated with a prototype pattern has moved to a completely new operation area (due to winners in close neighbourhood), but that the prototype pattern has not been updated (since the node itself was not a winner). From process modelling point of view also these prototypes contain useful information, as they are based on the latest information from the corresponding operating area. The latest data points (crosses in Fig. 16, right), are well represented by the prototype vectors.

Examining the RMSE's in Table 2, it is clear that the combined network outperforms the delay-coordinate approaches, providing a more robust and more accurate global non-linear SNN model. In on-line situations, a significant enhancement of the quality of the SNN training data could be achieved. The additional computational cost of the SOM/prototype layers of the combined network was insignificant, compared to the effort required for estimating the parameters of SNN models.

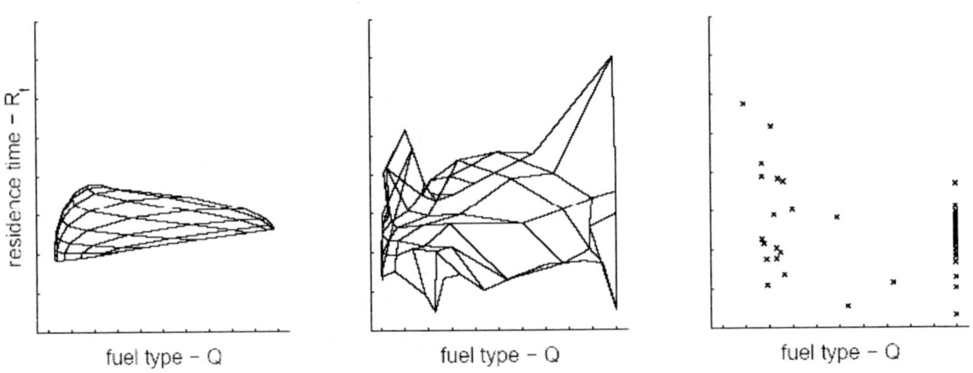

Fig. 16. Two-dimensional projection of the organization of the SOM and prototype nodes (taken from experiment CN II). Left: SOM nodes connected by solid lines. Middle: prototype patterns connected by solid lines. Right: input data patterns (dots). Recent patterns are shown as crosses

3.5. Discussion

Although the simulation example considered a static MISO model, the extension to the MIMO or dynamic cases is straightforward. Identification of dynamic time-series models using the combined network paradigm is possible, with the restriction that only non-recurrent model structures are utilised. The recurrent case is more complex, as the model output would then depend on previous model predictions, which would then need to be recalculated and stored after each parameter update. For feed-forward model types, such as the NARX (Non-linear Autoregressive eXogenous input) time series, the structure is readily applicable. It is also straightforward to include the model output or some auxiliary measurements into the SOM nodes, if that kind of clustering is desired. The intermediate SOM and prototype mappings can be easily utilised by the process operating personnel to support their decision-making. In practice, the 'true dimension' of an industrial process is often much less than that of the measurement space and a self-organising map may be used for revealing a reduced system dimension. One-, two-, or three-dimensional maps can also be used for graphical presentations, *e.g.*, of the past history of the process operation.

In power plants, hourly records of the most important process measurements are frequently consulted by the process operators. A typical use of these records is to consult how a certain operation point with given constraints was obtained earlier. The prototypes of the combined network could be readily used as a logbook. A computerized implementation would provide an easily-searchable, automatically updated look-up table, replacing (at least partially) the piles of paper, thus supporting the decision making of the operating personnel. A further potential application would combine these data (and the attached NO_x model) with a monitoring application and an optimisation procedure in order to provide optimal set points for the process controllers.

On-line identification of non-linear time-varying models of industrial processes is a difficult task, where a multitude of issues need to be taken into account. The combined network contributes to the problem of having insufficiently rich data for parameter estimation. Together with fault tolerant intelligent sensoring and techniques for parameter estimation under constraints, the suggested approach provides a useful method to be included in the tool-box of a process engineer.

4. Multivariable FBC Trajectory Following Control Using Genealogical Decision Trees

Advances in population-based evolutionary algorithms have introduced new tools for optimizing and controlling complex systems (see, *e.g.*, Zalzala and Fleming, 1997; Najim *et al.*, 2004). This section considers a novel algorithm that falls within the area of mathematical population genetics and interacting particle systems (Del Moral, 2004): the genealogical decision trees. These particle filter systems (Doucet *et al.*, 2001; Arulampalam *et al.*, 2002) belong to stochastic search models (Najim *et al.*, 2004). Their robustness and flexibility in terms of applicability to both linear and non linear systems, and for both continuous and batch processes, make them particularly attractive for applications of optimization and control of complex industrial processes.

Lately, genealogical decision trees have been introduced for solving open-loop tracking control problems (Ikonen *et al.*, 2004; 2005). In the following, a control approach based on the extensive use of a process model is considered. This approach is appropriate, *e.g.*, for processes for which there exist no on-line sensors for the controlled outputs, or if on-line analysers are very expensive or have high maintenance costs. Soft sensors and inferential control approaches have been dedicated to solve this type of process control problems where the outputs measurements are available only off-line. Soft sensors are based on models relating the desired controlled variable to some easily measured variables. The essence of the inferential control approach is in the prediction of the process outputs over the interval separating two successive measurements. As a consequence, the efficiency of soft sensors and inferential control approaches is limited by the adopted model for the sensor, or the characteristics of the predictor of the outputs. Similarly, the quality of the results of the genealogical decision tree-based optimization depends on the accuracy of the process model available.

4.1. Genealogical Decision Trees

The idea behind the genealogical decision tree control strategy consists of associating Gaussian distributions to both the norms of the control actions and the tracking errors. The resulting stochastic search model can be interpreted as simple evolutionary or genetic particle evolution model, which has a natural birth and death interpretation. Converge in probability can be shown. In the following, the control problem is formulated, followed by a description of the optimization and control algorithm under consideration (Ikonen *et al.*, 2004; 2005). The performance of the scheme is illustrated with a multivariable FBC problem related to trajectory following.

4.1.1. Problem Formulation

Consider any non-linear time-varying dynamic system described by the following state equations:

$$\begin{aligned} \mathbf{X}_n &= F_n(\mathbf{X}_{n-1}, \mathbf{U}_n) \\ \mathbf{Y}_n &= h_n(\mathbf{X}_n) \end{aligned} \qquad (6)$$

where $\mathbf{X}_n = [X_{n,1}, X_{n,2}, \ldots, X_{n,S}]^T \in \Re^S$, $\mathbf{U}_n = [U_{n,1}, U_{n,2}, \ldots, U_{n,P}]^T \in \Re^P$, and $\mathbf{Y}_n = [Y_{n,1}, Y_{n,2}, \ldots, Y_{n,Q}]^T \in \Re^Q$. Index $n = 1,2,\ldots,T$ represents the sampling instant. $\mathbf{X}_0$ represents the fixed initial state at instant $n = 0$. Let $\mathbf{A}_n$ and $\mathbf{B}_n$ be symmetric and semi-definite positive covariance matrices. The control objective in a finite horizon of length T is given by:

$$J_T(\mathbf{U}_1, \ldots, \mathbf{U}_T) = \sum_{n=1}^{T} \|\mathbf{U}_n\|_{\mathbf{A}_n}^2 + \sum_{n=1}^{T} \left\|\mathbf{Y}_n - \mathbf{Y}_n^{\text{ref}}\right\|_{\mathbf{B}_n}^2 \qquad (7)$$

where $\|\mathbf{a}\|_{\mathbf{b}}^2 = \mathbf{a}^T \mathbf{b}^{-1} \mathbf{a}$. Our objective is to find the sequence of control actions that will minimize this control objective for open-loop control.

4.1.2. Optimization of the Control Sequence

Let us associate Gaussian distributions to both the norms of the control actions and the tracking errors, *i.e.*,

$$\|\mathbf{U}_n\|_{\mathbf{A}_n}^2 \to \frac{1}{\sqrt{2\pi}} e^{\left(-\frac{\beta}{2}\mathbf{U}_n^T \mathbf{A}_n^{-1} \mathbf{U}_n\right)} \tag{8}$$

$$\left\|\mathbf{Y}_n - \mathbf{Y}_n^{\text{ref}}\right\|_{\mathbf{B}_n}^2 \to \frac{1}{\sqrt{2\pi}} e^{\left(-\frac{\beta}{2}\left(\mathbf{Y}_n - \mathbf{Y}_n^{\text{ref}}\right)^T \mathbf{B}_n^{-1}\left(\mathbf{Y}_n - \mathbf{Y}_n^{\text{ref}}\right)\right)} \tag{9}$$

The parameter β is similar to the inverse of the temperature in simulated annealing algorithms (Kirkpatrick *et al.*, 1983).

Let us consider the following scheme. At n = 1, the initial states $\hat{\mathbf{X}}_0^i$ are given by $\mathbf{X}_0$. At each instant n (n = 1,2,...,T), carry out N independent and identically distributed normal vectors $\mathbf{U}_n^i \sim \mathcal{N}(\mathbf{0}, \mathbf{A}_n)$, $i = 1,2,\ldots,N$. Using the model (F_n, h_n) these control values lead to:

$$\begin{aligned} \mathbf{X}_n^i &= F_n\left(\hat{\mathbf{X}}_{n-1}^i, \mathbf{U}_n^i\right) \\ \mathbf{Y}_n^i &= h_n\left(\mathbf{X}_n^i\right) \end{aligned} \tag{10}$$

In order to simplify the notations, let us introduce the following term:

$$p_n^i = \frac{e^{\left(-\frac{\beta}{2}\left(\mathbf{Y}_n^i - \mathbf{Y}_n^{\text{ref}}\right)^T \mathbf{B}_n^{-1}\left(\mathbf{Y}_n^i - \mathbf{Y}_n^{\text{ref}}\right)\right)}}{\sum_{j=1}^{N} e^{\left(-\frac{\beta}{2}\left(\mathbf{Y}_n^j - \mathbf{Y}_n^{\text{ref}}\right)^T \mathbf{B}_n^{-1}\left(\mathbf{Y}_n^j - \mathbf{Y}_n^{\text{ref}}\right)\right)}} \tag{11}$$

which can be interpreted as a probability measure. Let us then generate N independent and identically distributed random vectors $\hat{\mathbf{U}}_n^1, \hat{\mathbf{U}}_n^2, \ldots, \hat{\mathbf{U}}_n^N$ according to the distribution:

$$p_n(\mathbf{u}) = \sum_{i=1}^{N} p_n^i \delta_{\mathbf{U}_n^i}^i \tag{12}$$

where $\delta_{\mathbf{u}}^i$ is the Dirac measure at the control value $\mathbf{u} \in \Re^P$. In other words we have a resampling procedure where each random control $\hat{\mathbf{U}}_n^j$ takes the value $\mathbf{U}_n^i$ with probability equal to p_n^i. The implementation of these control actions leads to a population of states:

$$\hat{\mathbf{X}}_n^j = F_n\left(\hat{\mathbf{X}}_{n-1}^i, \mathbf{U}_n^i\right) \tag{13}$$

for $j = 1,2,...,N$. Using these as initial states, the procedure is repeated for all $n = 2,3,...,T$. A mechanization of the algorithm is given in the next subsection.

It is easy to see the relation between a standard Sampling Importance Resampling (SIR) particle filtering (Arulampalam *et al.*, 2002) scheme. The importance density $\mathcal{N}(\mathbf{0}, \mathbf{A}_n)$ is used to generate particles; the resampling is conducted at each sampling instant based on the criterion of the tracking error. Therefore it is intuitively clear that a cloud of sequences is efficiently simulated, that takes into account both the norm on the control action (importance density) and the norm on tracking errors (resampling). The optimal control sequence is found by taking the sequence, among the simulated ones, which minimizes the control objective J_T.

The above stochastic search model can be interpreted as simple genetic particle evolution model. These evolutionary models have a natural birth and death interpretation. More precisely, each controlled state $\hat{\mathbf{X}}_n^i$ results from the selection of a random control action, say $\hat{\mathbf{U}}_n^i = \mathbf{U}_n^j$, for some *j*. Therefore we have that $\hat{\mathbf{X}}_n^i = F_n\left(\hat{\mathbf{X}}_{n-1}^j, \mathbf{U}_n^j\right)$ and the state $\hat{\mathbf{X}}_{n-1}^j$ can be interpreted as the parent of the individual $\hat{\mathbf{X}}_n^i$. In the same way the parent individual $\hat{\mathbf{X}}_{n-1}^j$ results from the selection of a random control action, say $\hat{\mathbf{U}}_{n-1}^j = \mathbf{U}_{n-2}^k$, for some *k*. Running back in time, we can trace the complete ancestral line of the current individual, and that of the corresponding control actions. At $n = T$ the approximate optimal open loop control sequence is obtained as our solution to minimize the control objective.

4.1.3. Genealogical Decision Tree Algorithm

Let us assume that the plant model is given in the following form:

$$\begin{aligned} \mathbf{x}_{n+1} &= f(\mathbf{x}_n, \mathbf{u}_n) \\ \mathbf{y}_n &= g(\mathbf{x}_n) \end{aligned} \tag{14}$$

The following pseudo-code implements the algorithm:

for $n = 1{:}T$

for $i = 1{:}N$

if $n = 1$, Initialize $\mathbf{x}^i, \mathbf{y}^i$ and J_T^i; end

Generate random control action $\mathbf{u}^i \sim \mathcal{N}(\mathbf{0}, \mathbf{A}_n)$

Store action to a list: $\mathbf{v}_n^i = \mathbf{u}^i$

Evaluate costs $J_U = \mathbf{U}_n^T \mathbf{A}_n^{-1} \mathbf{U}_n$;

$$J_Y = \left(\mathbf{Y}_n - \mathbf{Y}_n^{\text{ref}}\right)^T \mathbf{B}_n^{-1} \left(\mathbf{Y}_n - \mathbf{Y}_n^{\text{ref}}\right);\ J_T^i = J_T^i + J_Y + J_U$$

Compute $p_{\text{un}}^i = \exp\left(\frac{-\beta}{2} J_Y\right)$

end

for all $i = 1,2,\ldots,N$: Compute resampling probabilities $p^i = \frac{p_{\text{un}}^i}{\sum_{i=1}^{N} p_{\text{un}}^i}$;

for all $i = 1,2,\ldots,N$, Resample, *i.e.*, select $\hat{I}_i = k$ such that $\Pr(k = j) = p^j$

for all $i \in \hat{I}$, Evaluate model for next n: $\begin{array}{l} \mathbf{x}^i = f\left(\mathbf{x}^i, \mathbf{u}^i\right) \\ \mathbf{y}^i = g\left(\mathbf{x}^i\right) \end{array}$

for all $j = 1,2,\ldots,N$: Death and birth, *i.e.*, replace

$$\mathbf{x}^j, \mathbf{y}^j, \mathbf{v}_n^j \text{ and } J_T^j \text{ by } \mathbf{x}^{\hat{I}j}, \mathbf{y}^{\hat{I}j}, \mathbf{v}_n^{\hat{I}j} \text{ and } J_T^{\hat{I}j};$$

end

The optimal control sequence is $\mathbf{v}_T^{i*}$, where $i^* = \arg\min_i J_T^i$.

4.2. FBC Trajectory Following

Let us illustrate the behaviour of the genealogical decision tree approach in the design of an optimal control sequence for a fluidized bed combustion (FBC) power plant. The experiments were based on a model of a saw-dust fired FBC power plant (Ikonen and Kortela, 1994), consisting of seven differential equations describing the fuel inventories, temperatures and oxygen concentrations in the bed and freeboard of the furnace, as well as the generated superheated steam power. The discrete-time state-space form was obtained by solving the ordinary differential equations numerically one sampling time ahead.

A three-input three-output MIMO control problem was considered. The control inputs were chosen to be the fuel feed rate [kg/s], the primary air flow, and the secondary air flow [Nm^3/s], which are typically used in the control of this type of plant. The fuel inventory [kg], flue gas oxygen [Nm^3/Nm^3], and the superheated steam power [MW] were controlled. The superheated steam is the product of the process, to be further converted into district heat. The flue gas oxygen gives an indication of the main combustion conditions. The fuel inventory models the amount of unburned char in the bed, which, depending on the fuel and the furnace conditions, is more or less instantly available for combustion. The former are typically measured on-line, whereas the fuel inventory can not be measured during normal operation (see Ikonen and Kortela, 1994).

In simulations, the control objective consisted of two phases: a) An increase in fuel inventory from initial steady state 150 kg to 300 kg following a 10 minutes ramp trajectory; b) The increase of the load from 21 MW to 26 MW following a 10 minutes ramp trajectory. The flue gas oxygen was to be kept at a constant 3.1 %.

An incremental control was used in the optimization, *i.e.*, the control action applied to the plant was given by $\mathbf{v}(k)=\mathbf{u}(k)+\mathbf{v}(k-1)$ where $\mathbf{v}(k)$ was the plant input vector [fuel feed, primary air, secondary air], and the change in $\mathbf{v}(k)$, $\mathbf{u}(k)$, was optimized using the genealogical decision tree scheme.

In general, the following parameters need to be set (Ikonen *et al.*, 2004): $\mathbf{A}_n$, $\mathbf{B}_n$ (cost function weight matrices), N (number of particles, *i.e.* decisions lines in parallel), and β (inverse temperature). In the FBC simulations, the algorithm parameters were set to:

$$\mathbf{A}_n=\frac{1}{20}\begin{bmatrix}4^2 & 0 & 0\\ 0 & 4^2 & 0\\ 0 & 0 & 12^2\end{bmatrix},\ \mathbf{B}_n=\begin{bmatrix}50^2 & 0 & 0\\ 0 & 0.002^2 & 0\\ 0 & 0 & 1^2\end{bmatrix},\ \beta=1$$

based on the scales of the variables; the population size was set to $N = 10000$. Setting the control interval to 15 seconds resulted in trajectories of length $T = 130$.

In real applications, feed-back control is required to diminish the effect of disturbances acting on the process. A straightforward way to use the genealogical decision tree optimization is to add simple PI-controllers to compensate for disturbances. The general scheme is illustrated in Fig. 17.

In FBC the bed fuel inventory is not measurable on-line. Consequently, it is simplest not to include inferential control as a feed-back control (alternatively a state estimator would need to be constructed). Based on steady state analysis and simple physical justifications (see Ikonen and Najim, 2002), two PI-controller loops are selected; the power output is controlled via fuel feed rate and the flue gas oxygen with secondary air flow.

The incremental open-loop control sequence was optimized using the genealogical decision tree approach. The resulting sequence was then two-way filtered so as to remove high frequency components (jitter) and used for the open-loop control. Two additional PI-controllers were then used for feedback control of disturbances, tuning was based on the SIMC rules (Skogestad, 2003). Figure 18 shows the plant inputs and outputs in a typical simulation, where an unmeasured 10% downward step in feed char content occurs at t = 20 min. The control sequences are shown in the plots at left. The dotted lines show the open-loop term, the PI-corrected control actions are shown using solid lines.

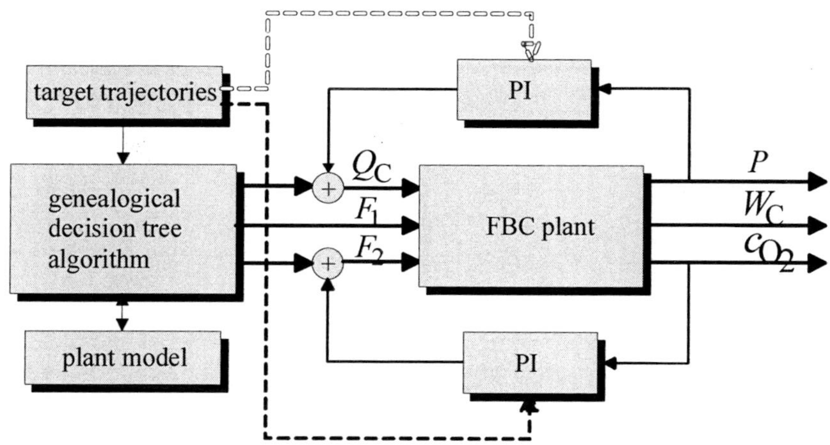

Fig. 17. FBC control with open-loop optimization and two feed-back loops. Notation: Q_C - fuel feed rate [kg/s], F_1 - primary air flow [Nm^3/s], F_2 secondary air flow [Nm^3/s], c_{O2} - oxygen concentration [Nm^3/Nm^3], W_C - fuel inventory [kg], P - combustion power [MW]

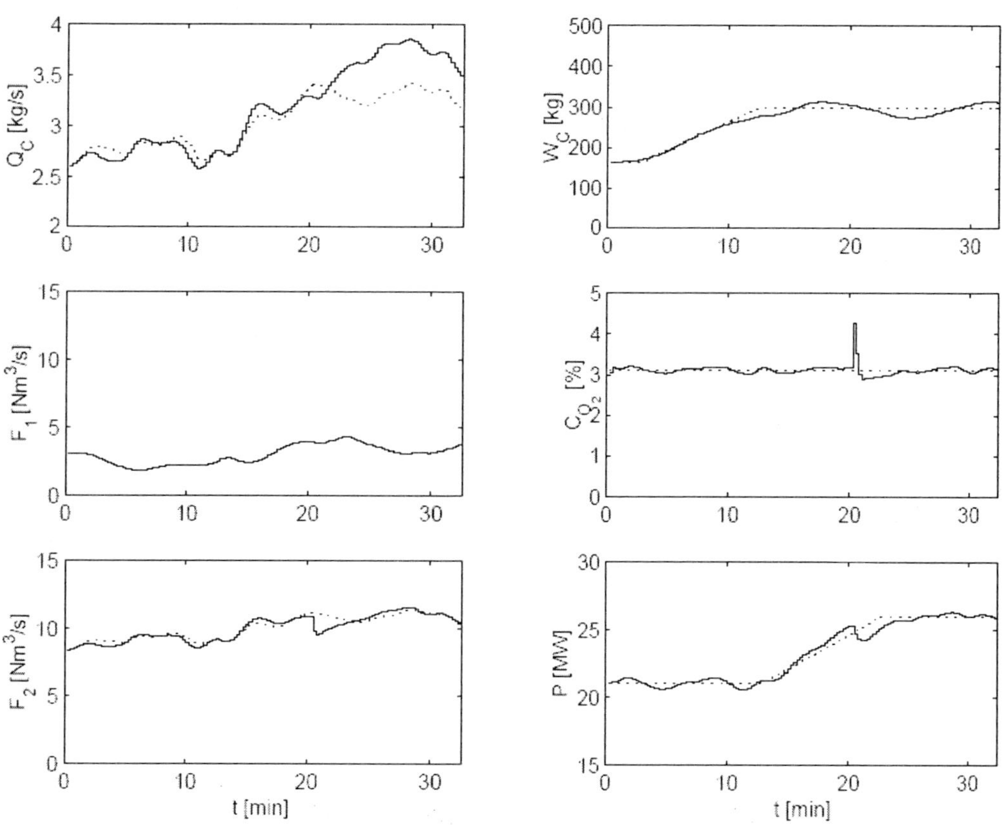

Fig. 18. Closed-loop simulation of trajectory control and a step disturbance in fuel quality. For notation, see caption of Fig. 17

The plots at the right show the two ramp trajectories: dotted lines indicate the desired trajectories, the solid lines show the outcomes of the controlled variables. The control actions of the manipulated variables appear realizable. The superheated steam power follows closely the desired trajectory. The plant model also gives an increased bed fuel inventory as desired. Despite the radical changes in the process, the flue gas oxygen is kept well in the vicinity of the given set point. Also the furnace temperatures remained between allowable limits (not shown in the figure). It can be concluded that the genealogical decision tree algorithm was able to find a feasible solution to the tracking problem. In addition, the unmeasured disturbance in feed char content is well compensated by the PI controllers. Overall, the simulations show good performance.

These simulations are illustrative of results that can be obtained in real situations and demonstrate the efficacy of this open-loop regulation and tracking control algorithm. The considered open-loop/closed-loop control strategy is well suited for a multitude of processes for which on-line sensors are not available, are too expensive, or can not be adapted for the measurement of different controlled variables. Indeed, models of these processes are often readily available, whereas the complex dynamics commonly exhibited by the models make the application of standard optimization routines unsatisfactory. The genealogical decision tree approach provides a flexible and robust optimization scheme for such dynamic problems.

5. CONCLUSION

In the development of learning control applications, a priority of importance can be emphasised:

(1) application (process, idea)
(2) control (regulation, optimisation)
(3) learning (adaptation, auto-tuning)

Needless to stress, process control is not just a matter of algorithmic signal manipulation. First of all, the idea of how to control the plant must be physically viable. This requires at least a basic understanding of the characteristics of the process (plant), the physical and chemical phenomena that occur, and which are to be affected by the manipulations. Second, the means for implementation and analysis of the control are to be selected. Plant control can be roughly categorized into three tasks: regulation (automatic control with feedback), control (servo behaviour with at least some degree of supervision by operators), and optimisation (initiated and implemented under supervision, often using complex constrained long-term or steady-state cost functions). Depending on these choices, different requirements are posed on the algorithms.

Learning can only be considered after a careful consideration and design of the first two stages. In regulation, the inclusion of learning is often limited, as the requirements of stability and robustness are essential (robustness – adaptation dilemma). Rather, an effort should be focused in keeping the control structures simple. In servo control, adaptation of simple models (under constraints) can be considered, resulting in adaptive control (auto-tuning, indirect/direct adaptive control, *e.g.*, predictive control.) In a complex system, proper design of servo controls includes taking into account interaction with other controls, lockings, *etc.* In optimisation (high level control), 'full scale' learning applications are more viable. At this level a sufficient variety of information is available, time is less critical, and supervision is provided by the human users. Also two-way interaction between users can benefit from the use of advanced techniques of artificial intelligence, *e.g.*, in improving the explanatory status of models using rule-based representations, or in data mining with applications in process monitoring, in simulation associated with numerical plant optimisation routines, and many more.

ACKNOWLEDGEMENTS

Many of the results discussed in this chapter have been prepared together with other researchers. In particular, we wish to thank Professor U. Kortela, Professor K. Najim and Professor P. Del Moral, as well as researchers I. Benyó, Z. Hímer, K. Leppäkoski, J. Mononen and M. Paloranta for their involvement.

Financial support from the Academy of Finland (projects No. 48545, 203231 and 50371) is gratefully acknowledged.

REFERENCES

Arulampalam M.S., Manskell S., Gordon N., Clapp T., 2002. A tutorial on particle filters for online nonlinear/non-Gaussian Bayesian tracking. *IEEE Transactions on Signal Processing*, Vol. 50, No. 2, 174-188.

Åström K.J., Bell R.D., 2000. Drum-boiler dynamics. *Automatica*, Vol. 36, No. 3, pp. 363-378.

Benyó I.,Paloranta M., Kovács J., Kortela U., 2005a. Advanced control of flue-gas oxygen content in fluidised bed boilers. *Control Engineering Practice* (to appear).

Benyó I, Kovács J., Mononen J., Kortela U., 2005b. Modelling of steam temperature dynamics of a superheater. *International Journal of Simulation*, Vol. 6, No. 6, pp. 3-9.

Benyó I., Kovács J., Paloranta M., Kortela U., 2006. Cascade generalized predictive controller: two in one; *International Journal of Control*, Vol 79, No 8, pp. 866-876.

Brown M., Harris C., 1994. *Neurofuzzy Adaptive Modelling and Control*, Prentice Hall, New York, 508p.

Del Moral P., 2004. *Feynman-Kac Formulae. Genalogical and Interacting Particle Systems with Applications*, Springer-Verlag, Berlin.

Donsi G., El-Sawi M., Formisani B., Sesti Osseo L., 1986. On the simulation of fluidized bed coal combustors. *Combustion and Flame*, Vol. 64, pp. 33-41.

Doucet A., de Freitas J.F.G, Gordon N.J., 2001. *Sequential Monte Carlo Methods in Practice.* Springer-Verlag, New York.

Henttonen J., Kojo I., Kortela U., 1992. Optimizing control of NOx and SO2 emissions in the FBC process. *Journal of the Institute of Energy*, Vol. 65, pp. 118-121.

Hímer Z., Wertz V., Kovács J., Kortela, U., 2004. Neuro-fuzzy model of flue gas oxygen content. *Proceedings of IASTED International Conference on Modelling, Identification and Control* (MIC 2004), Grindelwald, Switzerland. pp. 98-102.

Hupa M., Boström S., 1991. *Fluidized bed combustion: prospects and role*. Åbo Akademi, Department of Chemical Engineering, Report 91-7, Finland, 26 p.

Hupa M., 2005. Fluidised bed combustion. http://www.sedepiacenza.polimi.it/avvisi_eventi/file_recupero/Hupa.pdf

Hyötyniemi H., 1994. *Self-organizing artificial neural networks in dynamic systems modelling and control*. Helsinki University of Technology, Control Engineering Laboratory - Report 97. (Doctoral thesis).

Ikonen E., Kotajärvi H. (Editors), 1993. *Control concept for fluidized bed coal, peat and wood combustion: Optimization of flue gas emissions, control of process distrurbances and modelling of bed fuel inventory*. Systems Engineering Laboratory - Report C-8. March 1993. Oulu, Finland.

Ikonen E., Kortela U., 1994. Dynamic model for a bubbling fluidized bed coal combustor. *Control Engineering Practice*, Vol. 2, No. 6, pp. 1001–1006.

Ikonen E., Kortela U., 1995. On-line modelling using adaptive training prototypes with an application to the fluidized-bed combustion process. *IFAC Symposium on Power Plants and Power Systems* (SIPOWER '95), Mexico, II, pp. 113–118.

Ikonen E., 1996. *Algorithms for process modelling using fuzzy neural networks: A distributed logic processor approach*. Acta Universitatis Ouluensis, C95, Oulu, Finland (Doctoral thesis).

Ikonen E., Najim K., 1997. Use of learning automata in distributed fuzzy logic processor training. *IEE Proceedings - Control Theory and Applications*, Vol. 144, No. 3, pp. 255–262.

Ikonen E., Najim K., Kortela U., 2000. Neuro-fuzzy modelling of power plant flue-gas emissions. *Engineering Applications of Artificial Intelligence*, Vol. 13, No. 6, pp. 705–717.

Ikonen E., Najim K., 2001. Non-linear process modelling based on a Wiener approach. *Journal of Systems and Control Engineering – Proceedings of the Institution of Mechanical Engineers Part I*, Vol. 215, pp.15–27.

Ikonen E., Najim K., 2002. *Advanced Process Identification and Control*, Marcel Dekker Inc., New York, U.S.A., 310 p.

Ikonen E., Najim K., 2004. Process identification using distributed Wiener logic processors. *Differential Equations and Dynamical Systems*, Vol. 12, No. 1-2, pp. 171-194.

Ikonen E., Del Moral P., Najim K., 2004. A genealogical decision tree solution to optimal control problems. *Proceedings of the IFAC Workshop on Advanced Fuzzy/Neural Control* (AFNC'04), 16-17 September, 2004, Oulu, Finland, pp. 169-174.

Ikonen E., Najim K., Del Moral P., 2005. Application of genealogical decision trees for open-loop tracking control. *Proceedings of the 16th IFAC World Congress on DVD*, 4-8 July, 2005, Prague, Czech.

Jang R., Sun C., Mizutani E., 1997. *Neuro-fuzzy and Soft Computation*. Prentice Hall, New Jersey.

Karppanen E., 2000, *Advanced control of an industrial circulating fluidized bed boiler using fuzzy logic*. Acta Universitatis Ouluensis, C145, Oulu, Finland (Doctoral thesis).

Kirkpatrick S., Gelatt C., Vecchi M., 1983. Optimization by simulated annealing. Science, Vol. 220, pp. 671-680.

Klefenz G., 1986. *Automatic Control of Steam Power Plants*. 3rd Edition. Bibliographisches Institut.

Kohonen T., 1990. The self-organising map. *IEEE Proceedings*, Vol. 78, No. 9, pp. 1464–1480.

Kortela U., Lautala P., 1981. A new control concept for a coal power plant. *Proceedings of the 8th IFAC World Congress*, Kyoto, Japan.

Kovács J., 1998. *Model reference adaptive PIP control algorithms: Application to boiler drum control* (Doctoral thesis). Acta Universitatis Ouluensis Technica C112.

Kunii D., Levenspiel O., 1991. *Fluidization Engineering,* Butterworth-Heinemann, Boston, MA, U.S.A.

Leckner B., 1998. Fluidized bed combustion: mixing and pollutant limitation. *Progress in Energy and Combustion Science*, Vol. 24, pp. 31-61.

Leppäkoski K., Jouppila M, Kortela U., 1999. An expert system to reduce emissions in a full-scale BFB boiler. *Proceedings of the 14th World Congress of IFAC on CD-ROM*, Beijing, China.

Leppäkoski K., Paloranta M**,** Ikonen E., 2004. Comparing two nonlinear structures for secondary air process modelling. *Control 2004*, September 2004, University of Bath, UK.

Lindsley D., 2000. Power-plant control and instrumentation. *The control of boilers and HRSG systems*. IEE Control engineering series, London, UK.

Ljung L., Söderström S., 1983. *Theory and Practice of Recursive Identification.* MIT Press.

Mononen J., Hiltunen J, Kortela U., 1996. Developing a rule-based instruction system for the reduction of flue gas emission. *Proceedings of the IASTED Conference High Technology in the Power Industry*, 6 - 8 June 1996. Banff, Canada. Acta Press, Anaheim, USA, pp. 196-198.

Mononen J., Ikonen E., 2005. Modelling of an integrated superheater based on a Wiener approach. *Proceedings of the 16th IFAC World Congress on DVD*, 4-8 July, 2005, Prague, Czech.

Normey-Rico J.E., Bordons C., Camacho E F., 1997. Improving the robustness of dead-time compensating PI controllers, *Control Engineering Practice,* pp. 801-810.

Najim K., Ikonen E., 1999. Distributed logic processors trained under constraints using stochastic approximation techniques. *IEEE Transactions on Systems, Man and Cybernetics Part A: Systems and Humans*, Vol. 29, No. 4, pp. 421-426.

Najim K., Ikonen E., Aït-Kadi D., 2004. *Stochastic Processes: Estimation, Optimization and Analysis*. Kogan Page Science, London, U.K.

Paloranta M., Benyó I., Leppäkoski K., Kortela U., 2004. Time delay compensating control of flue-gas oxygen in FBB. *IASTED Conference on Power and Energy Systems on CD-ROM*, 28 November - 1 December 2004, Clearwater Beach, Florida, USA.

Pedrycz, W., 1993. Fuzzy neural networks and neurocomputations. *Fuzzy Sets and Systems*, Vol. 56, pp. 1-28.

Pedrycz W., 1994. Logic-based neurons: Extensions, uncertainty representation and development of fuzzy controllers. *Fuzzy Sets and Systems*, Vol. 66, pp. 251-266.

Pedrycz W., Lam P., Rocha A., 1995. Distributed fuzzy system modelling. *IEEE Transactions on Systems, Man and Cybernetics*, Vol. 25, pp. 627-636.

Ruoppila V., Sorsa T., Koivo H., 1993. Recursive least-squares approach to self-organizing maps. *IEEE Int. Conf. on Neural Networks*, San Francisco, CA, USA, pp. 1480–1485.

Skogestad S., 2003. Simple analytic rules for model reduction and PID controller tuning. *Journal of Process Control*, Vol. 13, pp. 291-309.

Zalzala A.M.S., Fleming P.J., 1997. *Genetic Algorithms in Engineering*, IEE Control Engineering Series.

In: Artificial Intelligence in Energy and Renewable Energy Systems ISBN 1-60021-261-1
Editor: Soteris Kalogirou, pp. 439-457

Chapter 13

APPLICATION OF ARTIFICIAL INTELLIGENCE TECHNIQUES FOR THE REDUCTION OF ENERGY CONSUMPTION OF REFRIGERATION SYSTEMS

Savvas A. Tassou, Debarati Datta and Ian N. Grace
School of Engineering and Design
Brunel University, UK

ABSTRACT

Refrigeration is part of everyday life and can be found in the home, car, office and industry. It is used very widely for food preservation, thermal comfort, food and drinks processing and chemical manufacturing. Since the inception of mechanical refrigeration towards the end of the 19th century two principal refrigeration cycles have emerged, the vapour compression and the vapour absorption cycle. The former cycle has dominated domestic and commercial applications due to its higher efficiency, smaller size and lower cost for the same refrigeration capacity. In recent years, environmental impacts from the working fluids of vapour compression systems, energy consumption and materials used in the construction of refrigeration equipment have increased pressure on the refrigeration industry to reduce the impacts of refrigeration systems through better design and control. This chapter presents results of investigations into the application of artificial intelligence techniques, based on Artificial Neural Networks, for the utilisation of data already collected from many large supermarkets for energy prediction, performance diagnosis and control. The results demonstrate that artificial intelligence can be used effectively to reduce both the energy consumption of refrigeration systems and predict deterioration in system performance due to gradual refrigerant leakage.

Keywords: Refrigeration, artificial intelligence; energy prediction, defrosting, leak detection.

ABBREVIATIONS

AI	Artificial Intelligence
ANN	Artificial Neural Networks
COP	Coefficient of Performance
FDD	Fault Detection and Diagnosis
LT	Low Temperature
HT	High Temperature
MLP	Multi-Layered Perceptron
RH	Relative Humidity
RMS	Root Mean Square

1. INTRODUCTION

Retail food stores are large consumers of energy, which in the industrialised countries, amounts to 3-5 % of total electricity consumption. Retail food stores are also responsible for the emission of large quantities of refrigerants to the environment, which annually are estimated to be between 10 – 30 % of total system refrigerant charge.

In most developed economies retail food stores, because of the size of their electrical energy consumption, qualify to join the competitive power market in which power can be purchased on a half-hourly basis. The half-hourly rate paid by the supermarket owner to the supplier is largely depended on his ability to estimate accurately the maximum half-hourly demand and the competitive rates offered by various suppliers. If the actual consumption exceeds the predicted value, the purchaser is penalized for the extra supply needed by paying at a higher than the negotiated rate. The ability, therefore, to predict the power consumption every half hour as accurately as possible will facilitate negotiations on electricity tariffs with the suppliers and will also enable the control of maximum demand by shifting some of the load to periods of reduced demand.

Most large retail food stores are equipped with central monitoring and control systems to primarily satisfy food hygiene regulations. These systems monitor and control the temperature in the refrigerated display cabinets within specified limits and control the centralised refrigeration systems (packs) to balance the load on the cabinets with the refrigeration capacity of the packs. The control functions performed are fairly simple and in the vast majority of cases, the data collected remain unutilised due to the unavailability of automated data mining and diagnostic systems for this application. These data, however, provide the opportunity not only to characterize the various energy consuming processes in the supermarket but also to relate the consumption patterns to fuel pricing and tariff structures and thereby develop advanced control techniques to minimize maximum electrical demand, energy consumption and fuel costs. It may be possible to perform these tasks on-line by employing adaptive control through predictive neural networks (ANNs).

The evaporator coils of refrigeration systems normally operate with surface temperatures below 0°C. Air circulating over the coils, in coming into contact with the cold surface, condenses and eventually freezes on the coil. With passage of time, the frost thickens and gradually the performance of the refrigeration system deteriorates as the thermal resistance between the warmer air and the cold refrigerant inside the evaporator increases due to the insulating effect of frost. For finned-tube evaporators frost formation also restricts the airflow through the coil thereby decreasing the cooling capacity further. Hence, periodic defrosting of cooling coils operating below 0°C is essential.

The defrost cycle in supermarket refrigeration systems is usually controlled by a pre-set time cycle as it is simple to install and easy to service. It is widely acknowledged, however, that timed defrost may cause a number of unnecessary defrost cycles and this reduces the energy efficiency of refrigeration systems as well as the accuracy of temperature control of the cabinets. Implementing defrost on demand should reduce the number of defrost cycles and thus has the potential to lead to energy savings.

A number of demand defrost techniques have been applied over the years which include: air pressure differential sensing across the evaporator (Ciricillo, 1985); sensing the temperature difference between the air and the evaporator surface (Buick *et al.* 1978) and adaptive defrost timer (Heinzen, 1988). Other methods which attempted to measure the frost thickness directly for defrost initiation included monitoring the resonant frequency of an acoustic oscillator installed on the evaporator (Llewelyn, 1984); measuring the thermal conductivity of ice (Llewelyn, 1984); and fibre optic sensors to detect the presence of ice (Paone and Rossi, 1991).

None of the above methods has gained acceptance by the refrigeration industry in general and the food retail industry in particular, due to excessive capital cost and reliability problems associated with complex and unreliable sensing methods. They are also system specific and so far their application has been restricted to domestic refrigerators and heat pumps. These disadvantages may be overcome through the development of a demand defrost control method based on Artificial Intelligence.

Refrigeration systems, particularly those in large retail food stores, have the potential to leak because pressures in the system are usually many times higher than atmospheric. Loss of refrigerant from industrial and commercial refrigeration systems can occur (Tassou and Grace, 2005):

a) due to gradual leakage from joints or seals, that can remain undetected for long periods of time, until sufficient refrigerant has been lost to adversely affect the operation of the refrigeration system,
b) through catastrophic damage which can occur when mechanical failure such as accidental rapture of a pipe or joint takes place and results in a significant loss of refrigerant charge in a short period of time, and
c) during servicing when some refrigerant can be accidentally vented to gain access to a section of pipe or a given piece of equipment for repair.

In many large systems, it is estimated that over 20% of the refrigerant charge could be released to the atmosphere each year, contributing to the depletion of the ozone layer and to global warming (Department of Environment, 1992; Hummel *et. al.,* 1992). Refrigerant loss also contributes to the reduction of the operating efficiency of the system, leading to increased power consumption and greenhouse gas emissions, higher maintenance costs and eventual system failure.

Various methods of refrigerant leak detection exist which are primarily based on fixed or portable refrigerant sensors. These methods have a number of limitations which include:

i) the requirement for careful location of the sensors,
ii) inability to detect slow refrigerant leaks which are the most common cause of refrigerant loss from refrigeration systems,
iii) inability to operate in well ventilated environments e.g. well ventilated plant rooms and outdoor location of equipment, and
iv) sensors are specific to only one or a small group of refrigerants.

It is well accepted that there is an optimum refrigerant charge for each system with the performance dropping at undercharged and overcharged conditions. The degree of degradation is a function of the type and design of the system but can be as high as 45% reduction in system capacity and COP for a 15% charge reduction. The performance of the system can also degrade due to other faults which may include `sticking' of the expansion valve, incorrect setting on the superheat on an expansion valve, fouling in the heat exchangers, compressor wear and many others (Fazad and O'Neil, 1991).

Given the limitations of existing refrigerant leak detection methods, attempts have been made to develop methods based on real time performance monitoring in conjunction with mathematical or black-box models of the system (Comstock *et al.,* 2002; Grimelius *et al.,* 1995). Although mathematical models can produce accurate predictions of system performance, they have the disadvantage of being relatively slow and highly specific to each individual unit. In recent years, attention has focused on the use of real-time system performance predictions based on Artificial Neural Networks (Stylianou and Nikanpour, 1996; Peitsman and Bakker, 1996; Bailey and Kreider, 2003). This chapter reports the work carried out by the authors on the development of a refrigerant leak detection and fault diagnosis system based on artificial intelligence and real-time performance monitoring of the refrigeration system (Tassou and Grace, 2005; Rueda *et al.*, 2005). The chapter also addresses the use of Artificial Intelligence Techniques for electrical power demand prediction and implementation of defrost on demand.

2. Power Demand Prediction

A typical layout of a supermarket, consists of a large number of display cabinets located around the perimeter of the sales area and walk-in boxes are placed in close proximity to the cases where products to be stocked are stored. In recent years, the retail industry has moved in favour of remote refrigeration systems, which provide the flexibility of installing the compressors and condensers at a centralized plant area, usually at the back of the store or mezzanine. The refrigeration pipe runs are usually installed under the floor of the sales area. In the machine room, the refrigeration compressors are mounted on bases or racks depending on the type of compressor system employed. Along with the compressors, all necessary piping, valves, and electrical components needed for the operation and control of the compressors are also mounted on the base or racks. Air-cooled condensers used in conjunction with the multiple compressor systems are installed remotely from the compressors, usually on the roof of the machine room (Tassou and Datta, 1999).

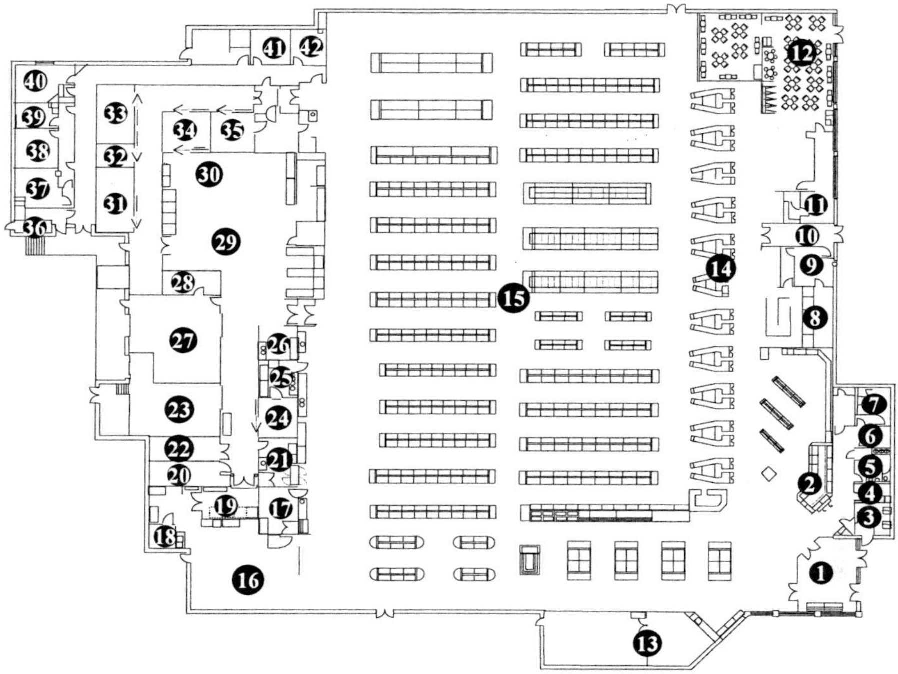

Zone No.	Description	Zone No.	Description
1	Entrance / Exit	22	Wine Store
2	Cigarette Kiosk	23	Refuse Store
3	ATM Room	24	Delicatessen Chiller
4	Security Room	25	Delicatessen Preparation Room
5	Male Toilets	26	Distress Goods Storage
6	Disable Toilets	27	Receiving Area
7	Female Toilets	28	Office
8	Office Counter	29	Warehouse
9	Administration Office	30	Produce Preparation
10	Side Exit	31	Dairy Chiller
11	Coffee Preparation Area	32	Ice Cream Freezer
12	Café / Restaurant	33	Frozen Food Freezer
13	Dry Clean Preparation / Counter	34	Produce Chiller
14	Check Out Area	35	Meat Chiller
15	Sales Area	36	Gas Monitoring
16	Bakery Preparation Area	37	Cash Office
17	Wash Up Room	38	Computer Room
18	Bakery Freezer	39	Manager Office
19	Flour Storage Area	40	Training Room
20	Electrical Switch Room	41	Training Manager Office
21	Chicken Preparation Area	42	General Office

Fig.1. A typical schematic layout of a supermarket

The compressors in large supermarkets are normally arranged in two groups. Low temperature (LT) serving the frozen food cabinets and low temperature cold rooms and high temperature (HT) serving the chilled food cabinets packs.

All field data employed in the investigations reported in this chapter were based on a Safeway retail food store situated in Airdrie, Scotland. A schematic diagram of the store is shown in Fig. 1. This store was equipped with a central monitoring and control system which monitors the temperatures in the display cabinets in the store and controls the refrigeration packs. The parameters monitored included (Datta and Tassou, 1998):

- temperature and relative humidity in the store
- external air temperature and humidity
- solar irradiance
- total electrical power consumption of the store
- electrical power consumption of the refrigeration packs
- gas consumption
- underfloor heating flow and return temperatures, and
- instantaneous store occupancy (shopping activity)

All the above data was logged at 15-minute intervals through a modem on a personal computer at Brunel University for analysis. Figure 2 shows a typical daily variation of the external and internal conditions and the total electrical load of the store. Although the temperature in the store is controlled by the heating and ventilating system, the relative humidity is allowed to float. As can be seen from Fig. 2 this variation is fairly small. The external relative humidity during the winter months is high and remains fairly constant whereas the external temperature rises during the day and drops during the night in a sinusoidal fashion as expected.

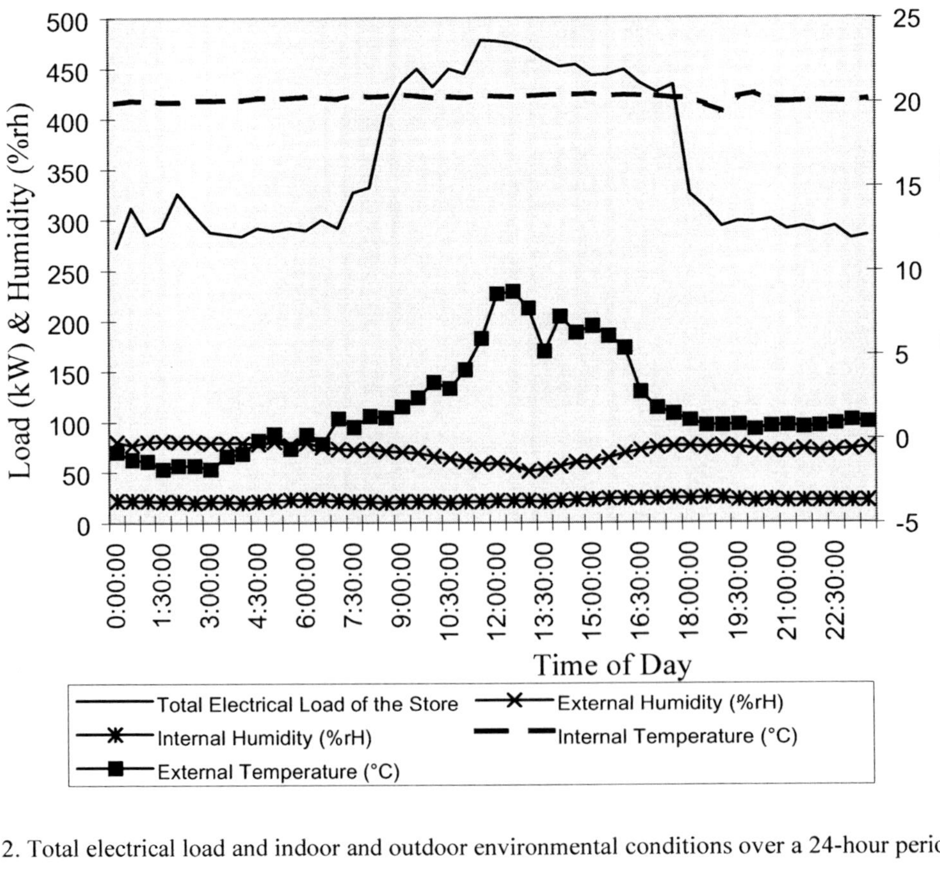

Fig. 2. Total electrical load and indoor and outdoor environmental conditions over a 24-hour period

From Fig. 2 it can also be seen that during the night there is a base load on the store comprised mainly of the refrigeration load which includes the compressor packs, the condenser fans and the display cabinet evaporator fans. On store opening in the morning the load increases sharply due to the store lights and other ancillary equipment being switched on and then drops again down to the base value on store closing in the evening.

Preliminary analysis of the data collected showed that there was a very good correlation between the time of day and day of the week with the occupancy of the store, which was measured with an infra-red sensor which triggered a counter once the light beam was interrupted. The sensor was placed at the main entrance of the store. It was also found that solar radiation had very little effect on the energy consumption of the store due to complete shading of external windows from direct solar radiation. For this reason and the relatively high cost of reliable occupancy and solar radiation meters, these parameters were not included in the final energy predictors developed. In the investigation presented here a multi-layered perceptron network (MLP), was found to be the most effective in predicting the energy consumption of the supermarket.

2.1. Total Electrical Power Demand Prediction

The input variables employed for the predictions of the total power consumption were: day of the week, time of the day, outdoor temperature and humidity and indoor temperature and humidity. A number of MLP network parameters were considered for the prediction of the total electrical power demand of the supermarket and different network configurations were tested and compared in terms of RMS error, Mean Absolute Error and Mean Absolute Percentage Error. A time series based MLP network with 14 inputs (outdoor humidity in three time steps, indoor humidity in three time steps, outdoor temperature in three time steps, indoor temperature in three time steps and power in past two time steps), 7 nodes on a single hidden layer, and one output (the total electrical demand) was found to produce best results in terms of prediction accuracy. The most important parameters were the outdoor temperature and the indoor relative humidity. Both of these parameters influence directly the power consumption of the refrigeration systems through their impact on the building load and condenser pressure and frosting and defrosting losses respectively. The performance for the 14:7:1 (input: hidden: output neurons) time series MLP network is illustrated in Fig. 3. It can be seen that the network can predict the variations in power demand quite accurately with an RMS error of 20 kW, a mean absolute error of 15 kW and a mean absolute percentage error of around 4%. It can also be seen that the neural network can predict quite accurately the power demand during the days when the store was closed during Christmas.

2.2. Refrigeration System Electrical Power Consumption

In a retail food store, refrigeration is the major contributor to the total electrical demand. Reliable and efficient operation of the refrigeration equipment is also very critical to the attainment of the required conditions for safe food storage and display. For this reason, we considered whether neural networks could be used to predict the energy demand of the refrigeration systems separately from the total electrical demand so that the predictions could be used for fault detection and diagnosis purposes. As with the prediction of the total electrical demand, different network configurations and input parameters were investigated. The best network was found to be a MLP of 7:4:1 configuration with 7 input variables, 4 nodes on a single hidden layer and one output variable (the mean daily power demand in kW determined from the total daily consumption in kWh divided by 24).

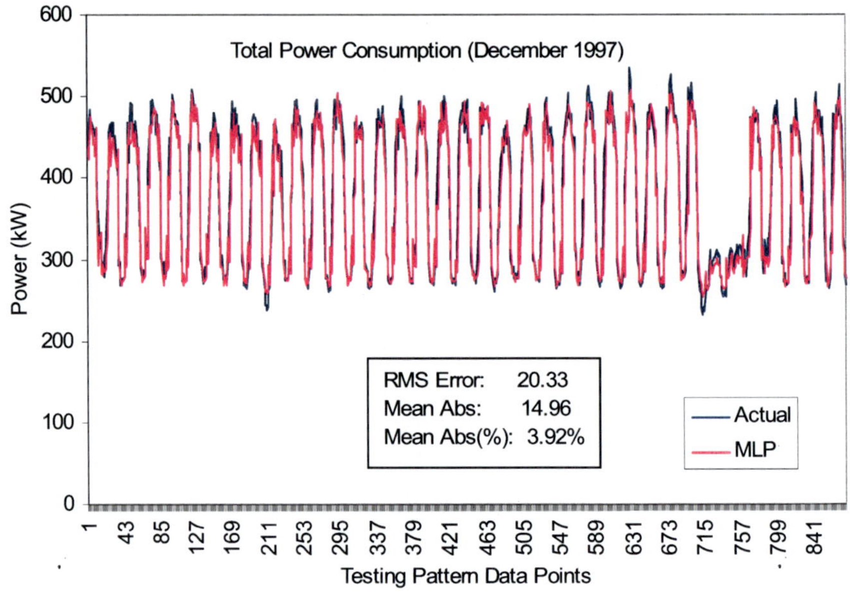

Fig. 3. Comparison of the target value and MLP output for network configuration 14:7:1 with conjugate training algorithm and hyperbolic tangent (tanh) activation function

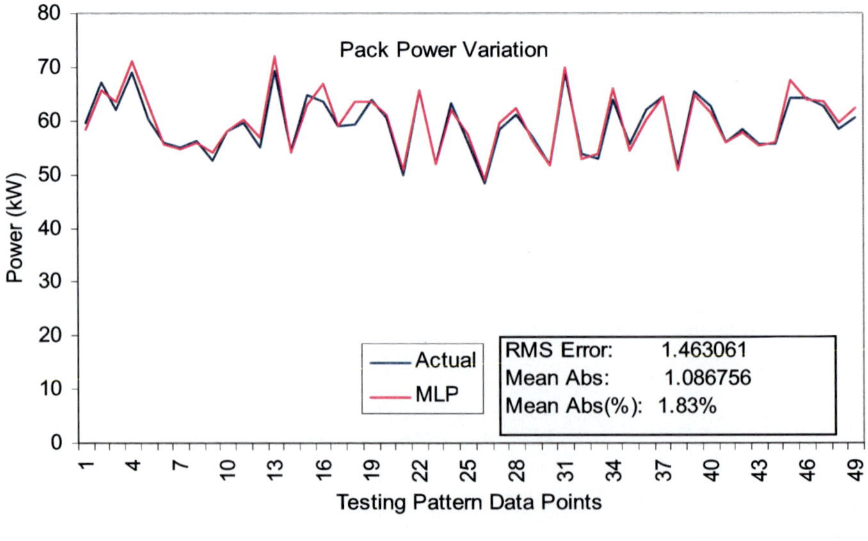

Fig. 4. Comparison of the target value and MLP output for network configuration 7:4:1 with conjugate training algorithm and hyperbolic tangent (tanh) activation function for pack electrical energy consumption

The variation of the predicted and actual power demand over a period of 49 days for the high temperature refrigeration pack in the store is shown in Fig. 4. It can be seen that the neural network can predict quite accurately the daily fluctuations in power demand with an RMS error of 1.5 kW, a mean absolute error of 1.0 kW and a mean absolute percentage error of 1.8%.

3. Defrost Control

3.1. Experimental Investigations

To identify suitable control parameters for the development of a defrost on demand controller based on artificial intelligence (AI), tests were carried out at controlled conditions in an environmental chamber at Brunel University and field measurements in two supermarkets. From the field measurements, which involved collecting and analysing environmental data and quantities of condensate collected after defrost, the environment within a supermarket was characterised. Tests in the laboratory were used to establish the performance of four different types of cabinets over a wide range of conditions and to monitor in detail the influence of frost formation on their performance characteristics. However, only one display cabinet is considered in this chapter, a high/medium temperature display cabinet 2.5 m long and 3.7 kW cooling capacity at +4°C. The cabinet was tested at controlled conditions as shown in Fig. 5. Plastic containers filled with water and food substitute materials were used to load the cabinet. The parameters measured were:

- evaporator air on and air off temperature,
- evaporator surface temperature,
- refrigerant temperature and pressure at evaporator inlet and outlet,
- compressor suction and discharge pressure and temperature,
- condenser inlet and outlet pressure and temperature,
- thermostatic expansion valve inlet and outlet pressure and temperature,
- mass flow of refrigerant, and
- space temperature and humidity.

In addition, the product temperature at various positions in the cabinet was monitored. These measurements were logged every minute onto a computer through a data-acquisition system. The air off velocity was also recorded every minute using a hot-wire anemometer.

The cabinet was operated overnight before testing to bring the products down to required temperature and to achieve steady state test conditions. Prior to all tests, a defrost cycle was implemented to ensure that the evaporator was clear of frost. The test period was considered to be between the end of the first defrost and start of the following defrost cycle. During each test period, the temperature and relative humidity in the test chamber were maintained at a constant preset value.

Experiments were conducted over a range of space temperatures at constant relative humidity and over a range of space relative humidities at constant temperature. Tests were also carried out for varying cooling time between defrost cycles.

Fig. 5. Display Cabinet under test in the Environmental Chamber

3.2. ANN Model Development and Validation

From the experiments conducted in the laboratory, vast amounts of data were obtained and used to train, validate and test artificial neural networks (ANNs) for the prediction of the air flow rate through the evaporator coil. The coil blockage due to frost accumulation was measured indirectly by measuring the air velocity through the coil using a hot-wire anemometer.

A commercial software package (NConnect from SPSS Inc.) was used for network training and evaluation. Table 1 shows the range of input and output data used. As can be seen, continuous variables often have dynamic ranges, which differ by orders of magnitude and hence a suitable normalisation need to be applied so that all transformed variables cover the same range. This ensures that each input field is used equally by the neural model. The most appropriate transformation is the *zero mean unit variance* transformation, which is a *linear* transformation applied to each input variable independently.

Table 1. Range of input and output data

Input parameters	Range	Input / Output parameters	Range
Space temperature	14 to 25°C	Air off temperature	-6 to 3°C
Space relative humidity	25 to 60% RH	Air off temperature at the backpanel	-6 to 4°C
Evaporating temperature	-10 to –3°C	Duration of the cooling cycle	0 to 800 min
Air on temperature	4 to 13°C	Air off velocity	0.9 to 2.9m/s

Multiple MLP networks were trained and evaluated using different network configurations and parameter settings as listed in Table 2. When choosing the optimal MLP, the selection is based on the lowest error obtained on the validation set and the testing set. The results show that the MLP 4:3:1 configuration achieved the lowest value of RMS error of

0.132 on the validation dataset. Also, the generalisation performance of 4:3:1 was better on the test dataset. So this configuration was chosen for implementation.

Table 2. Input Parameters for different network configurations

Configuration	Input parameters
7:4:1	Cooling time (mins), refrigerant temperature at evaporator inlet, air temperature on the evaporator, off the evaporator and the backpanel (°C), space temperature (°C) and space relative humidity (%RH).
6:4:1	Cooling time (mins), refrigerant temperature at evaporator inlet, air temperature on the evaporator, off the evaporator and the backpanel (°C) and space relative humidity (%RH).
5:3:1	Cooling time (mins), refrigerant temperature at evaporator inlet, air temperature on the evaporator, off the evaporator (°C) and space relative humidity (%RH).
4:3:1(A)	Cooling time (mins), refrigerant temperature at evaporator inlet, space temperature (°C) and space relative humidity (%RH).
4:3:1(B)	Cooling time (mins), air temperature on the evaporator, off the evaporator and the backpanel (°C).

Table 3. Neural network training, validation and testing results

Configuration	Training error	Validation error	Testing error	Max. error	r^2
7:4:1	0.291	0.151	0.053	0.169	0.99
6:4:1	0.145	0.204	0.079	0.265	0.99
5:3:1	0.160	0.190	0.088	0.269	0.99
4:3:1(A)	0.171	0.214	0.089	0.286	0.99
4:3:1(B)	0.292	0.132	0.046	0.148	1.00

Figure 6 along with Table 3, show the performance of the network on the test sets, which are typical of the neural networks that were trained in the project. They show that neural networks can be trained to accurately predict the air flow deterioration through the evaporator coil. The training time depends on the input/output patterns, the construction of the network (number of layers, the number of nodes, training algorithm and the use of activation function), error tolerance and the processing speed of the computer used.

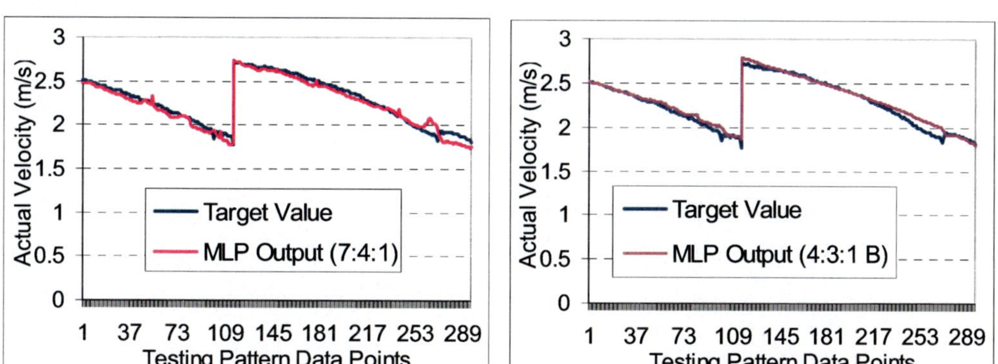

Fig. 6. Comparison of actual value and MLP network output for 7:4:1 and 4:3:1 (B) configurations

3.3. System Implementation

In order to implement the developed control strategy, a software-based controller was developed using JAVA, Fig. 7.

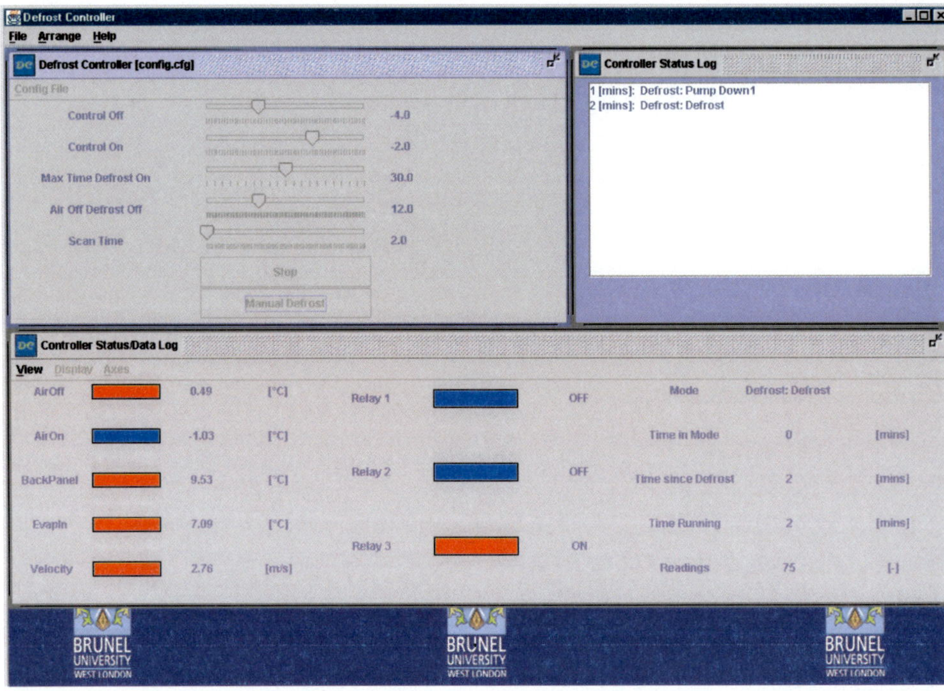

Fig. 7. Front end for the software-based cabinet demand defrost controller

The software-based controller is very flexible and has monitoring, data-logging, neural network for estimation, control and fault detection and diagnosis (FDD) capabilities. It is a cabinet controller which controls both the cooling and the defrost cycle. The software scans and stores data through a hardware system – the Measurement System in this case. For the cooling cycle the software looks at the cabinet temperature which is the average of the air off and air on temperature and switches the liquid line valve on and off depending on the set-temperature. For the defrost cycle the scanned values of the air on, air off and back-panel temperature along with time are used by the static-weight neural network module of the controller to estimate the air-off velocity. When the estimated velocity reduces to a predetermined value of 1.5m/s, which is a defined value in this case, the controller initiates a defrost cycle. The defrost cycle is terminated when the air off temperature reaches 12°C or the defrost duration is 30 minutes whichever occurs first.

3.4. Discussion

In laboratory tests, the demand defrost controller developed has shown defrost energy savings of 25% for space conditions of 22°C and 55% RH and 50% for space conditions of 22°C and 35% RH when electric defrost heaters were used. The savings result from the reduction of the number of defrost cycles at conditions of low space temperature and humidity, and reduction of the cabinet cooling (pull-down) load after each defrost. Therefore, even high temperature cabinets that use off-cycle rather than electric defrost can also benefit from a reduction in the number of defrost cycles per day. By using the demand defrost control method developed, the number of defrost cycles per day was reduced from 4 at 22°C and 65% RH to 3 at 22°C and 45% RH, and to 2 at 22°C and 35% RH, without affecting the product temperature.

4. Leak Detection

This section of the chapter reports on the development of a refrigerant leak detection and fault diagnosis system based on artificial intelligence and real-time performance monitoring of a small capacity refrigeration system.

4.1. Methodology

The results presented are based on a 4.0 kW nominal cooling capacity liquid-liquid system with a horizontal scroll compressor and plate heat exchangers. The FDD system developed is designed to operate on a range of refrigeration equipment including small air-to-air and large commercial chillers. The system consists of a number of discrete subsections or modules, which segregate the various functions and allow maximum flexibility by promoting module independence. The modules include (Tassou and Grace, 2005):

- data acquisition,
- fault-free performance prediction,
- residual calculation,
- steady-state detection,
- condition diagnosis, and
- user interfacing.

The system is able to operate in three modes – 'on-line mode' connected to the refrigeration system, 'off-line mode' for the processing of an existing data file and 'test mode' with simulated static inputs from the refrigeration system. In order to promote flexibility in the implementation of the system, the various modules are designed to achieve a low coupling between them and high cohesion of functions within each module. This enables modules to be added, removed or replaced with minimal modification to the remainder of the system and for the system to be configured as a distributed solution if required, i.e., individual modules could be implemented on remote machines. This allows each module to be connected via, for example, an interface, allowing data to pass from one unit to another irrespective of platform, operating system, programming language or physical location. This facilitates a flexible and cost effective solution by allowing detection and diagnosis for multiple units via a single central machine located remotely.

The prediction module uses a set of ten artificial neural networks (ANNs) to predict the fault-free steady-state condition of ten refrigeration system parameters (one network for each parameter). The ANNs receive the current coolant inlet temperatures for the evaporator and condenser as input data and each ANN calculates the expected fault-free condition of one of the ten parameters. The ten predicted parameters consist of the coolant outlet temperatures for the two heat exchangers and eight refrigerant conditions in the vapour compression cycle, including the suction and discharge pressures. The ANNs are trained off-line with steady-state data for the refrigeration system using a separate training module specifically developed as part of this project.

The residual calculator receives both the current measurement data and the fault-free predictions from the data acquisition and prediction modules respectively. The set of residuals, or differences between the current conditions and those predicted for fault-free operation, are generated by this module. The module also has provision for the calculation of derived residuals, i.e., further parameters that are derived from one or more of the ten existing residuals.

The steady-state detector simply monitors one parameter against a ‘rolling window’ of previous values. The parameter chosen is the one which exhibits the slowest response to a step change in conditions and is thus the last to achieve steady-state (Rueda *et. al.*, 2005).

The diagnosis module receives the set of residuals from the residual calculator, assuming that the steady-state detector has indicated that the system is operating under steady-state conditions. For each residual, a numerical range is allocated for five descriptive categories – “Very High”, “High”, “OK”, “Low” and “Very Low.” On receiving the set of current residuals, the diagnosis module uses the expert system to attempt to match the residual patterns to the expert system rules using the descriptive categories. Any rules that are fired modify the diagnosis to reflect the expert knowledge contained within the rules. If no rules are fired indicating a faulty state, a final ‘catch all’ rule is fired and the diagnosis is modified to indicate that the system is operating normally.

4.2. System Validation and Application

The system was validated using the 4.0 kW test rig outlined above. The system was operated across a range of `normal’ operating conditions and data gathered were used to provide a steady-state, fault-free ‘map’ of the system performance. The fault-free data was then used to train ten ANNs in the predictor module to enable prediction of the system conditions for given coolant inlet temperatures.

The test rig was then operated over a range of conditions and refrigerant charge levels, both above and below the design charge value. The data from these tests were used to establish the performance of the system at both over-and under-charge conditions. Over-charging can occur in practice in systems without liquid-line receivers and refrigerant leakage can be simulated by system under-charging. The residual calculator module compared the measurement data to the fault-free prediction data from the predictor module and generated the associated residuals.

The steady-state detector checked for steady-state conditions in the measurement data and forwarded the residuals to the diagnosis expert system. The expert system checked the residual patterns against the patterns in the set of rules and formed a diagnosis. The graphical user interface displayed all the data from the data acquisition, predictor and residual modules and the diagnosis was shown together with an estimate of the system charge level.

Figure 8 shows the variation of the Coefficient of Performance (COP) of the system with refrigerant charge. It can be seen that the COP increases as the refrigerant charge increases and reaches a maximum value of approximately 2.65 at a refrigerant charge of 1.2 kg. It can also be seen that the COP of this system is relatively constant at its maximum across a broad range of charge levels and that discrimination is required at charge levels differing by more than 33% from the nominal.

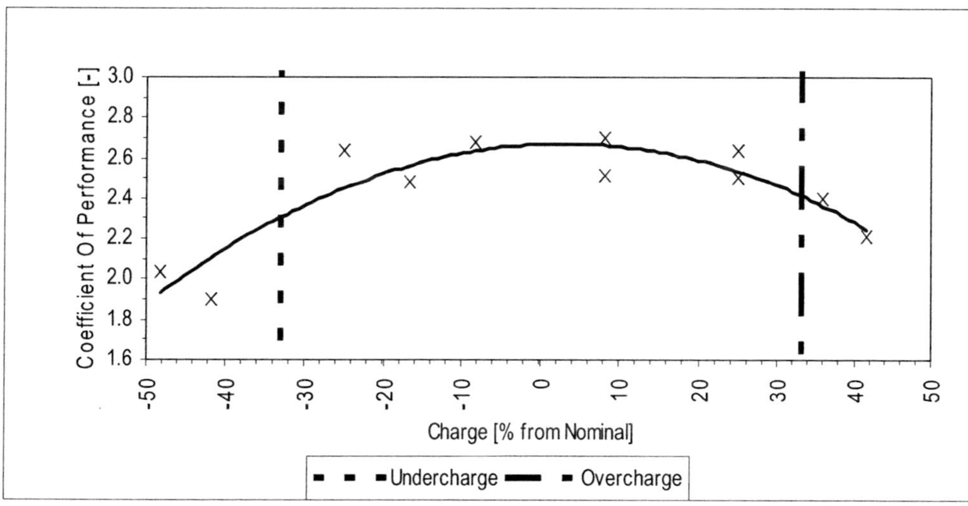

Fig. 8. Coefficient of performance (COP) against system charge

The ability of the system to diagnose four different states for the refrigeration system is shown in Table 4. The four states are "non-steady-state", "under charge", "normal" and "over charge."

Table 4. System diagnoses for validation data

Actual condition	Diagnosis (% of actual condition)			
	non-steady-state	under charge	normal	over charge
non-steady-state	100.0	0.0	0.0	0.0
under charge	0.0	99.9	0.1	0.0
normal	0.0	0.2	94.2	5.6
over charge	0.0	0.0	0.0	100.0

The "non-steady-state" diagnosis means that the measurement data is not sufficiently stable for a valid diagnosis to be made, "under charge" means that there is insufficient charge in the system, either through incorrect maintenance or through system leakage, "normal" means that there is no fault diagnosis and the system is operating correctly and "over charge" indicates excess charge in the system.

Table 4 shows that the system was able to distinguish between steady-state and transient conditions without error for the validation data and that all "over charge" conditions were detected without misdiagnosis. For the under charge and normal operation conditions, there was some small error in the diagnoses produced with some under charge conditions given as "normal" and some normal conditions given as "over charge". This is to be expected given the arbitrary nature of the distinction between under charge and normal, and between normal and over charge. For the validation dataset, 96.1% of the approximately 4,000 system conditions were correctly diagnosed.

Figure 9 compares the system diagnosis and actual charge for a test run using the rules in Table 5. The use of patterns of residuals enables not only charge levels to be detected but also any faults for which other distinct patterns can be discerned. This allows the system to be extended to diagnose a whole range of other faults simply by adding rules reflecting the diagnosis for each residual pattern.

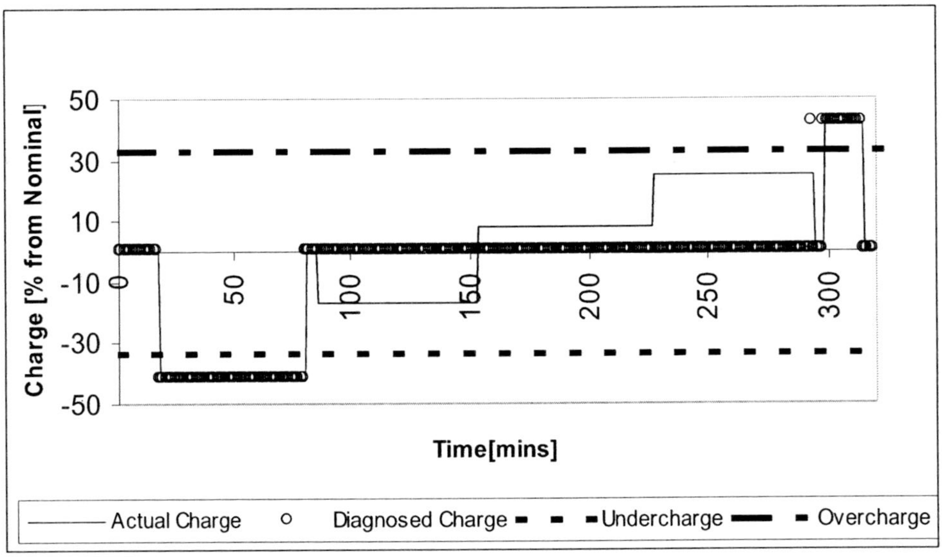

Fig. 9. System diagnosis and actual refrigerant charge against time

Table 5. Rules for discrimination of charge level

Diagnosis	Estimated Charge Level	Rule
under charge	-33%	IF (rE1Out is "Very High") THEN (diagnosis is "under charge")
over charge	+33%	IF (rDisp is "Very High") THEN (diagnosis is "over charge")

"rE1Out" denotes the residual for the evaporator coolant outlet temperature
"rDisp" indicates the residual for the compressor discharge pressure.

5. CONCLUSIONS

This chapter has demonstrated how artificial intelligence techniques can be employed for electrical power demand prediction, leak detection and fault diagnosis, and implementation of defrost on demand on commercial refrigeration systems.

From the analysis of the data collected and the consideration of different neural network architectures it has been established that time series MLP networks can effectively be used for the prediction of the total electrical power demand of supermarkets. The most important parameters to be used as input data were identified to be the current external and internal temperature and relative humidity and their values in the two previous time steps and the values of the total electrical power demand in the previous two time steps.

It has also been shown that MLP networks can be used to predict the energy consumption of the refrigeration plant. For this application, a network with input parameters the day of the week, suction and discharge pressure, external and internal temperature and relative humidity was found to produce best predictions.

A demand defrost controller based on Neural Networks has been developed and demonstrated in laboratory trials. It has been demonstrated that the controller can reduce the number of defrost cycles per day and produce significant energy savings. The controller can also reduce the variations in temperature that the products usually cycle through which should lead to improvements in their quality and shelf life.

The refrigerant leak detection system developed and demonstrated is capable of detecting modest leaks in system charge in any location, thus warning of any release of refrigerant into the environment and indicating operation at sub-optimal COP levels. The system is able to use instrumentation and control equipment already fitted to most commercial chillers and is capable of being implemented either within existing microprocessor control systems or as a distributed technology thus allowing remote central processing if required. This results in a minimal additional cost per chiller unit.

ACKNOWLEDGEMENTS

The work presented in this chapter was funded by the Engineering and Physical Sciences Research Council (EPSRC) of the UK, the Food Technology Unit of DEFRA (Department of Environment, Food and Rural Affairs) of the UK and a number of Industrial and commercial organisations. The authors are thankful for this funding and support.

REFERENCES

Bailey M.B., Kreider J.F., 2003. Creating an automated chiller fault detection and diagnosis tool using a data fault library, *ISA Transactions*, Vol. 42, pp. 486- 495.

Buick T.R., McMullan J.T., Morgan R., Murray R.B., 1978. Ice detection in heat pumps and coolers, *Energy Research*, Vol. 2, pp. 85-98.

Ciricillo, S.F., 1985. Heat pump de-icing – Controlling for energy conservation and costs, *Proceedings of CLIMA 2000, Congress on Heating, Ventilation and Air-Conditioning*, pp. 87-92.

Comstock M.C., Beaun J.E., Groll E.A., 2002. A survey of common faults for chillers. *ASHRAE Transactions*, Vol. 102, No. 1, pp. 819-825.

Datta, D., Tassou, S.A., 1998. Artificial neural network based electrical load prediction for food retail stores. *Journal of Applied Thermal Engineering*, Vol. 18, pp. 1121–1128.

Department of the Environment, 1992. *CFCs in the UK refrigeration and air conditioning industries. Usage and scope for substitution*, HMSO, London.

Farzad M., O'Neal D.L., 1991. System performance characteristics of an air conditioner over a range of charging conditions, *International Journal of Refrigeration*, Vol. 14, pp. 321-328.

Grimmelius H.T., Klein W.J., Been, G., 1995. On-line failure diagnosis for compression refrigeration plants. *International Journal of Refrigeration*, Vol. 18, pp. 31-41.

Heinzen, R.A., 1988. How adaptive defrost maintains refrigeration system efficiency, *Australian Refrigeration, Air Conditioning and Heating*, Vol. 42, No. 4, pp.14-16.

Hummel K.E., Nelson T.P., Thompson P.A., 1992. Survey of the use and emissions of chlorofluorocarbons from large chillers, *ASHRAE Transactions*, Paper No. 3536(RP-603), pp. 416-421.

Llewelyn, D.S., 1984. A significant advance in defrost control, *International Journal of Refrigeration*, Vol. 7, No. 5, pp. 334-335.

Peitsman H.C., Bakker V.E., 1996. Application of black-box models to HVAC systems for fault detection. *ASHRAE Technical Data Bulletin on Fault Detection and Diagnosis for HVAC Systems*, pp. 69-81.

Paone, N., Rossi, G., 1991. Fiber-optic ice sensors for refrigerators, *Proceedings of SPIE (Society of Optical Engineers) Fiber-Optic Sensors, Engineering and Applications*, pp. 129-139.

Rueda E., Tasou S.A., Grace, I.N., 2005. Fault detection and diagnosis in liquid chillers, *Proceedings Institution of Mechanical Engineers, Part E*: Journal of *Process Mechanical Engineering*, Vol. 219, No. E2, pp. 117-126.

Stylianou M., Nikanpour D., 1996. Performance monitoring, fault detection and diagnosis of reciprocating chillers. *ASHRAE Technical Data Bulletin on Fault Detection and Diagnosis for HVAC Systems,* pp. 56-68.

Tassou, S.A., Datta, D., 1999. Effectiveness and comparison of artificial neural network architectures in the prediction of electrical loads in supermarkets. *Proceedings of Twentieth International Congress of Refrigeration on CD-ROM,* International Institute of Refrigeration, 19-24 September 1999, Sydney, Australia, Paper No. 139.

Tassou, S.A., Grace, I.N., 2005. Fault diagnosis and refrigerant leak detection in vapour compression refrigeration systems, *International Journal of Refrigeration*, Vol. 28, No. 5, pp. 680-688.

INDEX

A

B

C

D

I

J

K

L

M

N

O

P

Q

R

S

T

U

V

W

Y